LABORATORY MANUAL FOR MAJORS GENERAL BIOLOGY

James W. Perry
University of Wisconsin, Fox Valley

David Morton
Frostburg State University

Joy B. Perry
University of Wisconsin, Fox Valley

Edited by
Mark Manteuffel
Washington University, St. Louis

Contributions to Animal Diversity
by Ronald E. Barry, Jr.

And Evolutionary Agents
by James H. Howard

 Cengage

Australia • Brazil • Canada • Mexico • Singapore • United Kingdom • United States

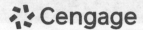

Laboratory Manual for Majors General Biology
James W. Perry, David Morton,
and Joy B. Perry

Executive Editor: Yolanda Cossio

Assistant Editor: Elizabeth Momb

Editorial Assistant: Samantha Arvin

Technology Project Manager: Kristina Razmara

Marketing Manager: Amanda Jellerichs

Marketing Assistant: Katy Malatesta

Marketing Communications Manager: Linda Yip

Project Manager, Editorial Production:
Michelle Cole

Creative Director: Rob Hugel

Art Director: John Walker

Print Buyer: Judy Inouye

Permissions Editor: Bob Kauser

Production Service: ICC Macmillan Inc.

Text Designer: Carolyn Deacy

Photo Researcher: Terri Wright

Copy Editor: Steven Summerlight

Cover Designer: William Stanton

Cover Image: Pete Oxford/Getty Images

Compositor: ICC Macmillan Inc.

For product information and technology assistance, contact us at
**Cengage Customer & Sales Support, 1-800-354-9706
or support.cengage.com.**

For permission to use material from this text or product, submit all
requests online at **www.copyright.com**.

Library of Congress Control Number: 2008927240

ISBN-13: 978-0-495-11505-2
ISBN-10: 0-495-11505-3

Cengage
200 Pier 4 Boulevard
Boston, MA 02210
USA

Cengage is a leading provider of customized learning solutions
with employees residing in nearly 40 different countries and sales in more
than 125 countries around the world. Find your local representative at:
www.cengage.com.

To learn more about Cengage platforms and services, register or access
your online learning solution, or purchase materials for your course,
visit **www.cengage.com.**

Printed in the United States of America
Print Number: 14   Print Year: 2022

Contents

Preface

Greetings from the authors! We're happy that you are examining the results of our efforts to assist you and your students. We believe you'll find the information below a valuable introduction to this laboratory manual.

Audience

This manual is designed for students at the college majors level. You'll find that the exercises support any biology text used in a majors course.

Features of This Edition

Inquiry Experiments and the Methods of Science

We realize that the best way to learn science is by doing science. Thus, in this edition, you will find numerous **inquiry-based experiments.** Some are extensions of a preceding activity, while others stand alone. Each follows the cognitive techniques whose foundations are laid in the first exercise, "Scientific Method."

Throughout the manual, we place more emphasis on testing predictions generated from hypotheses. We strongly believe that all biology students—benefit by repeating the logical thought processes of science.

New and Updated Exercises

As always, we strive to provide students with exciting, relevant activities and experiments that allow them to explore some of the rapidly developing areas of biological knowledge. We have added an additional animal diversity exercise and re-ordered and re-grouped some of the microbe, plant and invertebrate phyla.

Updated Taxonomy

An attempt to provide accurate systematic information is like trying to hit a moving target. As new information floods in, our best understanding of taxonomy and systematic relationships sometimes seems to change daily. The taxonomy in our exercises is completely updated and reflects the most widely accepted information at the time of publication.

What's Important to Us

In preparing this lab manual, we paid particular attention to pedagogy, clarity of procedures and terminology, illustrations, and practicality of materials used.

Pedagogy

The exercises are written so the conscientious student can accomplish the objectives of each exercise with minimal input from an instructor. As suggested by the publisher, the procedure sections of the exercises are more detailed and step-by-step than in other manuals. Instructions follow a natural progression of thought so the instructor need not conduct every movement.

We attempted to make each portion of the exercise part of a continuous flow of thought. Thus, we do not wait until the post-lab questions to ask students to record conclusions when it is more appropriate to do so within the body of the procedure. Answers to in-lab questions are to be found in the *Instructor's Manual* as well as online.

Terms required to accomplish objectives are **boldface.** Scientific names and precautionary statements, or those needing emphasis, are *italic*.

The use of scientific names is deemphasized when it is not relevant to understanding the subject. However, these names generally do appear in parentheses because the labels on many prepared microscope slides bear only the scientific name.

Format

Each exercise includes:

1. Objectives: a list of desired outcomes.

2. Introduction: to stimulate student interest, indicate relevance, and provide background.

3. Materials: a list for each portion of the exercise so a student can quickly gather the necessary supplies. Materials are listed "Per student," "Per pair," "Per group," and "Per lab room."

4. Procedures: including safety notes, illustrations of apparatus, figures to be labeled, drawings to be made, tables to record data, graphs to draw, and questions that lead to conclusions. The procedures are listed in easy-to-follow numbered steps.

5. Pre-lab Questions: ten multiple-choice questions that the student should be able to answer after reading the exercise, but prior to entering the laboratory.

6. Post-lab Questions: questions that draw on knowledge gained from doing the exercise and that the student should be able to answer after finishing the exercise. These post-lab questions assess **recall** (preparing students for lab practical assessments), **understanding,** and **application.**

Practical Post-lab Questions

Virtually all courses use laboratory practical examinations. We explain to our students the difference between lecture-type questions, which they need to read and provide an answer based on the written word, and practical-type questions, for which a response depends on observation.

We believe the post-lab questions should draw on the knowledge gained by observation. Consequently, we've incorporated as many illustrations as possible into the post-lab questions. These illustrations typically are similar, but not identical, to those in the procedures. Thus, they assess the student's ability to use knowledge gained during the exercise in a new situation.

Post-lab questions are identified by exercise section. This allows the instructor to easily assign or use only those questions relevant to portions of exercises performed by students. The questions also have been revised and are directly tied to the learning outcomes expected from each exercise.

Flexible Quiz Options

Each exercise has a set of pre-lab questions. We have found through nearly 80 combined years of experience that students left to their own initiative typically come to laboratory unprepared to do the exercise. Few read the exercise beforehand. One solution to this problem is to incorporate some sort of graded pre-exercise activity. At the same time, we recognize that grading a large number of lab papers each week can put an unreasonable burden on instructors. Consequently, we decided on a multiple-choice format, which is easy to grade but still accomplishes the pedagogical goal.

In our own courses, we have students take a pre-lab quiz consisting of the questions in their lab manual in scrambled order to discourage memorization. These scrambled quizzes are reproduced in our *Instructor's Manual.* Our quiz takes about five minutes of lab time and counts as a portion of the lab grade, thus rewarding students for preparation.

Other instructors have told us they use the pre-lab questions to assess learning after the exercise has been completed. We encourage you to be creative with the manual; do what you like best.

Lab Length and Exercise Options

We realize there is wide variation in the amount of time each instructor devotes to laboratory activities. To provide maximum flexibility for the instructor, the procedure portions of the exercises are divided by major headings, and the approximate time it takes to perform each portion indicated. Once the introduction has been studied, portions of the procedures can be deleted or conducted as demonstrations without sacrificing the pedagogy of the exercise as a whole.

It's our experience that if the lecture section covers the topic prior to the lab, students find the exercise much more relevant and understandable. We strive to create this situation in our courses, and thus no time is spent on a lecture-style introduction in the lab itself before the exercise begins. Therefore, we have two to three full hours for real scientific investigation and need to delete very little material to complete most exercises in the time allotted.

Illustrations

Perhaps our illustrations are more noticeable than anything else as you thumb through the manual. We continued our incorporation of a generous number of high-quality color illustrations, including everything a student needs visually to accomplish the objectives of each exercise. While there is no need for students to purchase supplemental publications, students may find the *Photoatlas for Biology*, ISBN 0-534-23556-5, to be a useful reference for this course and other biology courses.

Most illustrations of microscopic specimens are labeled to provide orientation and clarity. A few are unlabeled but are provided with leaders for students to attach labels. In other cases, more can be gained by requiring the student to do simple drawings. Space is included in the manual for these, with boxes for drawings of macroscopic specimens and circles for microscopic specimens.

Materials

Most of the equipment and supplies used in the exercises are readily available from biological and laboratory supply houses. Many others can be collected from nature or purchased in local supermarket, discount or office supply stores. (Our department budgets are not large!) We've attempted to keep instrumentation as simple and inexpensive as possible.

Anyone who lives in a temperate climate knows it may be necessary to adjust the sequence of the laboratory exercises to accommodate seasonal availability of certain materials. However, we provided alternatives, including the use of preserved specimens wherever possible, to avoid this problem.

Instructor's Manual

There is no need to worry, "Where can I get that?" or "How do I prepare this?" Our *Instructor's Manual* includes:

- Material and equipment lists for each exercise
- Procedures to prepare reagents, materials, and equipment
- Scheduling information for materials needing advance preparation
- Approximate quantities of materials needed
- Answers to in-text questions
- Answers for pre-lab questions
- Answers for post-lab questions
- Tear-out sheets of pre-lab questions, in scrambled order from those in the lab manual, for those who wish to duplicate them for quizzes
- Vendors and ordering information for supplies

And in the End . . .

There are very few things in life that are perfect. We don't suppose that this lab manual is one of them. We hope your students will enjoy the exercises. We **know** they will learn from them. Perhaps you and they will find places where rephrasing will make the activity better. Please contact us with your opinions and any ideas you wish to share; encourage your students to do likewise.

James W. Perry
Department of Biological
Sciences
University of Wisconsin
Fox Valley
1478 Midway Road
Menasha, WI 54952-1297
(920) 832-2610
james.perry@uwc.edu

David Morton
Department of Biology
Frostburg State University
Frostburg, MD 21532-1099
(301) 689-4355
dmorton@frostburg.edu

Joy B. Perry
Department of Biological
Sciences
University of Wisconsin
Fox Valley
1478 Midway Road
Menasha, WI 54952-1297
(920) 832-2653
joy.perry@uwc.edu

About the Editor

Mark Manteuffel holds a doctorate in ecology and evolutionary biology from the University of Miami. He is currently a faculty member at Washington University and St. Louis Community College, where he involves students in research learning experiences and teaches courses in biology, ecology, and environmental science. He teaches field courses around the world that focus on the conservation of biodiversity, sustainable development, and ecosystem function and health.

Professor Manteuffel also develops innovative interdisciplinary undergraduate and graduate science programs that cover global education and environmental sustainability. This focus led him to work with colleagues from the University of Missouri, Columbia, converting laboratory investigations into more inquiry-based and student-directed biological investigations. This work is currently ongoing, funded by National Science Foundation grant DEB-0618817, *Connecting Undergraduates to the Enterprise of Science* (CUES).

The inquiry-based lab modules in this edition of the laboratory manual are an extension of Professor Manteuffel's work on the CUES grant. These modules are designed to introduce students to the process and content of science. Students read and critique the lab modules' scientific reports and extend the research reported. The goal of these modules is for students to learn not only content matter, but also the process of science through their own direct experience. They design their own research that will extend the research reported in the modules. The intent is that students will benefit from learning science by doing science, thereby increasing their ability to read and interpret scientific literature.

To the Student

Welcome! You are about to embark on a journey through the cosmos of life. You will learn things about yourself and your surroundings that will broaden and enrich your life. You will have the opportunity to marvel at the microscopic world, to be fascinated by the cellular events occurring in your body at this very moment, and to gain an appreciation for the environment, including the marvelous diversity of the plant and animal world.

We offer a number of suggestions to make your college experience in biology a pleasant one. We have taken the first step toward that goal; we have written a laboratory guide that is user friendly. You will be able to hear the authors speaking as though we were there to share your experience. The authors share a personal belief that the more comfortable we make you feel, the more likely you will share our enthusiasm for biology. One thing we all must realize is that we are citizens of "spaceship Earth." The fate of our spaceship is largely in your hands because you are the decision makers of the future. As has been so aptly stated, "We inherited the earth from our parents and grandparents, but we are only the caretakers for our children and grandchildren."

As caretakers, we need to be informed about the world around us. That's why we enroll in colleges and universities with the hope of gaining a liberal education. In doing so, we establish a basis on which to make educated decisions about the future of the planet. Each exercise in this manual contains a lesson in life that is of a more global nature than the surroundings of your biology laboratory.

To enhance your biology education, take the initiative to give yourself the best possible advantage. Don't miss class. Read your text assignment routinely. And, read the laboratory exercise before you come to the lab.

Each exercise in the manual is organized in the same way:

1. Objectives tell exactly what you should learn from the exercise. If you wish to know what will be on the exam, consult the objectives for each exercise.

2. The Introduction provides background information for the exercise and is intended to stimulate your interest.

3. The Materials list for each portion of the exercise allows you to determine at a glance whether you have all the necessary supplies needed to do the activity.

4. The Procedure for each section, in easy-to-follow step-by-step fashion, describes the activity. Within the procedure, spaces are provided to make required drawings. Questions are posed with space for answers, asking you to draw conclusions about an activity you are engaged in. You'll find a lot of illustrations, most of which are labeled and others which are not but have leaders for you to attach labels. The terms to be used as labels are found in the procedure and in a list accompanying the illustration. We believe it best for you to sometimes make a simple drawing, and have inserted boxes or circles for your sketches. Where appropriate, tables and graphs are present to record your data.

5. Pre-lab questions can be answered easily by simply reading the exercise. They're meant to "set the stage" for the lab period by emphasizing some of the more salient points.

6. Post-lab questions are intended to be answered after the laboratory is completed. Some are straightforward interpretations of what you have done, while others require additional thought and perhaps some research in your textbook. In fact, some have no "right" or "wrong" answer at all!

It is our experience that students are much too reluctant to ask questions for fear of appearing stupid. Remember, there is no such thing as a stupid question. Speak up! Think of yourselves as "basic learners" and your instructors as "advanced learners." Interact and ask questions so that you and your instructors can further your/their respective educations.

Laboratory Supplies and Procedures

Materials and Supplies Kept in the Lab at All Times

The following materials will always be available in the lab room. Familiarize yourself with their location prior to beginning the exercises.

- Compound light microscopes
- Dissection microscopes
- Glass microscope slides
- Coverslips

- Lens paper
- Tissue wipes
- Plastic 15-cm rulers
- Dissecting needles
- Razor blades
- Assorted glassware-cleaning brushes

- Detergent for washing glassware
- Distilled water
- Hand soap
- Paper towels
- Safety equipment (see separate list)

Laboratory Safety

None of the exercises in this manual are inherently dangerous. Some of the chemicals are corrosive (causing burns to the skin) and others are poisonous if ingested or inhaled in large amounts. Contact with your eyes by otherwise innocuous substances may result in permanent eye injury. **Remember, once your sight is lost, it's probably lost forever.** Locate the following safety items and then study the list of basic safety rules.

1. Eyewash bottle or eye bath

   Should any substance be splashed in your eyes, wash them thoroughly.

2. Fire extinguisher

   Read the directions for use of the fire extinguisher.

3. Fire blanket

   Should someone's clothing catch fire, wrap the blanket around the individual and roll the person on the floor to smother the flames.

4. First-aid kit

   Minor injuries such as small cuts can be treated effectively in the lab. Open the first-aid kit to determine its contents.

5. Safety goggles

   Eye protection should be worn during the more experimental exercises.

Safety Rules

1. Do not eat, drink, or smoke in the laboratory.
2. Wash your hands with soap and warm water before leaving the laboratory.
3. When heating a test tube, point the mouth of the tube away from yourself and other people.
4. Always wear shoes in the laboratory.
5. Keep extra books and clothing in designated places so your work area is as uncluttered as possible.

6. If you have long hair, tie it back when in the laboratory.
7. Read labels carefully before removing substances from a container. Never return a substance to a container.
8. Discard used chemicals and materials into appropriately labeled containers. Certain chemicals should not be washed down the sink; these will be indicated by your instructor.

Caution: Report all accidents and spills to your instructor immediately!

Instructions for Washing Laboratory Glassware

1. Place contents to be discarded in proper waste container as described in exercise.
2. Rinse glassware with tap water.
3. Add a small amount of glassware cleaning detergent.
4. Scrub using an appropriately sized brush.

5. Rinse with tap water until detergent disappears.
6. Rinse three times with distilled water (dH$_2$O).
7. Allow to dry in inverted position on drying rack (if available).

When glassware is clean, dH$_2$O sheets off rather than remaining on the surface in droplets.

Microscopy

OBJECTIVES

After completing this exercise, you will be able to

1. define *magnification, resolving power, contrast, field of view, parfocal, parcentral, depth of field, working distance;*

2. describe how to care for a compound light microscope;

3. recognize and give the function of the parts of a compound light microscope;

4. accurately align a compound light microscope;

5. correctly use a compound light microscope;

6. make a wet mount;

7. correctly use a dissecting microscope;

8. describe the usefulness of the phase-contrast, transmission electron, and scanning electron microscopes;

9. use your skills to enjoy a fascinating world unavailable to the unaided eye.

INTRODUCTION

A microscope contains lenses (for example, transparent glass), which focus radiation (such as light rays) emanating from a specimen to produce an image of that specimen and on a surface sensitive to the radiation (like the retina, the light-sensitive layer of the eye). Table 1-1 presents the three most important properties of lenses and their images.

TABLE 1-1 Important Lens and Image Properties

Property	Definition
Magnification	The amount that the image of an object is enlarged—for example, 100×.
Resolving power	The extent to which object detail in an image is preserved during the magnifying process.
Contrast	The degree to which image details stand out against their background.

1.1 Compound Light Microscope *(About 80 min.)*

The lens of a normal unaided eye can project onto the retina a focused image of an object held no closer than about 10 cm. At this distance, details separated by 0.1 mm are visible. Most cells and related structures are smaller than this, and a light microscope is needed to see them. A microscope placed between the eye and a specimen (usually a section or thin object(s) mounted on a glass slide) acts to bring the specimen very close to the eye so that you can see its details. Ultimately, the greater the proportion of the retina covered by the final image of the specimen, the greater its magnification. It does this by producing a series of magnified images.

Magnification without enough resolving power is referred to as empty, and with a light microscope, the maximum useful magnification is about 1000 times the diameter of the specimen (1000×). Above this value, additional details are missing. Furthermore, adequate contrast is needed to see the details preserved in an image. Dyes are usually added to sections of biological specimens to increase contrast.

Like automobiles, models of compound light microscopes abound, and these instruments have numerous accessories. Typical examples are shown in Figure 1-1, and one is diagrammed in Figure 1-2. If your microscope differs significantly from Figure 1-2, your instructor will give you an unlabeled diagram. If the instructor assigns you a specific microscope for your lab work, record its identification code in the second column of the first row in Table 1-2.

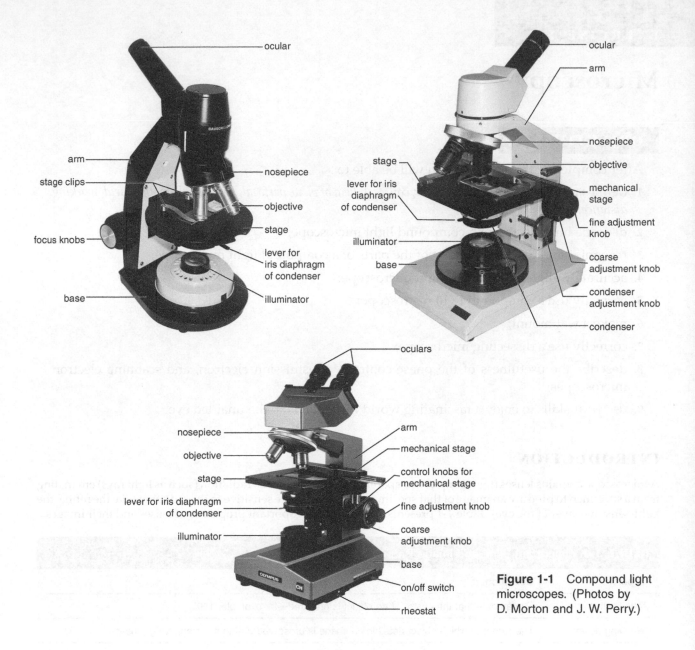

ocular

arm

stage clips

focus knobs

base

nosepiece

objective

stage

lever for
iris diaphragm
of condenser

illuminator

ocular

arm

nosepiece

objective

mechanical
stage

fine adjustment
knob

coarse
adjustment knob

condenser
adjustment knob

condenser

stage

lever for iris
diaphragm
of condenser

illuminator

base

oculars

nosepiece

objective

stage

lever for iris diaphragm
of condenser

illuminator

arm

mechanical stage

control knobs for
mechanical stage

fine adjustment knob

coarse
adjustment knob

base

on/off switch

rheostat

Figure 1-1 Compound light
microscopes. (Photos by
D. Morton and J. W. Perry.)

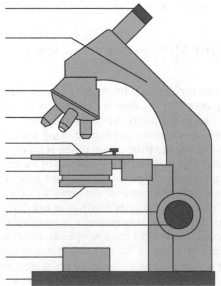

Figure 1-2 Compound light
microscope.
Labels: ocular, objective,
arm, base, illuminator,
condenser, lever for iris
diaphragm of condenser,
stage, stage clip, coarse
adjustment knob, fine
adjustment knob, nosepiece

TABLE 1-2 Characteristics of My Microscope

Characteristic	Description	Function
Code		identification of my microscope
Light source		
Condenser		
Stage		
Focusing knobs		
Objectives		
Ocular(s)		

MATERIALS

Per student:

- compound light microscope
- lens paper
- lint-free cloth (optional)
- unlabeled diagram of the compound light microscope model used in your course (optional)
- prepared slides with
  a whole mount of stained diatoms
  Wright-stained smear of mammalian blood
  mounted letter *e*
- index card

- prepared slides with
  crossed colored threads coded for thread order
  unstained fibers

Per student group (4):

- bottle of lens-cleaning solution (optional)
- dropper bottle of immersion oil (optional)

Per lab room:

- labeled chart of a compound light microscope

PROCEDURE

A. Care of a Compound Light Microscope

1. To carry a microscope to and from your lab bench, grasp the **arm** (Figure 1-1) with your dominant hand and support the **base** (Figure 1-1) with the other hand, always keeping the microscope upright. *Do not try to carry anything else at the same time.* Label the arm and the base on Figure 1-2 or on the diagram given to you by your instructor.

> ### Caution
> *Never wipe a glass lens with anything other than lens paper.*

2. Remove the dust cover and clean the exposed parts of the optical system. First, blow off any loose dust that may be on the ocular and then gently brush off any remaining dust with a piece of lens paper.
   If the part is still dirty, breathe on the lens and gently polish it with a rotary motion, using a fresh piece of lens paper. If the part is still dirty, and *with your instructor's approval,* clean the lens with a piece of lens paper moistened with lens-cleaning solution.

3. Always remember that your microscope is a precision instrument. Never force any of its moving parts.
4. It is just as difficult to see clearly through a dirty slide as through a dirty microscope. Clean dirty slides with a lint-free cloth or with lens paper before using.
5. At the end of an exercise, make sure the last slide has been removed from the stage and *rotate the nosepiece so that the low-power objective is in the light path.* If your instrument focuses by moving the body tube, turn the coarse adjustment so that it is racked all the way down. If your microscope has an electric cord, neatly fold it up on itself and tie it with a plastic strap or rubber band. Otherwise, wind the cord around the base of the arm of the microscope.
6. Replace the dust cover before returning your microscope to the cabinet.

Now that you know how to care for your microscope, remove the instrument assigned to you from the cabinet and place it on your lab bench. Use Figure 1-1 and the chart on the wall of your lab room to identify the various parts of your microscope. Read each step below and manipulate the parts *only where indicated*. Before you start, make sure the shortest objective is in the light path.

1. *Light source.* The compound microscope uses transmitted light to illuminate a transparent specimen usually mounted on a glass slide. Newer microscopes have a built-in illuminator (Figure 1-1). Locate the **illuminator**, the *off/on switch*, and perhaps also a *rheostat*, which is used to vary the intensity of the light. On some models, the switch and the rheostat are combined. Turn on the light source and look through the ocular. If the illuminator has a rheostat, adjust the intensity so that the light is not too bright.
   (a) Label the illuminator on Figure 1-2 or on your instructor's diagram.
   (b) Describe the light source in Table 1-2, then state its function.

2. *Condenser.* For maximum resolving power, a **condenser**—with a *condenser lens* and *iris diaphragm*—focuses the light source on the specimen so that each of its points is evenly illuminated. The **lever for the iris diaphragm** of the condenser is used to open and close the condenser. Establish whether there is a condenser adjustment knob (Figure 1-1) to set the height of the condenser. *Do not turn the knob; you will learn how to use it later.*

   There may be a *filter holder* under the condenser with a blue or frosted glass disk. Many microscope manufacturers believe that blue light is more pleasing to the eye because, when used with an incandescent bulb, it produces a color balance similar to daylight conditions. Also, theoretically at least, blue light gives better resolving power because of its shorter wavelength. The frosted glass disk scatters light and can be useful in producing even illumination at low magnifications.
   (a) Label the condenser and lever for the iris diaphragm of the condenser on Figure 1-2 or on your instructor's diagram.
   (b) Describe the condenser and its related parts in Table 1-2, then state its function.

3. *Stage.* Either a pair of *stage clips* or a *mechanical stage* holds a specimen mounted on a glass slide in place suspended over a central hole (Figure 1-1).
   (a) If your microscope has a mechanical stage, skip to part b. If your microscope has stage clips, place a prepared slide of stained diatoms under their free ends. Never remove the stage clips, because they make it easier to move a slide in small increments. Skip b and do part c next.
   (b) Position a prepared slide of stained diatoms on the stage by releasing the tension on the spring-loaded movable arm of the mechanical stage (Figure 1-1a). There are two knobs to the right or left of the stage: one to move the specimen forward and backward, the other to move it laterally (Figure 1-1). Label the stage and stage clips (or mechanical stage) on Figure 1-2 or on your instructor's diagram.

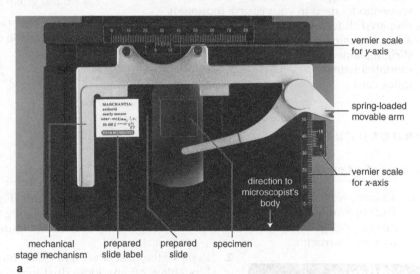

vernier scale for *y*-axis

spring-loaded movable arm

direction to microscopist's body

vernier scale for *x*-axis

mechanical stage mechanism | prepared slide label | prepared slide | specimen

a

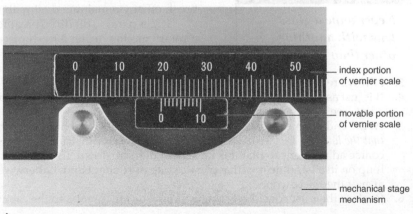

index portion of vernier scale

movable portion of vernier scale

mechanical stage mechanism

b

(Photos by J. W. Perry.)

Figure 1-3 (a) Mechanical stage. (b) Vernier scale on mechanical stage of compound light microscope. The correct reading is 19.6 mm.

(c) On most mechanical stages, each direction has a vernier scale so that you can easily locate interesting fields again and again. A vernier scale consists of two scales running side by side, a long one in millimeters and a short one, 9 mm in length and divided into 10 equal subdivisions. To take a reading, note the whole number on the long scale coinciding with or just below the zero line of the short scale. If the whole number of the long scale and the zero of the short scale coincide, the first place after the decimal point is zero. Otherwise, the first place after the decimal point is the value of the line on the short scale that coincides (or nearly coincides) with one of the next nine lines after the whole number on the long scale. For example, the correct reading of the vernier scale in Figure 1-3b is 19.6 mm.

(d) Describe the stage and its related parts in Table 1-2, then state its function.

4. *Focusing knobs* (Figure 1-1). The **coarse focus adjustment knob** is for use with the lower-power objectives, whereas the **fine focus adjustment knob** is for critical focusing, especially with the higher-power objectives. On most modern microscopes, you move the stage of the instrument up and down to focus the specimen. Modern microscopes usually have a *preset focus lock,* which stops the stage at a particular height. After setting this lock, you can lower the stage with the coarse focus knob, to facilitate changing of the specimen, and then raise it to focusing height without fear of colliding the specimen against the objective. There may also be a *focus tension adjustment knob,* usually located inside of the left-hand coarse focus knob.

(a) Turn the coarse focus knob. Do you turn the knob toward you or away from you to bring the slide and objective closer together?

_____

(b) Label the coarse and fine focus adjustment knobs on Figure 1-2 or on your instructor's diagram.

(c) Describe the focusing knobs in Table 1-2, then state their function. Note if a preset focus lock or focus adjustment knob is present.

5. *Objectives.* The compound light microscope has at least two magnifying lenses, the objective and the ocular (Figure 1-1). The objective scans the specimen. Most microscopes have several objectives mounted on a revolving **nosepiece.** The magnifying power of each objective is labeled on its side. Usually included are these objectives: a 4× *low-power* or scanning, a 10× *medium-power* (Figure 1-4b), an about 40× *high-dry,* and perhaps an about 100× *oil-immersion objective* (Figure 1-4c). The other number often labeled on the side of nosepiece objectives is the **numerical aperture** (NA). The larger the numerical aperture, the greater the resolving power and useful magnification.

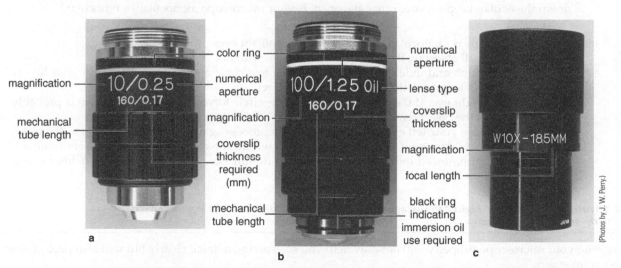

Figure 1-4 (a) 10× objective removed from microscope. (b) 100× oil immersion objective removed from microscope. (c) Ocular removed from microscope.

Objectives are **parfocal.** That is, once an objective has been focused, you can rotate to another one and the image will remain in coarse focus, requiring only slight movement of the fine focus knob. Objectives are also **parcentral,** meaning that the center of the field of view remains about the same for each objective. The **field of view** is the circle of light you see when looking into the microscope.

Objectives have different lengths; the lower-power objectives are shorter than the higher-power ones. That is, the working distance of objectives decreases with magnification. **Working distance** is the space between the objective lens and the slide. Therefore, the higher the power of the objective in use, the closer the objective is to the slide—and the more careful you must be.

(a) Record the magnifying power and NA of the objectives on your microscope in Table 1-3. If your instrument does not have a particular objective, indicate that it is not present (NP).

TABLE 1-3 Objectives Present on My Compound Light Microscope

Objective	Objective Magnifying Power (ObMP)	Total Magnifying Power (ObMP × OcMP = ____×)	Numerical Aperture (NA)
Low-power			
Medium-power			
High-dry			
Oil-immersion			

    (b) Label the nosepiece and objective on Figure 1-2 or on your instructor's diagram.
    (c) Describe the objectives in Table 1-2, then state their function.

6. *Ocular.* The magnifying lens you look into is called an **ocular** (Figure 1-4a). Oculars are generally 10×.
    (a) Since each objective has a different magnifying power, the total magnification is calculated by multiplying the magnifying power of the ocular by that of the objective in use. What is the ocular magnification power (OcMP) of the ocular(s) on your microscope?

           _____×
        Calculate the total magnification for each of your microscope's ocular/objective combinations, then record them in Table 1-3.
    (b) Label the ocular on Figure 1-2 or on your instructor's diagram.
    (c) Describe the ocular in Table 1-2, then state its function.
    (d) Your microscope will have one or two oculars mounted on a *monocular* or *binocular head*, respectively. There may be a pointer mounted in an ocular so that you can easily show a specimen detail to your instructor or another student. For a monocular microscope, it's best to use your dominant eye to look down the ocular, keeping your other eye open. Is your microscope monocular or binocular?

    _____
    (e) If your microscope is monocular, determine your dominant eye:
- Look at a small object on the far wall of your room with both eyes open.
- Form the thumb and index finger of one hand into a circle and place this circle in your line of sight, at arm's length, so that it surrounds the object.
- Close your right eye. If the object shifts out of the circle to your left, your right eye is probably dominant. If the object remains in the circle, your left eye is probably dominant.
- This time close your left eye and go through the process again. If the object shifts to the right, your left eye is dominant. If the object remains within the circle, your right eye is dominant. The more pronounced the shift, the greater the dominance. If there is no shift, neither eye is dominant.

C. Aligning a Compound Light Microscope with In-Base Illumination and a Condenser with an Iris Diaphragm

Aligning your microscope properly will not only help you see specimen detail clearly but will also protect your eyes from strain.

1. Rotate the nosepiece until the medium-power objective is in the light path. Open the iris diaphragm.
2. If it is not already there, place the prepared slide of stained diatoms on the stage; center and carefully focus on it. *Skip steps 3 and 4 if your microscope is monocular. Skip step 5 if your microscope does not have a control to adjust the height of the condenser.*
3. If your microscope is binocular, adjust the interpupillary distance. Hold a different ocular tube with each hand and, while looking at the specimen, pull the tubes apart or push them together until you see one field of view. After making this adjustment, read and record the number off the scale.
    My interpupillary distance is _____.
    From now on, you can set the interpupillary distance at this number.
4. Now compensate for any difference in diopter between the lenses of each eye:
    (a) *If there is one diopter adjustment ring around the left ocular tube,* cover your left eye with an index card and focus your microscope using the fine focus knob. Now uncover the left eye and cover your right one. Use the diopter adjustment ring to bring the specimen into focus.

(b) *If both ocular tubes have a diopter adjustment ring,* set the left one to the same number as the interpupillary distance, cover your right eye with an index card, and focus on the specimen. Then uncover the right eye and cover your left one. Use the diopter adjustment ring on the right ocular tube to bring the specimen into focus.

5. Place a sharp point (pencil, dissecting needle, or some similar object) on top of the illuminator and bring the silhouette into sharp focus by adjusting the height of the condenser.

   (a) *If the ocular on your microscope is removable (and with the permission of your instructor),* carefully slide it out and put the ocular open end down on a piece of lens paper in a safe place. Then, while looking down the ocular tube, adjust the iris diaphragm until the edge of the aperture lies just inside the margin of the back lens element of the objective (Figure 1-5). Replace the ocular.

   (b) *If the ocular cannot be removed,* close the condenser diaphragm and then open it until there is no further increase in brightness. Now close it again, stopping when you see the brightness begin to diminish.

6. If your microscope has a rheostat, adjust the illumination to a level that lets you see specimen detail and that is comfortable for your eyes. To maintain the same illumination at higher magnifications, you will have to increase its intensity.

7. For best results, repeat steps 5 and 6 each time you use a different objective.

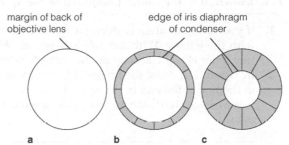

Figure 1-5 Correct setting for condenser iris diaphragm: Drawing **b** is correct. In **a**, you cannot see the edge of the iris diaphragm. In **c**, the diaphragm has been closed too much.

D. Using Different Magnifications

It's safest to observe a specimen on a slide first with low power and then, step by step, with higher-power objectives. This way you avoid colliding the objective against the slide or vice versa. Also, it's easier to use a lower-power objective to locate a specific specimen detail. This is why the low-power objective is sometimes called the scanning objective. Since the magnification is in diameters, the area of the field of view decreases dramatically with increasing magnification (Figure 1-6). It is just as easy to lose a specimen detail at higher magnifications, but once a specimen detail is lost, it's always easier to find it again if you switch to a lower-power objective.

Now follow these steps to use each objective.

1. Rotate the low-power objective into the light path.
2. If it is not already there, place a prepared slide of stained diatoms on the stage, securing it with either the stage clips or the movable arm of the mechanical stage.
3. Look through the ocular. Bring the diatoms into focus using the *coarse focus knob*. Adjust the illumination as described in steps 5 and 6 of Section C. At this magnification, the diatoms appear small. Center a diatom by moving the slide.
4. Rotate the nosepiece so that the medium-power objective is in the light path. Adjust the illumination. Focus the diatom.
5. Rotate the nosepiece so that the high-dry objective is in the light path. Adjust the illumination. Focus the diatom using the *fine focus-adjustment knob*.
6. If your microscope has an oil-immersion objective (and with the permission of your instructor), replace the prepared slide of diatoms with one with a smear of mammalian blood and repeat steps 1–5. Center and focus on a white blood cell. (See Exercise 25.) Most of the cells are red blood cells stained a light pink. A few cells are white blood cells with prominent blue-stained nuclei. Immersion oil, because it has optical properties similar to glass, increases resolving power and useful magnification. To use an oil-immersion objective:

   (a) Rotate the nosepiece so that the light path is midway between the high-dry and oil-immersion objectives.

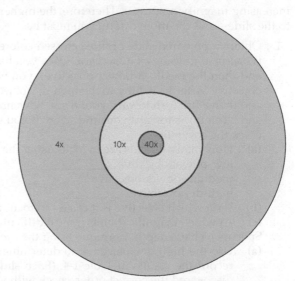

Figure 1-6 Illustration of the decreasing area of the field of view when a 4×, 10×, and 40× objective is used with a 10× ocular. The actual area of each circle has been enlarged 7.5×.

(b) Place a small drop of immersion oil onto the coverslip, using the circle of light above the specimen as a reference point.

(c) Rotate the nosepiece so that the oil-immersion objective is in the oil.

(d) Adjust the illumination and focus the white blood cell with the fine focus knob.

(e) After examining the white blood cell, rotate the oil-immersion objective out of the light path. Carefully wipe the oil from the oil-immersion objective with lens paper.

(f) Remove the slide from the stage and wipe the oil from the coverslip with lens paper.

E. Orientation of the Image Compared to the Specimen

1. If you have not already done so, remove the prepared slide of diatoms and replace it with a prepared slide with the letter *e*. With the medium-power objective in the light path, position the slide with the specimen (the letter *e*) right side up on the stage. Center the *e* in the field of view, and carefully bring it into focus.

2. In Figure 1-7, draw the image of the *e* as you see it through the ocular. Record the total magnification used in the line at the end of the legend.

   Is the image right side up or upside down compared to the specimen?

   _____

   Compared to the specimen, is the image backward as well as upside down?
   (yes or no) _____

   In summary, the image is *inverted* with respect to the specimen.

3. Move the specimen to the right while watching it through the microscope. In which direction does the image move?

   _____

4. Move the specimen away from you. In which direction does the image move?

   _____

5. Remove the slide and put it away.

Figure 1-7 Drawing of the letter *e* as seen through the ocular (_____×).

F. Depth of Field

The **depth of field** is the distance through which you can move the specimen and still have it remain in focus. Remember, the working distance—the space between the objective lens and the coverslip—decreases with increasing magnifying power. Therefore, the higher the power of the objective in use, the closer the objective is to the slide—and the more careful you must be.

1. Obtain a prepared slide of three crossed colored threads. *This exercise requires care, since you are probably not yet adept at focusing on a specimen.* Once you have the threads in focus (using first the low-power objective and then the medium-power objective), you need only use the fine focus knob to focus with the high-dry objective. After switching to the high-dry objective, try rotating the fine focus knob ½ turn away from you and then a full turn toward you. If you have not found the plane of focus, next try 1½ turns away from you and 2 full turns toward you, and so on. If you work deliberately, you will find the plane of focus and won't crack the coverslip.

   (a) How many threads are in focus using the
   low-power objective? _____
   medium-power objective? _____
   high-dry objective? _____

   (b) With which objective is it easiest to focus a specimen? _____

   (c) At which magnification is it most difficult to focus a specimen? _____

2. Specimens have depth. Continue using the prepared slide of three crossed colored threads.

   (a) Use the high-dry objective to determine the order of the three threads mounted on the slide and record the results in Table 1-4. (Each slide label has a code on it. When you believe that you have discovered the correct order, check with your instructor to find out if you are correct.)

   (b) Focusing carefully with the fine focus knob, move from the bottom to the upper thread. Did you move the knob away from or toward you? _____

3. Remove and put away the slide.

4. Viewing sections of three-dimensional structures makes interpretation of the original shape quite difficult (Figure 1-8).

TABLE 1-4 Order of Threads	
Location	Color
Closest to slide	
Middle	
Closest to coverslip	

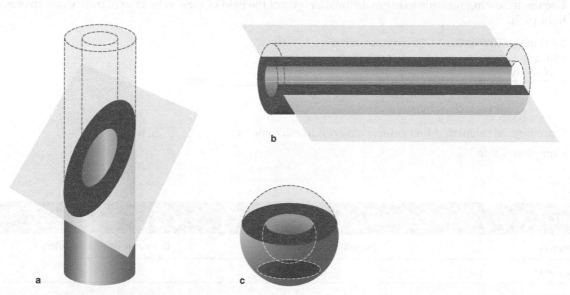

Figure 1-8 Difficulties encountered in interpreting the three-dimensional shape of objects from sections. Compare the transverse (cross) section of the cylinder with the distorted oblique section (**a**) and the longitudinal section (**b**). A similar shape to the transverse section of a cylinder results when a hollow ball is sectioned (**c**). Sectioning the hollow ball through the wall results in a solid shape.

(a) Examine a slide of the cortex of the mammalian kidney with your compound microscope. Complete the hypotheses as to the three-dimensional shape of the structures labeled renal corpuscles and nephron tubules.

The shape of renal corpuscles is _____.

The shape of nephron tubules is _____.

(b) Check the illustration in Exercise 22 to see if your hypotheses are acceptable.

G. Using the Iris Diaphragm to Improve Contrast

1. Place a specimen of unstained fibers on the stage. Locate and focus on these fibers using the medium-power objective. Make sure the condenser and iris diaphragm are correctly set.
2. Close the iris diaphragm.
Does this procedure increase or decrease contrast? _____.
Although this procedure is useful when viewing specimens with low contrast, it should be used only as a last resort because resolving power is also decreased.
3. Remove and put away the slide.

H. Units of Measurement

The basic metric unit of length at the light-microscopic level is the micrometer (μm). An even smaller unit, the nanometer (nm) is often used at the electron-microscopic level.

1000 μm = 1 mm
1000 nm = 1 μm

How many nanometers are there in 1 mm? _____ nm
How many millimeters are there in 1 nm? _____ mm

I. Determining the Diameter of the Field of View

1. Rotate the low-power objective into the light path.

   What is the total magnification? _____×

2. Place a transparent 15-cm ruler on the stage.

3. What is the diameter of the field of view? _____mm

4. Repeat step 3 with the medium-power objective in the light path.

   The total magnification is _____×.

   The diameter of the field of view is _____mm.

5. Use the following formula to estimate the diameter of the field of view when the high-dry objective is in the light path.

$$\frac{\begin{array}{c}\text{total magnification}\\\text{using low-power}\\\text{objective}\end{array} \times \begin{array}{c}\text{mm counted}\\\text{with that}\\\text{objective}\end{array}}{\begin{array}{c}\text{total magnification}\\\text{of high-dry objective}\end{array}} = \underline{\hspace{2cm}} \text{ mm}$$

   Once you've calculated this value, convert it to micrometers: _____μm

6. Complete Table 1-5.

TABLE 1-5 Diameter of Field of View		
Objective	**Magnifying Power**	**Diameter of Field of View**
Low-power		
Medium-power		
High-dry		
Oil-immersion		

7. You will use this information to estimate size by observing the percentage of the diameter of the field of view taken up by the specimen or part of the specimen. Using the medium-power objective, estimate the percentage of the diameter of the field of view covered by the letter *e*. The approximate diameter of the letter *e* is

$$\frac{\text{percent} \times \text{diameter of field of view (mm)}}{100\%} = \underline{\hspace{2cm}} \text{ mm}$$

8. If your microscope is equipped with an ocular micrometer, your instructor may provide directions on how to accurately measure specimen details.

1.2   How to Make a Wet Mount *(About 10 min.)*

In the mid-seventeenth century, Robert Hooke used a microscope to discover tiny, empty compartments in thin shavings of cork. He named them cells. Repeating this historic observation is a good way to learn how to prepare a wet mount.

MATERIALS

Per student:

- compound microscope, lens paper, a bottle of lens-cleaning solution (optional), a lint-free cloth (optional)
- cork

- razor blade
- glass microscope slide
- glass coverslip
- dissecting needle

Per student group (4):

■ dropper bottle of distilled water (dH₂O)

PROCEDURE

1. Carefully use a razor blade to cut a number of *very thin shavings* from a cork stopper. Place them on a glass microscope slide.
2. Gently add a drop of distilled water.
3. Place one end of a glass coverslip to the right or left of the specimen so that the rest of the slip is held at a 45° angle over the specimen (Figure 1-9a).

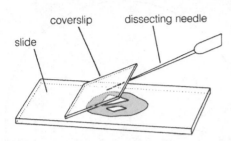
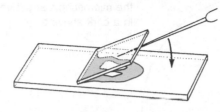

Figure 1-9 How to make a wet mount.

4. Slowly lower the coverslip with a dissecting needle so as not to trap air bubbles (Figure 1-9b).
5. Observe the wet mount, first at low magnification and then with higher power. Air may be trapped either in the cork or as free bubbles (Figure 1-10). Trapped air will appear dark and refractive around its edges. This effect is due to sharply bending rays of light. Draw what you see in Figure 1-11. Note the total magnification used to make the drawing.
6. Clean and replace the slide and coverslip as indicated by your instructor.

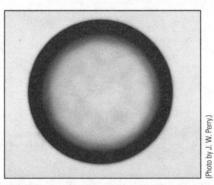

Figure 1-10 Free air bubble (250×).

1.3   Microscopic Observations *(About 20 min.)*

Examining the microscopic world is both challenging and fun. Most of the macroscopic world has been explored, but the microscopic world is barely touched. Yet the microbes in it are essential to our very existence. So be an explorer and see what you can discover!

MATERIALS

Per student:

■ compound microscope, lens paper, a bottle of lens-cleaning solution (optional), a lint-free cloth (optional)
■ glass microscope slide
■ glass coverslip

Per student group (4):

■ pond water or some other mixed culture in a dropper bottle
■ dropper bottle of Protoslo®

Per lab room:

■ reference books for the identification of micro-organisms

PROCEDURE

1. Obtain a drop of pond water or other mixed culture from the bottom of the bottle.
2. Add a drop of Protoslo®. This methyl cellulose solution slows down any swimming microorganisms.
3. Make a wet mount (Figure 1-9).

4. Observe the wet mount with your compound microscope. Start at the upper-left corner of the coverslip and scan the wet mount with the low-power objective. When you find something interesting, focus on it and switch to the medium-power objective and then, if necessary, the high-dry objective.
5. Draw what you find on Figure 1-12 and note the total magnification.
6. Attempt to identify it using the resource books provided you by your instructor. If successful, write its name under your drawing.
7. Clean and replace the slide and coverslip as indicated by your instructor.
8. Put away your compound microscope.

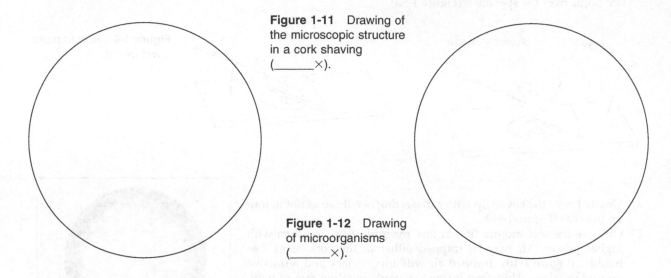

Figure 1-11 Drawing of the microscopic structure in a cork shaving (_____×).

Figure 1-12 Drawing of microorganisms (_____×).

| 1.4 | **Dissecting Microscope** (About 10 min.) |

Dissecting microscopes (Figure 1-13) have a large working distance between the specimen and the objective lens. They are especially useful in viewing larger specimens (including thicker slide-mounted specimens) and in manipulating the specimen (when dissection of a small structure or organism is required, for example).

The large working distance also allows for illumination of the specimen from above (reflected light) as well as from below (transmitted light). Reflected light shows up surface features on the specimen better than transmitted light does.

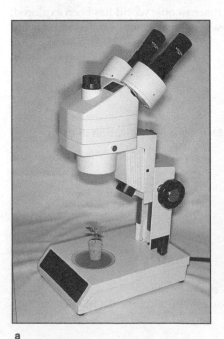

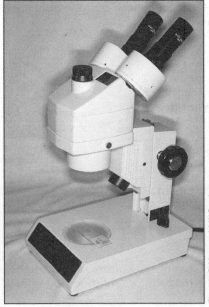

Figure 1-13 Dissecting microscope (**a**) reflected light, (**b**) transmitted light.

(Photos by D. Morton and J. W. Perry.)

a   b

MATERIALS

Per student group:

- dissecting microscope
- specimens appropriate for viewing with the dissecting microscope (for example, a prepared slide with a whole mount of a small organism, bread mold, an insect mounted on a pin stuck in a cork, a small flower)

PROCEDURE

1. Under a dissecting microscope, view one or more of the specimens provided by your instructor. What is the magnification range of this microscope?

   _____× to _____×

2. Is the image of the specimen inverted as in the compound light microscope? (yes or no) _____

3. Describe the type of illumination used by your dissecting microscope. Is there a choice?

   _____

   _____

1.5   Other Microscopes *(About 10 min.)*

In future exercises you will examine pictures taken with other types of microscopes (Figure 1-14). Some will be of living cells taken with a phase-contrast microscope or similar instrument, including those using the Nomarski process. Others will be of very thin-sectioned, heavy metal–stained specimens taken with a transmission electron microscope or TEM (Figure 1-15c and d). Still others will be of precious metal–coated surfaces produced by signals from the scanning electron microscope or SEM (Figure 1-16). Table 1-6 summarizes the technology and use of these microscopes.

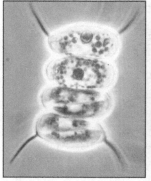 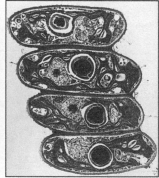 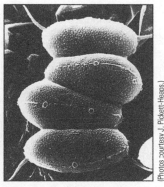

a phase contrast   **b** Nomarski process   **c** transmission electron   **d** scanning electron

(Photos courtesy J. Pickett-Heaps.)

Figure 1-14   How different types of microscopes reveal detail in cells of the green alga *Scenedesmus.*

TABLE 1-6 Other Microscopes		
Microscope	**Technology**	**Use**
Phase-contrast and Nomarski process	Converts phase differences in light to differences in contrast.	Observation of low contrast specimens (often living).
Transmission electron	Increases resolving power by using electrons in a vacuum and magnetic lenses instead of light and glass lenses, respectively.	Preservation of greater specimen detail allows for magnifications up to 1,000,000× or more (usually dead materials).
Scanning electron	Forms TV-like picture from a secondary electron signal, which is emitted from surface points excited by a thin beam of electrons drawn across the surface in a raster pattern.	Investigation of the fine structure of surfaces (usually dead specimens).

a

c

path of light rays (bottom to top) to eye

Ocular lens enlarges primary image formed by objective lenses.

prism that directs rays to ocular lens

Objective lenses (those closest to specimen) form the primary image. Most compound light microscopes have several.

stage (holds microscope slide in position)

Condenser lenses focus light rays onto specimen.

illuminator

microscope base housing source of illumination

b

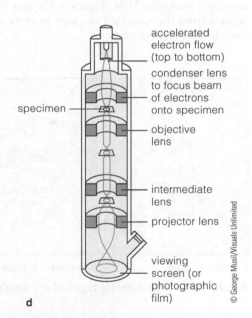

accelerated electron flow (top to bottom)

condenser lens to focus beam of electrons onto specimen

specimen

objective lens

intermediate lens

projector lens

viewing screen (or photographic film)

© George Musil/Visuals Unlimited

d

Figure 1-15 External view (**c**) and diagram (**d**) of a transmission electron microscope. Similar illustrations of a compound light microscope (**a, b**) are provided for comparing the similarities and differences of these microscopes. (After Starr and Taggart, 2001; (**a**) Leica Microsystems, Inc., Deerfield, IL; (**c**) George Musil/Visuals Unlimited.)

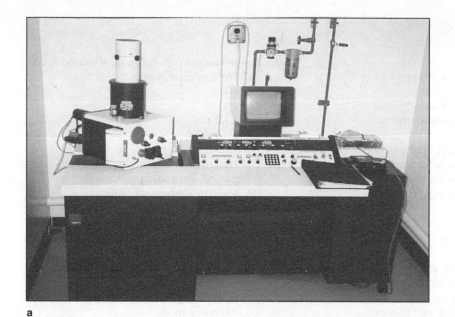
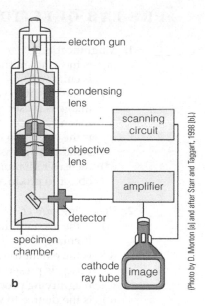

a

b

(Photo by D. Morton (a) and after Starr and Taggart, 1998 (b).)

- electron gun
- condensing lens
- scanning circuit
- objective lens
- amplifier
- detector
- specimen chamber
- cathode ray tube — image

Figure 1-16 External view (**a**) and diagram (**b**) of a scanning electron microscope.

MATERIALS

Per student group:

- photographs of TEM micrographs (negatives) or digital images
- photographs of SEM negatives or digital images

PROCEDURE

1. Examine some photographs of TEM micrographs (negatives) or digital images. The darker areas are more electron-dense in the specimen than the lighter areas.
2. Now look at some photographs of SEM negatives or digital images. The lighter areas correspond to the emission of greater numbers of secondary electrons from that part of the specimen; the darker areas emit less.
3. What type of microscope (compound light, dissecting, phase-contrast, TEM, or SEM) would you use to examine the specimens listed in Table 1-7?

TABLE 1-7 Microscope Use	
Specimen	**Microscope**
Living surface of the finger	
Dye-stained slide of a section of the finger	
Gold-coated bacteria on a single cell of the finger	
Unstained section of a biopsy from the finger	
Heavy metal–stained, very thin section of the finger	

_____ 1. Magnification
   (a) is the amount that an object's image is enlarged
   (b) is the extent to which detail in an image is preserved during the magnifying process
   (c) is the degree to which image details stand out against their background
   (d) focuses radiation emanating from an object to produce an image

_____ 2. Resolving power
   (a) is the amount that an object's image is enlarged
   (b) is the extent to which detail in an image is preserved during the magnifying process
   (c) is the degree to which image details stand out against their background
   (d) focuses radiation emanating from an object to produce an image

_____ 3. A lens
   (a) is the amount that an object's image is enlarged
   (b) is the extent to which detail in an image is preserved during the magnifying process
   (c) is the degree to which image details stand out against their background
   (d) focuses radiation emanating from an object to produce an image

_____ 4. Contrast
   (a) is the amount that an object's image of an object is enlarged
   (b) is the extent to which detail in an image is preserved during the magnifying process
   (c) is the degree to which image details stand out against their background
   (d) focuses radiation emanating from an object to produce an image

_____ 5. The maximum useful magnification for a light microscope is about
   (a) 100×
   (b) 1000×
   (c) 10,000×
   (d) 100,000×

_____ 6. The two image-forming lenses of a compound light microscope are
   (a) the condenser and objective
   (b) the condenser and ocular
   (c) the objective and ocular
   (d) none of these choices

_____ 7. Dyes are usually added to sections of biological specimens to increase
   (a) resolving power
   (b) magnification
   (c) contrast
   (d) all of the above

_____ 8. If the magnification of the two image-forming lenses are both 10×, the total magnification of the image will be
   (a) 1×
   (b) 10×
   (c) 100×
   (d) 1000×

_____ 9. The distance through which a microscopic specimen can be moved and still have it remain in focus is called the
   (a) field of view
   (b) working distance
   (c) depth of field
   (d) magnification

_____ 10. Electron microscopes differ from light microscopes in that
   (a) electrons are used instead of light
   (b) magnetic lenses replace glass lenses
   (c) the electron path has to be maintained in a high vacuum
   (d) a, b, and c are all true

EXERCISE 1

Microscopy

POST-LAB QUESTIONS

1.1 Compound Light Microscope

1. What is the function of the following parts of a compound light microscope?

   a. condenser lens

   b. iris diaphragm

   c. objective

   d. ocular

2. In order, list the lenses in the light path between a specimen viewed with the compound light microscope and its image on the retina of the eye.

3. What happens to contrast and resolving power when the aperture of the condenser (that is, the size of the hole through which light passes before it reaches the specimen) of a compound light microscope is decreased?

4. What happens to the field of view in a compound light microscope when the total magnification is increased?

5. Describe the importance of the following concepts to microscopy.
   a. magnification

   b. resolving power

   c. contrast

6. Which photomicrograph of unstained cotton fibers was taken with the iris diaphragm closed? _____

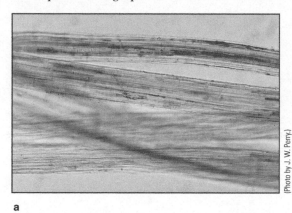

a

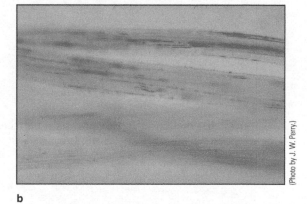

b

7. Describe how you would care for and put away your compound light microscope at the end of lab.

1.2 How to Make a Wet Mount

8.   Describe how to make a wet mount.

1.4 Dissecting Microscope, 1.5 Other Microscopes

9.   A camera mounted on a _____ microscope took this photo of a cut piece of cork.

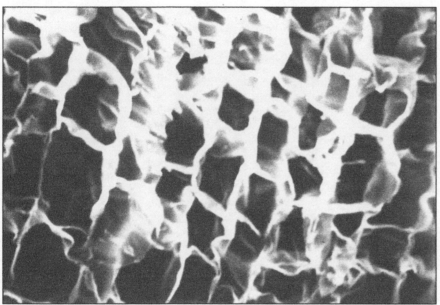

(516×).

Food for Thought

10.   Why were humans unaware of microorganisms for most of their history?

Macromolecules and You: Food and Diet Analysis

OBJECTIVES

After completing this exercise, you will be able to

1. define *macromolecule, vitamin, mineral, carbohydrate, monosaccharide, disaccharide, polysaccharide, lipid, protein, amino acid, calorie;*

2. describe the basic structures of carbohydrates, lipids, glycerides, and proteins;

3. recognize positive and negative tests for carbohydrates, lipids, and proteins;

4. identify the roles that carbohydrates, lipids, proteins, minerals, and vitamins play in the body's construction and metabolism;

5. test food substances to determine the presence of biologically important macromolecules;

6. identify common dietary sources of nutrients;

7. identify the components and relative proportions of a healthy diet with respect to nutrients and calories.

INTRODUCTION

Humans are omnivores, animals who consume a wide variety of food items from several trophic levels. As with all consumers, our food items provide us with the energy stored in the chemical bonds of the food molecules, plus the raw carbon-based materials from which cellular components are built.

Food nutrients include minerals and vitamins, plus the larger molecules known as carbohydrates (which include sugars, starches, and cellulose), lipids (fats and oils), and proteins. As it happens, these last nutrients are three of the four major groups of biological **macromolecules,** large organic molecules of which all cells are made. (The fourth group is the nucleic acids that store and control the genetic instructions within a cell. These crucial macromolecules are not usually utilized by the body as nutrients, and so will not be considered in this exercise.)

Carbohydrates, lipids, and proteins supply the materials from which our cells and tissues are constructed and the energy (measured in kilocalories) to run our metabolic processes. **Vitamins** are necessary organic molecules that our bodies do not construct from other molecules; we require vitamins in our diets in only small amounts. **Minerals** are required inorganic (noncarbon-containing) nutrients such as calcium and potassium. Good health, then, depends upon ingesting the proper balance and quantities of these major nutrients.

In this exercise, you will test for the presence of carbohydrates, lipids, and proteins, and then use your knowledge to determine the macromolecular composition of various common food products. You will also analyze your own diet and compare it to a diet recommended for maintaining good health.

2.1   Identification of Macromolecules

You will use some simple tests for carbohydrates, lipids, and proteins in a variety of substances, including food products. Most of the reagents used are not harmful; however, observe all precautions listed and perform the experiments only in the proper location as identified by your instructor.

A. Carbohydrates

A **carbohydrate** is a simple sugar or a larger molecule composed of multiple sugar units. Carbohydrates are composed of carbon, hydrogen, and oxygen. Those composed of a single sugar molecule are called **monosaccharides.** Examples include ribose and deoxyribose (components of our genetic material), fructose (sometimes called fruit sugar), and glucose (a sugar that commonly serves as the most immediate source of cellular energy needs; see Figure 2-1). Monosaccharides are easily used within cells as energy sources.

Monosaccharides can be bonded together. If two monosaccharides are joined, they form a **disaccharide.** Examples of disaccharides are sucrose (common table sugar), maltose (found in many seeds), and lactose (milk sugar). If more than two monosaccharides are bonded together, the resulting large carbohydrate molecule is called a **polysaccharide.**

Animals, including humans, store glucose in the form of glycogen (Figure 2-2), a highly branched polysaccharide chain of glucose molecules. Starch (Figure 2-2), a key glucose-storing polysaccharide in plants, is an important food molecule for humans. Plant cells are surrounded by a tough cell wall made of another chain formed of glucose molecules, cellulose (Figure 2-2). Our digestive system does not break cellulose apart very well. Fibrous plant materials thus provide bulky cellulose fiber, which is necessary for good intestinal health.

Carbohydrate digestion occurs in humans as digestive enzymes produced in the salivary glands, pancreas, and lining of the small intestine break polysaccharides and disaccharides into monosaccharides such as glucose. The simple sugars are small enough to be absorbed into intestinal cells and then carried throughout the bloodstream.

Many small carbohydrate molecules react upon heating with a copper-containing compound called Benedict's reagent, changing the reagent color from blue to orange or red. A disaccharide may or may not react with Benedict's solution, depending upon how the bonding of the component monosaccharides took place.

Other tests are available for some polysaccharides, the most common of which is Lugol's test for starch. In this test, a dilute solution of potassium iodide reacts with the starch molecule to form a deep blue (nearly black) color product.

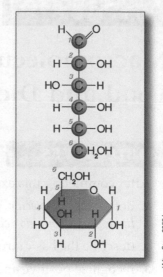

(After Starr, 2000.)

Figure 2-1 Structural diagrams of glucose.

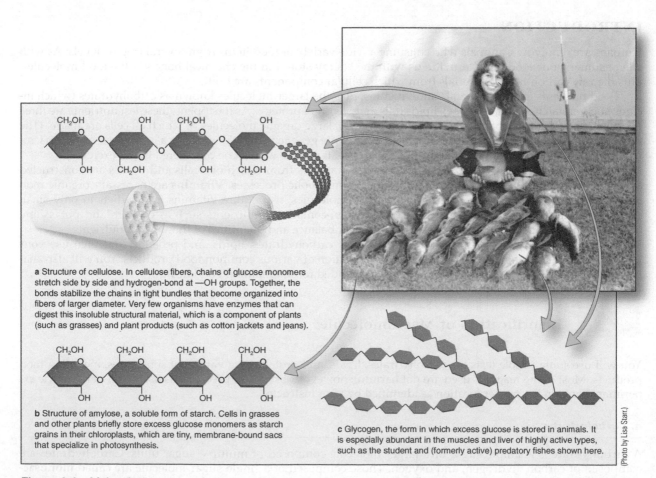

a Structure of cellulose. In cellulose fibers, chains of glucose monomers stretch side by side and hydrogen-bond at —OH groups. Together, the bonds stabilize the chains in tight bundles that become organized into fibers of larger diameter. Very few organisms have enzymes that can digest this insoluble structural material, which is a component of plants (such as grasses) and plant products (such as cotton jackets and jeans).

b Structure of amylose, a soluble form of starch. Cells in grasses and other plants briefly store excess glucose monomers as starch grains in their chloroplasts, which are tiny, membrane-bound sacs that specialize in photosynthesis.

c Glycogen, the form in which excess glucose is stored in animals. It is especially abundant in the muscles and liver of highly active types, such as the student and (formerly active) predatory fishes shown here.

(Photo by Lisa Starr.)

Figure 2-2 Molecular structure of starch, cellulose, and glycogen, and their typical locations in a few organisms.

A.1. Test for Sugars Using Benedict's Solution *(About 20 min.)*

MATERIALS

Per student group (4):

- china marker
- 11 test tubes in test tube rack
- test tube clamp
- hot plate *or* ring stand with wire gauze support and Bunsen burner
- 250- or 400-mL beaker with boiling beads or stones
- stock bottles* of
    Benedict's solution
    distilled water (dH₂O)
    glucose solution

- maltose solution
- lactose solution
- sucrose solution
- fructose solution
- starch solution
- lemon juice
- unsweetened orange juice
- colorless nondiet soda
- colorless diet soda
- vortex mixer (optional)

PROCEDURE

1. Half-fill the beaker with tap water and apply heat with a hot plate or burner to bring the water to a gentle boil.
2. Using the china marker, number the test tubes 1–11.
3. Pipet 2 mL of the correct stock solutions into the test tubes as described in Table 2-1. (If you don't know how to use the Pi-pump, check with your instructor before proceeding.)
4. Add 2 mL of Benedict's solution to each test tube, agitate the mixture by shaking the tubes from side to side or with a vortex mixer, if available, and record the color of the mixture in Table 2-1 in the column "Initial Color."
5. Heat the tubes in the boiling water bath for 3 minutes. Remove the tubes with the test tube clamp, and record any color changes that have taken place in the column "Color After Heating." Also record your conclusions regarding the presence or absence of simple carbohydrates.

TABLE 2-1	Benedict's Test for Sugars			
Tube	Contents	Initial Color	Color After Heating	Conclusion
1	Distilled water			
2	Glucose			
3	Maltose			
4	Lactose			
5	Sucrose			
6	Starch			
7	Fructose			
8	Lemon juice			
9	Orange juice			
10	Nondiet soda			
11	Diet soda			

---

*Each stock bottle should have its own 2- or 2-mL pipet fitted with a Pi-pump.

A.2. Test for Starch Using Lugol's Solution (About 15 min.)

MATERIALS

Per student group (4):

- china marker
- 7 test tubes in test tube rack
- dropper bottle containing Lugol's solution
- stock bottles* of
    distilled water (dH₂0)
    glucose solution

- maltose solution
- lactose solution
- sucrose solution
- starch solution
- cream
- vortex mixer (optional)

PROCEDURE

1. Using the china marker, number the test tubes 1–7.
2. Pipet 3 mL of the correct stock solutions into the test tubes as described in Table 2-2.
3. Record the color of each solution in the column "Initial Color" in Table 2-2.
4. Add 9 drops of Lugol's solution to each test tube, agitate the mixture by shaking the tubes from side to side or with a vortex mixer, if available, and record the color of the mixture in the column "Color After Adding Lugol's Solution." Also record your conclusions regarding the presence or absence of starch in each test solution.

TABLE 2-2 Lugol's Test for Starch

Tube	Contents	Initial Color	Color After Adding Lugol's Solution	Conclusion
1	Distilled water			
2	Glucose			
3	Maltose			
4	Lactose			
5	Sucrose			
6	Starch			
7	Cream			

5. Hypothesize about why you got the result you did with the Lugol's test for each substance:

dH₂O _____

glucose _____

maltose _____

lactose _____

sucrose _____

starch _____

Does cream contain starch? _____

B. Lipids

Lipids are oily or greasy compounds insoluble in water, but dissolvable in nonpolar solvents such as ether or chloroform. Those lipids having fatty acids—hydrocarbon chains with a carboxyl (—COOH) group at one end—combine with glycerol to form glycerides, a rich source of stored energy. Lipids with three fatty acid chains attached to a glycerol backbone are called triglycerides (Figure 2-3).

---

*Each stock bottle should have its own 2-mL pipet fitted with a Pi-pump.

Lipids provide long-term energy storage in cells and are very diverse. Lipid digestion occurs primarily in the small intestine where bile produced by the liver breaks lipid globules into smaller droplets, and then pancreatic enzymes break large lipid molecules into smaller components for absorption. The lipid components are then transported throughout the body in lymph, the fluid that bathes the tissues.

Substances we think of as fats and oils are examples of lipids. One of the most simple tests for lipids is to determine whether they leave a grease spot on a piece of uncoated paper, such as a grocery sack. A test commonly used to identify fats and oils in microscopic preparations—Sudan IV—can also be used to indicate their presence in a test tube.

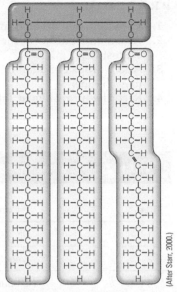

Figure 2-3 A triglyceride.

(After Starr, 2000.)

B.1. Uncoated Paper Test (About 20 min.)

MATERIALS

Per student group (4):

- china marker
- dropper bottles of
    distilled water (dH₂O)
    vegetable oil
- stock bottles* of
    hamburger juice
    onion juice

- colorless nondiet soft drink
- cream
- dropper bottle of Sudan IV
- vortex mixer (optional)
- piece of uncoated paper (grocery bag)

PROCEDURE

1. Place a drop of distilled water and a drop of vegetable oil on a piece of grocery bag. Set aside to dry for 10 minutes.
2. After 10 minutes, describe the appearance of each spot on the paper.

    distilled water _____

    vegetable oil _____

3. Place a drop of each test substance on a labeled spot on the grocery bag piece. Set aside to dry for 10 minutes.
4. After 10 minutes, describe the appearance of each spot on the paper.

    hamburger juice _____

    onion juice _____

    colorless
    nondiet soft drink _____

    cream _____

5. What can you infer about the lipid content of each substance?

    _____

    _____

    _____

B.2. Sudan IV Test (About 20 min.)

MATERIALS

Per student group (4):

- china marker
- dropper bottles of
    distilled water (dH₂O)
    vegetable oil
- 6 test tubes in test tube rack
- stock bottles* of
    hamburger juice
    onion juice

- colorless nondiet soft drink
- cream
- dropper bottle of Sudan IV
- vortex mixer (optional)

---

*Each stock bottle should have its own 2-mL pipet fitted with a Pi-pump.

PROCEDURE

1. With a china marker, number the test tubes 1–6.
2. Add 3 mL of dH_2O to each test tube and then 3 mL of each substance indicated in Table 2-3 to the appropriate test tube.
3. Add 9 drops of Sudan IV to each tube, agitate by shaking the tube side, to side and then add 2 mL of dH_2O to each tube.
4. Record your results in Table 2-3.

TABLE 2-3 Test for Lipids with Sudan IV

Tube	Contents	Observations After Addition of Sudan IV	Conclusion
1	Distilled water		
2	Vegetable oil		
3	Hamburger juice		
4	Onion juice		
5	Soft drink		
6	Cream		

Which test, the uncoated paper test or the Sudan IV test, do you think is most sensitive to small quantities of lipids?

_____

What are the limitations of these tests?

_____

_____

C. Proteins

Proteins are a diverse group of macromolecules with a wide range of functions in the human body. Many are structural components of muscle, bone, hair, fingernails, and toenails, among other tissues. Others are enzymes that speed up cellular reactions that would otherwise take years to occur. Movement of structures within cells (such as occurs during cell division) and of sperm cells is also associated with proteins.

All proteins are chains of **amino acids,** whose general structure is illustrated in Figure 2-4.

Though there are only about 20 amino acids, innumerable kinds of proteins result from different amino acid sequences. A protein begins to form when two or more amino acids are linked together by peptide bonds (bonding between the amino group of one amino acid and the acid group of another), forming a polypeptide chain (Figure 2-5). Subsequently, the chain folds and twists, with links formed between adjacent parts of the chain. Humans cannot synthesize about half of the amino acids; these essential amino acids must be obtained through our foods.

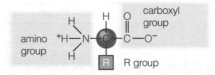

Figure 2-4 General structure of an amino acid. The R group consists of one or more atoms that make each kind of amino acid unique. (After Starr, 2000.)

Protein digestion begins in the stomach lining and continues in the small intestine, where various enzymes break protein molecules first into protein fragments, and then into amino acids, which are absorbed across the small intestine wall. The amino acids are then transported throughout the body in the blood.

Several tests are used for proteins, including Biuret reagent, which indicates the presence of peptide bonds. The greater the number of peptide bonds, the more intense the color reaction with Biuret.

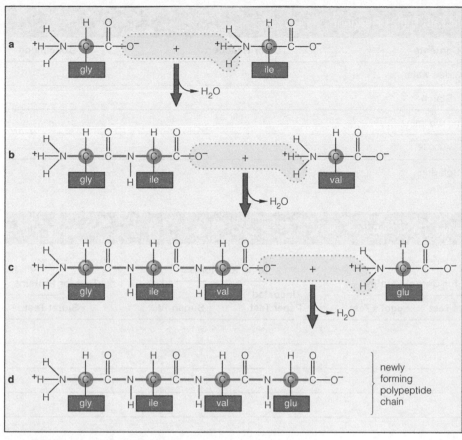

(a) The first two amino acids shown are glycine (gly) and isoleucine (ile). They are at the start of the sequence for one of two polypeptide chains that make up the protein insulin in cattle.

(b) Through a condensation reaction, the isoleucine becomes joined to the glycine by a peptide bond. A water molecule forms as a by-product of the reaction.

(c) A peptide bond forms between the isoleucine and valine (val), another amino acid, and water again forms.

(d) Remember, DNA specifies the order in which the different kinds of amino acids follow one another in a growing polypeptide chain. In this case, glutamate (glu) is the fourth amino acid specified.

(After Starr, 2000.)

Figure 2-5 Peptide bond formation during protein synthesis.

C.1. Biuret Test for Proteins with Biuret Reagent (About 15 min.)

MATERIALS

Per student group (4):

- china marker
- 5 test tubes in test tube rack
- stock bottles* of
  Biuret reagent
  distilled water (dH₂O)

  starch solution
  egg albumin
  glucose solution
  colorless soft drink
- vortex mixer (optional)

PROCEDURE

1. Using the china marker, number the test tubes 1–5.
2. Pipet 1 mL of the correct stock solutions into the test tubes as described in Table 2-4.
3. Add 10 drops of Biuret reagent to each test tube and agitate the mixture by shaking the tubes from side to side or with a vortex mixer, if available.
4. Wait 2 minutes and then record the color of the mixture in Table 2-4, in the column "Color Reaction." Also enter your conclusion about the presence of protein in the substance.

D. Testing Unknown Food Substances (About 30 min.)

Your instructor will provide you with several samples of unidentified food substances.

PROCEDURE

1. Following the previous procedures, perform tests for carbohydrates (Benedict's test and Lugol's solution test), lipids (uncoated paper and Sudan IV tests), and proteins (Biuret reagent) on each item.
2. Record your test results (+ or – for each test) in Table 2-5.

---

*Each stock bottle should have its own 2- or 2-mL pipet fitted with a Pi-pump.

TABLE 2-4 Test for Protein with Biuret Reagent

Tube	Contents	Color Reaction	Conclusion
1	Distilled water		
2	Starch		
3	Egg albumin		
4	Glucose		
5	Soft drink		

TABLE 2-5 Tests for Unknown Food Substances

Unknown Sample Designation	Tests for Carbohydrates		Tests for Lipids		Test for Proteins
	Benedict's Test	Lugol's Test	Uncoated Paper Test	Sudan IV	Biuret Test

Note: After completing all experiments, take your dirty glassware to the sink and wash it following the directions given in *"Instructions for Washing Laboratory Glassware,"* page x. Invert the test tubes in the test tube rack so they drain. Tidy up your work area, making certain all equipment used in this exercise is there for the next class.

2.2 The Food Pyramid and Diet Analysis

The foods we eat are usually complex mixtures of various macromolecules, vitamins, and minerals that supply energy and raw materials to the body. However, food must be digested before most nutrients are available within cells. Within the digestive system, macromolecules are hydrolyzed to smaller components (proteins broken down to amino acids, for example). These smaller molecules are absorbed across the intestinal wall, transported throughout the body, and used to build and power cells. We truly *are* what we eat.

The U.S. Department of Agriculture issued its "MyPyramid" food guide in 2005 to provide daily dietary guidelines for average people of good health. Several nutrition recommendations underlie MyPyramid: moderation in portion sizes (many American portions are 2 to 3 times larger than recommended); balance among food groups to ensure that all key nutrients are included; and emphasis on daily exercise for a multitude of health benefits.

The major food groups are the grains, vegetables, fruits, dairy and soy, meat and beans, and vegetable oils. There is also an allotment of "discretionary calories" for each person. Why is it important to include each of these in our diets? What do we gain from each type of food?

Grains are high in the complex carbohydrates (starches and cellulose) that provide ready energy sources. The USDA recommends that you "make half your grains whole," meaning that *at least* half your grain intake should be whole grains, which supply important vitamins and minerals, plus the fiber needed for a healthy digestive system. Whole grains include oatmeal, brown rice, and foods whose first ingredients include the words *whole grain* or *whole wheat*.

Vegetables and **fruits** are especially important sources of the vitamins and minerals needed for a healthy diet, as well as complex carbohydrates and **phytochemicals**, a catchall term for a diverse group of molecules whose benefits we are just beginning to understand. Diets that include plenty of vegetables and fruits help prevent a variety of illnesses, including heart disease, stroke, and certain types of cancer. To meet your requirements for these important foods, mix it up. Eat more vegetables and fruits of every color and variety, especially the vegetables that provide high volume and high nutritional value with low calorie impact (unless they're deep-fried).

Low-fat **dairy** and **soy** products are important sources of calcium and provide significant amounts of proteins. Protein-rich, vitamin-D-fortified dairy and soy foods, along with weight-bearing exercise, build and maintain strong bones.

Smaller quantities are recommended of the "meat and bean" group, which includes poultry, fish, and nuts. These protein-rich foods supply essential amino acids, vitamins, and minerals, and also help maintain the "full" feeling between meals. Choose beans, fish, and lean meats and poultry, along with lower quantities of calorie-rich nuts, rather than fatty or deep-fried choices with unhealthy lipid levels.

Vegetable oils have been promoted to their own food group in MyPyramid, and include monounsaturated vegetable oils (olive oil and canola oil, for example) and polyunsaturated oils (corn oil, soybean oil). These oils are considered healthy in moderation, unlike the solid saturated fats (sour cream, butter, palm oil) and especially the harmful trans fats (partially hydrogenated vegetable oils) that are found in many processed and restaurant foods. Aim for zero trans fats in your diet.

Discretionary calories can be used to eat more foods from any food group or to eat higher-calorie forms of a food or beverage with added sugars or fats. Such foods are high in calorie-dense lipids and the simple sugars that are the "lighter fluid" (quickly used energy source) of metabolism. While some lipids are essential nutrients, other food groups on the pyramid provide those lipids in abundance. Foods that provide abundant fats and simple sugars usually contain few other nutrients and so are not recommended dietary components.

The dietary guidelines of MyPyramid vary depending upon your gender and activity level, with more calories of food energy needed to power your cells as your activity level increases. Use Table 2-6 to determine your recommended average daily calorie allowance.

TABLE 2-6 Approximate Daily Calorie Requirements

FEMALES	Sedentary (less than 30 min. moderate activity daily)	Moderate (30 to 60 min. moderate activity daily)	Active (more than 60 min. moderate activity daily)
Ages 16 to 18	1800	2000	2400
Ages 19 to 30	2000	2200	2400
Ages 31 to 50	1800	2000	2200
MALES			
Ages 16 to 18	2400	2800	3200
Ages 19 to 30	2400	2800	3000
Ages 31 to 50	2200	2600	3000

Use Table 2-7 to find out recommendations for consumption of each food group for your calorie level.

Follow the MyPyramid guidelines and eat a variety of foods to enjoy a healthy diet that provides balanced nutrients without excess calories.

A. Food Diary

In this activity you will analyze *your* diet, so that you can see in which areas you maintain good nutrition, and which areas need improvement.

PROCEDURE (to be completed before class, about 15 minutes per day)
For 2 days during the week preceding this activity, keep a complete diary of *every* food and drink item you consume other than water, unsweetened coffee or tea, or other zero-calorie beverages. In Table 2-8, record what you eat and drink, and how much (the portion size: ounces, cups, teaspoons, and so on). Enter this data in the first two columns. Be as honest, specific, and descriptive as you can. Also keep track separately of which items of processed foods contain trans fats (this information will be found on the food label). Do this for *each* meal and snack.

TABLE 2-7 Recommended Daily Food Group Intakes

Calorie level	1800	2000	2200	2400	2600	2800	3000	3200
Grains	6 oz	6 oz	7 oz	8 oz	9 oz	10 oz	10 oz	10 oz
Vegetables	2 cups (c)	2½ c	3 c	3 c	3½ c	3½ c	4 c	4 c
Fruits	1½ c	2 c	2 c	2 c	2 c	2½ c	2½ c	2½ c
Dairy & Soy	3 c	3 c	3 c	3 c	3 c	3 c	3 c	3 c
Meat & Beans	5 oz	5½ oz	6 oz	6½ oz	6½ oz	7 oz	7 oz	7 oz
Vegetable Oils	5 tsp	6 tsp	6 tsp	7 tsp	8 tsp	8 tsp	10 tsp	11 tsp
Discretionary calories*	195	267	290	362	410	426	512	650

*Discretionary calories are those "extra" calories, usually from fats and sugars, that remain when all the other food group portions and nutrients are consumed.

Eat your usual diet. Don't change your eating habits for this exercise.

It's often difficult to determine portion sizes. To do this accurately, record the information on food product labels regarding serving size and calorie content of each item whenever possible. You also may want to use a measuring cup for this activity. For nonlabeled foods, use the following guidelines:

Portion	Approximate Size of Item
1 ounce	one slice processed cheese; two dice
3 ounces meat	one deck of playing cards or computer mouse
½ cup	one tennis ball or one ice cream scoop
1 serving fruit	one baseball (*not* the larger softball)
1 cup	volume of a woman's fist
1 teaspoon	top half of thumb
1 ounce grains	½ cup oatmeal or rice or 1 cup cereal flakes or ½ English muffin or 1 slice commercial bread

TABLE 2-8 Food Diary

| Food or Beverage Item | Total Portion Size | Food Group Servings in Each Food Item | | | | | | Calories |
		Grains (ounces)	Vegetables (cups)	Fruits (cups)	Dairy & Soy (cups)	Meat & Beans (ounces)	Oils (teaspoons)	
2-day total:								
2-day average:								

MATERIALS

Per student group (4):

- food diary data
- diet analysis books and/or computers with diet analysis software
- colored paper squares
- glue or tape

PROCEDURE

1. Complete the food diary table (Table 2-8) in class if necessary, listing everything you've ingested, including both the food group and the number of servings. Then determine how each food item is allocated as servings of the various MyPyramid food groups: grains, vegetables, fruits, dairy and soy, meat and bean, and vegetable oils. You can find portion information for common foods in online sources and/or the books provided in your classroom.

2. Determine and record the calorie content of each item using the resources provided in class.

3. Average the food group servings and calorie content over the 2 days of data collection, and use the averaged data for the rest of this activity.

4. Construct a food pyramid to visualize the MyPyramid recommendations for your gender and activity level. Use the colored paper blocks, with each block corresponding to one serving unit (ounce, cup, or teaspoon, depending on the food group involved). Glue or tape the pyramid to the bottom of page 33 in the following order, from bottom up: orange = grain group servings (ounces); blue = vegetable group servings (cups); green = fruit group servings (cups); yellow = dairy and soy group servings (cups); brown = meat and bean group servings (ounces); red = vegetable oil servings (teaspoons); and purple = discretionary calories (100 calories per block).

5. Now construct your **personal food pyramid** in the same way on the next page, using 2-day average data from your food diary.

6. Answer the following questions AFTER you have completed your 2-day food diary and pyramids:

    (a) What is your typical level of physical activity (sedentary, moderate, active)? _____

    (b) What is your approximate daily calorie requirement? _____

    (c) How does your caloric intake compare with the recommendations for your gender and activity level?

    _____

    _____

    _____

    _____

    (d) What food groups are you eating too much of? What specific kinds of nutrient molecules are you thus overconsuming? What are the potential consequences of this excess for your health?

    _____

    _____

    _____

    _____

    (e) What food groups are you not eating enough of? What kinds of specific nutrient molecules are thus underrepresented in your diet? What are the potential consequences of this deficiency for your health?

    _____

    _____

    _____

    _____

    (f) What proportion of your grain servings included whole grains? _____
    How many different *kinds* of vegetables and fruits did you consume? _____
    How many food items contained saturated or trans fats? _____

    (g) Given all the above information, describe two *reasonable* changes you could make to improve your diet and your health.

    _____

    _____

    _____

    _____

_____ 1. A carbohydrate consists of
   (a) amino acid units
   (b) one or more sugar units
   (c) lipid droplets
   (d) glycerol

_____ 2. A protein is made up of
   (a) amino acid units
   (b) one or more sugar units
   (c) lipid droplets
   (d) Biuret solution

_____ 3. Benedict's solution is commonly used to test for
   (a) proteins
   (b) certain carbohydrates
   (c) nucleic acids
   (d) lipids

_____ 4. Glycogen is
   (a) a polysaccharide
   (b) a storage carbohydrate
   (c) found in human tissues
   (d) all of the above

_____ 5. To test for starch, one would use
   (a) Benedict's solution
   (b) uncoated paper
   (c) Sudan IV
   (d) Lugol's solution

_____ 6. Rich sources of stored energy that are dissolvable in organic solvents are
   (a) carbohydrates
   (b) proteins
   (c) glucose
   (d) lipids

_____ 7. Rubbing a substance on uncoated paper should reveal if it is a
   (a) lipid
   (b) carbohydrate
   (c) protein
   (d) sugar

_____ 8. Proteins consist of
   (a) monosaccharides linked in chains
   (b) amino acid units
   (c) polysaccharide units
   (d) condensed fatty acids

_____ 9. Biuret reagent will indicate the presence of
   (a) peptide bonds
   (b) proteins
   (c) amino acids units linked together
   (d) all of the above

_____ 10. The largest number of food servings in your daily diet should be from
   (a) meats and beans
   (b) dairy products
   (c) vegetables and fruits
   (d) vegetable oils

Name _____ Section Number _____

EXERCISE 2

Macromolecules and You: Food and Diet Analysis

POST-LAB QUESTIONS

A.1. Test for Sugars Using Benedict's Solution

1. Let's suppose you are teaching science in a part of the world without ready access to a doctor and you're worried that you may have developed diabetes. (Diabetics are unable to regulate blood glucose levels, and glucose accumulates in blood and urine.) What could you do to gain an indication of whether or not you have diabetes?

2. The test tubes in the photograph contain Benedict's solution and two unknown substances that have been heated. What do the results indicate?

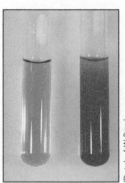
(Photo by J. W. Perry.)

3. How could you verify that a soft-drink can contains diet soda rather than soda sweetened with fructose?

A.2. Test for Starch Using Lugol's Solution

4. Observe the photomicrograph accompanying this question. This section of a potato tuber has been stained with Lugol's iodine solution. When you eat french fries, the potato material is broken down in your small intestine into what small subunits?

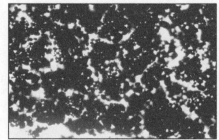
(Photo by J. W. Perry.)

B.2. Sudan IV Test

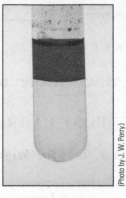

(Photo by J. W. Perry)

5. The test tube in the photograph contains water at the bottom and another substance that has been stained with Sudan IV at the top. What is the macromolecular composition of this stained substance?

6. You are given a sample of an unknown food. Describe how you would test it for the presence of lipids.

7. You wish to test the same unknown food for the presence of sugars. Describe how you would do so.

Food for Thought

8. What is the purpose of the distilled water sample in each of the chemical tests in this exercise?

9. Many health food stores carry enzyme preparations that are intended to be ingested orally (by mouth) to supplement existing enzymes in various organs like the liver, heart, and muscle. Explain why these preparations are unlikely to be effective as advertised.

10. A young child grows rapidly, with high levels of cell division and high energy requirements. If you were planning the child's diet, which food groups would you emphasize, and why? Which food groups would you deemphasize, and why?

Analysis of the Macromolecules Found in Various Items Commonly Available in Grocery Stores

Abstract

Food labels provide consumers with nutritional information that is necessary for people to make healthy food choices. The nutrition facts box that is commonly found on product packaging contains information on the number of servings in the package, serving size, and amount per serving of total fat, saturated fat, cholesterol, total carbohydrates, simple sugars, dietary fiber, protein, and other nutrients of major health concern. However, this information is useful only if it is accurate.

   In this study, I chose to examine the accuracy of nutritional information provided in the nutrition facts box of five common packaged food products from a local supermarket. I used chemical tests to assess the presence of simple sugars, starch, protein, and fats. I discovered that two of the five food products tested did not have accurate nutritional information. The significance of these findings is discussed and further research is suggested.

Introduction

The U.S. Department of Agriculture (USDA) publishes the periodic *Dietary Guidelines for Americans*. These guidelines are designed to promote good health and reduce the risk of major chronic disease (USDA 1995). The food guide pyramid, also published by the USDA, provides instructions in following these guidelines (USDA 1992). To track the quality of consumers' diets, the USDA uses the Healthy Eating Index (HEI), which is based on a 10-component system of five food groups, four nutrients, and a measure of variety in food intake (Kennedy et al. 1995). By using the HEI, consumers can discover how well their diets conform to the dietary guidelines and the food guide pyramid.

   Consumers that use the HEI must rely on nutrient labels to determine the amounts of carbohydrates, proteins, fats and certain other key nutrients they consume. Nutrition labels contain information on the number of servings in the package, serving size, and the amount per serving of saturated fat, cholesterol, sugars, dietary fiber, protein, and other nutrients of major health concern. The Office of Nutritional Products, Labeling, and Dietary Supplements, under regulations from the U.S. Food and Drug Administration of the Department of Health and Human Services and the Food Safety and Inspection Service of the U.S. Department of Agriculture, is responsible for making sure that food labels offer complete, useful, and accurate nutrition information (U.S. Food and Drug Administration 2007). Nutrient content claims must adhere to these regulations to ensure that consumers have the information they need to make healthful food choices. This is especially true for consumers who may be allergic to certain additives.

   My goal was to analyze the accuracy of the nutrition facts box for various packaged products sold at a local supermarket. I hypothesized that if the product nutrient label showed that fat, sugars, starch, and protein were present in the product, then these macromolecules would test positive in samples of the product analyzed in the laboratory.

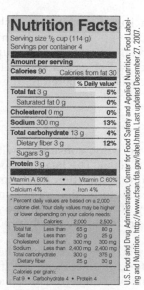

Nutrition facts box.

Materials and Methods

I obtained the five common packaged products listed in Table 1 at a local supermarket. My sample size included three separate packages for each different packaged item. I placed a one-half cup sample of the product in two cups of water and liquefied as much material as possible using a food processor. I then poured the resulting liquid into a test tube.

TABLE 1 Results of chemical tests for biologically important macromolecules in common packaged food items

	Benedict's Test	Lugol's Test	Biuret Test	Sudan IV Test
Sugarless cookies	−	+	−	+
Oatmeal cookies	+	+	+	+
Canned black beans	−	+	+	−
Canned red beans	+	+	+	+
Canned nuts	+	−	+	+
Control	−	−	−	−

To test for the presence of simple sugars, I used the Benedict's test. I expected the products that listed simple sugars on their nutrition facts labels would have solutions that changed from blue to green (or reddish brown if they contained a high amount of simple sugars) when heated with Benedict's reagent. I added a dropperful of Benedict's solution to each test tube and heated the test tubes in a boiling water bath for three minutes. Data were recorded in Table 1.

To test for the presence of starch, a complex carbohydrate, I used the iodine test (also known as Lugol's test). I expected the products that listed starch on their nutrition facts labels would have solutions that changed from a yellowish-brown color to a bluish-black color when iodine was added. Data were recorded in Table 1.

To test for the presence of proteins, I used the Biuret test. I expected the products that listed protein on their nutrition facts labels would have solutions that turned a violet color when the Biuret test reagent was added. Data were recorded in Table 1.

To test for the presence of fats, I used the Sudan IV test. I expected the products that listed fat on their nutrition facts labels would have solutions that interacted with the Sudan IV dye to develop two separate liquid layers, with a red color occurring at the interface of the lipid and the Sudan IV dye. Data were recorded in Table 1.

Results

The results of the chemical analysis of the packaged food items are shown in Table 1. The sugarless cookies, oatmeal cookies, and canned black beans had observed test results in accordance with their expected results based on their nutrition facts labels. Therefore, the hypothesis was supported for these three packaged products. The canned beans and the canned nuts had observed results that were not in accordance with their expected results based on their nutrition facts labels. Therefore, the hypothesis was rejected for these two packaged products.

Discussion

The USDA's *Dietary Guidelines for Americans* provide consumers with science-based recommendations for exercise and dietary intake for health (USDA 2005). The health-conscious consumer who wishes to follow these guidelines needs accurate information from nutrition labels to make informed choices. The nutrition facts box is an important and valuable source of information for helping consumers choose the foods that best contribute to their health (National Academy of Sciences 2004). The results of my study, however, revealed that two of five common packaged food items failed to accurately report their nutrient content in the nutrient facts box.

The significance of the results of this study in terms of consumer health choices cannot be overemphasized. Truth in labeling is a serious issue, especially for consumers with specific diseases or allergic conditions. Carbohydrates, especially glucose levels, are of paramount concern for individuals with diabetes or for those who are at risk of cardiovascular disease (CVD) (Jacobs et al. 1998). Also of concern to health-conscious consumers are lipids, especially saturated fats, because of their association with CVD development in certain types of individuals (Hu et al. 1997).

Based on the results of this study, further investigations are warranted that examine commonly available food products from local grocers. Other studies could be designed that analyze the daily meals that individual's consume in terms of the macromolecules present in their diet. Pet food products could also be included in future analyses. Finally, with recent concerns over contaminants in food imported from abroad (Buzby 2002), chemical tests for other molecules could be examined.

Citations

Buzby, J. C. 2002. Effects of food safety perceptions on food demand and global trade. In A Regmi (Ed.), *Changing Structure of Global Food Consumption and Trade*. Market and Trade Economics Division, Economic Research Service, USDA, Agriculture and Trade Report. WRS-01-1. Washington, DC: U.S. Government Printing Office.

Hu, F. B., M. J. Stampfer, J. E. Manson, E. B. Rimm, A. Wolk, G. A. Colditz, C. H. Hennekens, and W. C. Willett. 1997. Dietary fat intake and the risk of coronary heart disease in women. *New England Journal of Medicine* 337: 1491–9.

Jacobs, D. R. Jr., K. A. Meyer, L. H. Kushi, and A. R. Folsom. 1998. Whole-grain intake may reduce the risk of ischemic heart disease death in postmenopausal women: The Iowa Women's Health Study. *American Journal of Clinical Nutrition* 68: 248–57.

Kennedy, E. T., J. Ohls, S. Carlson, and K. Fleming. 1995. The healthy eating index: Design and applications. *Journal of the American Dietetic Association* 95: 1103–8.

National Academy of Sciences. 2004. *Dietary reference intakes: Guiding principles for nutrition labeling and fortification* (free Executive Summary). Washington, DC: National Academies Press. http://www.nap.edu/catalog/10872.html.

U.S. Department of Agriculture (USDA). 1992. *The food guide pyramid*. Washington, DC: U.S. Government Printing Office.

———. 1995. *Nutrition and your health: Dietary guidelines for Americans*. Washington, DC: U.S. Government Printing Office.

———. 2998. Dietary guidelines. March. Washington, DC: U.S. Government Printing Office. www.mypyramid.gov/guidelines/index_print.html.

U.S. Food and Drug Administration. 2007. *Food labeling and nutrition*. Center for Food Safety and Applied Nutrition. www.cfsan.fda.gov/label.html.

Structure and Function of Living Cells

OBJECTIVES

After completing this exercise, you will be able to

1. define *cell, cell theory, prokaryotic, eukaryotic, nucleus, cytomembrane system, organelle, multinucleate, cytoplasmic streaming, sol, gel, envelope;*

2. list the structural features shared by all cells;

3. describe the similarities and differences between prokaryotic and eukaryotic cells;

4. identify the cell parts described in this exercise;

5. state the function for each cell part;

6. distinguish between plant and animal cells;

7. recognize the structures presented in **boldface** in the procedure sections.

INTRODUCTION

Structurally and functionally, all life has one common feature: All living organisms are composed of **cells.** The development of this concept began with Robert Hooke's seventeenth-century observation that slices of cork were made up of small units. He called these units "cells" because their structure reminded him of the small cubicles that monks lived in. Over the next 100 years, the **cell theory** emerged. This theory has three principles: (1) All organisms are composed of one or more cells; (2) the cell is the basic *living* unit of organization; and (3) all cells arise from preexisting cells.

Although cells vary in organization, size, and function, all share three structural features: (1) All possess a **plasma membrane** defining the boundary of the living material; (2) all contain a region of **DNA** (deoxyribonucleic acid), which stores genetic information; and (3) all contain **cytoplasm,** everything inside the plasma membrane that is not part of the DNA region.

With respect to internal organization, there are two basic types of cells, **prokaryotic** and **eukaryotic.** Study Table 3-1, comparing the more important differences between prokaryotic and eukaryotic cells. The Greek word *karyon* means "kernel," referring to the nucleus. Thus, *prokaryotic* means "before a nucleus," while *eukaryotic* indicates the presence of a "true nucleus." Prokaryotic cells typical of bacteria, cyanobacteria, and archaea are believed to be similar to the first cells, which arose on Earth 3.5 billion years ago. Eukaryotic cells, such as those that comprise the bodies of protists, fungi, plants, and animals, probably evolved from prokaryotes.

This exercise will familiarize you with the basics of cell structure and the function of prokaryotes (prokaryotic cells) and eukaryotes (eukaryotic cells).

3.1   Prokaryotic Cells *(About 20 min.)*

MATERIALS

Per student:

- dissecting needle
- compound microscope
- microscope slide
- coverslip

Per student pair:

- distilled water (dH$_2$O) in dropping bottle

Per student group (table):

- culture of a cyanobacterium (either *Anabaena* or *Oscillatoria*)

Per lab room:

- 3 bacterium-containing nutrient agar plates (demonstration)
- 3 demonstration slides of bacteria (coccus, bacillus, spirillum)

TABLE 3-1 Comparison of Prokaryotic and Eukaryotic Cells

Characteristic	Cell Type	
	Prokaryotic	**Eukaryotic**
Genetic material	Located within cytoplasm, not bounded by a special membrane Consists of a single molecule of DNA	Located in **nucleus,** a double membrane-bounded compartment within the cytoplasm Numerous molecules of DNA combined with protien Organized into chromosomes
Cytoplasmic structures	Small ribosomes Photosynthetic membranes arising from the plasma membrane (in some representatives only)	Large ribosomes **Cytomembrane system,** a system of connected membrane structures **Organelles,** membrane-bounded compartments specialized to perform specific functions
Kingdoms represented	Bacteria Archaea	Protista Fungi Plantae Animalia

PROCEDURE

1. Observe the culture plate with bacteria growing on the surface of a nutrient medium. Can you see the individual cells with your naked eye?

   _____

2. Observe the microscopic preparations of bacteria on *demonstration* next to the culture plate. The three slides represent the three different shapes of bacteria. Which objective lenses are being used to view the bacteria?

   _____

3. Can you discern any detail within the cytoplasm?

   _____

   In the space provided in Figure 3-1, sketch what you see through the microscope. Record the magnification you are using in the blank provided in the figure caption. Then record the approximate size of the bacterial cells. (Return to Exercise 1 if you've forgotten how to estimate the size of an object being viewed through a microscope.)

4. Study Figure 3-2, a three-dimensional representation of a bacterial cell. Now examine the electron micrograph of the bacterium *Escherichia coli* (Figure 3-3). Locate the **cell wall,** a structure chemically distinct from the wall of plant cells but serving the same primary function to contain and protect the cell's contents.

5. Find the **plasma membrane,** which is lying flat against the internal surface of the cell wall and is difficult to distinguish.

6. Look for two components of the **cytoplasm: ribosomes,** electron-dense particles (they appear black) that give the cytoplasm its granular appearance, and a relatively electron-transparent region (appears light) containing fine threads of DNA called the **nucleoid.**

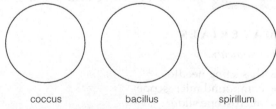

coccus   bacillus   spirillum

Figure 3-1 Drawing of several bacterial cells (_____×). Approximate size = _____ μm.

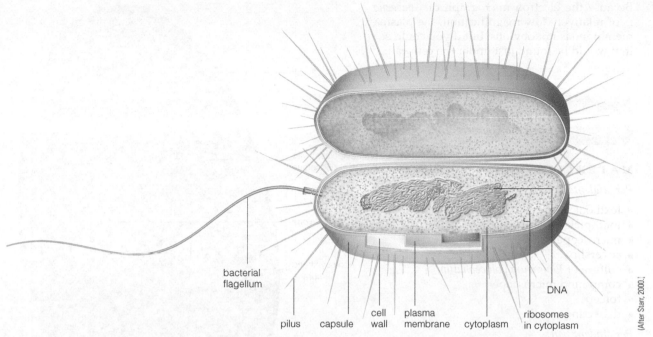

bacterial flagellum

pilus   capsule   cell wall   plasma membrane   cytoplasm   DNA   ribosomes in cytoplasm

(After Starr, 2000.)

Figure 3-2   Three-dimensional representation of a bacterial cell as seen with the electron microscope.

Another type of prokaryotic cell is exemplified by cyanobacteria, such as *Oscillatoria* and *Anabaena*. Cyanobacteria (sometimes called blue-green algae) are commonly found in water and damp soils. They obtain their nutrition by converting the sun's energy through photosynthesis.

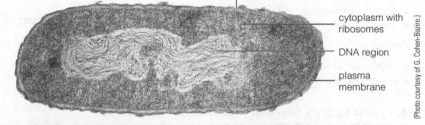

cell wall

cytoplasm with ribosomes

DNA region

plasma membrane

(Photo courtesy of G. Cohen-Bazire.)

Figure 3-3   Electron micrograph of the bacterium *Escherichia coli* (28,300×).

7. With a dissecting needle, remove a few filaments from the cyanobacterial culture, placing them in a drop of water on a clean microscope slide.

8. Place a coverslip over the material and examine it first with the low-power objective and then using the high-dry objective (or oil-immersion objective, if your microscope is so equipped).

9. In the space provided in Figure 3-4, sketch the cells you see at high power. Estimate the size of a *single* cyanobacterial cell and record the magnification you used to make your drawing.

10. Now examine the electron micrograph of *Anabaena* (Figure 3-5), which identifies the **cell wall, cytoplasm,** and **ribosomes.** The cyanobacteria also possess membranes that function in photosynthesis. Identify these **photosynthetic membranes,** which look like tiny threads within the cytoplasm.

11. Look at the captions for Figures 3-3 and 3-5. Judging by the magnification of each electron micrograph, which cell is larger, the bacterium *E. coli* or the cyanobacterium *Anabaena*?

_____

_____

_____

Figure 3-4   Drawing of several prokaryotic cells of a cyanobacterium (_____×). Approximate size = _____ μm.

Because the electron micrograph of *Anabaena* is of relatively low magnification, the plasma membrane is not obvious, but if you could see it, it would be found just under the cell wall.

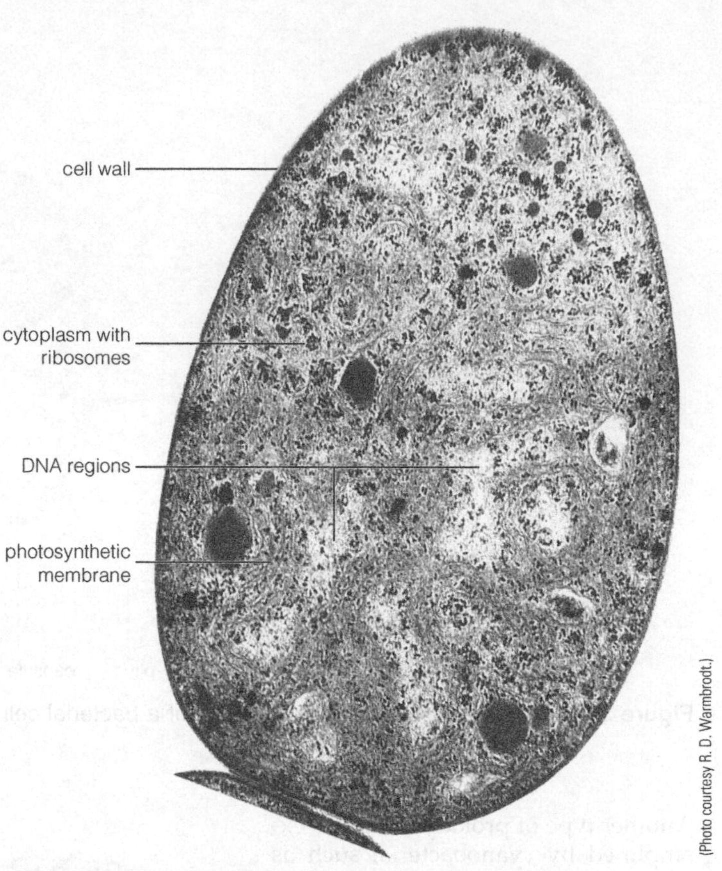

cell wall

cytoplasm with ribosomes

DNA regions

photosynthetic membrane

(Photo courtesy R. D. Warmbrodt.)

Figure 3-5 Electron micrograph of *Anabaena* (11,600×).

3.2	Eukaryotic Cells

(About 75 min.)

MATERIALS

Per student:

- textbook
- toothpick
- microscope slide
- coverslip
- culture of *Physarum polycephalum*
- compound microscope
- forceps
- dissecting needle

Per student pair:

- methylene blue in dropping bottle
- distilled water (dH_2O) in dropping bottle

Per student group (table):

- *Elodea* in water-containing culture dish
- onion bulb
- tissue paper

Per lab room:

- model of animal cell
- model of plant cell

A. Protist Cells as Observed with the Light Microscope

The slime mold *Physarum polycephalum* is in the Kingdom Protista. *Physarum* is a unicellular organism, so it contains all the metabolic machinery for independent existence.

PROCEDURE

1. Place a plain microscope slide on the stage of your compound microscope. This will serve as a platform on which you can place a culture dish.
2. Now obtain a petri dish culture of *Physarum*, remove the lid, and place it on the platform. Observe initially with the low-power objective and then with the medium-power objective. Place a coverslip over part of the organism before rotating the high-dry objective into place. (This prevents the agar from getting on the lens.)

Physarum is **multinucleate,** meaning that more than one nucleus occurs within the cytoplasm. Unfortunately, the nuclei are tiny; you won't be able to distinguish them from other granules in the cytoplasm.

3. Locate the **plasma membrane,** which is the outer boundary of the cytoplasm. Once again, the resolving power of your microscope is not sufficient to allow you to actually view the membrane.
4. Watch the cytoplasm of the organism move. This intracellular motion is known as **cytoplasmic streaming.** Although not visible with the light microscope without using special techniques, contractile proteins called **microfilaments** are believed responsible for cytoplasmic streaming.
5. Note that the outer portion of the cytoplasm appears solid; this is the **gel** state of the cytoplasm. Notice that the granules closer to the interior are in motion within a fluid; this portion of the cytoplasm is in the **sol** state. Movement of the organism occurs as the sol-state cytoplasm at the advancing tip pushes against the plasma membrane, causing the region to swell outward. The sol-state cytoplasm flows into the region, converting to the gel state along the margins.
6. In Figure 3-6, sketch the portion of *Physarum* that you have been observing and label it.

As you might predict, temperature affects many cellular processes. You may have observed that snakes and insects, being ectotherms (animals that gain heat from the environment and unlike humans, not primarily from metabolic activities), in nature are relatively sluggish during cold weather. Is the same true for other organisms, like the slime mold?

This simple experiment addresses the hypothesis that *cold slows cytoplasmic streaming in Physarum polycephalum*. Before starting this experiment, you may wish to review the discussion in Exercise 1, "The Scientific Method."

MATERIALS

Per experimental group:

- culture of *Physarum polycephalum*
- compound microscope
- container with ice *or* refrigerator
- timer *or* watch with second hand
- Celsius thermometer

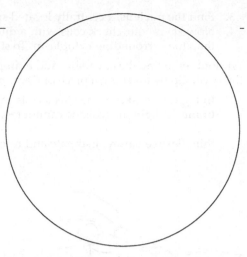

Figure 3-6 Drawing of a portion of *Physarum* (_____×).

PROCEDURE

1. Place the *Physarum* culture on the stage of your compound microscope as described in Section A.
2. Time the duration of cytoplasmic streaming in one direction and then in the other direction. Do this for five cycles of back-and-forth motion. Calculate the average duration of flow in either direction. Record the temperature and your observations in Table 3-2.
3. Remove your culture from the microscope's stage, replace the cover, and place it and the thermometer in a refrigerator or atop ice for 15 minutes.
4. While you are waiting, in Table 3-2 write a prediction for the effect on the duration of cytoplasmic streaming by reducing the temperature of the culture of *Physarum polycephalum*.
5. After 15 minutes have elapsed, remove the culture from the cold treatment, record the temperature of the experimental treatment, and repeat the observations in step 2.
6. Record your observations and make a conclusion in Table 3-2, accepting or rejecting the hypothesis.

TABLE 3-2	Effect of Temperature on Cytoplasmic Streaming	
Prediction:		
Temperature (°C)	**Time**	**Observations and Duration of Directional Flow (sec.)**

A logical question to ask at this time is *why* temperature has the effect you observed. If you perform Exercise 5, "Enzymes: Catalysts of Life," you may be able to make an educated guess (another hypothesis).

C. Animal Cells Observed with the Light Microscope

PROCEDURE

1. *Human cheek cells.* Using the broad end of a clean toothpick, gently scrape the inside of your cheek. Stir the scrapings into a drop of distilled water on a clean microscope slide and add a coverslip. Dispose of used toothpicks in the jar containing alcohol.
2. Because the cells are almost transparent, decrease the amount of light entering the objective lens to increase the contrast. (See Exercise 1.) Find the cells using the low-power objective of your microscope; then switch to the high-dry objective for detailed study.

3. Find the **nucleus,** a centrally located spherical body within the **cytoplasm** of each cell.
4. Now stain your cheek cells with a dilute solution of methylene blue, a dye that stains the nucleus darker than the surrounding cytoplasm. To stain your slide, follow the directions illustrated in Figure 3-7.

Without removing the coverslip, add a drop of the stain to one edge of the coverslip. Then draw the stain under the coverslip by touching a piece of tissue paper to the *opposite* side of the coverslip.

5. In Figure 3-8, sketch the cheek cells, labeling the **cytoplasm, nucleus,** and the location of the **plasma membrane.** (A light microscope cannot resolve the plasma membrane, but the boundary between the cytoplasm and the external medium indicates its location.) Many of the cells will be folded or wrinkled due to their thin, flexible nature. Estimate and record in your sketch the size of the cells.

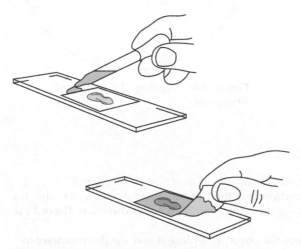

Figure 3-7 Method for staining specimen under coverslips of microscope slide.

Figure 3-8 Drawing of human cheek cells (_____×).
Approximate size = _____μm.
Labels: cytoplasm, nucleus, plasma membrane

D. Animal Cells as Observed with the Electron Microscope

Studies with the electron microscope have yielded a wealth of information on the structure of eukaryotic cells. Structures too small to be seen with the light microscope have been identified. These include many **organelles,** structures in the cytoplasm that have been separated ("compartmentalized") by enclosure in membranes. Examples of organelles are the nucleus, mitochondria, endoplasmic reticulum, and Golgi bodies. Although the cells in each of the six kingdoms have some peculiarities unique to that kingdom, electron microscopy has revealed that all cells are fundamentally similar.

PROCEDURE

1. Study Figure 3-9, a three-dimensional representation of an animal cell.
2. With the aid of Figure 3-9, identify the parts on the model of the animal cell that is on *demonstration*.
3. Figure 3-10 is an electron micrograph (EM) of an animal cell (kingdom Animalia). Study the electron micrograph and, with the aid of Figure 3-9 and any electron micrographs in your textbook, label each structure listed.
4. Pay particular attention to the membranes surrounding the nucleus and mitochondria. Note that these two are each bounded by *two* membranes, which are commonly referred to collectively as an **envelope.**
5. Using your textbook as a reference, list the function for the following cellular components:

   (a) plasma membrane _____

   _____

   (b) cytoplasm _____

   _____

   (c) nucleus (the plural is *nuclei*) _____

   _____

   (d) nuclear envelope _____

   _____

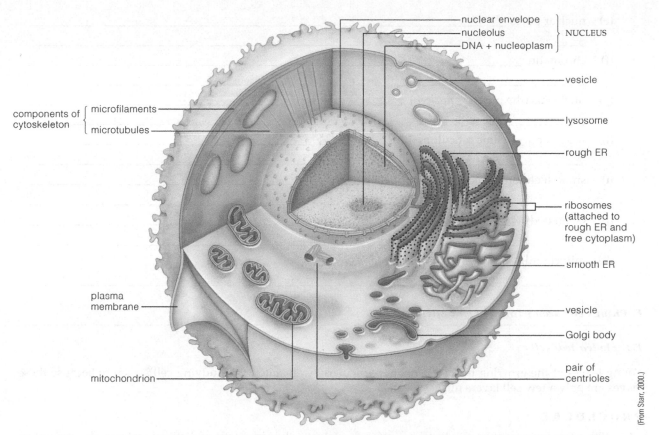

Figure 3-9 Three-dimensional representation of an animal cell as seen with the electron microscope.

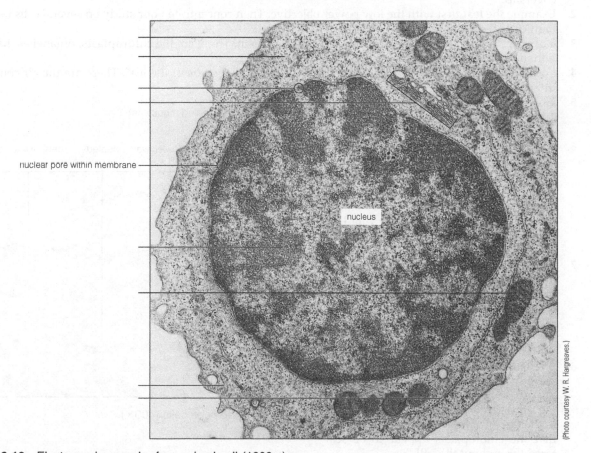

nuclear pore within membrane

nucleus

Figure 3-10 Electron micrograph of an animal cell (1600×).
Labels: plasma membrane, cytoplasm, nuclear envelope, nuclear pore, chromatin, rough ER, smooth ER, Golgi body, mitochondrion

(e) nuclear pores _____

_____

(f) chromatin _____

_____

(g) nucleolus (the plural is *nucleoli*) _____

_____

(h) rough endoplasmic reticulum (RER) _____

_____

(i) smooth endoplasmic reticulum (SER) _____

_____

(j) Golgi body _____

_____

(k) mitochondrion (the plural is *mitochondria*) _____

_____

E. Plant Cells Seen with the Light Microscope

E.1. Elodea leaf cells

Young leaves at the growing tip of *Elodea* are particularly well suited for studying cell structure because these leaves are only a few cell layers thick.

PROCEDURE

1. With a forceps, remove a single young leaf, mount it on a slide in a drop of distilled water, and cover with a coverslip.
2. Examine the leaf first with the low-power objective. Then concentrate your study on several cells using the high-dry objective. Refer to Figure 3-11.
3. Observe the abundance of green bodies in the cytoplasm. These are the **chloroplasts,** organelles that function in photosynthesis and that are typical of green plants.
4. Locate the numerous dark lines running parallel to the long axis of the leaf. These are the air-containing **intercellular spaces.**
5. Find the **cell wall,** a structure distinguishing plant from animal cells, visible as a clear area surrounding the cytoplasm.
6. After the cells have warmed a bit, notice the **cytoplasmic streaming** taking place. Movement of the chloroplasts along the cell wall is the most obvious visual evidence of cytoplasmic streaming. Microfilaments (much too small to be seen with your light microscope) are responsible for this intracellular motion.
7. Remember that you are looking at a three-dimensional object. In the middle portion of the cell is the large, clear **central vacuole,** which can take up from 50% to 90% of the cell interior. Because the vacuole in *Elodea* is transparent, it cannot be seen with the light microscope.
8. The chloroplasts occur in the cytoplasm surrounding the vacuole, so they will appear to be in different locations, depending on where you focus in the cell. Focus in the upper or lower surface and observe that the chloroplasts appear to be scattered throughout the cell.

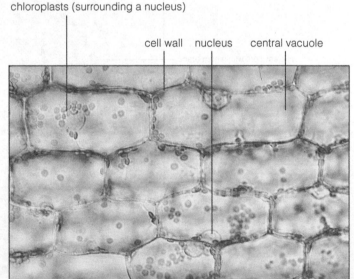

chloroplasts (surrounding a nucleus)

cell wall   nucleus   central vacuole

(Photo by J. W. Perry.)

Figure 3-11 *Elodea* cells (400×).

9. Now focus in the center of the cell (by raising or lowering the objective with the fine focus knob), and note that the chloroplasts lie in a thin layer of cytoplasm along the wall.
10. Locate the **nucleus** within the cytoplasm. It will appear as a clear or slightly amber body that is slightly larger than the chloroplasts. (You may need to examine several cells to find a clearly defined nucleus.)
11. Describe the three-dimensional shape of the *Elodea* leaf cell.

_____

_____

12. What are the shapes of the chloroplasts and nucleus? _____

_____

13. Now add a drop of methylene blue stain to make the cell wall more obvious. Add the stain as shown in Figure 3-7.
14. Look for the very, very tiny **mitochondria.** (If you have an oil-immersion lens on your microscope, you should use that lens.)
15. Compare the size of the mitochondria to chloroplasts:

_____

_____

E.2. Onion scale cells

1. Make a wet mount of a colorless scale of an onion bulb, using the technique described in Figure 3-12. The *inner* face of the scale is easiest to remove, as shown in Figure 3-12d.
2. Observe your preparation with your microscope, focusing first with the low-power objective. Continue your study, switching to the medium-power and finally the high-dry objective. Refer to Figure 3-13.
3. Identify the **cell wall** and **cytoplasm.**
4. Find the **nucleus,** a prominent sphere within the cytoplasm.
5. Examine the nucleus more carefully at high magnification. Within it, find one or more nucleoli (the singular is *nucleolus*). Nucleoli are rich in a nucleic acid known as RNA (ribonucleic acid), while the nucleus as a whole is largely DNA (deoxyribonucleic acid), the genetic material.
6. You may see numerous **oil droplets** within the cytoplasm, visible in the form of granulelike bodies. These oil droplets are a form of stored food material. You may be surprised to learn that onion scales are actually leaves! Which cellular components present in *Elodea* leaf cells are absent in onion leaf cells?

_____

7. If you are using the pigmented tissue from a red onion, you should see a purple pigment located in the vacuole. In this case, the cell wall appears as a bright line.
8. In Figure 3-14, sketch and label several cells from onion scale leaves.

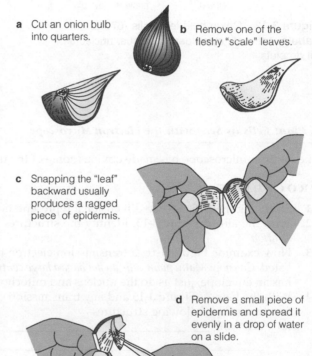

a Cut an onion bulb into quarters.

b Remove one of the fleshy "scale" leaves.

c Snapping the "leaf" backward usually produces a ragged piece of epidermis.

d Remove a small piece of epidermis and spread it evenly in a drop of water on a slide.

d Gently lower a coverslip to prevent trapping air bubbles. Examine with your microscope. Add more water to the edge of the coverslip with an eye dropper if the slide begins to dry.

Figure 3-12 Method for obtaining onion scale cells.

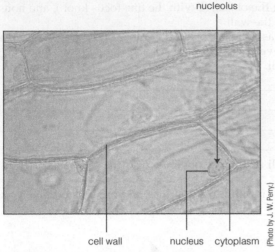

nucleolus

cell wall   nucleus   cytoplasm

(Photo by J. W. Perry.)

Figure 3-13 Onion bulb leaf cells (67×).
Labels: cell wall, cytoplasm, nucleus, nucleolus,
oil droplets

Figure 3-14 Drawing of onion scale cells
(_____×).
Labels: cell wall, cytoplasm, nucleus

F. Plant Cells as Seen with the Electron Microscope

The electron microscope has made obvious some of the unique features of plant cells.

PROCEDURE

1. Study Figure 3-15, a three-dimensional representation of a typical plant cell.
2. With the aid of Figure 3-15, identify the structures present on the model of a plant cell that is on *demonstration.*
3. Now examine Figure 3-16, a transmission electron micrograph from a corn leaf. Label all of the structures listed. *Caution: Many plant cells do not have a large central vacuole. This is one of them.* Notice that the chloroplast has an envelope, just as do the nucleus and mitochondria.
4. With the help of Figure 3-15 and any transmission electron micrographs and text in your textbook, list the function of the following structures.
    (a) cell wall _____
    _____
    (b) chloroplast _____
    _____
    (c) vacuole _____
    _____
    (d) vacuolar membrane _____
    _____
    (e) plasma membrane _____
    _____
    (f) cytoplasm _____
    _____
    (g) nucleus _____
    _____

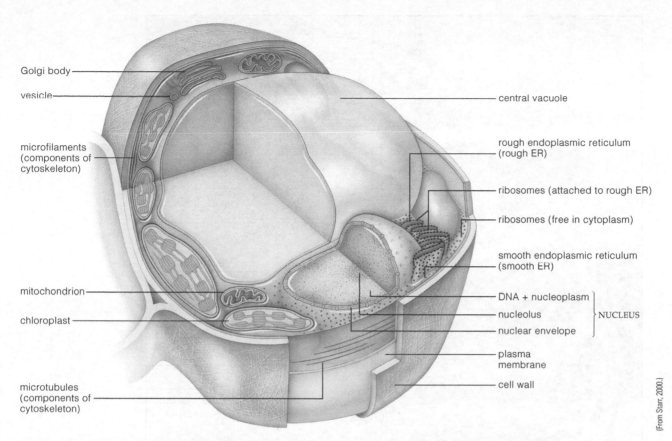

Golgi body

vesicle

microfilaments
(components of
cytoskeleton)

mitochondrion

chloroplast

microtubules
(components of
cytoskeleton)

central vacuole

rough endoplasmic reticulum
(rough ER)

ribosomes (attached to rough ER)

ribosomes (free in cytoplasm)

smooth endoplasmic reticulum
(smooth ER)

DNA + nucleoplasm
nucleolus } NUCLEUS
nuclear envelope

plasma
membrane

cell wall

(From Starr, 2000.)

Figure 3-15 Three-dimensional representation of a plant cell as seen with the electron microscope.

(h) nuclear envelope _____

_____

(i) nuclear pore _____

_____

(j) chromatin _____

_____

(k) nucleolus _____

_____

(l) rough endoplasmic reticulum (RER) _____

_____

(m) smooth endoplasmic reticulum (SER) _____

_____

(n) Golgi body _____

_____

(o) mitochondrion _____

_____

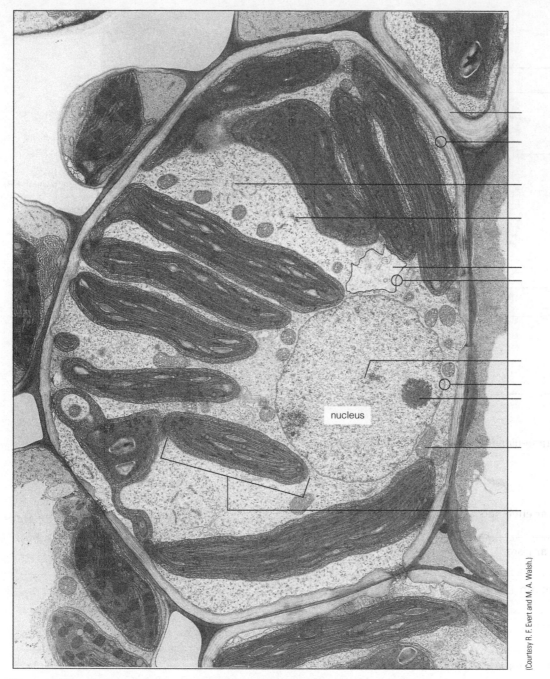

nucleus

(Courtesy R. F. Evert and M. A. Walsh.)

Figure 3-16 Electron micrograph of a corn leaf cell (2700×).
Labels: cell wall, chloroplast, vacuole, vacuolar membrane, plasma membrane, nuclear envelope, chromatin, nucleolus, endoplasmic reticulum (ER), Golgi body, mitochondrion

_____ 1. The person who first used the term *cell* was
   (a) Darwin
   (b) Leeuwenhoek
   (c) Hooke
   (d) Watson

_____ 2. All cells contain
   (a) a nucleus, plasma membrane, and cytoplasm
   (b) a cell wall, nucleus, and cytoplasm
   (c) DNA, plasma membrane, and cytoplasm
   (d) mitochondria, plasma membrane, and cytoplasm

_____ 3. Prokaryotic cells *lack*
   (a) DNA
   (b) a true nucleus
   (c) a cell wall
   (d) none of the above

_____ 4. The word *eukaryotic* refers specifically to a cell containing
   (a) photosynthetic membranes
   (b) a true nucleus
   (c) a cell wall
   (d) none of the above

_____ 5. A bacterium is an example of a
   (a) prokaryotic cell
   (b) eukaryotic cell
   (c) plant cell
   (d) all of the above

_____ 6. Methylene blue
   (a) is used to kill cells that are moving too quickly to observe
   (b) renders cells nontoxic
   (c) is a portion of the electromagnetic spectrum used by green plant cells
   (d) is a biological stain used to increase contrast of cellular constituents

_____ 7. Components typical of plant cells but not of animal cells are
   (a) nuclei
   (b) cell walls
   (c) mitochondria
   (d) ribosomes

_____ 8. A central vacuole
   (a) is found only in plant cells
   (b) may take up between 50% and 90% of the cell's interior
   (c) is both of the above
   (d) is none of the above

_____ 9. The intercellular spaces between plant cells
   (a) contain air
   (b) are responsible for cytoplasmic streaming
   (c) are nonexistent
   (d) contain chloroplasts

_____ 10. An envelope
   (a) surrounds the nucleus
   (b) surrounds mitochondria
   (c) consists of two membranes
   (d) does all of the above

EXERCISE 3

Structure and Function of Living Cells

POST-LAB QUESTIONS

3.1 Prokaryotic Cells

1. Did all living cells that you saw in lab contain mitochondria?

2. Below is a high-magnification photomicrograph of an organism you observed in this exercise. Each rectangular box is a single cell. What organelle is absent from each cell that makes it "prokaryotic?"

(750×).

3.2 Eukaryotic Cells

3. Is it possible for a cell to contain more than one nucleus? Explain.

4. When students are asked to distinguish between an animal cell and a plant cell, they typically answer that plant cells contain chloroplasts and animal cells do not. If you were the professor reading that answer, what sort of credit would you give and why?

5. Describe a major distinction between most plant cells and animal cells.

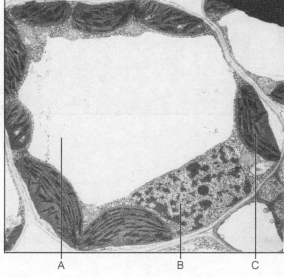

6. Observe the electron micrograph to the right. Is the cell prokaryotic or eukaryotic?

Identify the labeled structures.

A. _____

B. _____

C. _____

(4800×).

7. Look at the photomicrograph to the right, which was taken with a technique that gives a three-dimensional impression. Identify the structures labeled A, B, and C.

A. _____

B. _____

C. _____

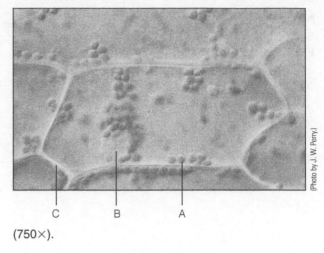

(Photo by J. W. Perry.)

8. Is the electron micrograph below a plant or an animal cell?

_____

Identify structures labeled A and B.

A. _____

B. _____

C   B   A

(750×).

9. What are the numerous "wavy lines" within the cell (labeled C)? _____

_____

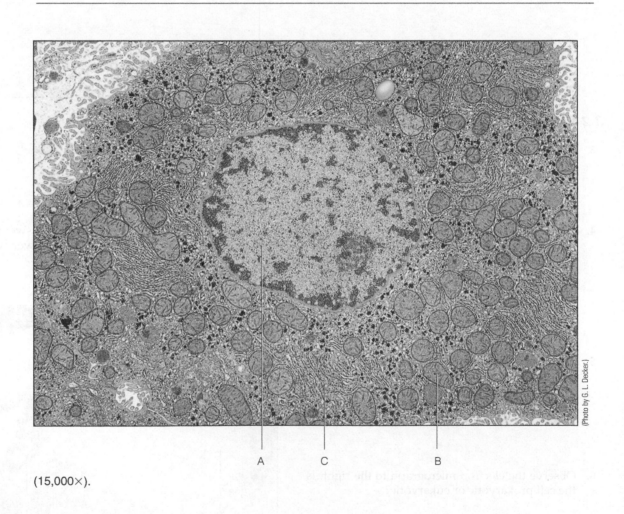

A   C   B

(Photo by G. L. Decker.)

(15,000×).

10. What structure(s) found in plant cells is (are) primarily responsible for cellular support?

Food for Thought

11. What structural differences did you observe between prokaryotic and eukaryotic cells?

12. Are the cells in the electron micrograph below prokaryotic or eukaryotic? How do you know?

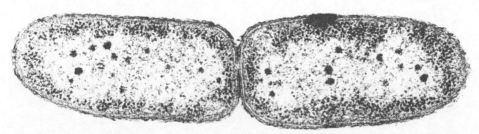

(Photo by J. J. Cardamone, Jr., University of Pittsburgh/BPS.)

Diffusion, Osmosis, and the Functional Significance of Biological Membranes

After completing this exercise, you will be able to

1. define *solvent, solute, solution, selectively permeable, diffusion, osmosis, concentration gradient, equilibrium, turgid, plasmolyzed, plasmolysis, turgor pressure, tonicity, hypertonic, isotonic, hypotonic;*

2. describe the structure of cellular membranes;

3. distinguish between diffusion and osmosis;

4. determine the effects of concentration and temperature on diffusion;

5. describe the effects of hypertonic, isotonic, and hypotonic solutions on red blood cells and *Elodea* leaf cells.

INTRODUCTION

Water is a great environment. Earthly life is believed to have originated in the water. Without it, life as we know it would cease to exist. Recently, the discovery of water in meteorites originating within our solar system has fueled speculation that life may not be unique to earth.

Living cells are made up of 75–85% water. Virtually all substances entering and leaving cells are dissolved in water, making it the **solvent** most important for life processes. The substances dissolved in water are called **solutes** and include such substances as salts and sugars. The combination of a solvent and dissolved solute is a **solution.** The cytoplasm of living cells contains numerous solutes, like sugars and salts, in solution.

All cells possess membranes composed of a phospholipid bilayer that contains different kinds of embedded and surface proteins. Look at Figure 4-1 to get an idea of the complexity of a cellular membrane.

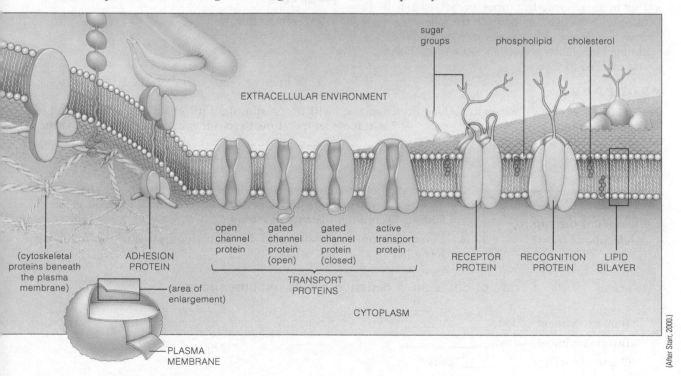

Figure 4-1 Artistic rendering of cutaway view of part of the plasma membrane.

Membranes are boundaries that solutes must cross to reach the cellular site where they will be utilized in the processes of life. These membranes regulate the passage of substances into and out of the cell. They are **selectively permeable,** allowing some substances to move easily while completely excluding others.

The most simple means by which solutes enter the cell is **diffusion,** the movement of solute molecules from a region of high concentration to one of lower concentration. Diffusion occurs without the expenditure of cellular energy. Once inside the cell, solutes move through the cytoplasm by diffusion, sometimes assisted by cytoplasmic streaming.

Water (the solvent) also moves across the membrane. **Osmosis** is the movement of *water* across selectively permeable membranes. Think of osmosis as a special form of diffusion, one occurring from a region of higher *water* concentration to one of lower *water* concentration.

The difference in concentration of like molecules in two regions is called a **concentration gradient.** Diffusion and osmosis take place *down* concentration gradients. Over time, the concentration of solvent and solute molecules becomes equally distributed, the gradient ceasing to exist. At this point, the system is said to be at **equilibrium.**

Molecules are always in motion, even at equilibrium. Thus, solvent and solute molecules continue to move because of randomly colliding molecules. However, at equilibrium there is no *net change* in their concentrations.

This exercise introduces you to the principles of diffusion and osmosis.

Note: If Sections 4.2 and 4.3 are to be done during this lab period, start them before doing any other activity in this exercise.

4.1   Experiment: Rate of Diffusion of Solutes (*About 10 min.*)

Solutes move within a cell's cytoplasm largely because of diffusion. However, the rate of diffusion (the distance diffused in a given amount of time) is affected by such factors as temperature and the size of the solute molecules. In this experiment, you will discover the effects of these two factors in gelatin (the substance of Jell-O®), a substance much like cytoplasm and used to simulate it in this experiment.

MATERIALS

Per student:

■ metric ruler

Per student group (table):

■ 1 set of 3 screw-cap test tubes, in rack, each half-filled with 5% gelatin, to which the following dyes have been added: potassium dichromate, aniline blue, Janus green; labeled with each dye and marked "5°C"

■ 1 set of 3 screw-cap test tubes, in rack, as above but marked "Room Temperature"

Per lab room:

■ 5°C refrigerator

PROCEDURE

Two sets of three screw-cap test tubes have been half-filled with 5% gelatin; and 1 mL of a dye has been added to each test tube. Set 1 is in a 5°C refrigerator; set 2 is at room temperature. Record the time at which your instructor tells you the experiment was started: _____

1. Remove set 1 from the refrigerator and compare the distance the dye has diffused in corresponding tubes of each set.

Caution

Be certain the cap to each tube is tight!

2. Invert and hold each tube vertically in front of a white sheet of paper. Use a metric ruler to measure how far each dye has diffused from the gelatin's surface. Record this distance in Table 4-1.

3. Determine the *rate* of diffusion for each dye by using the following formula:

rate of diffusion = distance ÷ elapsed time (hours)

Time experiment ended: _____

Time experiment started: _____

Elapsed time: _____ hours

TABLE 4-1 Effect of Temperature on Diffusion Rates of Various Solutes

Solute (dye)	Set 1 (5°C)		Set 2 (Room Temp.)	
	Distance (mm)	Rate	Distance (mm)	Rate
Potassium dichromate (MW = 294)[a]				
Janus green (MW = 511)				
Aniline blue (MW = 738)				

[a]MW = molecular weight, a reflection of the mass of a substance. To determine MW, add the atomic weights of all elements in a compound.

Which of the solutes diffused the slowest (regardless of temperature)? _____

Which diffused the fastest? _____

What effect did temperature have on the rate of diffusion? _____

Make a conclusion about the diffusion of a solute in a gel, relating the rate of diffusion to the molecular weight of the solute and to temperature.

_____

_____

Note: Return set 1 to the refrigerator.

4.2   Experiment: Osmosis *(About 20 min. for setup)*

Osmosis occurs when different concentrations of water are separated by a selectively permeable membrane. One example of a selectively permeable membrane within a living cell is the plasma membrane. In this experiment, you will learn about osmosis using dialysis membrane, a selectively permeable cellulose sheet that permits the passage of water but obstructs passage of larger molecules. If you examined the membrane with a scanning electron microscope, you would see that it is porous; it thus prevents molecules larger than the pores from passing through the membrane.

MATERIALS

Per student group (4):

- four 15-cm lengths of dialysis tubing, soaking in dH_2O
- eight 10-cm pieces of string or waxed dental floss
- ring stand and funnel apparatus (Figure 4-2)
- 25-mL graduated cylinder
- 4 small string tags
- china marker
- four 400-mL beakers

Per student group (table):

- dishpan half-filled with dH_2O
- paper toweling
- balance

Per lab room:

- source of dH_2O (at each sink)
- 15% and 30% sucrose solutions
- scissors (at each sink)

PROCEDURE

Work in groups of four for this experiment.

1. Obtain four sections of dialysis tubing, each 15 cm long, that have been presoaked in dH_2O. Recall that the dialysis tubing is permeable to water molecules but not to sucrose.
2. Fold over one end of each tube and tie it tightly with string or dental floss.
3. Attach a string tag to the tied end of each bag and number them 1–4.

4. Slip the open end of the bag over the stem of a funnel (Figure 4-2). Using a graduated cylinder* to measure volume, fill the bags as follows:

Bag 1. 10 mL of dH_2O Bag 3. 10 mL of 30% sucrose
Bag 2. 10 mL of 15% sucrose   Bag 4. 10 mL of dH_2O

5. As each bag is filled, force out excess air by squeezing the bottom end of the tube.
6. Fold the end of the bag and tie it securely with another piece of string or dental floss.
7. Rinse each filled bag in the dishpan containing dH_2O; gently blot off the excess water with paper toweling.
8. Weigh each bag to the nearest 0.5 g.
9. Record the weights in the column marked "0 min." in Table 4-2.
10. Number four 400-mL beakers with a china marker.
11. Add 200 mL of dH_2O to beakers 1–3.
12. Add 200 mL of 30% sucrose solution to beaker 4.
13. Place bags 1–3 in the correspondingly numbered beakers.
14. Place bag 4 in the beaker containing 30% sucrose.
15. After 15 minutes, remove each bag from its beaker, blot off the excess fluid, and weigh each bag.
16. Record the weight of each bag in Table 4-2.
17. Return the bags to their respective beakers immediately after weighing.
18. Repeat steps 15–17 at 30, 45, and 60 minutes from time zero.

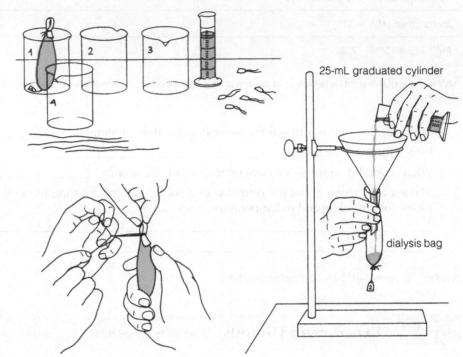

Figure 4-2 Method for filling dialysis bags.

At the end of the experiment, take the bags to the sink, cut them open, pour the contents down the drain, and discard the bags in the wastebasket. Pour the contents of the beakers down the drain and wash them according to the instructions given on page x.

Make a *qualitative* statement about what you have observed. _____

_____

_____

Was the direction of *net* movement of water in bags 2–4 into or out of the bags? _____

_____

Which bag gained the most weight? Why? _____

_____

_____

---

*Be sure to rinse the cylinder if it has been used to measure sucrose.

TABLE 4-2 Change in Weight as a Consequence of Osmosis

Bag	Bag Contents	Beaker Contents	Bag Weight (g)					Weight Change (g)
			0 min.	15 min.	30 min.	45 min.	60 min.	
1	dH$_2$O	dH$_2$O						
2	15% sucrose	dH$_2$O						
3	30% sucrose	dH$_2$O						
4	dH$_2$O	30% sucrose						

4.3 Experiment: Selective Permeability of Membranes *(About 15 min. for setup)*

Dialysis tubing is a selectively permeable material that provides a means to demonstrate the movement of substances through cellular membranes.

MATERIALS

Per student group (4):

- 1 25-cm length of dialysis tubing, soaking in dH$_2$O
- two 10-cm pieces of string or waxed dental floss
- bottle of 1% soluble starch in 1% sodium sulfate (Na$_2$SO$_4$)
- dishpan half-filled with dH$_2$O
- 400-mL graduated beaker
- ring stand and funnel apparatus (Figure 4-2)
- bottle of 1% albumin in 1% sodium chloride (NaCl)
- 8 test tubes
- test tube rack
- china marker

- 25-mL graduated cylinder
- iodine (I$_2$KI) solution in dropping bottle
- 2% barium chloride (BaCl$_2$) in dropping bottle
- 2% silver nitrate (AgNO$_3$) in dropping bottle
- Biuret reagent in dropping bottle
- albustix reagent strips (optional)
- scissors

Per lab room:

- series of 4 test tubes in test tube rack demonstrating positive tests for starch, sulfate ion, chloride ion, protein

PROCEDURE

Work in groups of four.

1. Obtain a 25-cm section of dialysis tubing that has been soaked in dH$_2$O.
2. Fold over one end of the tubing and tie it securely with string or dental floss to form a leakproof bag (Figure 4-2).
3. Slip the open end of the bag over the stem of a funnel and fill the bag approximately half full with 25 mL of a solution of 1% soluble starch in 1% sodium sulfate (Na$_2$SO$_4$).
4. Remove the bag from the funnel; fold and tie the open end of the bag.
5. Rinse the tied bag in a dishpan partially filled with dH$_2$O.
6. Pour 200 mL of a solution of 1% albumin (a protein) in 1% sodium chloride (NaCl) into a 400-mL beaker.
7. Place the bag into the fluid in the beaker.
8. Record the time: _____
9. With a china marker, label eight test tubes, numbering them 1–8.
10. Seventy-five minutes after the start of the experiment, pour 20 mL of the *beaker contents* into a *clean* 25-mL graduated cylinder.
11. Decant (pour out) 5 mL from the graduated cylinder into each of the first four test tubes.

12. Perform the following tests, recording your results in Table 4-3. Your instructor will have a series of test tubes showing positive tests for starch, sulfate and chloride ions, and proteins. You should compare your results with the known positives.

   (a) *Test for starch.* Add several drops of iodine solution (I_2KI) from the dropper bottle to test tube 1. If starch is present, the solution will turn blue-black.

   (b) *Test for sulfate ion.* Add several drops of 2% barium chloride ($BaCl_2$) from the dropper bottle to test tube 2. If sulfate ions (SO^{-4}) are present, a white precipitate of barium sulfate ($BaSO_4$) will form.

   (c) *Test for chloride ion.* Add several drops of 2% silver nitrate ($AgNO_3$) from the dropper bottle to test tube 3. A milky-white precipitate of silver chloride ($AgCl$) indicates the presence of chloride ions (Cl^2).

   (d) *Test for protein.* Add several drops of Biuret reagent from the dropper bottle to test tube 4. If protein is present, the solution will change from blue to pinkish-violet. The more intense the violet hue, the greater the quantity of the protein.

   An alternative method for determining the presence of protein is the use of albustix reagent strips. Presence of protein is indicated by green or blue-green coloration of the paper.

13. Wash the graduated cylinder, using the technique described on page x.

14. Thoroughly rinse the bag in the dishpan of dH_2O.

15. Using scissors, cut the bag open and empty the contents into the 25-mL graduated cylinder.

16. Decant 5-mL samples into each of the four remaining test tubes.

17. Perform the tests for starch, sulfate ions, chloride ions, and protein on tubes 5–8, respectively.

18. Record the results of this series of tests in Table 4-4.

To which substances was the dialysis tubing permeable?

_____

What physical property of the dialysis tubing might explain its differential permeability?

_____

19. Discard contents of test tubes and beaker down sink drain. Wash glassware by using the technique described on page x.

20. Discard dialysis tubing in wastebasket.

TABLE 4-3	Results of Tests for Substances in Beaker[a]	
	At Start of Experiment	**After 75 min.**
Starch	−	
Sulfate ion	−	
Chloride ion	+	
Albumin	+	

[a]Contents of beaker: (+) = presence, (−) = absence.

TABLE 4-4	Results of Tests for Substances in Dialysis Bag[a]	
	At Start of Experiment	**After 75 min.**
Starch	+	
Sulfate ion	+	
Chloride ion	−	
Albumin	−	

[a]Contents of dialysis bag: (+) = presence, (−) = absence.

Plant cells are surrounded by a rigid cell wall, composed primarily of the glucose polymer, cellulose. Recall from Exercise 6 that many plant cells have a large central vacuole surrounded by the vacuolar membrane. The vacuolar membrane is selectively permeable. Normally, the solute concentration within the cell's central vacuole is greater than that of the external environment. Consequently, water moves into the cell, creating **turgor pressure,** which presses the cytoplasm against the cell wall. Such cells are said to be **turgid.** Many nonwoody plants (like beans and peas) rely on turgor pressure to maintain their rigidity and erect stance.

In this experiment, you will discover the effect of external solute concentration on the structure of plant cells.

MATERIALS

Per student:

- forceps
- 2 microscope slides
- 2 coverslips
- compound microscope

Per student group (table):

- *Elodea* in tap water
- 2 dropping bottles of dH_2O
- 2 dropping bottles of 20% sodium chloride (NaCl)

PROCEDURE

1. With a forceps, remove two young leaves from the tip of an *Elodea* plant.
2. Mount one leaf in a drop of distilled water on a microscope slide and the other in 20% NaCl solution on a second microscope slide.
3. Place coverslips over both leaves.
4. Observe the leaf in distilled water with the compound microscope. Focus first with the medium-power objective and then switch to the high-dry objective.
5. Label the photomicrograph of turgid cells (Figure 4-3).
6. Now observe the leaf mounted in 20% NaCl solution. After several minutes, the cell will have lost water, causing it to become **plasmolyzed.** (This process is called **plasmolysis.**) Label the plasmolyzed cells shown in Figure 4-4.

Figure 4-3   Turgid *Elodea* cells (400×). (Photo by J. W. Perry.)
Labels: cell wall, chloroplasts in cytoplasm, central vacuole

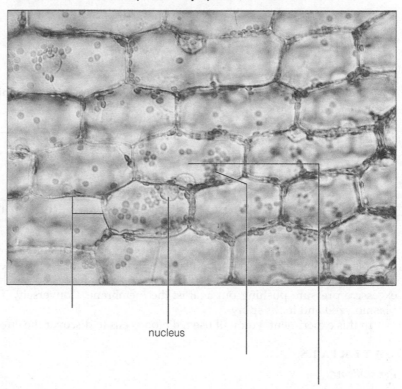

nucleus

Tonicity describes one solution's solute concentration compared to that of another solution. The solution containing the lower concentration of solute molecules than another is **hypotonic** *relative to the second solution.* Solutions containing equal concentrations of solute are **isotonic** to each other, while one containing a greater concentration of solute relative to a second one is **hypertonic.**

Were the contents of the vacuole in the *Elodea* leaf in distilled water hypotonic, isotonic, or hypertonic compared to the dH_2O? _____

Was the 20% NaCl solution hypertonic, isotonic, or hypotonic relative to the cytoplasm? _____

_____

If a hypotonic and a hypertonic solution are separated by a selectively permeable membrane, in which direction will the water move? _____

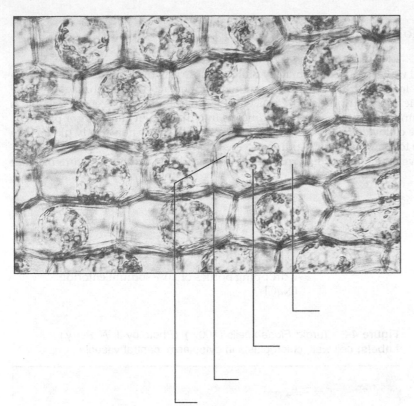

Figure 4-4 Plasmolyzed *Elodea* cells (400×). (Photo by J. W. Perry.) **Labels:** cell wall, chloroplasts in cytoplasm, plasma membrane, space (between cell wall and plasma membrane)

Name two selectively permeable membranes that are present within the *Elodea* cells and that were involved in the plasmolysis process.

1. _____

2. _____

4.5 Experiment: Osmotic Changes in Red Blood Cells *(About 15 min.)*

Animal cells lack the rigid cell wall of a plant. The external boundary of an animal cell is the selectively permeable plasma membrane. Consequently, an animal cell increases in size as water enters the cell. However, since the plasma membrane is relatively fragile, it ruptures when too much water enters the cell. This is because of excessive pressure pushing out against the membrane. Conversely, if water moves out of the cell, it becomes plasmolyzed and looks spiny.

In this experiment, you will use red blood cells to discover the effects of osmosis in animal cells.

MATERIALS

Per student:

■ compound microscope

Per student group (4):

■ 3 clean screw-cap test tubes
■ test tube rack
■ metric ruler
■ china marker
■ bottle of 0.9% sodium chloride (NaCl)
■ bottle of 10% NaCl
■ bottle of dH$_2$O

■ 3 disposable plastic pipets
■ 3 clean microscope slides
■ 3 coverslips

Per student group (table):

■ bottle of sheep blood (in ice bath)

Per lab room:

■ source of dH$_2$O

PROCEDURE

Work in groups of four for this experiment, but do the microscopic observations individually.

1. Observe the scanning electron micrographs in Figure 4-5.

Figure 4-5a illustrates the normal appearance of red blood cells. They are biconcave disks; that is, they are circular in outline with a depression in the center of both surfaces. Cells in an isotonic solution will appear like these blood cells.

Figure 4-5b shows cells that have been plasmolyzed. (In the case of red blood cells, plasmolysis is given a special term, *crenation*; the blood cell is said to be *crenate*.)

Figure 4-5c represents cells that have taken in water but have not yet burst. (Burst red blood cells are said to be *hemolyzed*, and of course they can't be seen.) Note their swollen, spherical appearance.

2. Obtain three clean screw-cap test tubes.
3. Lay test tubes 1 and 2 against a metric ruler and mark lines indicating 5 cm *from the bottom of each tube*.
4. Fill each tube as follows:
   Tube 1: 5 cm of 0.9% sodium chloride (NaCl)
            5 drops of sheep blood
   Tube 2: 5 cm of 10% NaCl
            5 drops of sheep blood
5. Lay test tube 3 against a metric ruler and mark lines indicating 0.5 cm and 5 cm *from the bottom of the tube*.
6. Fill tube 3 to the 0.5-cm mark with 0.9% NaCl, and to the 5-cm mark with dH_2O. Then add 5 drops of sheep blood. Enter the contents of each tube in the appropriate column of Table 4-5.

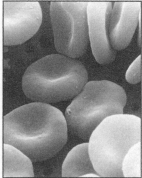
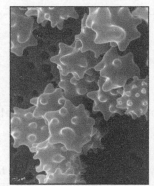
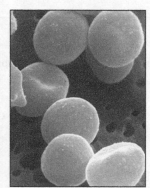

a Red blood cells in an isotonic solution ("normal")

b Red blood cells in a hypertonic solution ("crenate")

c Red blood cells in a hypotonic solution

Figure 4-5 Scanning electron micrographs of red blood cells. (Photos from M. Sheetz, R. Painter, and S. Singer. Reproduced from *The Journal of Cell Biology*, 1976, 70:193, by copyright permission of the Rockefeller University Press and M. Sheetz.)

7. Replace the caps and mix the contents of each tube by inverting several times (Figure 4-6a).
8. Hold each tube flat against the printed page of your lab manual (Figure 4-6b). *Only if the blood cells are hemolyzed should you be able to read the print.*
9. In Table 4-5, record your observations in the column "Print Visible?"
10. Number three clean microscope slides.
11. With three *separate* disposable pipets, remove a small amount of blood from each of the three tubes. Place 1 drop of blood from tube 1 on slide 1, 1 drop from tube 2 on slide 2, and 1 drop from tube 3 on slide 3.
12. Cover each drop of blood with a coverslip.
13. Observe the three slides with your compound microscope, focusing first with the medium-power objective and finally with the high-dry objective. (Hemolyzed cells are virtually unrecognizable; all that remains are membranous "ghosts," which are difficult to see with the microscope.)
14. In Figure 4-7, sketch the cells from each tube. Label the sketches, indicating whether the cells are normal, plasmolyzed (crenate), or hemolyzed.

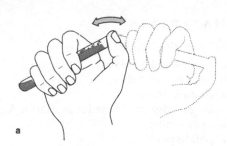

a

b

(After Abramoff and Thomson, 1982.)

Figure 4-6 Method for studying effects of different solute concentrations on red blood cells.

15. Record the microscopic appearance in Table 4-5.
16. Record the relative tonicity of the sodium chloride solutions you added to the test tubes in Table 4-5.

Why do red blood cells burst when put in a hypotonic solution whereas *Elodea* leaf cells do not?

_____

_____

_____

Figure 4-7 Microscopic appearance of red blood cells in different solute concentrations (_____×).
Labels: normal, plasmolyzed (crenate), hemolyzed

After completing all experiments, take your dirty glassware to the sink and wash it as directed on page x. Invert the test tubes in the test tube rack so they drain. Reorganize your work area, making certain all materials used in this exercise are present for the next class.

TABLE 4-5 Effect of Salt Solutions on Red Blood Cells

Tube	Contents	Print Visible?	Microscopic Appearance of Cells	Tonicity of External Solution[a]
1				
2				
3				

[a]With respect to that inside the red blood cell at the start of the experiment.

4.6 Experiment: Determining the Concentration of Solutes in Cells
(About 20 min. for setup)

If you've done the previous experiments of this exercise, you now know that water flows into or out of cells in response to the concentration of solutes within the cells. But you might logically ask at this point how much solute is present in a typical cell. While the answer varies from cell to cell, a simple experiment enables you to determine the osmotic concentration in the cells of a potato tuber.

MATERIALS

Per student group (4):

- five 250-mL beakers
- large potato tuber
- china marker
- single-edge razor blades *or* paring knife
- metric ruler
- potato peeler

Per table:

- balance
- paper toweling
- bottles containing solutions of 0.15, 0.20, 0.25, 0.30, and 0.35 M sucrose

PROCEDURE

1. With the china marker, label the five 250-mL beakers with the concentrations of sucrose solution.
2. Pour about 100 mL of each solution into its respective beaker.
3. Peel the potato and then cut it into five 3-cm cubes (3 cm on each side).
4. Without delay, weigh each cube to the nearest 0.01 g. Record the weights in Table 4-6.
5. Place one cube in each beaker and allow it to remain there for a minimum of 30 minutes, longer if time is available.

6. After the experimental period has elapsed, remove each cube, one at a time, and blot it lightly but thoroughly with the paper toweling.
7. Weigh each cube and record its final weight in Table 4-6. Then calculate and record the weight loss or gain.
8. Calculate the percent change in weight by dividing the *initial weight* by the *final weight*.

TABLE 4-6 Determining the Solute Concentration in Potato Tuber Cells

Solution	Weight		Change	Percent Change
	Initial	Final		
0.15 M				
0.20 M				
0.25 M				
0.30 M				
0.35 M				

The cube with the lowest percentage of weight loss or gain is in a solution that most closely approximates the solute concentration of the cells within the potato tuber. Of course, most of the solute within the tuber is in the form of starch, and our experimental solution is sucrose. The results of this experiment indicate that the *concentration* of the solute, but not the *type* of solute, is important for osmosis to occur.

What was the approximate concentration of solute in the potato tuber? _____

_____

Which concentration resulted in the *greatest* percentage change? _____

_____

Make a statement that relates the amount of water loss or gain to the concentration of the solute. _____

_____

_____

_____

_____ 1. If one were to identify the most important compound for sustenance of life, it would probably be
  (a) salt
  (b) $BaCl_2$
  (c) water
  (d) I_2KI

_____ 2. A solvent is
  (a) the substance in which solutes are dissolved
  (b) a salt or sugar
  (c) one component of a biological membrane
  (d) selectively permeable

_____ 3. Diffusion
  (a) is a process requiring cellular energy
  (b) is the movement of molecules from a region of higher concentration to one of lower concentration
  (c) occurs only across selectively permeable membranes
  (d) is none of the above

_____ 4. Cellular membranes
  (a) consist of a phospholipid bilayer containing embedded proteins
  (b) control the movement of substances into and out of cells
  (c) are selectively permeable
  (d) are all of the above

_____ 5. An example of a solute would be
  (a) Janus green B
  (b) water
  (c) sucrose
  (d) both a and c

_____ 6. Dialysis membrane is
  (a) selectively permeable
  (b) used in these experiments to simulate cellular membranes
  (c) permeable to water but not to sucrose
  (d) all of the above

_____ 7. Specifically, osmosis
  (a) requires the expenditure of cellular energy
  (b) is diffusion of water from one region to another
  (c) is diffusion of water across a selectively permeable membrane
  (d) is none of the above

_____ 8. Which of the following reagents does *not* fit with the substance being tested for?
  (a) Biuret reagent protein
  (b) $BaCl_2$ starch
  (c) $AgNO_3$ chloride ion
  (d) albustix protein

_____ 9. When the cytoplasm of a plant cell is pressed against the cell wall, the cell is said to be
  (a) turgid
  (b) plasmolyzed
  (c) hemolyzed
  (d) crenate

_____ 10. If one solution contains 10% NaCl and another contains 30% NaCl, the 30% solution is
  (a) isotonic
  (b) hypotonic
  (c) hypertonic
  (d) plasmolyzed, with respect to the 10% solution

EXERCISE 4

Diffusion, Osmosis, and the Functional Significance of Biological Membranes

POST-LAB QUESTIONS

4.1 Experiment: Rate of Diffusion of Solutes

1. You want to dissolve a solute in water. Without shaking or swirling the solution, what might you do to increase the rate at which the solute would go into solution? Relate your answer to your method's effect on the motion of the molecules.

4.2 Experiment: Osmosis

2. If a 10% sugar solution is separated from a 20% sugar solution by a selectively permeable membrane, in which direction will there be a net movement of water?

3. Based on your observations in this exercise, would you expect dialysis membrane to be permeable to sucrose? Why?

4.4 Experiment: Plasmolysis in Plant Cells

4. You are having a party and you plan to serve celery, but your celery has gone limp, and the stores are closed. What might you do to make the celery crisp (turgid) again?

5. Why don't plant cells undergo osmotic lysis?

6. This drawing represents a plant cell that has been placed in a solution.
   a. What *process* is taking place in the direction of the arrows? What is happening at the cellular level when a wilted plant is watered and begins to recover from the wilt?

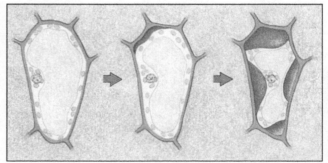

(After Starr and Taggart, 1989.)

   b. Is the solution in which the cells have been placed hypotonic, isotonic, or hypertonic relative to the cytoplasm?

4.5 Experiment: Osmotic Changes in Red Blood Cells

7. A human lost at sea without fresh drinking water is effectively lost in an osmotic desert. Why would drinking salt water be harmful?

Food for Thought

8. How does diffusion differ from osmosis?

9. Plant fertilizer consists of numerous different solutes. A small dose of fertilizer can enhance plant growth, but overfertilization can kill the plant. Why might overfertilization have this effect?

10. What does the word *lysis* mean? (*Now* does the name of the disinfectant Lysol® make sense?)

Enzymes: Catalysts of Life

OBJECTIVES

After completing this exercise, you will be able to

1. define *catalyst, enzyme, activation energy, enzyme–substrate complex, substrate, product, active site, denaturation, cofactor;*

2. explain how an enzyme operates;

3. recognize benzoquinone as a brown substance formed in damaged plant tissue;

4. indicate the substrates for the enzyme catechol oxidase;

5. describe the effect of temperature on the rate of chemical reactions in general and on enzymatically controlled reactions in particular;

6. describe the effect that an atypical pH may have on enzyme action;

7. indicate how a cofactor might operate and identify a cofactor for catechol oxidase.

INTRODUCTION

Life as we know it is impossible without enzymes. The energy required by your muscles simply to open your lab manual would take years to accumulate without enzymes. Due to the presence of enzymes, the myriad chemical reactions occurring in your cells at this very moment are being completed in a fraction of a second rather than the years or even decades that would be otherwise required.

Enzymes are proteins that function as biological catalysts. A **catalyst** is a substance that lowers the amount of energy necessary for a chemical reaction to proceed. You might think of this so-called **activation energy** as a mountain to be climbed. Enzymes decrease the size of the mountain, in effect turning it into a molehill (Figure 5-1).

By lowering the activation energy, an enzyme affects the *rate* at which reaction occurs. Enzyme-boosted reactions may proceed from 100,000 to 10 million times faster than they would without the enzyme.

In an enzyme-catalyzed reaction, the reactant (the substance being acted upon) is called the **substrate**. Substrate molecules combine with enzyme molecules to form a temporary **enzyme–substrate complex**. **Products** are

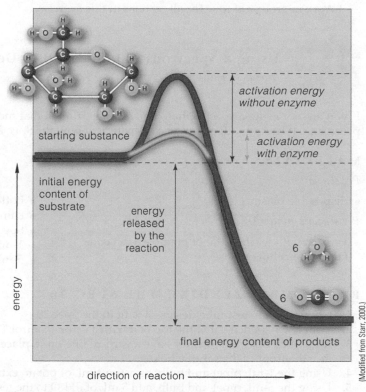

Figure 5-1 Enzymes and activation energy.

(Modified from Starr, 2000.)

formed, and the enzyme molecule is released unchanged. Thus, the enzyme is not used up in the process and is capable of catalyzing the same reaction again and again. This can be summarized as follows:

$$\text{substrate} \xrightarrow{\text{enzyme}} \text{enzyme–substrate complex} \longrightarrow \text{products} + \text{enzyme}$$

Before we proceed, let's visualize an enzyme. Look at Figure 5-2. Using common, everyday items, think of an enzyme as a key that unlocks a lock. Imagine that you have a number of locks, and all use the same key. You would only need one key to unlock them.

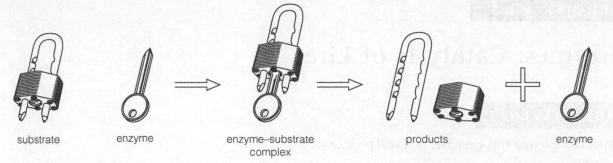

substrate   enzyme   enzyme–substrate complex   products   enzyme

Figure 5-2 Action of an enzyme.

Changing the shape of the key just a tiny bit may still allow the key to function in the lock, but you may have to fumble with the key a bit to get the lock open. Changing the key more results in an inability to open the lock. Similarly, a change in the shape of an enzyme alters its function. We will examine a number of factors in this exercise to determine their effects on enzyme action, including

1. temperature
2. pH (hydrogen-ion concentration of the environment)
3. specificity (how discriminating the enzyme is in catalyzing different potential substrates)
4. cofactor necessity (the need for a metallic ion for enzyme activity)

Although thousands of enzymes are present within cells, we'll examine only one: catechol oxidase (also known as tyrosinase).

Work in groups of four for all sections in this exercise.

5.1   Using a Spectronic 20 (Spec 20) to Determine Color Changes
(About 15 min.)

If you are not using a spectrophotometer, an instrument that measures color change, skip the following 11-step procedure. (See Appendix 2 for an explanation of how the Spec 20 works.)

MATERIALS

Per student:

■ disposable plastic gloves

Per student group (4):

■ ice bath with wash bottle of potato extract containing catechol oxidase

■ bottle of dH₂O
■ china marker
■ Spec 20 spectrophotometer
■ 1-mL pipet and bulb
■ 5-mL pipet and bulb

PROCEDURE: ZEROING THE SPEC 20

1. Obtain a clean test tube designed to fit in the Spec 20.
2. With a wax pencil, label the top of it with a C for "control."
3. If the tube does not already have a vertical line on it, place one on it with the wax pencil. This mark must face the front of the sample holder.
4. Using the 1-mL pipet and bulb, measure 1 mL of potato extract into test tube C.
5. Using the 5-mL pipet and bulb, add 5 mL of dH₂O to the test tube.
6. Place your gloved thumb over the mouth of the test tube and invert the tube to mix the contents.
7. Adjust the wavelength knob (top, right) of the Spec 20 to 540 nm.
8. Rotate the lower *left* absorbance adjustment so the needle reads infinity (∞) on the bottom scale.
9. Clean the surface of the tube by wiping it with a tissue and insert the tube into the Spec 20 sample holder with the vertical mark facing front (toward you).
10. Rotate the lower *right* absorbance adjustment so the needle reads zero.
11. You have now zeroed the Spec 20 for this exercise. Remove tube C and set it aside.

The previous procedure is done to account for the fact that the potato extract has color and absorbs light. In your experiments, you will be measuring the change in the amount of light absorbed by potato extract before and after various experimental treatments.

Formation and Detection of Benzoquinone *(About 20 min.)*

Catechol oxidase is an enzyme that catalyzes the production of benzoquinone and water from catechol:

$$\text{catechol} + \tfrac{1}{2} O_2 \xrightarrow{\text{catechol oxidase}} \text{benzoquinone} + H_2O$$

(substrates)   (enzyme in potato extract)   (product)

This is an oxidation reaction, with catechol and oxygen as the substrates. Hence, the enzyme gets its preferred name, *catechol oxidase*. (This suffix *-ase* is a tipoff that the substance is an enzyme.)

Catechol and catechol oxidase are present in the cells of many plants, although in undamaged tissue they are separated in different compartments of the cells. Injury causes mixing of the substrate and enzyme, producing benzoquinone, a brown substance.

You've probably noticed the brown coloration of a damaged apple or the blackening of an injured potato tuber (Figure 5-3). Benzoquinone inhibits the growth of certain microorganisms that cause rot.

In this section, you will form the product, benzoquinone, and establish a color intensity scale or absorbance standard that you will use in subsequent experiments.

(Photo by J. W. Perry.)

Figure 5-3   Potato showing browning due to benzoquinone production. The potato on the left was sliced immediately before this photo was taken. The one on the right had been cut and exposed to the oxygen in the air for several minutes before being photographed.

MATERIALS

Per student:

- disposable plastic gloves

Per student group (4):

- 3 test tubes
- test tube rack
- metric ruler
- ice bath with wash bottle of potato extract containing catechol oxidase
- wash bottle containing 1% catechol solution

- bottle of dH_2O
- china marker
- warmed up and zeroed Spec 20, optional; see Appendix 2

Per lab room:

- 40°C waterbath
- vortex mixer (optional)

PROCEDURE

1. With a china marker, label three test tubes 2_A, 2_B, and 2_C. Place your initials on each test tube for later identification.
2. Lay the test tubes against a metric ruler and mark lines on the tubes corresponding to 1 cm and 2 cm *from the bottom* of each tube.
3. Fill each tube as follows:
   Tube 2_A: 1 cm of potato extract containing catechol oxidase
           1 cm of 1% catechol solution
   Tube 2_B: 1 cm of potato extract containing catechol oxidase
           1 cm of dH_2O
   Tube 2_C: 1 cm of 1% catechol solution
           1 cm of dH_2O

> ### Caution
>
> *Some of the chemicals (catechol, hydroquinone) used in these experiments can be hazardous to your health if they are ingested or taken in through your skin. Wear disposable plastic gloves for all experiments.*

Note: Be certain to return potato extract to ice bath IMMEDIATELY after use in this and subsequent experiments.

4. Shake all tubes (using a vortex mixer if available).
5. Record the color of the solution in each tube in the "time 0" spaces of Table 5-1.
6. Place the tubes in a 40°C waterbath.
7. After 10 minutes, the catechol should be completely oxidized. Remove the tubes from the waterbath. Record the *color* of the substance in each test tube in Table 5-1.

   What you do next depends on whether you're using a spectrophotometer. Proceed with *either* step 8 *or* 9.
8. *Color-intensity method:* Consider the color of the product in tube 2_A to be a 5 on a color intensity scale of 0–5, while the color of the substance in tubes 2_B and 2_C is a 0. In Table 5-2, record the color intensity for tubes 2_B and 2_C as 0, and that of tube 2_A as 5.

   You will use this scale to make comparisons in Sections 5.3–5.6. Keep the contents in tubes 2_A, 2_B, and 2_C, and refer to them as you make comparisons in subsequent experiments.
9. *Spec 20 method:*
   (a) Pipet 6 mL of the contents of each tube into six separate Spec 20 tubes.
   (b) Clean off each tube.
   (c) Zero the Spec 20 using the contents of tube 2_C.
   (d) Insert tubes containing the contents of tube 2_A and then tube 2_B, determine the absorbance of each, and record them in Table 5-2.

TABLE 5-1	Formation and Detection of Benzoquinone: Record Color		
Time	Tube 2_A: Potato Extract and Catechol	Tube 2_B: Potato Extract and Water	Tube 2_C: Catechol and Water
0 min.			
10 min.			

What is the brown-colored substance that appeared in tube 2_A? _____

What is the substrate for the reaction that occurred in tube 2_A? _____

What is the product of the reaction in tube 2_A? _____

What substances do tubes 2_B and 2_C lack that account for the absence of the brown-colored substance?
2_B _____ 2_C _____

What is the purpose of tubes 2_B and 2_C? _____

TABLE 5-2	Color-Intensity Scale or Absorbance	
Intensity/Absorbance	Tube	Color of Product
	2_C	
	2_B	
	2_A	

5.3   Experiment: Enzyme Specificity *(About 20 min.)*

Generally, enzymes are substrate-specific, acting on one particular substrate or a small number of structurally similar substrates. This specificity is due to the three-dimensional structure of the enzyme. For the enzyme–substrate complex to form, the structure of the substrate must very closely complement that of the enzyme's

active site. The active site is a special region of the enzyme where the substrate binds. The active site has a small amount of moldability, so that the active site and substrate become fully complementary to each other, as shown in Figure 5-4.

Think again about the lock and key analogy (Figure 5-2). If the lock and key are not complementary, the lock won't open. But how exact a fit is necessary?

In this experiment, you will determine the ability of the enzyme catechol oxidase to catalyze the oxidation of two different but structurally similar substrates: catechol and hydroquinone. First, examine the chemical structure of each compound:

catechol   hydroquinone

You need not memorize these structural formulas, but do notice that both are ring structures with two hydroxyl (—OH) groups attached.

Keep this in mind as you do the next experiment, in which you will determine how specific (discriminating) catechol oxidase is for particular substrates.

MATERIALS

Per student group (4):

- 3 test tubes
- test tube rack
- metric ruler
- china marker
- wash bottle containing 1% catechol
- wash bottle containing 1% hydroquinone
- ice bath with wash bottle of potato extract containing catechol oxidase
- warmed up and zeroed Spec 20, optional; see Appendix 6

Per lab room:

- 40°C waterbath
- vortex mixer (optional)

PROCEDURE

1. With a china marker, label three clean test tubes 3_A, 3_B, and 3_C. Include your initials for identification.
2. Lay the test tubes against a metric ruler and mark lines indicating 1 cm and 2 cm *from the bottom* of each test tube.
3. Fill each tube as follows:
   Tube 3_A: 1 cm of potato extract containing catechol oxidase
   1 cm of 1% catechol
   Tube 3_B: 1 cm of potato extract containing catechol oxidase
   1 cm of 1% hydroquinone
   Tube 3_C: 1 cm of potato extract containing catechol oxidase
   1 cm of dH_2O
4. Gently shake the test tubes to mix the contents.
5. Compare the color intensity of the solution in each test tube *with the standards produced in Section 5.2* and record them at time 0 in Table 5-3.
6. Place the test tubes in a 40°C waterbath.

This experiment addresses the hypothesis that *the structure of a substrate determines how well an enzyme acts upon the substrate.*

7. While you wait for the experiment to run its course, write your prediction of the outcome in Table 5-3.
8. After 10 minutes, remove the test tubes from the water bath and examine them. Choose step 9 *or* 10 and proceed.
9. *Color-intensity method:* Record the color intensity (scale 0–5) of each tube's contents in Table 5-3.

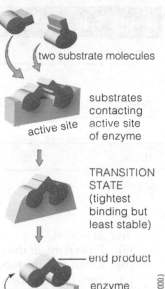

two substrate molecules

substrates contacting active site of enzyme

active site

TRANSITION STATE (tightest binding but least stable)

end product

enzyme unchanged by the reaction

(After Starr, 2000.)

Figure 5-4 Induced-fit model of enzyme–substrate interactions.

TABLE 5-3 Specificity of Catechol Oxidase for Different Substrates

Prediction:

Time	Relative Color Intensity or Absorbance		
	Tube 3_A: Catechol	Tube 3_B: Hydroquinone	Tube 3_C: dH_2O
0 min.			
10 min.			
Conclusion:			

10. *Spec 20 method:*
    (a) Pipet 6 mL of the contents of each tube into three labeled Spec 20 tubes.
    (b) Clean off each tube.
    (c) Zero the Spec 20 using the contents of tube 3_C.
    (d) Insert tubes containing the contents of tube 3_A and then tube 3_B, determine the absorbance of each, and record them in Table 5-3.
11. Upon which substrate does catechol oxidase work best, forming the most benzoquinone in the shortest amount of time?

    _____

12. Based on your knowledge of the structures of the two substrates, what apparently determines the specificity of catechol oxidase?

    _____

    _____

13. Why was tube 3_C included in this experiment?

    _____

    _____

14. Record your conclusion in Table 5-3, either accepting or rejecting the hypothesis.

5.4 Experiment: Effect of Temperature on Enzyme Activity *(25 min.)*

The rate at which chemical reactions take place is largely determined by the temperature of the environment. *Generally, for every 10°C rise in temperature, the reaction rate doubles.* Within a rather narrow range, this is true for enzymatic reactions also. However, because enzymes are proteins, excessive temperature alters their structure, destroying their ability to function. When an enzyme's structure is changed sufficiently to destroy its function, the enzyme is said to be **denatured.** Most enzymatically controlled reactions have an *optimum* temperature and pH—that is, one temperature and pH where activity is maximized.

   In this experiment, you will determine the temperature range over which the enzyme catechol oxidase is able to catalyze its substrate. You will also determine the best (optimum) temperature for the reaction.

MATERIALS

Per student group (4):

- 6 test tubes
- test tube rack
- metric ruler
- china marker
- wash bottle containing 1% catechol
- ice bath with wash bottle of potato extract containing catechol oxidase
- three 400-mL graduated beakers

- heat-resistant glove
- Celsius thermometer
- warmed up and zeroed Spec 20, optional; see Appendix 2

Per student group (table):

- hot plate *or* burner, tripod support, wire gauze, and matches or striker
- boiling chips

Per lab room:

- source of room-temperature water
- three waterbaths: 40°C, 60°C, 80°C
- vortex mixer (optional)

PROCEDURE

1. Half fill one 400-mL beaker with tap water. Add a few boiling chips and turn on the hotplate to the highest temperature setting, *or,* if your lab is equipped with burners, light the burner. Bring the water to a boil, then turn the heat down so that the water just continues to boil.
2. Put 150 mL of tap water into a second beaker and add ice to the water.
3. Half fill a third beaker with water from the source at room temperature.
4. With a china marker, label six test tubes 4_A–4_F. Include your initials for identification.
5. Lay the test tubes against a metric ruler and mark off lines indicating 1 cm and 2 cm *from the bottom* of each tube.
6. Fill each tube to the 1-cm mark with potato extract containing catechol oxidase.
7. **(a)** Place tube 4_A in the 400-mL beaker of ice water.
   Measure and record the water temperature: _____°C
   (b) Place tube 4_B in the 400-mL beaker containing room-temperature water.
   Room temperature: _____°C
   (c) Place tube 4_C in the 40°C waterbath.
   (d) Place tube 4_D in the 60°C waterbath.
   (e) Place tube 4_E in the 80°C waterbath.
   (f) Place tube 4_F in the 400-mL beaker containing boiling water.
   Temperature of boiling water: _____°C
8. Allow the test tubes to remain at the various temperatures for 5 minutes.
9. Remove the tubes and add catechol to the 2-cm line on each. Agitate the tubes (with a vortex mixer if available) to mix the contents.
10. *Color-intensity method:* In Table 5-4, record the relative color intensity (scale 0–5) of the solution in each tube, using the standard established in Section 5.2. Return each tube to its respective temperature bath immediately after recording.

This experiment addresses the hypothesis that *the temperature of a substrate and an enzyme determines the amount of product that is formed.*

Caution

Wear a heat-resistant glove when handling heated glassware.

11. While you wait for the experiment to run its course, write your prediction of the outcome of the experiment in Table 5-4.
12. Shake periodically (by hand) all tubes over the next 10 minutes.
13. After 10 minutes, remove the test tubes from the water baths. Choose step 14 *or* 15 and proceed.
14. *Color-intensity method:* Record the color intensity (scale 0–5) of each tube's contents in Table 5-4.

TABLE 5-4 Effect of Temperature on Enzyme Activity

Prediction:

Time	Relative Color Intensity or Absorbance					
	Tube 4_A	Tube 4_B	Tube 4_C	Tube 4_D	Tube 4_E	Tube 4_F
0 min.						
10 min.						
Conclusion:						

15. *Spec 20 method:*
    (a) Pipet 6 mL of the contents of each tube into six labeled Spec 20 tubes.
    (b) Clean off each tube.
    (c) Zero the Spec 20 using the contents of tube 2_C.
    (d) Insert tubes containing the contents of each tube, recording them in Table 5-4.
16. Plot the data from Table 5-4 for the 10-minute reading in Figure 5-5.
17. Over what temperature *range* is catechol oxidase active?

    _____

18. What is the *optimum* temperature for activity of this enzyme?

    _____

19. What happens to enzyme activity at very high temperatures?

    _____

    _____

    _____

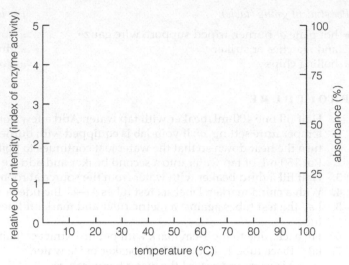

Figure 5-5 Effect of temperature on catechol oxidase activity.

20. Record your conclusion in Table 5-4, either accepting or rejecting the hypothesis.

5.5 Experiment: Effect of pH on Enzyme Activity *(25 min.)*

Another factor influencing the rate of enzyme catalysis is the hydrogen-ion concentration (pH) of the solution. Like temperature, pH affects the three-dimensional shape of enzymes, thus regulating their function. Most enzymes operate best when the pH of the solution is near neutrality (pH 7). Others, however, have pH optima in the acidic or basic range, corresponding to the environment in which they normally function.

In this experiment, you will determine the pH range over which the enzyme catechol oxidase is able to catalyze its substrate. You will also determine the optimum pH for the reaction.

MATERIALS

Per student group (4):

- 7 test tubes
- test tube rack
- metric ruler
- china marker
- wash bottle containing 1% catechol
- ice bath with wash bottle of potato extract containing catechol oxidase
- warmed up and zeroed Spec 20, optional; see Appendix 2

Per lab room:

- 40°C waterbath
- phosphate buffer series, pH 2–12 (2, 4, 6, 7, 8, 10, 12)
- vortex mixer (optional)

PROCEDURE

1. With a china marker, label seven test tubes 5_A–5_G. Include your initials for identification.
2. Lay the test tubes against a metric ruler and mark lines indicating 4 cm, 5 cm, and 6 cm *from the bottom* of each tube.
3. Take your test tubes to the location of the phosphate buffer series and fill each tube according to the following directions:

4. Return to your work area and add 1 cm of potato extract containing catechol oxidase to each of the seven tubes (thus bringing the total volume of each to the 5-cm mark). Agitate the tubes by hand.
5. Add 1% catechol to each of the seven tubes, bringing the total volume to the 6-cm mark. Agitate the contents of the tubes, using a vortex mixer if available.
6. In Table 5-5 at time 0, record the relative color intensity of each tube *immediately after adding the 1% catechol.*
7. Place the tubes in the 40°C waterbath.
8. Agitate the tubes periodically over the next 10 minutes.

Tube	Fill to the 4-cm Mark with Buffer of
5A	pH 2
5B	pH 4
5C	pH 6
5D	pH 7
5E	pH 8
5F	pH 10
5G	pH 12

This experiment addresses the hypothesis that *the pH of a substrate and an enzyme determines the amount of product that is formed.*

9. While you wait for the experiment to run its course, write your prediction of the outcome in Table 5-5.
10. After 10 minutes, remove the test tubes from the water baths. Choose step 11 *or* 12 and proceed.
11. *Color-intensity method:* Record the color intensity (scale 0–5) of each tube's contents in Table 5-5.

TABLE 5-5 Effect of pH on Enzyme Activity

Prediction:

	Relative Color Intensity or Absorbance						
Time	Tube 5A (pH 2)	Tube 5B (pH 4)	Tube 5C (pH 6)	Tube 5D (pH 7)	Tube 5E (pH 8)	Tube 5F (pH 10)	Tube 5G (pH 12)
0 min.							
10 min.							

Conclusion:

12. *Spec 20 method:*
   (a) Pipet 6 mL of the contents of each tube into seven labeled Spec 20 tubes.
   (b) Clean off each tube.
   (c) Zero the Spec 20 using the contents of tube 2C from Section 5.2.
   (d) Insert each tube, determine the absorbance, and record it in Table 5-5.
13. Plot the data from Table 5-5 for your 10-minute reading in Figure 5-6.
14. Over what pH *range* does catechol oxidase catalyze catechol to benzoquinone?
_____
15. What is the *optimum* pH for catechol oxidase activity?
_____
16. Record your conclusion in Table 5-5, either accepting or rejecting the hypothesis.

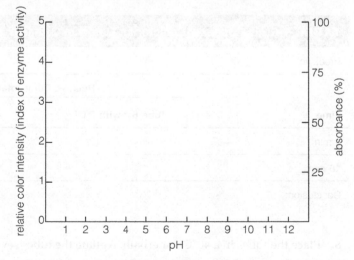

Figure 5-6 Effect of pH on catechol oxidase activity.

Some enzymatic reactions occur only when the proper *cofactors* are present. **Cofactors** are nonprotein organic molecules and metal ions that are part of the structure of the active site, making the formation of the enzyme–substrate complex possible.

In this experiment, you will use phenylthiourea (PTU), which binds strongly to copper, to remove copper ions. Thus, you'll be able to determine whether copper is a necessary cofactor necessary for producing benzoquinone from catechol.

MATERIALS

Per student group (4):

- 2 test tubes
- test tube rack
- metric ruler
- china marker
- ice bath with wash bottle of potato extract containing catechol oxidase
- wash bottle containing 1% catechol solution
- bottle of dH_2O
- china marker
- scoopula (small spoon)

- phenylthiourea crystals in small screw-cap bottle
- warmed up and zeroed Spec 20, optional; see Appendix 2

Per lab room:

- 40°C waterbath
- vortex mixer (optional)
- bottle of 95% ethanol (at each sink)
- tissues (at each sink)

PROCEDURE

1. With a china marker, label two test tubes 6_A and 6_B. Include your initials for identification.
2. Lay the test tube against a metric ruler and mark lines indicating 1 cm and 2 cm *from the bottom* of each tube.
3. Add potato extract containing catechol oxidase to the 1-cm mark of each test tube.
4. Using a scoopula, add five crystals of phenylthiourea (PTU) to tube 6_A. Do not add anything to tube 6_B.

Caution
PTU is poisonous.

5. Agitate the contents of both test tubes frequently by hand during the next 5 minutes.
6. Add 1% catechol to the 2-cm mark of both test tubes and agitate the contents of the tubes, using a vortex mixer if available.
7. *Color-intensity method:* In Table 5-5 at time 0, record the relative color intensities (scale of 0–5).

TABLE 5-6 Is Copper a Cofactor for Catechol Oxidase?		
Prediction:		
	Relative Color Intensity or Absorbance	
Time	Tube 6_A: with PTU	Tube 6_B: without PTU
0 min.		
10 min.		
Conclusion:		

8. Place the tubes in a 40°C waterbath. Agitate the tubes several times during the next 10 minutes.

This experiment addresses the hypothesis that *a cofactor is necessary for the action of the enzyme catecol oxidase.*

9. While you wait for the experiment to run its course, write your prediction of the outcome in Table 5-6.
10. After 10 minutes, remove the test tubes from the water baths. Choose step 11 *or* 12 and proceed.
11. *Color-intensity method:* Record the color intensity (scale 0–5) of each tube's contents in Table 5-6.

12. *Spec 20 method:*
    (a) Pipet 6 mL of the contents of each tube into seven separate Spec 20 tubes.
    (b) Clean off each tube.
    (c) Zero the Spec 20 using the contents of tube 2_C from Section 5.2.
    (d) Insert each tube in the Spec 20, determine the absorbance, and record it in Table 5-6.
13. Did benzoquinone form in tube 6_A? In tube 6_B? _____
14. From this experiment, what can you conclude about the necessity for copper for catechol oxidase activity?
    _____
15. What substance used in this experiment contained copper?
    _____
16. Record your conclusion in Table 5-5, either accepting or rejecting the hypothesis.

Note: **After completing all experiments, take your dirty glassware to the sink and wash it following directions on page x. Use 95% ethanol to remove the china marker. Invert the test tubes in the test tube racks so they drain. Tidy up your work area, making certain all materials used in this exercise are there for the next class.**

_____ 1. Enzymes are
   (a) biological catalysts
   (b) agents that speed up cellular reactions
   (c) proteins
   (d) all of the above

_____ 2. Enzymes function by
   (a) being consumed (used up) in the reaction
   (b) lowering the activation energy of a reaction
   (c) combining with otherwise toxic substances in the cell
   (d) adding heat to the cell to speed up the reaction

_____ 3. The substance that an enzyme combines with is
   (a) another enzyme
   (b) a cofactor
   (c) a coenzyme
   (d) the substrate

_____ 4. Enzyme specificity refers to the
   (a) need for cofactors for some enzymes to function
   (b) fact that enzymes catalyze one particular substrate or a small number of structurally similar substrates
   (c) effect of temperature on enzyme activity
   (d) effect of pH on enzyme activity

_____ 5. For every 10°C rise in temperature, the rate of most chemical reactions will
   (a) double
   (b) triple
   (c) increase by 100 times
   (d) stop

_____ 6. When an enzyme becomes denatured, it
   (a) increases in effectiveness
   (b) loses its requirement for a cofactor
   (c) forms an enzyme–substrate complex
   (d) loses its ability to function

_____ 7. An enzyme may lose its ability to function because of
   (a) excessively high temperatures
   (b) a change in its three-dimensional structure
   (c) a large change in the pH of the environment
   (d) all of the above

_____ 8. pH is a measure of
   (a) an enzyme's effectiveness
   (b) enzyme concentration
   (c) the hydrogen-ion concentration
   (d) none of the above

_____ 9. Catechol oxidase
   (a) is an enzyme found in potatoes
   (b) catalyzes the production of catechol
   (c) has as its substrate benzoquinone
   (d) is a substance that encourages the growth of microorganisms

_____ 10. The relative color intensity used in the experiments of this exercise
   (a) is a consequence of production of benzoquinone
   (b) is an index of enzyme activity
   (c) may differ depending on the pH, temperature, or presence of cofactors, respectively
   (d) is all of the above

EXERCISE 5

Enzymes: Catalysts of Life

POST-LAB QUESTIONS

5.4 Experiment: Effect of Temperature on Enzyme Activity

1. Eggs can contain bacteria such as *Salmonella*. Considering what you've learned in this exercise, explain how cooking eggs makes them safe to eat.

2. As you demonstrated in this experiment, high temperatures inactivate catechol oxidase. How is it that some bacteria live in the hot springs of Yellowstone Park at temperatures as high as 73°C?

3. Why do you think high fevers alter cellular functions?

4. Some surgical procedures involve lowering a patient's body temperature during periods when blood flow must be restricted. What effect might this have on enzyme-controlled cellular metabolism?

5. At one time, it was believed that individuals who had been submerged under water for longer than several minutes could not be resuscitated. Recently this has been shown to be false, especially if the person was in cold water. Explain why cold-water "drowning" victims might survive prolonged periods under water.

6. Explain what happens to catechol oxidase when the pH is on either side of the optimum.

7. What would you expect the pH optimum to be for an enzyme secreted into your stomach?

Food for Thought

8. Is it necessary for a cell to produce one enzyme molecule for every substrate molecule that needs to be catalyzed? Why or why not?

9. Explain the difference between *substrate* and *active site*.

10. The photo shows slices of two apples. The one on the left sat on the counter for 15 minutes prior to being photographed. The one on the right was sliced immediately prior to the photo being taken.

    a. Explain as thoroughly as possible what you see and why the two slices differ.

(Photo by J. W. Perry.)

    b. If you don't want a cut apple to brown, what can you do to prevent it?

The Digestive Action of Salivary Enzymes on Various Macromolecules

Abstract

Salivary glands secrete enzymes that begin the digestion of some foods within the mouth. Our objective was to determine which macromolecules from various foods the enzymes in saliva would digest. We hypothesized that salivary enzymes would begin digestion of certain carbohydrates and proteins. We used chemical tests to examine the digestive ability of salivary enzymes. Our results indicated that salivary enzymes were capable of digesting starch but were not able to digest glucose or protein. Extensions of these experimental findings are discussed.

Introduction

Enzymes are essential to living organisms. They are catalysts that facilitate and coordinate the metabolic, life-sustaining reactions within cells (Starr, Evers, and Starr 2008). For example, digestive system enzymes help break down food into molecules that are small enough to be absorbed by the body's cells. Enzymes within the cells complete the breakdown of molecules to even smaller units, freeing the energy held within chemical bonds so that it can be stored and later used by the cell (Starr, Evers, and Starr 2008).

To better understand the human digestive system and its enzymatic functions, we undertook a study to investigate the digestive properties of the enzymes contained in saliva. Our research question focused on the function of saliva: What macromolecules does saliva digest? Previous research on mammals has shown that some starches begin to be digested in the mouth (Pederson et al. 2002). To verify this research finding and to extend it by investigating proteins, we designed an experiment to examine the digestion of glucose, starch, and protein by salivary enzymes.

Our hypothesis was that saliva contains a broad spectrum of digestive enzymes that start to digest both simple and complex carbohydrates as well as proteins. This hypothesis led to the predictions that saliva will digest (1) glucose, a simple carbohydrate; (2) starch, a complex carbohydrate; and (3) protein.

Methods

General Procedure

We made separate solutions of glucose, potato starch, and egg albumin (protein). Each solution was contained in a separate test tube. We then collected saliva in test tubes by gently spitting into them until we had a sufficient quantity. For our control groups, we used test tubes with water and the particular macromolecule being tested. For our experimental groups, we used test tubes with water, the particular macromolecule, and saliva. We had one additional control, saliva and water, to verify that saliva and water did not test positive for any of the chemical tests used. We used five replicates of each control and experimental group. To allow for potential digestive action to take place in the experimental groups, we waited 10 minutes before conducting the chemical tests.

Test Procedures

To test for the presence of simple sugars, we used the Benedict's test. We expected that control glucose solutions would change from blue to green or reddish brown when heated with Benedict's reagent, but the experimental solutions with glucose and saliva would remain blue when tested if glucose was digested. We added a dropperful of Benedict's solution to each test tube and heated the test tubes in a boiling water bath for three minutes. Data were recorded in the accompanying table.

To test for the presence of starch, we used the iodine test. We expected that the control starch solutions would change from a yellowish-brown color to a bluish-black color when iodine was added, but the experimental solutions with starch and saliva would remain yellowish-brown if starch was digested. Data also were recorded in the table.

To test for the presence of proteins, we used the Biuret test. We expected the control protein solutions would turn a violet color when the Biuret test reagent was added, but the experimental solutions, with protein and saliva, would remain a light blue if protein were digested. These data were recorded in the table.

TABLE 1 Results of chemical tests for glucose, starch, and protein. NA = test not applicable.

Test Solution	Benedict's Test	Lugol's Test	Biuret Test
Control: water	–	–	–
Control: saliva	–	–	–
Control: glucose + water	+	NA	NA
Glucose + saliva	Expect + if saliva does not digest glucose	NA	NA
Control: Starch (amylose) + water	–	+	NA
Starch (amylose) + saliva	Expect + if digested into simple sugars	Expect—if digested	NA
Control: Egg albumin + water	NA	NA	+
Egg albumin + saliva	NA	NA	Expect—if protein digested

Results

The table contains the results of the chemical tests conducted on all control and experimental groups. The data indicate that saliva digests starch (amylose from potatoes) but is unable to digest glucose or protein. All replicates within a group tested identical.

Discussion

The results of our experimental analysis do not support our hypothesis. Saliva was able to digest potato starch but was unable to digest glucose or protein. Glucose, a small molecule that is highly soluble in water, may begin digestion only after being absorbed into cells (intracellular digestion). Perhaps the glycolytic enzymes are only found within cells and are not excreted for extracellular digestion as are other enzymes of the digestive system. Proteins, on the other hand, are known to be digested by enzymes that are excreted by cells lining the stomach and small intestine (Starr, Evers, and Starr 2008). As the results of this study suggest, proteins may begin digestion only after reaching the stomach, with mastication and no chemical digestion occurring in the mouth.

Future studies that explore the potential effects of temperature, pH, and salinity would add to our knowledge of the digestive action of saliva. Weak bonds, such as the hydrogen bonds that hold enzymes in their functional shapes, may be broken or disrupted by changes in temperature or salinity. Perhaps the old adage "starve a fever but feed a cold" is because the higher body temperatures experienced during a fever disrupt the hydrogen bonds of the body's enzymes, interfering with digestion. Salinity changes may also disrupt the three-dimensional shape of enzymes. Studies that examine the effect of foods high in salt are warranted. Changes in pH may also influence the digestive action of saliva. Beverages such as tea and coffee that are high in acidity may interfere with the digestive ability of saliva and impair digestion in the mouth.

Additional studies could be designed to investigate potential enzyme inhibitors, enzyme activators, and the influence of cofactors on the digestive action of saliva. Many drugs are enzyme inhibitors, as are some herbicides and pesticides. Could pesticide residue on foods reduce the digestive action of saliva? Also, many plants have polyphenolic allelochemicals, such as tannins, that may reduce the digestive action of saliva. A study by Robbins et al. (1987) found differences among herbivores in the ability of their saliva to neutralize tannins. A study to investigate the influence of various polyphenolic allelochemicals would be useful for extending our knowledge of salivary enzyme functioning.

To extend our understanding of enzyme functioning in general, studies could be designed to test the function of over-the-counter digestive enzyme supplements. Various experimental conditions could be created, such as those discussed above for enzymes in saliva, to study the effects of temperature, pH, salinity, enzyme inhibitors and activators, and polyphenolic allelochemicals on enzymatic functioning.

Citations

Pederson, A. M., A. Bardow, S. Beier Jenson, and B. Nauntofte. 2002. Saliva and gastrointestinal functions of taste, mastication, swallowing and digestion. *Oral Diseases* 8(3): 117–29.

Robbins, C. T., S. Mole, A. E. Hagerman, and T. A. Hanley. 1987. Role of tannins in defending plants against ruminants: Reduction in dry matter digestion. *Ecology* 68(6): 1606–15.

Starr, C., C. A. Evers, and L. Starr. 2008. *Biology: Concepts and application*, 7th ed. Belmont, CA: Cengage/Brooks Cole.

Photosynthesis: Capture of Light Energy

OBJECTIVES

After completing this exercise, you will be able to

1. define *photosynthesis, autotroph, heterotroph, chlorophyll, chromatogram, absorption spectrum, carotenoid;*

2. describe the role of carbon dioxide in photosynthesis;

3. determine the effect of light and carbon dioxide on photosynthesis;

4. determine the wavelengths absorbed by pigments;

5. identify the pigments in spinach chloroplast extract;

6. identify the carbohydrate produced in geranium leaves during photosynthesis;

7. identify the structures composing the chloroplast and indicate the function of each structure in photosynthesis.

INTRODUCTION

Photosynthesis is the process by which light energy converts inorganic compounds to organic substances with the subsequent release of elemental oxygen. It may very well be the most important biological event sustaining life. Without it, most living things would starve, and atmospheric oxygen would become depleted to a level incapable of supporting animal life. Ultimately, the source of light energy is the sun, although on a small scale we can substitute artificial light.

Nutritionally, two types of organisms exist in our world, autotrophs and heterotrophs. **Autotrophs** (*auto* means self, *troph* means feeding) synthesize organic molecules (carbohydrates) from inorganic carbon dioxide. The vast majority of autotrophs are the photosynthetic organisms that you're familiar with—plants, as well as some protistans and bacteria. These organisms use light energy to produce carbohydrates. (A few bacteria produce their organic carbon compounds chemosynthetically, that is, using chemical energy.)

By contrast, **heterotrophs** must rely directly or indirectly on autotrophs for their nutritional carbon and metabolic energy. Heterotrophs include animals, fungi, many protistans, and most bacteria.

In both autotrophs and heterotrophs, carbohydrates originally produced by photosynthesis are broken down by *cellular respiration* (Exercise 10), releasing the energy captured from the sun for metabolic needs.

The photosynthetic reaction is often conveniently summarized by the equation:

$$12H_2O + 6CO_2 \xrightarrow{\text{light energy}} 6O_2 + C_6H_{12}O_6 + 6H_2O$$

water   carbon   oxygen   glucose   water
         dioxide

Although glucose is often produced during photosynthesis, it is usually converted to another transport or storage compound unless it is to be used immediately for carbohydrate metabolism. In plants and many protistans, the most common storage carbohydrate is *starch*, a compound made up of numerous glucose units linked together. Starch is designated by the chemical formula $(C_6H_{12}O_6)_n$, where n indicates a large number. Most plants transport carbohydrate as sucrose.

The following experiments will acquaint you with the principles of photosynthesis.

6.1   Test for Starch (*About 10 min.*)

As indicated by the overall formula of photosynthesis, one end product is a carbohydrate (CH_2O). But a number of different carbohydrates have the empirical formula CH_2O. In this section, you will perform a simple test to visually distinguish between two different carbohydrates and water.

MATERIALS

Per student group (4):

- 1 dropper bottle each of
  iodine (I_2KI) solution
  starch solution

- glucose solution
  dH$_2$O
- depression (spot) plate

PROCEDURE

1. Place a couple drops of starch, glucose, and distilled water in three different depressions of the spot plate. Now add a drop of iodine solution to each.
2. Record your observations. How can you identify the presence of starch?

Observations: _____

_____

_____

6.2	Experiment: Effects of Light and Carbon Dioxide on Starch Production

(About 30 min.)

In this experiment, you will perform a test to determine the environmental conditions necessary for photosynthesis and starch production. This experiment addresses the hypothesis that *photosynthesis proceeds only in the presence of light and carbon dioxide.*

MATERIALS

Per student group (4):

- two 400-mL beakers
- square of aluminum foil
- hot plate in fume hood
- heat-resistant glove
- petri dish halves
- bottle of iodine solution
- bottle of 95% ethanol (EtOH)
- forceps

Per lab room:

- source of dH$_2$O
- Fast Plants™, 9–10 days old, grown for 4 days in three different environments:
  - I. normal conditions with both light and carbon dioxide
  - II. in dark, with normal carbon dioxide
  - III. in light, but with carbon dioxide removed

PROCEDURE

1. Carefully observe and record any differences in appearance among the three sets of plants in Table 6-1.
2. Write a prediction regarding starch presence, and thus photosynthesis activity, for each growing condition in Table 6-1.
3. Pigments present in the plants must be removed before a test for starch presence can be performed. Kill the plants and extract the pigments:
   - **(a)** With a china marker, label the beakers A (for alcohol) and dH$_2$O.
   - **(b)** Add about 150 mL distilled water to the dH$_2$O beaker, set it on the hot plate, and turn on the hot plate to the highest setting. Allow the water to come to a boil.
   - **(c)** Completely remove 1–2 plants of each treatment from its growing container. Wash off all soil from the roots. Keep plants of each treatment separate and labeled.
   - **(d)** Alcohol has a much lower boiling point than water, and so takes very little time to come to a boil. When the water is boiling, put about 150 mL of alcohol in the A beaker, set it also on the hot plate and bring to a boil. Keep the alcohol beaker covered with aluminum foil as much as possible throughout the lab to prevent excess evaporation.

Caution

Ethanol is highly flammable. Use only electric hot plates, never open flame. Also, never let a beaker boil dry. Add more liquid, or remove the beaker from the burner, and place it on a pad of folded paper towels.

   - **(e)** Place the plants from one treatment in the beaker of boiling water for about 1 minute. This kills the tissue and breaks down internal membranes.
   - **(f)** Use the long forceps to move the wilted plants from the water into the boiling alcohol. This will extract the photosynthetic pigments from the plant tissues. When the pigments have been extracted, the liquid will appear green, and the plant will appear to be mostly bleached.

Fast Plants™ Growing Condition	Appearance	Prediction	Starch Presence and Location
TABLE 6-1 Effect of Light and Carbon Dioxide on Starch Presence			
I. Normal conditions with both light and carbon dioxide			
II. In dark, with normal carbon dioxide			
III. In light, but with carbon dioxide removed			

(g) Remove the plants from the alcohol with forceps, and dip momentarily in the boiling water to soften.

4. Test the plants for the presence and localization of starches:
   (a) Place killed, depigmented plants in petri dishes filled with iodine solution.
   (b) Let the plants soak in the iodine solution for a couple minutes, rinse, and float in water in another petri dish in order to observe the pattern of staining.

5. Repeat the pigment extraction an d staining process for plants of the other two treatments, being careful to keep the treatments separate and identified.

6. Remove the A beaker from the hot plate, and turn the heat off.

7. In Figure 6-1, sketch a plant from each experimental treatment, shading in the portions that stained dark. Be careful to note where any dark staining occurs. Record your written observations in Table 6-1.

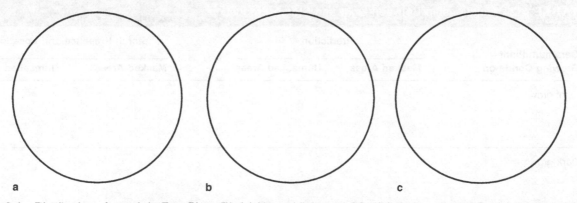

Figure 6-1   Distribution of starch in Fast Plants™. (a) Normal light and CO_2; (b) dark, normal CO_2; (c) light, no CO_2.

8. What does this staining pattern in plants of the three treatments indicate?
_____
_____

9. What conclusion can you draw about the effect of light on the presence of starch?
_____
_____

10. What conclusion can you draw about the effect of carbon dioxide on the presence of starch?

_____

_____

11. Write a conclusion either accepting or rejecting the hypothesis.

_____

6.3	Experiment: Relationship Between Light and Photosynthetic Products

(About 15 min.)

This experiment addresses the hypothesis that *light is necessary for photosynthesis to proceed.*

MATERIALS

Per student group (4):

- china marker
- two 400-mL graduated beakers
- hot plate in fume hood
- heat-resistant glove
- bottle of 95% ethanol
- forceps
- 2 petri dishes
- I₂KI solution in foil-wrapped stock bottle

Per lab room:

- source of dH₂O
- light-grown geranium plant or leaves of geranium plant with "masks"
- dark-grown geranium plant or leaves of geranium plant with "masks" (kept in dark place)

PROCEDURE

Work in groups of four.

1. Observe the two geranium plants available. One plant has been growing in bright light for several hours; the other has been kept in the dark for a day or more. Both leaves have had an area of the lamina (blade) masked by an opaque design.
2. Write a prediction regarding starch presence, and thus photosynthesis activity, for each growing condition and leaf treatment area in Table 6-2.

TABLE 6-2 Relationship Between Light and Starch Production

Geranium Plant Growing Condition	Prediction		Starch Presence and Location	
	Masked Areas	Unmasked Areas	Masked Areas	Unmasked Areas
Light-grown				
Dark-grown				

3. Select a leaf from one of the two geranium plants. Pigments present in the plants must be removed before a test for starch presence can be performed. Kill the leaf and extract the pigments:
   (a) With a china marker, label the beakers A (for alcohol) and dH₂O.
   (b) Add about 150 mL distilled water to the dH₂O beaker, set it on the hot plate, and turn on the hot plate to the highest setting. Allow the water to come to a boil.
   (c) Remove a treated leaf from each plant. Keep each separate and labeled. Remove the opaque cover from the leaf before proceeding.

(d) Alcohol has a much lower boiling point than water, and so takes very little time to come to a boil. When the water is boiling, put about 150 mL of alcohol in the A beaker, set it also on the hot plate, and bring to a boil. Keep the alcohol beaker covered with aluminum foil as much as possible throughout the lab to prevent excess evaporation.

(e) Place the leaf from one treatment in the beaker of boiling water for about 1 minute. This kills the tissue and breaks down internal membranes.

Caution

Ethanol is highly flammable. Use only electric hot plates, never open flame. Also, never let a beaker boil dry. Add more liquid, or remove the beaker from the burner, and place it on a pad of folded paper towels.

(f) Use the long forceps to move the wilted leaf from the water into the boiling alcohol. This will extract the photosynthetic pigments from the plant tissues. When the pigments have been extracted, the liquid will appear green, and the leaf will appear to be mostly bleached.

(g) Remove the leaf from the alcohol with forceps, dip momentarily in the boiling water to soften.

4. Test the plants for the presence and localization of starches:
   (a) Place killed, depigmented leaf in a petri dish filled with iodine solution.
   (b) Let the leaf soak in the iodine solution for a couple minutes, rinse, and float in water in another petri dish in order to observe the pattern of staining.

5. Show the distribution of the stain in the leaf by shading in and labeling Figure 6-2. In the blank provided in the legend for Figure 6-2, record the *substance* that I_2KI stains.

6. Repeat the pigment extraction and staining process for a leaf of the other treatment. Show the distribution of stain in this leaf by shading in and labeling Figure 6-2.

7. Remove the A beaker from the hot plate, and turn the heat off.

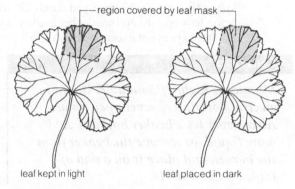

Figure 6-2 Distribution of the photosynthetic product _____.

What does the blue-black coloration of the leaf indicate? _____

_____

Why did the masked area fail to stain? _____

_____

Write a conclusion accepting or rejecting the hypothesis. _____

_____

6.4 Experiment: Necessity of Photosynthetic Pigments for Photosynthesis
(About 15 min.)

Coleus plants are widely planted ornamentals that are popular for their striking foliage color patterns. Observe the plants available in the lab and note their wide variety and attractiveness. This experiment addresses the hypothesis that *chlorophyll is necessary for photosynthesis to occur.*

MATERIALS

Per student group (4):

- colored pencils or pens
- two 400-mL beakers
- hot plate in fume hood
- heat-resistant glove
- petri dish halves
- bottle of iodine solution

- bottle of 95% ethanol (EtOH)
- forceps

Per lab room:

- source of dH_2O
- variegated *Coleus* plants

PROCEDURE

1. **Obtain a leaf of variegated *Coleus*.** In the left-hand circle, carefully sketch the leaf, indicating the distribution of each color on the leaf with colored pencils or pens. Green coloration is due to chlorophyll, the major photosynthetic pigment. Pink colors are caused by water-soluble anthocyanin pigments (not involved in photosynthesis), and yellows are formed by carotenoid pigments. Be sure that you look at both surfaces of the leaf in case pigment distribution differs.

   Make a prediction regarding starch presence and photosynthetic activity for each pigmentation area:

   _____

   _____

2. Kill and extract the pigments from the leaf:
   (a) With a china marker, label the beakers A (for alcohol) and dH$_2$O.
   (b) Add about 150 mL distilled water to the dH$_2$O beaker, set it on the hot plate, and turn on the hot plate to the highest setting. Allow the water to come to a boil.
   (c) Alcohol has a much lower boiling point than water, and so takes very little time to come to a boil. When the water is boiling, put about 150 mL of alcohol in the A beaker, set it also on the hot plate, and bring to a boil. Keep the alcohol beaker covered with aluminum foil as much as possible throughout the lab to prevent excess evaporation.

   > **Caution**
   >
   > *Ethanol is highly flammable. Use only electric hot plates, never open flame. Also, never let a beaker boil dry. Add more liquid, or remove the beaker from the burner, and place it on a pad of folded paper towels.*

   (d) Place the leaf from one treatment in the beaker of boiling water for about 1 minute. This kills the tissue and breaks down internal membranes.
   (e) Use the long forceps to move the wilted leaf from the water into the boiling alcohol. This will extract the photosynthetic pigments from the plant tissues. When the pigments have been extracted, the liquid will appear green, and the leaf will appear to be mostly bleached.
   (f) Remove the leaf from the alcohol with forceps, and dip momentarily in the boiling water to soften.

3. Test the leaf for the presence and localization of starch:
   (a) Place the killed, depigmented leaf in a petri dish filled with iodine solution.
   (b) Let the leaf soak in the iodine solution for a couple minutes, rinse, and float in water in another petri dish in order to observe the pattern of staining.

4. On the right-hand side of the space below, resketch the leaf, indicating the pattern of staining with iodine.

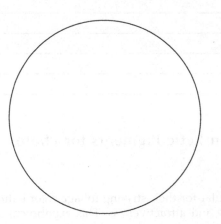 

5. How does the pattern of starch storage relate to the distribution of chlorophyll?

   _____

   _____

6. Write a conclusion either accepting or rejecting the hypothesis.

   _____

   _____

We tend to think of sunlight as being white. However, as you will see in this experiment, white light consists of a continuum of wavelengths. If we see light of just one wavelength, that light will appear colored.

When light hits a pigmented surface, some of the wavelengths are absorbed and others are reflected or transmitted. In this experiment, you will discover *which* wavelengths are absorbed, transmitted, or reflected by *particular* pigments, among them the photosynthetic pigment **chlorophyll**.

MATERIALS

Per lab room:

- several spectroscope setups (Figure 6-3a)
- sets of colored pencils (violet, blue, green, yellow, orange, red)
- colored filters (blue, green, red)
- small test tube containing pigment extract

Per student:

- hand-held spectroscope (optional; Figure 6-3b)

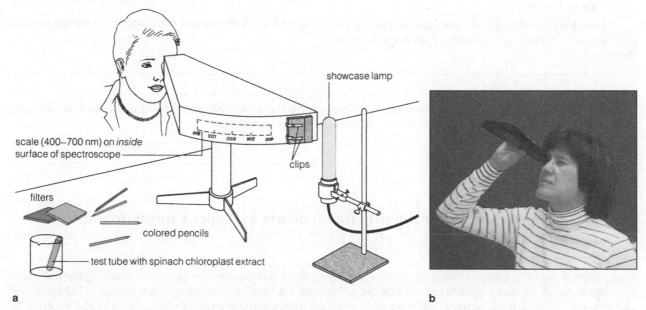

Figure 6-3   Use of a spectroscope. **(a)** Table-mounted. (After Abramoff and Thomson, 1982.) **(b)** Hand-held. (Photo by J. W. Perry.)

PROCEDURE

Work alone.

One way to separate light into its component parts is to view the light through a spectroscope. The spectroscope contains a prism that causes a spectrum of colors to form. A nanometer scale is imposed on the spectrum to indicate the wavelength of each component of white light.

1. Observe the spectrum of white light given off by an incandescent bulb through the spectroscope. With the colored pencils provided, record the positions of the colors violet, blue, green, yellow, orange, and red on the scale in Figure 6-4.
2. Observe the spectrum produced by the three colored filters using the spectroscope.

Which color or colors are absorbed when a red filter is placed between the light and the prism?

_____

_____

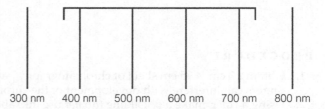

Figure 6-4   Spectrum of white light.

When a blue filter is used? _____

A green filter? _____

Make a general statement concerning the color of a pigment (filter) and the absorption of light by that pigment.

_____

_____

3. Now obtain a small test tube containing spinach chloroplast pigment extract and place it between the light source and the spectroscope. By adjusting the height of the tube so that the upper portion of the light passes through the pigment extract and the lower portion is white light, you can compare the absorption spectrum of the pigment extract with the spectrum of white light.

An **absorption spectrum** is a spectrum of light waves absorbed by a particular pigment. By contrast, the wavelengths that pass through the pigment extract and are visible in the spectroscope make up the *transmission spectrum* of the pigment.

4. Using the colored pencils, record the transmission spectrum of the chloroplast extract on the scale in Figure 6-5.

How does the absorption spectrum of the chloroplast extract compare with the absorption spectrum of the green filter?

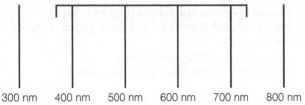

300 nm   400 nm   500 nm   600 nm   700 nm   800 nm

Figure 6-5 Transmission spectrum of chloroplast extract.

_____

_____

How might you explain the difference in absorption by the green filter and by the chloroplast pigment extract? (You might want to do Section 6.6 before answering this question.)

_____

_____

| 6.6 | Separation of Photosynthetic Pigments by Paper Chromatography |

(15 min.)

Paper chromatography allows substances to be separated from one another on the basis of their physical characteristics. A chloroplast pigment extract has been prepared for you by soaking spinach leaves in cold acetone and ethanol. Although the extract appears green, other pigments present may be masked by the chlorophyll. In this activity, you will use paper chromatography to separate any pigments present. Separation occurs due to the solubility of the pigment in the chromatography solvent and the affinity of the pigments for absorption to the paper surface. The finished product, showing separated pigments, is called a **chromatogram.**

MATERIALS

Per student:

■ chromatography paper, 3 cm × 15 cm sheet
■ metric ruler

Per student pair:

■ chloroplast pigment extract in foil-wrapped dropping bottle
■ chromatography chamber containing solvent
■ colored pencils (green, blue-green, yellow, orange)

PROCEDURE

1. Obtain a 3 cm × 15 cm sheet of chromatography paper. *Touch only the edges of the paper,* because oil from your fingers can interfere with development of the chromatogram.
2. Using a ruler, make a *pencil* line (do *not* use ink) about 2 cm from the bottom of the paper.
3. Load the paper by applying a droplet of the chloroplast pigment extract near the center of the pencil line. Allow the pigment spot to dry for about 30 seconds. Five to eight applications of extract on the same spot are necessary to get enough pigment for a good chromatogram. Be certain to allow the pigment to air-dry between applications.

4. Insert your "loaded" chromatography paper, spot-side down, into a chromatography chamber—a bottle containing a solvent consisting of 10% acetone in petroleum ether. The level of the solvent should cover the bottom of the strip but no portion of the pigment spot. Seal the chromatography chamber and allow the solvent to rise on the paper. (Two chromatograms can be inserted in a single bottle, but attempt to keep each separate.)

Caution

Avoid inhaling the solvent vapors. Keep the chamber tightly capped whenever possible.

5. Watch the separation take place over the next 10 minutes. When the solvent is within about 1 cm of the top of the paper, the separation is complete. Remove the strip, close the chromatography chamber, and allow the chromatogram to dry.
6. Using colored pencils, sketch the results in Figure 6-6, showing the relative position of the colors along the paper.
7. Beginning nearest the original pigment spot, identify and label the yellow-green pigment **chlorophyll b.**
8. Moving upward, find the blue-green **chlorophyll a,** two yellow-orange **xanthophylls** in the middle, and an orange **carotene** at the top. Xanthophylls and carotenes belong to the class of pigments called **carotenoids.**
9. You may preserve your chromatogram for future reference by keeping it in a dark place (for example, between the pages of your textbook). Light causes the chromatogram to fade.

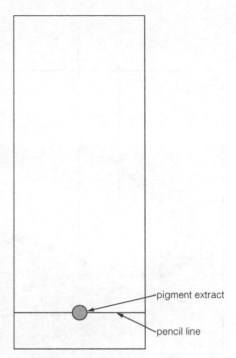

Figure 6-6 Chloroplast pigment chromatogram.
Labels: chlorophyll b, chlorophyll a, xanthophylls, carotene

What pigments are contained within the chloroplasts of spinach leaves? _____

_____

What common "vegetable" is particularly high in carotenes? _____
(Are you familiar with the medical condition called "carotenosis"? If not, go to the library and look it up in a medical dictionary.)

6.7	**Structure of the Chloroplast** *(15 min.)*

Work alone.

The chloroplast is the organelle concerned with photosynthesis. Study Figure 6-7, an artist's conception of the three-dimensional structure of a chloroplast.

Like the mitochondrion and the nucleus, the chloroplast is surrounded by two membranes. Within the **stroma** (semifluid matrix), identify the **thylakoid disks** stacked into **grana** (a single stack is a **granum**). The chloroplast pigment molecules are located on the surface of the thylakoid disks. Hydrogen ion buildup occurs within the interior of the disks. As these ions are expelled back into the stroma, ATP is formed. Within the stroma, the ATP is used to generate organic compounds. These compounds are converted to carbohydrates, lipids, and amino acids from carbon dioxide, water, and other raw materials.

Now examine Figure 6-8, a high-magnification electron micrograph of a chloroplast. With the aid of Figure 6-7, label the electron micrograph.

If the plant is killed and fixed for electron microscopy after being exposed to strong light, the chloroplasts will contain **starch grains.** Note the large starch grain present in this chloroplast. (Starch grains appear as ellipsoidal white structures in electron micrographs.)

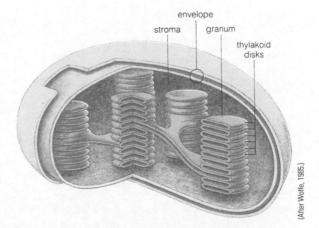

(After Wolfe, 1985.)

Figure 6-7 The arrangement of membranes and compartments inside a chloroplast.

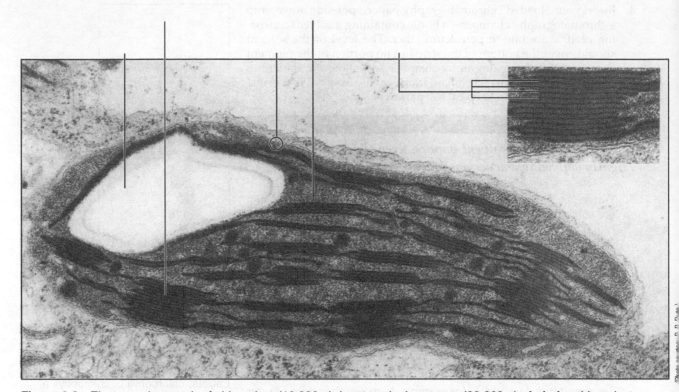

Figure 6-8 Electron micrograph of chloroplast (10,000×). Inset: a single granum (20,000×). **Labels:** chloroplast membrane, thylakoid disks, stroma, starch

_____ 1. The raw materials used for photosynthesis include
(a) O_2
(b) $C_6H_{12}O_6$
(c) $CO_2 + H_2O$
(d) CH_2O

_____ 2. A device useful for viewing the spectrum of light is a
(a) spectroscope
(b) volumeter
(c) chromatogram
(d) chloroplast

_____ 3. Products and byproducts of photosynthesis do NOT include
(a) O_2
(b) $C_6H_{12}O_6$
(c) CO_2
(d) H_2O

_____ 4. A paper chromatogram is useful for
(a) measuring the amount of photosynthesis
(b) determining the amount of gas evolved during photosynthesis
(c) separating pigments based on their physical characteristics
(d) determining the distribution of chlorophyll in a leaf

_____ 5. Which of the following pigments would you find in a geranium leaf?
(a) chlorophyll, xanthophyll, phycobilins
(b) chlorophyll a, chlorophyll b, carotenoids
(c) phycocyanin, xanthophyll, fucoxanthin
(d) carotenoids, chlorophylls, phycoerythrin

_____ 6. Which reagent would you use to determine the distribution of the carbohydrate stored in leaves?
(a) starch
(b) Benedict's solution
(c) chlorophyll
(d) I_2KI

_____ 7. An example of a heterotrophic organism is
(a) a plant
(b) a geranium
(c) a human
(d) none of the above

_____ 8. Organisms capable of producing their own food are known as
(a) autotrophs
(b) heterotrophs
(c) omnivores
(d) herbivores

_____ 9. Grana are
(a) the same as starch grains
(b) the site of ATP production within chloroplasts
(c) part of the outer chloroplast membrane
(d) contained within mitochondria and nuclei

_____ 10. The ultimate source of energy trapped during photosynthesis is
(a) CO_2
(b) H_2O
(c) O_2
(d) sunlight

EXERCISE 6

Photosynthesis: Capture of Light Energy

POST-LAB QUESTIONS

6.2 Experiment: Effects of Light and Carbon on Starch Production

1. Is starch stored in the leaves of some plants? Would you expect leaves in a temperate climate plant to be the primary area for long-term starch storage? Why or why not? What part(s) of a plant might be better-suited for long-term starch storage?

6.3 Experiment: Relationship Between Light and Photosynthetic Products

2. Examine the photo at right, which shows the location of starch in two geranium leaves treated in much the same way as you did in Section 9.3. Explain the results you see.

(Photo by J.W. Perry.)

6.4 Experiment: Necessity of Photosynthetic Pigments for Photosynthesis

3. Examine the photo at right of a *Coleus* leaf. Describe an experiment that would allow you to determine whether the deep purple portion of the leaf is photosynthesizing.

(Photo by J.W. Perry.)

6.5 Experiment: Absorption of Light by Chloroplast Extract

4. The photo at right was taken through a spectroscope. What color was the pigment extract used to produce this spectrum? What color(s) did this extract absorb?

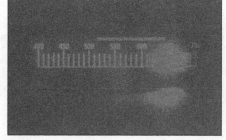
(Photo by J.W. Perry.)

5. Would you illuminate your house plants with a green light bulb? Why or why not?

6.7 Structure of the Chloroplast

6. Examine this electron micrograph of a chloroplast.

   a. Identify the stack of membranes labeled A.

   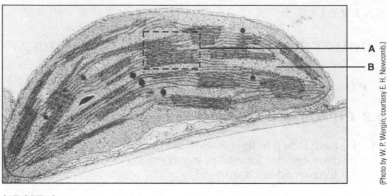

   (Photo by W. P. Wergin, courtesy E. H. Newcomb.)

   b. Identify the region labeled B.

   (17,257×).

   c. Would the production of organic compounds during the light-independent reactions occur in region B or on the membranes labeled A?

   d. Would you expect the plant in which this structure was found to have been illuminated with strong light immediately before it was prepared for electron microscopy? Why or why not?

Food for Thought

7. Numerous hypotheses have been proposed for the extinction of the dinosaurs. Recently, evidence has been found of the impact of a large meteor at about the time of this mass extinction. The amount of dust and debris put into the atmosphere upon impact, as well as atmospheric heating, would have been enormous. Using your knowledge of photosynthesis, speculate as to why the dinosaurs subsequently became extinct.

8. Explain the statement: "Without autotrophic organisms, heterotrophic life would cease to exist."

9. Why do you suppose that a chloroplast kept in darkness for some time prior to being fixed for electron microscopy does not contain starch?

10. With the results of the preceding experiments in mind, what might you do to increase the vigor of your house plants?

The Influence of CO_2 Levels on Photosynthesis

Abstract

Understanding the influence of global warming on photosynthetic rates is of paramount importance for predicting ecosystem energy budgets. Global warming involves many environmental factors, not the least of which are rising levels of greenhouse gases such as carbon dioxide (CO_2). The effect of rising CO_2 levels on photosynthesis was examined for one aquatic plant species, *Elodea*. An experimental group given a high concentration of CO_2 was compared to a control group with low levels of CO_2. I hypothesized that the higher CO_2 levels would lead to a lowering of the rate of photosynthesis. This was supported by the experiment. The rate of photosynthesis decreased in the experimental group compared to the control group. Further investigations of the influence of global warming on photosynthesis are suggested and discussed.

Introduction

The photosynthetic efficiency of terrestrial and aquatic species is of central importance for understanding the availability of food and energy within ecosystems. Currently, increases in global warming have led to widespread concern as to its effects on photosynthetic production systems (Chaves and Pereira 1992; Schimel 2006). As such, it is imperative to discover the influence of global warming on photosynthesis. The current study was undertaken to examine the influence of rising CO_2 levels on photosynthesis in an aquatic plant species, *Elodea*. I chose *Elodea* as my study organism because previous research has shown that its photosynthetic rates are reduced under acidic conditions of < pH 6 (Jones, Eaton, and Hardwick 1999). I chose CO_2 levels as my independent variable because this is a leading greenhouse gas contributing to global warming (Lashof and Ahyja 1990), and it may have a detrimental effect on photosynthesis within aquatic ecosystems because it chemically reacts with water, and high levels of CO_2 lead to acidification of aquatic systems (Doney 2006).

As rising CO_2 levels lead to lowering of the pH in aquatic systems, I hypothesized that high CO_2 would lead to a lowering of the pH in an aquatic system and influence the photosynthesis rates of aquatic plant species. I predicted that high levels of CO_2 would lead to a lowering of the photosynthesis rate that would be indicated by a decrease in oxygen production when compared to photosynthesis rates at lower levels of CO_2.

Materials and Methods

Overall Procedure

To measure the rate of photosynthesis, 15 milliliter (ml) centrifuge tubes were filled with water, and a 3 gram (g) sample of *Elodea* was placed inside each. The samples of *Elodea* leaves were matched as closely as possible (size, shape, number, etc). Next, in one-half of the centrifuge tubes, I blew CO_2 through a straw for 5 minutes, using a stopwatch and stopping the timer during pauses where I needed to inhale, covering the top of the centrifuge tube with my thumb to prevent CO_2 from escaping. A rubber stopper fitted with a capillary pipette inserted into a predrilled hole in the stopper was used to plug the centrifuge tubes once I was done. Care was taken to ensure that no air bubbles occurred in the centrifuge tubes once the rubber stopper was fitted. The water level that rose in the capillary pipette after the rubber stopper was fitted was initially marked. It was *assumed* that any water displaced up the capillary pipette column was the result of oxygen being produced by photosynthesis. The capillary tubes were then placed in front of a full-spectrum light source. A period of 15 minutes was allowed to elapse during which time the plants were allowed to acclimate. Then every 10 minutes, the water level was recorded until 30 minutes had elapsed.

Experimental Design

The experimental group for this investigation consisted of three centrifuge tubes with high CO_2 levels, and the control group consisted of three centrifuge tubes with low levels of CO_2. The independent variable was CO_2 levels, and the dependent variable was the displacement of water, which is an indirect measure of the rate of photosynthesis. Two additional controls were used to check the equipment design for the experiment. Two centrifuge tubes only held water and, two centrifuge tubes with just high CO_2 and water were used to ensure that no displacement of water occurred under these conditions.

The average displacement of water was calculated for the experimental and control groups, as well as the standard deviation and standard error of the mean. A t-test was conducted to compare the averages.

Results

The results for the influence of CO_2 levels on photosynthesis in *Elodea* support the hypothesis that high CO_2 levels lead to lower rates of photosynthesis. The *t*-test revealed a significant difference between the control group and the experimental group ($t = 3.0226$, df $= 4$, $p = 0.02$), therefore the hypothesis was accepted. As depicted in Figure 1 and Table 1, the results were in accordance with what was predicted.

The results for the two additional controls that were used to check the equipment design for the experiment indicated only a slight (0.5 mm) displacement of water on average (mean $= 0.5$, $SE = 0.2$). This increased the reliability of the data collected.

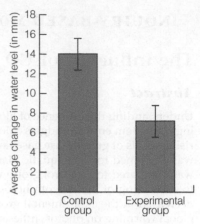

Figure 1 Change in water level (an indirect measure of oxygen production) at different concentrations of CO_2 for *Elodea*. Values are means $\pm$ SE.

TABLE 1 Summary Data Table		
Displacement of Water (mm)	**Control Group**	**Experimental Group**
Mean	14.00	7.17
Standard Deviation	2.78	2.75
Standard Error	1.61	1.59

Discussion

Photosynthesis was decreased in the experimental group because of the higher levels of CO_2. It was predicted that such higher levels would lead to acidification of the water and lead to reductions in photosynthesis. This was borne out in the experiment. The significance of the results of the current investigation is that even though CO_2 is a reactant molecule for photosynthesis, because of the nature its chemical reactions with water in aquatic systems, too much CO_2 actually decreases photosynthetic rates for pH-sensitive plant species, at least in the short term.

Global warming has serious implications for human societies because of the potential influence on photosynthesis in agricultural crops (Abelson 1992). In addition, rising CO_2 levels have led to increasing acidification of marine environments and the decline of several photosynthetic species such as corals and their symbiotic zooxanthelle (Scott 2006). Thus, it is imperative that we discover the influence of CO_2 levels not only on photosynthesis but also on changes in pH, changes in temperature, changes in sunlight availability, and, for aquatic systems, changes in salinity as well. Future studies that address the influence of these and other factors, as well as examining additional species of photosynthetic organisms, will aid in devising strategies for coping with the effects of global warming.

References

Abelson, P. H. 1992. Agriculture and climate change. *Science* 257: 9.

Chaves, M. M., and J. S. Pereira. 1992. Water stress, CO_2 and climate change. *Journal of Experimental Botany* 43(8): 1131–9.

Doney, S. C. 2006. The dangers of ocean acidification. *Scientific American* March 2006: 58–65.

Jones, J. I., J. W. Eaton, and K. Hardwick. 1999. The effect of changing environmental variables in the surrounding water on the physiology of *Elodea nuttallii*. *Aquatic Botany* 66(2): 115–29.

Lashof, D. A., and D. R. Ahyja. 1990. Relative contributions of greenhouse gas emissions to global warming. *Nature* 344: 529–31.

Schimel, David. 2006. Climate change and crop yields: Beyond Cassandra. *Science* 312: 188–9.

Respiration: Energy Conversion

OBJECTIVES

After completing this exercise, you will be able to

1. define *metabolism, reaction, metabolic pathway, respiration, ATP, aerobic respiration, alcoholic fermentation;*

2. give the overall balanced equations for aerobic respiration and alcoholic fermentation;

3. distinguish among the inputs, products, and efficiency of aerobic respiration and those of fermentation;

4. explain the relationship between temperature and goldfish respiration rate;

5. identify the structures and list the functions of each part of a mitochondrion.

INTRODUCTION

The first law of thermodynamics states that energy can neither be created nor destroyed, only converted from one form to another. Because all living organisms have a constant energy requirement, they have mechanisms to gather, store, and use energy. Collectively, these mechanisms are called **metabolism.** A single, specific reaction that starts with one compound and ends up with another compound is a **reaction,** and a sequence of such reactions is a **metabolic pathway.**

In Exercise 9, we investigated the metabolic pathways by which green plants capture light energy and use it to make carbohydrates such as glucose. Carbohydrates are temporary energy stores. The process by which energy stored in carbohydrates is released to the cell is **respiration.**

Both autotrophs and heterotrophs undergo respiration. Photoautotrophs such as plants utilize the carbohydrates they have produced by photosynthesis to build new cells and maintain cellular machinery. Heterotrophic organisms may obtain materials for respiration in two ways: by digesting plant material or by digesting the tissues of animals that have previously digested plants.

Several different forms of respiration have evolved. The specific respiration pathway used depends on the specific organism and/or environmental conditions. In this exercise, we will consider two alternative pathways: (1) **aerobic respiration,** an oxygen-dependent pathway common in most organisms; and (2) **alcoholic fermentation,** an ethanol-producing process occurring in some yeasts.

Perhaps the most important aspect to remember about these two processes is that aerobic respiration is by far the most energy-efficient. **Efficiency** refers to the amount of energy captured in the form of ATP relative to the amount available within the bonds of the carbohydrate. **ATP, adenosine triphosphate,** is the so-called universal energy currency of the cell. Energy contained within the bonds of carbohydrates is transferred to ATP during respiration. This stored energy can be released later to power a wide variety of cellular reactions.

For aerobic respiration, the general equation is

$$C_6H_{12}O_6 + 6O_2 \xrightarrow{\text{enzymes}} 6CO_2 + 6H_2O + 36ATP^*$$

glucose   oxygen   carbon dioxide   water   chemical energy

If glucose is completely broken down to CO_2 and H_2O, about 686,000 calories of energy are released. Each ATP molecule produced represents about 7500 calories of usable energy. The 36 ATP represent 270,000 calories of energy (36 × 7500 calories). Thus, aerobic respiration is about 39% efficient [(270,000/686,000) × 100%].

By contrast, fermentation yields only 2 ATP. Thus, these processes are only about 2% efficient [(2 × 7500/686,000) × 100%]. Obviously, breaking down carbohydrates by aerobic respiration gives a bigger payback than the other means.

---

*Depending on the tissue, as many as 38 ATP may be produced.

During the process of aerobic respiration, relatively high-energy carbohydrates are broken down in stepwise fashion, ultimately producing the low-energy products of carbon dioxide and water and transferring released energy into ATP. But what is the role of oxygen?

During aerobic respiration, the carbohydrate undergoes a series of oxidation–reduction reactions. Whenever one substance is oxidized (loses electrons), another must be reduced (accept, or gain, those electrons). The final electron acceptor in aerobic respiration is oxygen. Tagging along with the electrons as they pass through the electron transport process are protons (H$^+$). When the electrons and protons are captured by oxygen, water (H$_2$O) is formed:

$$2H^+ \ + \ 2e^- \ + \ \tfrac{1}{2} O_2 \longrightarrow H_2O$$

In the following experiments, we examine aerobic respiration in two sets of seeds.

A. Experiment: Carbon Dioxide Production (About 25 min. for setup, 1 3/4 hr to complete)

Seeds contain stored food material, usually in the form of some type of carbohydrate. When a seed germinates, the carbohydrate is broken down by aerobic respiration, liberating the energy (ATP) required for each embryo to grow into a seedling.

Two days ago, one set of dry pea seeds was soaked in water to start the germination process. Another set was not soaked. In this experiment, you will compare carbon dioxide production between germinating pea seeds, germinating pea seeds that have been boiled, and ungerminated (dry) pea seeds.

This experiment investigates the hypothesis that *germinating seeds produce carbon dioxide from aerobic respiration.*

MATERIALS

Per student group (4):

- 600-mL beaker
- hot plate *or* burner, wire gauze, tripod, and matches
- heat-resistant glove
- 3 respiration bottle apparatuses (Figure 7-1)
- china marker
- phenol red solution

Per lab room:

- germinating pea seeds
- ungerminated (dry) pea seeds

PROCEDURE

Work in groups of four.

1. Place about 250 mL of tap water in a 600-mL beaker, put the beaker on a heat source, and bring the water to a boil.
2. Obtain three respiration bottle setups (Figure 7-1). With a china marker, label one "Germ" for germinating pea seeds, the second "Germ-Boil" for those you will boil, and the third "Ungerm" for ungerminated seeds.
3. From the class supply, obtain and put enough germinating pea seeds into the two appropriately labeled respiration bottles to fill them approximately halfway. Fill the third bottle half full with ungerminated (dry) pea seeds.
4. Dump the germinating peas from the "Germ-Boil" bottle into the boiling waterbath; continue to boil for 5 minutes.

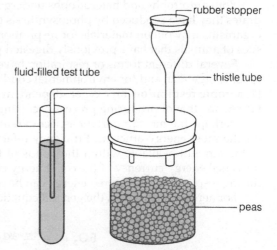

Figure 7-1 Respiration bottle apparatus.

After 5 minutes, turn off the heat source, put on a heat-resistant glove, and remove the water bath. Pour the water off into the sink and cool the boiled peas by pouring cold water into the beaker. Allow 5 minutes for the peas to cool to room temperature; then pour off the water. Now replace the peas into the "Germ-Boil" respiration bottle.
5. Fit the rubber stopper with attached glass tubes into the respiration bottles. Add enough water to the test tube to cover the end of the glass tubing that comes out of the respiration bottle. (This keeps gases from escaping from the respiration bottle.)
6. Insert rubber stoppers into the thistle tubes.
7. Set the three bottles aside for the next 1 1/4 hours and do the other experiments in this exercise.

8. Make a prediction about carbon dioxide production in each of the three bottles:

_____

_____

_____

Now start the next series of experiments while you allow this one to proceed.

9. After 1 1/4 hours, pour the water in each test tube into the sink and replace it with an equal volume of dilute phenol red solution. Phenol red solution, which should appear pinkish in the stock bottle, will be used to test for the presence of carbon dioxide (CO_2) within the respiration bottles. If CO_2 is bubbled through water, carbonic acid (H_2CO_3) forms:

$$CO_2 \;+\; H_2O \longrightarrow H_2CO_3$$

Phenol red solution is mostly water. When the phenol red solution is basic (pH > 7), it is pink; when it is acidic (pH < 7), the solution is yellow. The phenol red solution in the stock bottle is

_____ (color); therefore, the stock solution is

_____ (acidic/basic).

10. Put several hundred milliliters of tap water in the 600-mL beaker.
11. Remove the stopper plugging the top of the thistle tube and *slowly* pour water from the beaker into each thistle tube. The water will force out gases present in the bottles. If CO_2 is present, the phenol red will become yellow.
12. Record your observations in Table 7-1.

TABLE 7-1 CO_2 Evolution by Pea Seeds

Pea Seeds	Indicator Color (Phenol Red)	Conclusion (CO_2 Present or Absent)
Germinating—unboiled		
Germinating—boiled		
Ungerminated		

Which set(s) of seeds underwent respiration? _____

What happened during boiling that caused the results you found? (*Hint:* Think "enzymes.")

_____

Write a conclusion, accepting or rejecting the hypothesis.

_____

_____

B. Experiment: Oxygen Consumption (About 1 1/2 hr.)

One set of pea seeds has been soaked in water for the past 48 hours to initiate germination. In this section, you will measure oxygen consumption to answer the question, "Do germinating peas have a higher respiratory rate than ungerminated peas?"

MATERIALS

Per student group (4):

- volumeter (Figure 7-2)
- china marker
- 80 germinating pea seeds
- 80 ungerminated (dry) pea seeds
- glass beads

- nonabsorbent cotton
- metric ruler
- bottle of potassium hydroxide (KOH) pellets
- 1/4 teaspoon measure
- marker fluid in dropping bottle

PROCEDURE

Work in groups of four.

1. Obtain a volumeter set up as in Figure 7-2. Skip to step 6 if your instructor has already assembled the volumeters as described by steps 2–5.

2. Remove the test tubes from the volumeter. With a china marker, number the tubes and then fill as follows:
Tube 1: 80 germinating (soaked) pea seeds
Tube 2: 80 ungerminated (dry) pea seeds plus enough glass beads to bring the total volume equal to that in tube 1
Tube 3: Enough glass beads to equal the volume of tube 1

Both temperature and pressure affect gas pressure within a closed tube. Tube 3 serves as a thermobarometer and is used as a control to correct experimental readings to account for changes in temperature and barometric pressure taking place during the experiment.

3. Pack cotton *loosely* into each tube to a thickness of about 1.5 cm above the peas/beads.
4. Measure out 1 cubic centimeter (cm^3) (about $1/4$ teaspoon) of KOH pellets and pour them atop the cotton.

Potassium hydroxide absorbs CO_2 given off during aerobic respiration. Since the volumeter measures change in gas volume, any gas *given off* during respiration must be removed from the tube so an accurate measure of O_2 consumption can be made.

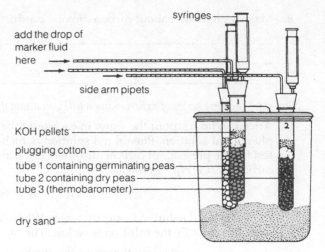

Figure 7-2 Volumeter.

Caution

Potassium hydroxide can cause burns. Do not get any on your skin or clothing. If you do, wash immediately with copious amounts of water.

5. Insert the stopper-syringe assembly in place.
6. Add a small drop of marker fluid to each side arm pipet by touching the dropper to the end of each. The drop should be taken into the side arm by capillary action. Gently withdraw the plunger of each syringe and adjust the position of the drop so it is between 0.80 and 0.90 cm^3 on the scale of the graduated pipet.
7. Adjust each side arm pipet so it is parallel to the table top. Wait 5 minutes before starting data collection.
8. In Table 7-2 at time 0, record the position of the marker droplet within each pipet. Record readings for each tube at 5-minute intervals for the next 60 minutes, keeping track of whether changes are positive (movement toward the test tube) or negative (movement away from the test tube).

If respiration is rapid and the marker drop moves too near the end of the scale to read, *carefully* use the syringe to readjust its position so it is between 0.80 and 0.90 cm^3 on the scale of the graduated pipet again. Note this in Table 7-2 and continue to record readings every 5 minutes.

9. To determine change in volume of gas within each tube at each sampling, subtract each subsequent reading from the previous reading.
10. Determine cumulative volume change (cumulative oxygen consumption) by adding each volume change to the previous volume-change measurement. The final figure represents the total oxygen consumption (in mL) in that tube.
11. At the end of the experiment, correct for any volume changes caused by changes in temperature or barometric pressure by using the reading obtained from the thermobarometer. If the thermobarometric marker moves *toward* the test tube (decrease in volume), *subtract* the volume change from the total oxygen consumption measurement of tubes 1 and 2. If the marker droplet moves *away* from the test tube (increase in volume), *add* the volume change to the last total oxygen consumption measurement for tubes 1 and 2.
12. In Figure 7-3, graph the consumption of oxygen over time. Use a + for data points of germinating peas, a • for dry peas.

How do the respiratory rates for germinating and nongerminating seeds compare?

_____

_____

How do you account for this difference?

_____

_____

It takes 820 cm^3 of oxygen to completely oxidize 1 g of glucose. How much glucose are the 80 germinating peas consuming per hour? (Recall that 1 cm^3 = 1 mL.)

TABLE 7-2 Respiratory Rate as Measured by Oxygen Consumption

Time (min.)	Tube 3: Thermobarometer		Tube 1: Germinating Peas			Tube 2: Dry Peas		
	Reading	Total Change in Volume	Reading	Total Change in Volume	Total Oxygen Consumption	Reading	Total Change in Volume	Total Oxygen Consumption
0		0		0	0		0	0
5								
10								
15								
20								
25								
30								
35								
40								
45								
50								
55								
60								

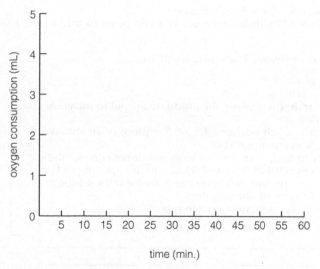

Figure 7-3 Oxygen consumption by germinating and nongerminating pea seeds.

7.2   Fermentation *(About 20 min. to set up, 1¹/₂ hr. to complete)*

Despite relatively low energy yield, fermentation provides sufficient energy for certain organisms to survive. Alcoholic fermentation by yeast is the basis for the baking, wine-making, and brewing industries. It's been said that yeast and alcoholic fermentation made Milwaukee famous.

The chemical equation for this process is

$$C_6H_{12}O_6 \longrightarrow 2CH_3CH_2OH + CO_2 + 2ATP$$

glucose ethanol carbon dioxide energy

Starch (amylose), a common storage carbohydrate in plants, is a polymer consisting of a chain of repeating glucose ($C_6H_{12}O_6$) units. The polymer has the chemical formula $(C_6H_{12}O_6)n$,[*] where n represents a large number. Starch is broken down by the enzyme amylase into individual glucose units. To summarize:

$$(C_6H_{12}O_6)n \xrightarrow{\text{amylase}} C_6H_{12}O_6 + C_6H_{12}O_6 + C_6H_{12}O_6 + \cdots$$

starch glucose

This section demonstrates the action of yeast cells on carbohydrates.

MATERIALS

Per student group (4):

- china marker
- three 50-mL beakers
- 25-mL graduated cylinder
- bottle of 10% glucose
- bottle of 1% starch
- 0.5% amylase in bottle fitted with graduated pipet
- 3 glass stirring rods

- 1/4 teaspoon measure (optional)
- 3 fermentation tubes
- 15-cm metric ruler

Per lab room:

- 0.5-g pieces of fresh yeast cake
- scale and weighing paper (optional)
- 37°C incubator

PROCEDURE

Work in groups of four.

1. Using a china marker, number three 50-mL beakers.
2. With a *clean* 25-mL graduated cylinder, measure out and pour 15 mL[†] of the following solutions into each beaker:

Note: Wash the graduated cylinder between solutions.

    Beaker 1: 15 mL of 10% glucose
    Beaker 2: 15 mL of 1% starch
    Beaker 3: 15 mL of 1% starch; then, using the graduated pipet to measure, add 5 mL of 0.5% amylase.
3. Wait 5 minutes and then to each beaker add a 0.5-g piece of fresh cake yeast. Stir with *separate* glass stirring rods.
4. When each is thoroughly mixed, pour the contents into three correspondingly numbered fermentation tubes (Figure 7-4). Cover the opening of the fermentation tube with your thumb and invert each fermentation tube so that the "tail" portion is filled with the solution.
5. Write a prediction about gas production in each tube.

    _____

    _____

    _____

6. Place the tubes in a 37°C incubator.
7. At intervals of 20, 40, and 60 minutes after the start of the experiment, remove the tubes and, using a metric ruler, measure the distance from the tip of the tail to the fluid level. Record your results in Table 7-3. Calculate the volume of gas evolved using the formula below Table 7-3. (If time is short, do your calculations later.)

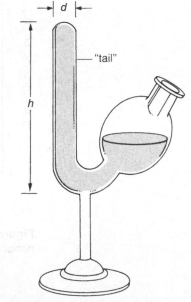

Figure 7-4 Fermentation tube.

---

[*]A number of carbohydrates share this same chemical formula but differ slightly in the arrangement of their atoms. These carbohydrates are called *structural isomers.*

[†]The amount of fluid needed to fill the fermentation tube depends on its size. Your instructor may indicate the required volume.

TABLE 7-3	Evolution of Gas by Yeast Cells				
		Distance from Tip of Tube to Fluid Level (mm)			
Tube	Solution	20 min.	40 min.	60 min.	Volume of Gas Evolved (mm³)
1	10% glucose + yeast				
2	1% starch + yeast				
3	1% starch + yeast + amylase				

To calculate the volume of gas evolved, use the following equation: $V = \pi r^2 h$, where $\pi = 3.14$, r = radius of tail of fermentation tube [$r = \frac{1}{2} d$ (diameter)], h = distance from top of tail to level of solution.

Did your results conform to your predictions? If not, speculate on reasons why this might be so.

_____

_____

What gas accumulates in the tail portion of the fermentation tube?

_____

7.3   Ultrastructure of the Mitochondrion

1. Study Figure 7-5b, which shows the three-dimensional structure of a mitochondrion, the respiratory organelle of all living eukaryotic cells. The mitochondrion has frequently been referred to as the "power-house of the cell," because most of the cell's chemical energy (ATP) is produced here.

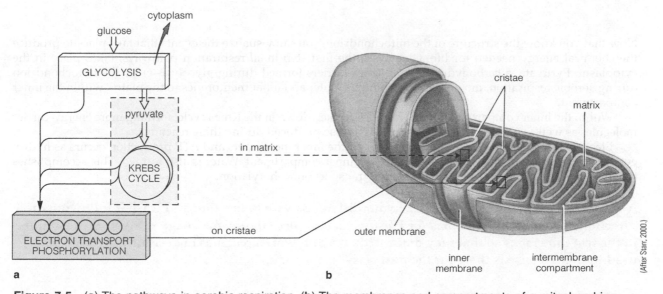

Figure 7-5 **(a)** The pathways in aerobic respiration. **(b)** The membranes and compartments of a mitochondrion.

2. Now observe Figure 7-6, a high-magnification electron micrograph of a mitochondrion. Identify and label the **outer membrane** separating the organelle from the cytoplasm.

3. Note the presence of an inner membrane, folded into fingerlike projections. Each projection is called a **crista** (the plural is *cristae*). The folding of the inner membrane greatly increases the surface area on which many of the chemical reactions of aerobic respiration take place. Label the crista.

4. Identify and label the **outer compartment,** the space between the inner and outer membranes. The outer compartment serves as a reservoir for hydrogen ions.

5. Finally, identify and label the **inner compartment** (filled with the *matrix*), the interior of the mitochondrion.

6. Study Figure 7-5a, a diagram of the pathways in aerobic respiration.

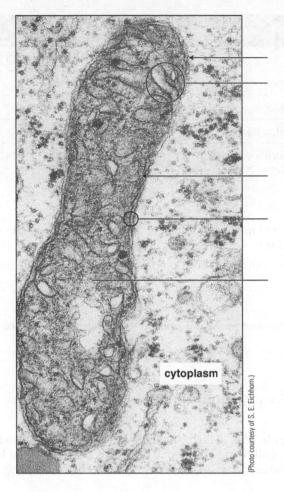

cytoplasm

(Photo courtesy of S. E. Eichhorn.)

Figure 7-6 Transmission electron micrograph of a mitochondrion (18,600×).
Labels: outer membrane, inner membrane, crista, intermembrane compartment, matrix

Now that you know the structure of the mitochondrion, you can visualize the events that take place to produce the chemical energy needed for life. Glycolysis, the first step in *all* respiration pathways, takes place in the cytoplasm. Pyruvate, a carbohydrate, and energy carriers formed during glycolysis enter the mitochondrion during aerobic respiration, moving through both the outer and inner membranes to the matrix within the inner compartment.

Within the inner compartment, the pyruvate is broken down in the Krebs cycle, forming more energy-carrier molecules as well as CO_2. A small amount of ATP is also produced during these reactions.

Electron transport molecules are embedded on the inner membrane, and ATP production occurs as hydrogen ions cross from the outer compartment to the inner compartment. Water is also formed. This accomplishes the third portion of aerobic respiration, electron transport phosphorylation.

Note: **After completing all labs, take your dirty glassware to the sink and wash it following the directions given in** *"Instructions for Washing Laboratory Glassware,"* **page x. Invert the test tubes in the test tube racks so that they drain. Tidy up your work area, making certain all equipment used in this exercise is there for the next class.**

1. A metabolic pathway is
   (a) a single, specific reaction that starts with one compound and ends up with another
   (b) a sequence of chemical reactions that are part of the metabolic process
   (c) a series of events that occur only in autotrophs
   (d) all of the above

2. The "universal energy currency" of the cell is
   (a) O_2
   (b) $C_6H_{12}O_6$
   (c) ATP
   (d) H_2O

3. Products of aerobic respiration include
   (a) glucose
   (b) oxygen
   (c) carbon dioxide
   (d) starch

4. "Efficiency" of a respiration pathway refers to the
   (a) number of steps in the pathway
   (b) amount of CO_2 produced relative to the amount of carbohydrate entering the pathway
   (c) amount of H_2O produced relative to the amount of carbohydrate entering the pathway
   (d) amount of ATP energy produced relative to the energy content of the carbohydrate entering the pathway

5. The purpose of the thermobarometer in a volumeter is to
   (a) judge the amount of O_2 evolved during respiration
   (b) determine the volume changes as a result of respiration
   (c) indicate oxygen consumption by germinating pea seeds
   (d) indicate volume changes resulting from changes in temperature or barometric pressure

6. Phenol red is used in the experiments as
   (a) an O_2 indicator
   (b) a CO_2 indicator
   (c) a sugar indicator
   (d) an enzyme

7. As temperatures rise, the body temperatures of _____ organisms rise.
   (a) ectothermic
   (b) endothermic
   (c) autotrophic
   (d) heterotrophic

8. Which of the following enzymes breaks down starch into glucose?
   (a) kinase
   (b) maltase
   (c) fructase
   (d) amylase

9. Oxygen is necessary for life because
   (a) photosynthesis depends on it
   (b) it serves as the final electron acceptor during aerobic respiration
   (c) it is necessary for glycolysis
   (d) of all of the above

10. Yeast cells undergoing alcoholic fermentation produce
    (a) ATP
    (b) ethanol
    (c) CO_2
    (d) all of the above

Name _____ Section Number _____

Respiration: Energy Conversion

POST-LAB QUESTIONS

7.1 Aerobic Respiration

1. Explain the role of the following components in the experiment on carbon dioxide production (page 106).

   germinating pea seeds   _____

   ungerminated (dry) pea seeds   _____

   germinating, boiled pea seeds   _____

   phenol red solution   _____

2. If you performed the experiment on oxygen consumption (page 107) without adding KOH pellets to the test tubes, what results would you predict? Why?

7.2 Fermentation

3. Sucrose (table sugar) is a disaccharide composed of glucose and fructose. Glycogen is a polysaccharide composed of many glucose subunits. Which of the following fermentation tubes would you expect to produce the greatest gas volume over a 1-hour period? Why?

   Tube 1: glucose plus yeast

   Tube 2: sucrose plus yeast

   Tube 3: glycogen plus yeast

4. Bread is made by mixing flour, water, sugar, and yeast to form a dense dough. Why does the dough rise? What gas is responsible for the holes in bread?

7.3 Ultrastructure of the Mitochondrion

6. Examine the electron micrograph of the mitochondrion on the right.

   a. What portions of aerobic respiration occur in region b?

   _____

   b. What substance is produced as hydrogen ions cross from the space between the inner and outer membranes into region b?

   _____

   c. What portion of cellular respiration takes place in the cytoplasm *outside* of this organelle?

   _____

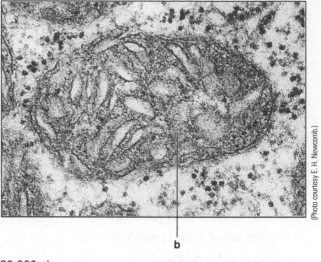

(Photo courtesy E. H. Newcomb.)

b

(20,000×).

Food for Thought

7. Oxygen is used during aerobic respiration. What biological process is the source of the oxygen?

8. Compare aerobic respiration and fermentation in terms of
   a. efficiency of obtaining energy from glucose

   b. end products

9. How would you explain this statement: "The ultimate source of our energy is the sun"?

10. The first law of thermodynamics seems to conflict with what we know about ourselves. For example, after strenuous exercise we run out of "energy." We must eat to replenish our energy stores. Where has that energy gone? What form has it taken?

The Effect of Temperature on Goldfish Respiration

Abstract

Currently, the Earth is experiencing rapidly changing environmental conditions because of global warming. This has led to concern over the effects on aquatic organisms because oxygen solubility decreases with increases in water temperature. Our study examined the metabolic consequences of changes in water temperature in goldfish, *Carassius auratus*. We hypothesized that respiratory rates of goldfish would increase with increases in temperature. We found that goldfish respiration rates nearly doubled when water temperature was raised from 22° C to 27°C, supporting our hypothesis. Both the significance of increasing environmental temperatures on metabolic rates of organisms and future avenues of research are discussed.

Introduction

Aerobic respiration is a metabolic process that requires oxygen. Cells use the oxygen during the degradative pathways in which energy is harvested from organic molecules (Starr, Evers, and Starr 2008). Organisms use this energy, which is stored temporarily in adenosine triphosphate (ATP) molecules, to do cellular activities. A crucial function of aerobic organisms is using cellular energy to drive muscles such as those used in breathing. The breathing process functions to obtain oxygen from the environment and deliver it to the mitochondria of cells, where the main energy-releasing metabolic pathways take place.

Changing environmental conditions can alter the availability of oxygen to organisms, especially in aquatic environments. The solubility of oxygen in water decreases as temperature increases. Global warming (Carpenter et al. 1992) and local warming by the effluents of power plants (Hill and Magnuson 1990; Schindler 1998) have altered oxygen concentrations in aquatic ecosystems. Aquatic organisms may respond to these changes in the environment by altering their respiration rates. Ectotherms are especially vulnerable to changing environmental conditions and may respond by making behavioral changes to adjust to these changes in the environment.

Our investigation focused on the behavioral responses of goldfish, *Carassius auratus*, to changes in temperature of their aquatic environment. We hypothesized that goldfish would increase their respiration rates as the temperature of their environment increased. We reasoned that oxygen availability would decrease as water temperature increased, leading to increases in goldfish respiratory rates as they attempted to meet their oxygen needs.

Materials and Methods

Study Organism

We obtained 20 size-matched male goldfish, *Carassius auratus*, from a local pet shop. We housed 10 fish each in two aquariums equipped with filters and aerators. Both aquariums were maintained at 22°C with a 12-hour to 12-hour light-to-dark cycle using 15-watt full-spectrum aquarium lighting. Fish were fed twice a day with commercial goldfish food for at least a week before the experimental trials began.

Goldfish respire by drawing oxygen-laden water into their mouth and forcing it over the respiratory surface area in the gills (Burggren 1982). This "air gulping" can be observed as goldfish open and close their mouths. In the gills, the oxygen is removed and carbon dioxide diffuses from the blood across the gill filaments and into the water. The gill cover opens as the now CO_2-laden water is forced out. Goldfish respiratory rate can thus be measured by observing the number of times their gill covers open and close (i.e. 'beat') over the course of 1 minute (gill beat rate = respiratory rate).

Experimental Design

For our experiment, we set up four aquariums containing identical conditions to the aquariums where fish were maintained (described above), except that the experimental group of aquariums had water temperature maintained at 27°C. Goldfish were individually netted and randomly assigned to either the experimental (27°C) or the control (22°C) conditions. Only one fish at a time was placed in either the control or the experimental aquarium. Once placed in the assigned aquarium, fish were allowed to acclimate for 1 minute. We then measured gill beat rate for 1 minute. Recorded data are shown in Table 1.

Data Analysis

We used a *t*-test to analyze the effect of temperature on goldfish respiratory rates under the two temperature conditions.

TABLE 1 Operculum (gill cover) beat rate (beats per minute) for goldfish in control and experimental aquariums

Individual Goldfish	Control Group at 22° C	Experimental Group at 27° C
1	61	128
2	58	135
3	65	127
4	60	131
5	61	127
6	57	138
7	63	133
8	55	140
9	59	126
10	66	129
Mean	60.5	131.4
Standard deviation	3.4	4.9
Standard error of the mean	1.1	1.5

Results

The results of the *t*-test indicated a highly significant difference between respiratory rates of goldfish in the control versus the experimental temperature conditions ($t = -37.2$, df = 18, $p < 0.0001$; see Figure 1). Individual data are shown in Table 1. As we predicted, goldfish respiratory rates increased at higher water temperatures. We therefore accepted our hypothesis.

Discussion

Our study revealed that goldfish approximately double their rate of respiration at 27° C compared to 22° C. This may entail energetic consequences for the fish. More energy allocated to metabolism and respiration means less energy allocated to growth, feeding, or reproduction (Buckley, Rodda, and Jetz 2008). Fish that are unable to behaviorally thermoregulate or adjust to temperature changes by some other means, such as morphological (Sollid, Weber, and Nilsson 2005) or biochemical adjustments (Goldspink 1995), may experience growth reductions and other fitness-related costs (Hill and Magnuson 1990). It is therefore imperative that we continue to investigate the potential effects of global climate warming on organisms, especially metabolic and respiratory responses.

Further investigations that would extend our research would be useful. We used 10 male goldfish of roughly the same size for this experiment. Perhaps body size has an influence on how organisms respond to temperature changes. Are juveniles more affected than larger adults or vice versa? What is the affect of temperature on females that are brooding young versus nonbrooders? Future studies could expand the focus to look at how temperature affects different life stages of goldfish: eggs, hatchlings, juveniles, adults. Another area of investigation involves the temperature range over which effects take place. Does respiration change smoothly and gradually, or do effects occur at specific temperatures?

The importance of studies focused on elucidating the impact of thermal conditions on plants and animals will increase as global warming continues (Root et al. 2003). As the results of our study have shown, even small changes of 5° C may have large energetic consequences for organisms, at least in the short term.

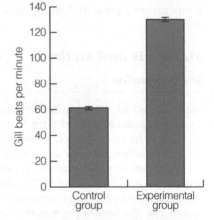

Figure 1 Average gill beat rate for goldfish held in aquariums with water at two different temperatures: control group at 22° C and experimental group at 27° C (values are means ±*SE*)

Citations

Buckley, L. B., G. H. Rodda, and W. Jetz. 2008. Thermal and energetic constraints on ectotherm abundance: A global test using lizards. *Ecology* 89(1): 48–55.

Burggren, W. W. 1982. "Air gulping" improves blood oxygen transport during aquatic hypoxia in the goldfish *Carassius auratus*. *Physiological Zoology* 55: 327–34.

Carpenter, S. R., S. G. Fisher, N. B. Grimm, and J. F. Kitchell. 1992. Global change and freshwater ecosystems. *Annual Review of Ecology and Systematics* 23: 119–39.

Goldspink, G. 1995. Adaptation of fish to different environmental temperatures by qualitative and quantitative changes in gene expression. *Journal of Thermal Biology* 20: 167–74.

Hill, D. K., and J. J. Magnuson. 1990. Potential effects of global climate warming on the growth and prey consumption of Great Lakes fish. *Transactions of the American Fisheries Society* 119(2): 265–75.

Root, T. L., J. T. Price, K. R. Hall, S. H. Schneider, C. Rosenzweig, and J. A. Pounds. 2003. Fingerprints of global warming on wild animals and plants. *Nature* 421(6918): 57–60.

Schindler, D. W. 1998. Widespread effects of climatic warming on freshwater ecosystems in North America. *Hydrological Processes* 11(8): 1043–67.

Sollid, J., R. E. Weber, and G. E. Nilsson. 2005. Temperature alters the respiratory surface area of crucian carp *Crassius carassius* and goldfish, *Crassius auratus*. *Journal of Experimental Biology* 208: 1109–16.

Starr, C., C. A. Evers, and L. Starr. 2008. Biology: *Concepts and application*, 7th ed. Belmont, CA: Cengage/Brooks Cole.

Mitosis, Meiosis, and Cytokinesis

OBJECTIVES

After completing this exercise, you will be able to

1. define *fertilization, zygote, DNA, chromosome, mitosis, cytokinesis, nucleoprotein, sister chromatid, centromere, meristem, meiosis, homologue (homologous chromosome), diploid, haploid, gene pair, allele, gamete, ovum, sperm, fertilization, locus, synapsis, zygote, genotype,* and *nondisjunction;*

2. identify the stages of the cell cycle;

3. distinguish between mitosis and cytokinesis as they take place in animal and plant cells;

4. identify the structures involved in nuclear and cell division (those in **boldface**) and describe the role each plays;

5. indicate the differences and similarities between meiosis and mitosis;

6. describe the basic differences between the life cycles of higher plants and higher animals;

7. describe the process of meiosis, and recognize events that occur during each stage;

8. discuss the significance of crossing over, segregation, and independent assortment;

9. identify the meiotic products in male and female animals;

10. describe the process of nondisjuction and chromosome number abnormalities in resulting gametes and zygotes.

INTRODUCTION

"All cells arise from preexisting cells." This is one tenet of the cell theory. It's easy to understand this concept if you think of a single-celled *Amoeba* or bacterium. Each cell divides to give rise to two entirely new individuals, and it is fascinating that each of us began life as *one* single cell and developed into this astonishingly complex animal, the human. Our first cell has *all* the hereditary information we'll ever get.

In higher plants and animals, **fertilization,** the fusion of egg and sperm nuclei, produces a single-celled **zygote.** The zygote divides into two cells, these two into four, and so on to produce a multicellular organism. During cell division, each new cell receives a complete set of hereditary information and an assortment of cytoplasmic components.

Recall from Exercise 6 that there are two basic cell types, prokaryotic and eukaryotic. The genetic material of both consists of **DNA (deoxyribonucleic acid).** In prokaryotes, the DNA molecule is organized into a single circular **chromosome.** Prior to cell division, the chromosome duplicates. Then the cell undergoes **fission,** the splitting of a preexisting cell into two, with each new cell receiving a full complement of the genetic material.

In eukaryotes, the process of cell division is more complex, primarily because of the much more complex nature of the hereditary material. Here the chromosomes consist of DNA and proteins complexed together within the nucleus. Cell division is preceded by duplication of the chromosomes and usually involves two processes: **mitosis** (nuclear division) and **cytokinesis** (cytoplasmic division). Whereas mitosis results in the production of two nuclei, both containing identical chromosomes, cytokinesis ensures that each new cell contains all the metabolic machinery necessary for sustenance of life.

In this exercise, we consider only what occurs in eukaryotic cells.

Dividing cells pass through a regular sequence of events called the cell cycle (Figure 8-1). Notice that the majority of the time is spent in interphase and that actual nuclear division—mitosis—is but a brief portion of the cycle.

Interphase is comprised of three parts (Figure 8-1): the G1 period, during which cytoplasmic growth takes place; the S period, when the DNA is duplicated; and the G2 period, when structures directly involved in mitosis are synthesized.

Unfortunately, because of the apparent relative inactivity that early microscopists observed, interphase was given the misnomer "resting stage." In fact, we know now that interphase is anything but a resting period. The cell is producing new DNA, assembling proteins from amino acids, and synthesizing or breaking down carbohydrates. In short, interphase is a very busy time in the life of a cell.

Like mitosis, meiosis is a process of nuclear division. During mitosis, the number of chromosomes in the daughter nuclei remains the same as that in the parental nucleus. In meiosis, however, the genetic complement is halved, resulting in daughter nuclei containing only one-half the number of chromosomes as the parental nucleus. Thus, while mitosis is sometimes referred to as an *equational division,* meiosis is often called *reduction division.* Moreover, while mitosis is completed after a single nuclear division, two divisions, called meiosis I and meiosis II, occur during meiosis. Table 8-1 summarizes the differences between mitosis and meiosis.

In the body cells of most eukaryotes, chromosomes exist in pairs called homologues (homologous chromosomes); that is, there are two chromosomes that are physically similar and contain genetic information for the same traits. To visualize this, press your palms together, lining up your fingers. Each "finger pair" represents one pair of homologues.

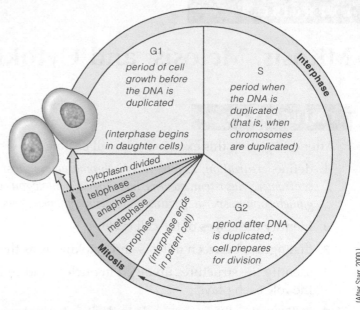

Figure 8-1 The eukaryotic cell cycle.

TABLE 8-1 Comparison of Mitosis and Meiosis

Mitosis	Meiosis
Equational division: Amount of genetic material remains constant	Reduction division: Amount of genetic material is halved
Completed in one division cycle	Requires two division cycles for completion
Produces two genetically identical nuclei	Produces two to four genetically different nuclei
Generally produces cells not directly involved in sexual reproduction	Ultimately produces cells used for sexual reproduction

When both homologues are in the *same* nucleus, the nucleus is **diploid** (2n); when only one of the homologues is present, the nucleus is **haploid** (n). If the parental nucleus normally contains the diploid (2n) chromosome number before meiosis, all four daughter nuclei contain the haploid (n) number at the completion of meiosis.

The reduction in chromosome number is the basis for sexual reproduction. In animals, the cells containing the daughter nuclei produced by meiosis are called **gametes: ova** (singular is *ovum*) if the parent is female, **sperm** cells if male. As you probably know, gametes are produced in the gonads—ovaries and testes, respectively. In fact, this is the *only* place where meiosis occurs in higher animals. Figure 8-2 shows where meiosis occurs in humans, while Figure 8-3 shows the life cycle of a higher animal.

Note when meiosis occurs—during gamete production. During **fertilization** (the fusion of a sperm nucleus with an ovum nucleus), the diploid chromosome number is restored as the two haploid gamete nuclei fuse to form the **zygote,** the first cell of the new diploid generation.

What about plants? Do plants have sex? Indeed they do. However, the

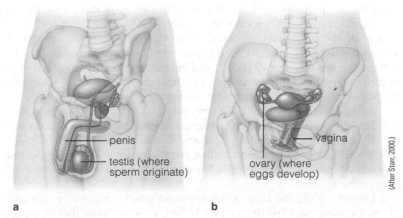

Figure 8-2 Gamete-producing structures in humans. (**a**) Human male, (**b**) human female.

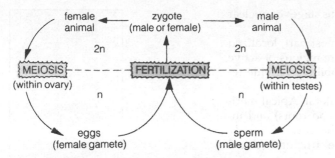

Figure 8-3 Life cycle of higher animals.

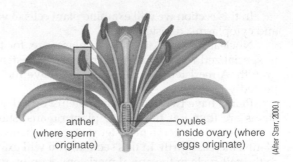

(After Starr, 2000.)

Figure 8-4 Gamete-producing structures in a flowering plant.

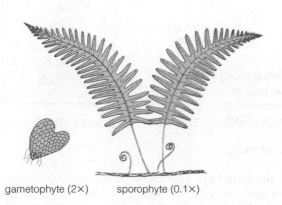

gametophyte (2×)   sporophyte (0.1×)

Figure 8-5 Gametophyte and sporophyte phases of the same fern species. Note size differences.

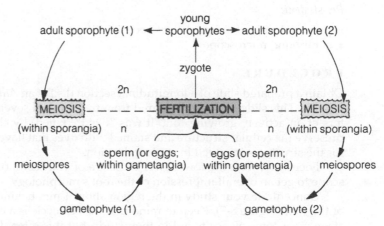

Figure 8-6 Life cycle of a plant.

plant life cycle is a bit more complex than that of animals. Plants of a single species have two completely different body forms. The primary function of one is the production of gametes. This plant is called a *gametophyte* ("gamete-producing plant") and it is haploid. Because the entire plant is haploid, gametes are produced in specialized organs by mitosis. The other body form, a *sporophyte*, is diploid. This diploid sporophyte has specialized organs in which meiosis occurs, producing haploid spores (hence the name *sporophyte*, "spore-producing plant"). When spores germinate and produce more cells by mitosis, they grow into haploid gametophytes, completing the life cycle.

Figure 8-4 shows the structures in a typical flower that produce sperm and eggs.

Examine Figure 8-5, which shows the gametophyte and sporophyte of a fern plant. Remember, the gametophyte and sporophyte are different, free-living stages of the *same* species of fern. Now look at Figure 8-6, which diagrams a typical plant life cycle. Again, note the consequence of meiosis. In plants, it results in the production of **spores,** not gametes.

You should understand an important concept from these diagrams: *Meiosis always halves the chromosome number. The diploid chromosome number is eventually restored when two haploid nuclei fuse during fertilization.*

Understanding meiosis is an absolute necessity for understanding the patterns of inheritance in Mendelian genetics. Gregor Mendel, an Austrian monk, spent years deciphering the complexity of simple genetics. Although he knew nothing of genes and chromosomes, he noted certain patterns of inheritance and formulated three principles, now known as Mendel's principles of recombination, segregation, and independent assortment. The following activities will demonstrate the events of meiosis and the genetic basis for Mendel's principles.

| 8.1 | The Cell Cycle in Plant Cells: Onion Roots *(About 45 min.)* |

During much of a cell's life, each DNA–protein complex, the **nucleoprotein,** is extended as a thin strand within the nucleus. In this form, it is called **chromatin.** Prior to the onset of nuclear division, the genetic material duplicates itself. As nuclear division begins, the chromatin condenses. The two identical condensed nucleoproteins are called **sister chromatids** and are attached at the **centromere.** The centromere gives the appearance of dividing each chromatid into two "arms." Collectively, the two attached sister chromatids are referred to as a *duplicated chromosome.*

In this section we will examine plant cells in various stages of nuclear and cytoplasmic division.

Nuclear and cell divisions in plants are, for the most part, localized in specialized regions called meristems. **Meristems** are regions of active growth. A meristem contains cells that have the capability to divide repeatedly.

Plants have two types of meristems: apical and lateral. Apical meristems are found at the tips of plant organs (shoots and roots) and increase length. Lateral meristems, located beneath the bark of woody plants, increase girth. In this section, you will examine structures related to the cell cycle in the apical meristem of onion roots.

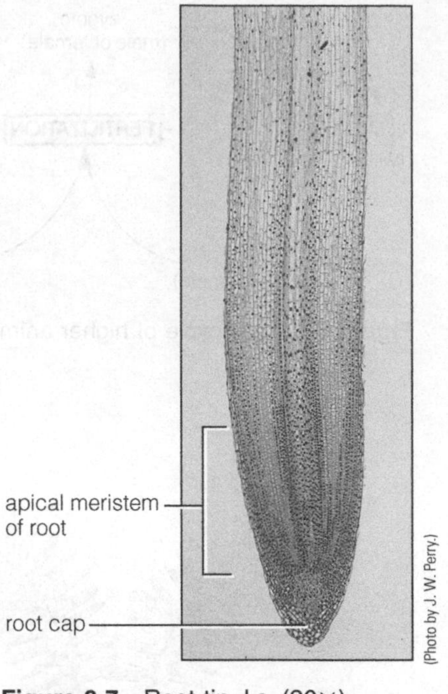

MATERIALS

Per student:

- prepared slide of onion, *Allium,* root tip mitosis
- compound microscope

PROCEDURE

Obtain a prepared slide of a longitudinal section (l.s.) of an *Allium* (onion) root tip. This slide has been prepared from the terminal several millimeters of an actively growing root. It was "fixed" (killed) by chemicals to preserve the cellular structure and stained with dyes that have an affinity for the structures involved in nuclear division.

Focus first with the low-power objective of your compound microscope to get an overall impression of the root's morphology.

Concentrate your study in the region about 1 mm behind the actual tip. This region is the apical meristem of the root (Figure 8-7). Keep in mind that the cell cycle is a continuous cycle; we separate events into different stages as a convenience to aid in their study, but it is often difficult to say definitively when one phase begins and another ends.

apical meristem of root

root cap

(Photo by J. W. Perry.)

Figure 8-7   Root tip, l.s. (20×).

A. Interphase and Mitosis

1. **Interphase.** Use the medium-power objective to scan the apical meristem. Note that most of the nuclei are in interphase.

    Switch to the high-dry objective, focusing on a single interphase cell. Note the distinct **nucleus,** with one or more **nucleoli,** and the **chromatin** dispersed within the bounds of the **nuclear envelope.** Label these features in cell 1 of Figure 8-8.

2. **Mitosis.**

    (a) *Prophase.* During **prophase** the chromatin condenses, rendering the duplicated chromosomes visible as threadlike structures. At the same time, microtubules outside the nucleus are beginning to assemble into **spindle fibers.** Collectively, the spindle fibers make up the **spindle,** a three-dimensional structure widest in the middle and tapering to a point at the two **poles** (opposite ends of the cell). *You will not see the spindle during prophase.*

    Find a nucleus in prophase. Draw and label a prophase nucleus in cell 2 of Figure 8-8.

    The transition from prophase to metaphase is marked by the fragmentation and disappearance of the nuclear envelope. At about the same time, the nucleoli disappear.

    (b) *Metaphase.* When the nuclear envelope is no longer distinct, the cell is in **metaphase.** Identify a metaphase cell by locating a cell with the duplicated chromosomes, each consisting of two **sister chromatids,** lined up midway between the two poles. This imaginary midline is called the **spindle equator.** (You will not be able to distinguish the chromatids.) The spindle has moved into the space the nucleus once occupied. The microtubules have become attached to the chromosomes at the **kinetochores,** groups of proteins that form the outer faces of the centromeres. Find a cell in metaphase. Label cell 3 of Figure 8-8.

    (c) *Anaphase.* During **anaphase** sister chromatids of each chromosome separate, each chromatid moving toward an opposite pole.

    Find an early anaphase cell, recognizable by the slightly separated chromatids. Notice that the chromatids begin separating at the centromere. The last point of contact before separation is complete is at the ends of the "arms" of each chromatid. Although incompletely understood, the mechanism of chromatid separation is based on action of the spindle-fiber microtubules. Once separated, each chromatid is referred to as an individual daughter chromosome. Note that now the chromosome consists of a *single* chromatid.

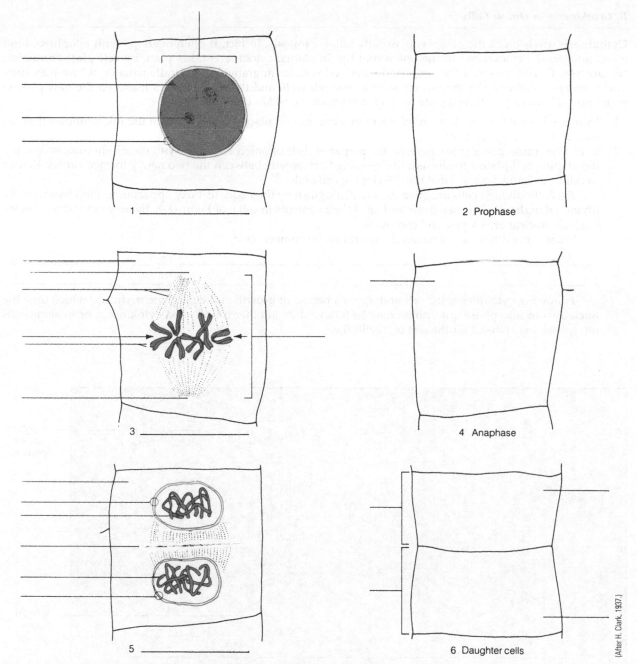

Figure 8-8 Interphase, mitosis, and cytokinesis in onion root tip cells.
Labels: interphase, cytoplasm, nucleus, nucleolus, chromatin, nuclear envelope, metaphase, spindle fibers, spindle, pole, spindle equator (between arrows), sister chromatids, telophase and cell plate formation, chromosome, cell plate, daughter cell
(*Note:* Some terms are used more than once.)

Find a late anaphase cell and draw it in cell 4 of Figure 8-8.

(d) *Telophase.* When the daughter chromosomes arrive at opposite poles, the cell is in **telophase.** The spindle disorganizes. The chromosomes expand again into chromatin form, and a nuclear envelope re-forms around each newly formed daughter nucleus.

Find a telophase cell and label individual chromosomes, nuclei, and nuclear envelopes on cell 5 of Figure 8-8.

Cytokinesis, division of the cytoplasm, usually follows mitosis. In fact, it often overlaps with telophase. Find a cell undergoing cytokinesis in the onion root tip. In plants, cytokinesis takes place by **cell plate formation** (Figure 8-9). During this process, Golgi body–derived vesicles migrate to the spindle equator, where they fuse. Their contents contribute to the formation of a new cell wall, and their membranes make up the new plasma membranes. In most plants, cell plate formation starts in the *middle* of the cell.

1. Examine Figure 8-9, an electron micrograph showing cell plate formation. Note the microtubules that are part of the spindle apparatus.

2. Find a cell undergoing cytokinesis on the prepared slide of onion root tips. With your light microscope, the developing **cell plate** appears as a line running horizontally between the two newly formed nuclei. Return to cell 5 of Figure 8-8 and label the developing cell plate.

   Recently divided cells are often easy to distinguish by their square, boxy appearance. Find two recently divided **daughter cells;** then draw and label their contents in cell 6 of Figure 8-8. Include cytoplasm, nuclei, nucleoli, nuclear envelopes, and chromatin.

   What is the difference between chromatin and chromosomes?

   _____

   _____

   Following cytokinesis, the cell undergoes a period of growth and enlargement, during which time the nucleus is in interphase. Interphase may be followed by another mitosis and cytokinesis, or in some cells interphase may persist for the rest of a cell's life.

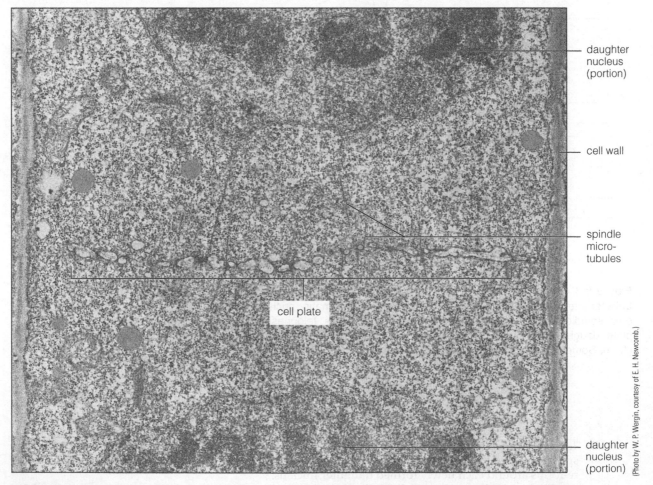

Figure 8-9 Transmission electron micrograph of cytokinesis by cell plate formation in a plant cell (2000×).

An onion root tip preparation can be thought of as a snapshot capturing cells in various phases of the cell cycle at a particular moment in time. The frequency of occurrence of a cell cycle phase is directly proportional to the length of a phase. You can therefore estimate the amount of time each phase takes by tallying the proportions of cells in each phase.

The length of the cell cycle for cells in actively dividing onion root tips is approximately 24 hours, with mitosis lasting for about 90 minutes.

1. Examine the meristem region of an onion root tip slide. Count the number of cells in each of the stages of mitosis plus interphase in one field of view. Repeat this procedure for other fields of view until you count 100 cells. Record your data in Table 8-2.
2. Now calculate the time spent in each stage based on a 24-hour cell cycle by dividing the number of cells in each stage by the total number of cells counted to determine percent of total cells in each stage.
3. Multiply the fraction obtained in step 2 by 24 to determine duration.

Although your calculations are only a rough approximation of the time spent in each stage, they do illustrate the differences in duration of each stage in the cell cycle.

TABLE 8-2 Determining Duration of Cell Cycle Phases

Phase	Number Seen	% of Total	Duration (hrs)
Interphase			
Prophase			
Metaphase			
Anaphase			
Telophase			
Total			

8.2   The Cell Cycle in Animal Cells: Whitefish Blastula *(About 20 min.)*

Fertilization of an ovum by a sperm produces a zygote. In animal cells, the zygote undergoes a special type of cell division, *cleavage*, in which no increase in cytoplasm occurs between divisions. A ball of cells called a blastula is produced by cleavage. Within the blastula, repeated nuclear and cytoplasmic divisions take place; consequently, the whitefish blastula is an excellent example in which to observe the cell cycle of an animal.

Note a key difference between plants and animals: Whereas plants have meristems where divisions continually take place, animals do not have specialized regions to which mitosis and cytokinesis are limited. Indeed, divisions occur continually throughout many tissues of an animal's body, replacing worn-out or damaged cells.

With several important exceptions, mitosis in animals is remarkably like that in plants. These exceptions will be pointed out as we go through the cell cycle.

MATERIALS

Per student:

- prepared slide of whitefish blastula mitosis
- compound microscope

PROCEDURE

Obtain a slide labeled "whitefish blastula." Scan it with the low-power objective and then at medium power. This slide has numerous thin sections of a blastula. Select one section (Figure 8-10) and then switch to the high-dry objective for detailed observation.

As you examine the slides, draw the cells to show the correct sequence of events in the cell cycle of whitefish blastula.

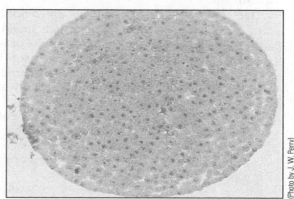

Figure 8-10   Section of a whitefish blastula (75×).

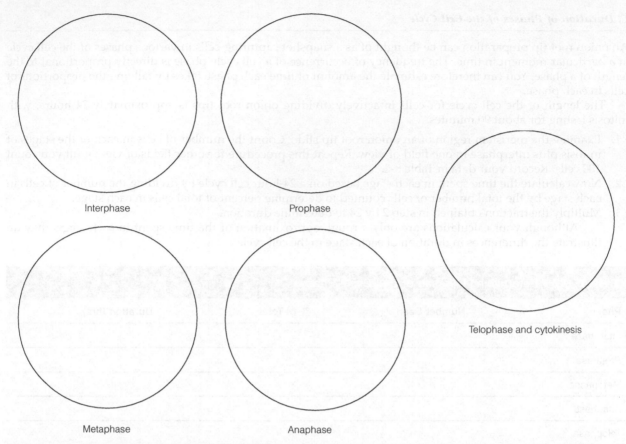

Interphase

Prophase

Telophase and cytokinesis

Metaphase

Anaphase

Figure 8-11 Drawings of cell cycle stages in whitefish blastula.
Labels: cytoplasm, nucleus, plasma membrane, spindle, chromosomes, spindle equator, sister chromatids, daughter nuclei, chromatin, furrow
(*Note:* Some terms are used more than once.)

A. Interphase and Mitosis

1. **Interphase.** Locate a cell in **interphase**. As you observed in the onion root tip, note the presence of the nucleus and chromatin within it. Note also the absence of a cell wall.

    Draw an interphase cell above the word "Interphase" in Figure 8-11 and label the cytoplasm, nucleus, and plasma membrane.

2. **Mitosis.**

    (a) *Prophase.* The first obvious difference between mitosis in plants and animals is found in **prophase.** Unlike the onion cells, those of whitefish contain **centrioles** (Figure 8-12). As seen with the electron microscope, centrioles are barrel-shaped structures consisting of nine radially arranged triplets of microtubules.

    One pair of centrioles was present in the cytoplasm in the G1 stage of interphase. These centrioles duplicated during the S stage of interphase. Subsequently, one new and one old centriole migrated to each pole.

    Although the centrioles are too small to be resolved with your light microscope, you can see

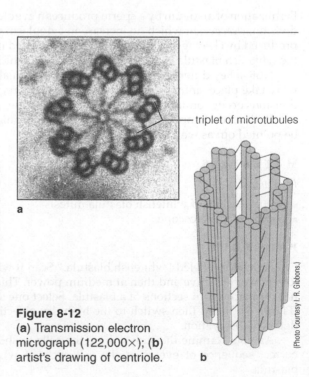

triplet of microtubules

a

Figure 8-12
(a) Transmission electron micrograph (122,000×); **(b)** artist's drawing of centriole.

b

(Photo Courtesy I. R. Gibbons.)

a starburst pattern of spindle fibers that appear to radiate from the centrioles. Other microtubules extend between the centrioles, forming the **spindle** (Figure 8-13). The chromosomes become visible as the chromatin condenses.

Find a prophase cell, identifying the spindle and starburst cluster of fibers about the centriole.

Draw the prophase cell in the proper location on Figure 8-11. Label the spindle, chromosomes, cytoplasm, and the position of the plasma membrane.

(b) *Metaphase.* As was the case in plant cells, during **metaphase** the spindle fiber microtubules become attached to the **kinetochore** of each centromere region, and the duplicated chromosomes (each consisting of two **sister chromatids**) line up on the **spindle equator.** Locate a metaphase cell.

Draw the metaphase cell in the proper location on Figure 8-11. Label the chromosomes on the spindle equator, spindle, and plasma membrane.

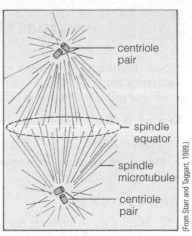

Figure 8-13 Spindle apparatus in animal cell.

(c) *Anaphase.* Again similar to that observed in plant cells, **anaphase** begins with the separation of sister chromatids into individual (daughter) chromosomes. Observe a blastula cell in anaphase.

Draw the anaphase cell in the proper location on Figure 8-11. Label the separating sister chromatids, spindle, cytoplasm, and plasma membrane.

(d) *Telophase.* **Telophase** is characterized by the arrival of the individual (daughter) chromosomes at the poles. A nuclear envelope forms around each daughter nucleus. Find a telophase cell.

Is the spindle still visible? _____

Is there any evidence of a nuclear envelope forming around the chromosomes?_____

Draw the telophase cell in Figure 8-11. Label daughter nuclei, chromatin, cytoplasm, and plasma membrane.

B. Cytokinesis in Animal Cells

A second major distinction between cell division in plants and animals occurs during cytoplasmic division. Cell plates are absent in animal cells. Instead, cytokinesis takes place by **furrowing.**

To visualize how furrowing takes place, imagine wrapping a string around a balloon and slowly tightening the string until the balloon has been pinched in two. In life, the animal cell is pinched in two, forming two discrete cytoplasmic entities, each with a single nucleus. Figure 8-14 illustrates the cleavage furrow in animal cell.

Find a cell in the blastula undergoing cytokinesis. The telophase cell that you drew in Figure 8-11 may also show an early stage of cytokinesis. Label the cleavage furrow if it does.

    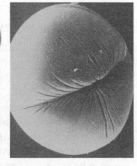

a Mitosis is completed and the bipolar spindle is starting to disassemble.

b At the former spindle equator, a ring of actin filaments attached to the plasma membrane contracts.

c The diameter of the contractile ring continues to shrink and pull the cell surface inward.

d The contractile mechanism continues to operate until the cytoplasm is partitioned.

e

Figure 8-14 (a–d) Cytoplasmic division of an animal cell. **(e)** Scanning electron micrograph of the cleavage furrow at the plane of the former spindle equator.

Simulating Mitosis (*About 20 min.*)

Understanding chromosome movements is crucial to understanding mitosis. You can simulate mitosis with a variety of materials. This is a simple activity, but a valuable one. It will be especially helpful when comparing the events of mitosis with those of meiosis in the next exercise.

MATERIALS

Per student pair:

- 44 pop beads each of two colors
- 8 magnetic centromeres

PROCEDURE

1. Build the components for two pairs of chromosomes by assembling strings of pop beads as follows:
    (a) Assemble two strands of pop beads with eight pop beads of one color on each arm, with a magnetic centromere connecting the two arms.
    (b) Repeat step a, but use pop beads of the second color.
    (c) Assemble two more strands of pop beads, but with three pop beads of one color on each arm.
    (d) Repeat step c, using pop beads of the second color.
       You should have four long strings, two of each color, and four short strings, with two of each color. Each pop bead string should have a magnetic centromere at its midpoint. Note that pop bead strings can attach to each other at the magnetic centromere. Each pop bead string represents a single molecule of DNA plus proteins.
2. Place **one** of each kind of strand in the center of your workspace, which represents the nucleus. You have created a nucleus with four "chromosomes," two long and two short.
3. Manipulate these model chromosomes through the phases of the cell cycle, beginning in the G1 phase of interphase and proceeding through the rest of interphase, mitosis, and cytokinesis.

Demonstrations of Meiosis Using Pop Beads (*About 75 min.*)

MATERIALS

Per student pair:

- 44 pop beads each of two colors (red and yellow, for example)
- 8 magnetic centromeres
- marking pens
- 8 pieces of string, each 40 cm long
- meiotic diagram cards similar to those used here
- colored pencils

Per student group (table):

- bottle of 95% ethanol to remove marking ink
- tissues

PROCEDURE

Work in pairs.

Within the nucleus of an organism, each chromosome bears **genes,** which are units of inheritance. Genes may exist in two or more alternative forms called **alleles.** Each homologue bears *genes* for the same traits; these are the **gene pairs.** However, the homologues may or may not have the same *alleles.* An example will help here.

Suppose the trait in question is flower color and that a flower has only two possible colors, red or white (Figure 8-15a, b). The gene is coding (providing the information) for flower color. There are two homologues in the same nucleus, so each bears the gene for flower color. *But,* on one homologue, the *allele* might code for red flowers, while the allele on the other homologue might code for white flowers

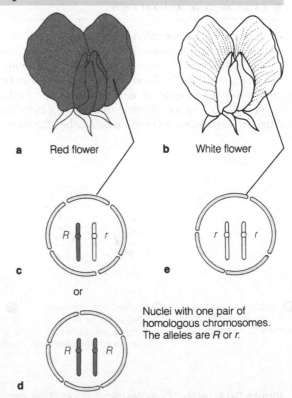

a   Red flower   b   White flower

c   or

d

e   Nuclei with one pair of homologous chromosomes. The alleles are *R* or *r*.

Figure 8-15 Chromosomal control of flower color. (c–e) show a nucleus with one pair of homologous chromosomes. The alleles for flower color are *R* or *r.*

(Figure 8-15c). There are two other possibilities. The alleles on *both* homologues might be coding for red flowers (Figure 8-15d), or they *both* might be coding for white flowers (Figure 8-15e). Note that these three possibilities are mutually exclusive.

1. Build the components for two pairs of homologous chromosomes by assembling strings of pop beads as follows:
    (a) Assemble two strands of pop beads with eight pop beads of one color on each arm, with a magnetic centromere connecting the two arms.
    (b) Repeat step a, but use pop beads of the second color.
    (c) Assemble two more strands of pop beads, but with three pop beads of one color on each arm.
    (d) Repeat step c, using pop beads of the second color.
    You should have four long strings, two of each color, and two short strings, also two of each color. Each pop-bead string should have a magnetic centromere at its midpoint by which pop-bead strings can attach to each other. Each pop-bead string represents a single molecule of DNA plus proteins, with each bead representing a gene.
2. Place **one** of each kind of strand in the center of your workspace, which represents the interphase nucleus of a cell that will undergo meiosis. You have created a nucleus with four "chromosomes," two long and two short. The long strands represent one homologous pair, and the short strands represent a second homologous pair of chromosomes.
    We start by assuming that these chromosomes represent the diploid condition. The two colors represent the origin of the chromosomes: One homologue (color _____) came from the male parent, and the other homologue (color _____) came from the female parent.
3. The four single-stranded chromosomes represent four unduplicated chromosomes. Now simulate DNA duplication during the S-phase of interphase (Figure 8-1), whereby each DNA molecule and its associated proteins are copied exactly. The two copies, called sister chromatids, remain attached to each other at their centromeres (Figure 8-16). During chromosome replication, the genes also duplicate. Thus, alleles on sister chromatids are identical.

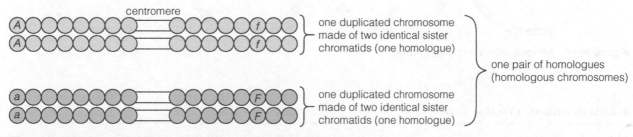

Figure 8-16 One pair of homologous pop-bead chromosomes.

How many sister chromatids are there in a duplicated chromosome? _____

How many chromosomes are represented by four sister chromatids? _____ By eight? _____

What is the diploid number of the starting (parental) nucleus? (*Hint:* Count the number of homologues to obtain the diploid number.) _____

4. As mentioned previously, genes may exist in two or more alternative forms, called alleles. The location of an allele on a chromosome is its **locus** (plural: *loci*). Using the marking pen, mark two loci on each long chromatid with letters to indicate alleles for a common trait. Suppose the long pair of homologous chromosomes codes for two traits, skin pigmentation and the presence of attached earlobes in humans. We'll let the capital letter *A* represent the allele for normal pigmentation and a lowercase *a* the allele for albinism (the absence of skin pigmentation); *F* will represent free earlobes and *f* attached earlobes. A suggested marking sequence is illustrated in Figure 8-16.
5. Let's assign a gene to our second homologous pair of chromosomes, the short pair. We'll suppose this gene codes for the production of an enzyme necessary for metabolism. On one homologue (consisting of two chromatids) mark the letter *E*, representing the allele causing enzyme production. On the other homologue, *e* represents the allele that interferes with normal enzyme production.
6. Obtain a meiotic diagram card like the one in Figure 8-17. Manipulate your model chromosomes through the stages of meiosis described below, locating the chromosomes in the correct diagram circles (representing nuclei) as you go along. Reference to Figure 8-17 will be made at the proper steps. *DO NOT* draw on the meiotic diagram cards.

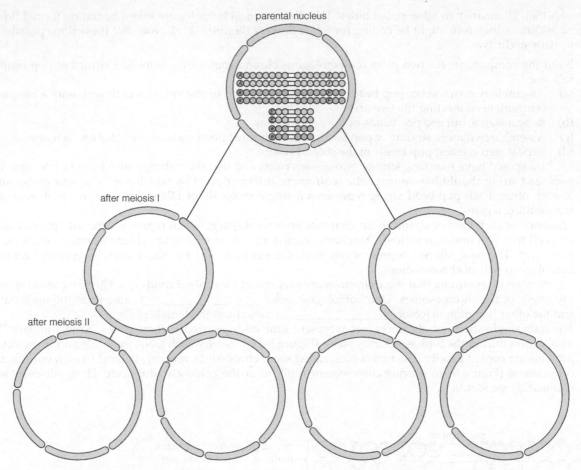

parental nucleus

after meiosis I

after meiosis II

Figure 8-17 Meiosis without crossing over.

A. Meiosis without Crossing Over

Although crossing over is a nearly universal event during meiosis, we will first work with a simplified model to illustrate chromosomal movements and separations during meiosis. Refer to Figure 8-18 as you manipulate your model.

1. **Late interphase.** During interphase, the nuclear envelope is intact and the chromosomes are randomly distributed throughout the nucleoplasm (semifluid substance within the nucleus). All duplicated chromosomes (eight chromatids) should be in the parental nucleus, indicating that DNA duplication has taken place. The sister chromatids of each homologue should be attached by their magnetic centromeres, but the four homologues should be separate. Your model nucleus contains a diploid number 2n = 4.

    The pop-bead chromosomes should appear during interphase in the parental nucleus as shown in Figure 8-17. Be sure to mark the location of the alleles. Use different pencil or pen colors to differentiate the homologues on your drawings.

2. **Meiosis I.** During meiosis I, homologues are separated from each other into different nuclei. Daughter nuclei created are thus haploid.

    (a) *Prophase I.* During the first prophase, the parental nucleus contains four duplicated homologous chromosomes, each comprised of two sister chromatids joined at their centromeres. The chromatin condenses to form discrete, visible chromosomes. The homologues pair with each other. This pairing is called *synapsis.* Slide the two homologues together.

    Twist the chromatids about one another to simulate synapsis.
    The nuclear envelope disorganizes at the end of prophase I.

    (b) *Metaphase I.* Homologous chromosomes now move toward the spindle equator, the centromeres of each homologue coming to lie *on either side of the equator.* Spindle fibers, consisting of aggregations of microtubules, attach to the centromeres. One homologue attaches to microtubules extending from one pole, and the other homologue attaches to microtubules extending from the opposite spindle pole.

To simulate the spindle fibers, attach one piece of string to each centromere. Then lay the free ends of strings from two homologues toward one spindle pole and the ends of the other homologues toward the opposite pole.

(c) *Anaphase I.* During anaphase I, the homologous chromosomes separate, one homologue moving toward one pole, the other toward the opposite pole. The movement of the chromosomes is apparently the result of shortening of some spindle fibers and lengthening of others. Each homologue is still in the duplicated form, consisting of two sister chromatids.

Pull the two strings of one homologous pair toward its spindle pole and the other toward the opposite spindle pole, separating the homologues from one another. Repeat with the second pair of homologues.

(d) *Telophase I.* Continue pulling the string spindle fibers until each homologue is now at its respective pole. The first meiotic division is now complete. You should have two nuclei, each containing two chromosomes (one long and one short) consisting of two sister chromatids.

Draw your pop-bead chromosomes as they appear after meiosis I on the two nuclei labeled "after meiosis I" of Figure 8-17. Depending on the organism involved, an interphase (interkinesis) and cytokinesis may precede the second meiotic division, *or* each nucleus may enter directly into meiosis II. The chromosomes decondense into chromatin form.

It is important to note here that DNA synthesis *does not* occur following telophase I (between meiosis I and meiosis II).

Before meiosis II, the spindle is rearranged into two spindles, one for each nucleus.

3. **Meiosis II.** During meiosis II, sister chromatids are separated into different daughter nuclei. The result is four haploid nuclei.

(a) *Prophase II.* At the beginning of the second meiotic division, the sister chromatids are still attached by their centromeres. During prophase II, the nuclear envelope disorganizes, and the chromatin recondenses.

(b) *Metaphase II.* Within each nucleus, the duplicated chromosomes align with the equator, the centromeres lying *on the equator*. Spindle fiber microtubules attach the centromeres of each chromatid to opposite spindle poles.

Your string spindle fibers should be positioned so that the two spindle fiber strings from sister chromatids lie toward opposite poles. Note that each nucleus contains only *two* duplicated chromosomes (one long and one short) consisting of *two* sister chromatids each.

(c) *Anaphase II.* The sister chromatids separate, moving to opposite poles. Pull on the string until the two sister chromatids separate. After the sister chromatids separate, each is an individual (not duplicated) daughter chromosome.

(d) *Telophase II.* Continue pulling on the string spindle fibers until the two daughter chromosomes are at opposite poles. The nuclear envelope re-forms around each chromosome and the chromosomes decondense back into chromatin form. Four daughter nuclei now exist. Note that each nucleus contains two individual unduplicated chromosomes (each formerly a chromatid) originally present within the parental nucleus. These nuclei and the cells they're in generally undergo a differentiation and maturation process to become gametes (in animals) or spores (in plants).

Draw your pop-bead chromosomes as they appear after meiosis II in the "gamete nuclei" of Figure 8-17. Your diagram should indicate the genetic (chromatid) complement *before* meiosis and *after* each meiotic division, *not* the stages of each division.

Remember that meiosis takes place in both male and female organisms. (See Figure 8-3.)

If the parental nucleus was from a male, what is the gamete called? _____

If female? _____

Is the parental nucleus diploid or haploid? _____

Are the nuclei produced after the *first* meiotic division diploid or haploid? _____

Are the nuclei of the gametes diploid or haploid? _____

What is the **genotype** of each gamete nucleus after meiosis II? (The genotype is the genetic composition of an organism, or the alleles present. Another way to ask this question is, What alleles are present in each gamete nucleus? Write these in the format: *AFE*, *afe*, and so on.)

_____

If you answered the preceding questions correctly, you might logically ask, "If the chromosome number of the gametes is the same as that produced after the first meiotic division, why bother to have two separate divisions? After all, the genes present are the same in both gametes and first-division nuclei."

There are two answers to this apparent paradox. The first, and perhaps the most obvious, is that the second meiotic division ensures that a *single* chromatid (nonduplicated chromosome) is contained within each gamete. After gametes fuse, producing a zygote, the genetic material duplicates prior to the zygote's undergoing mitosis.

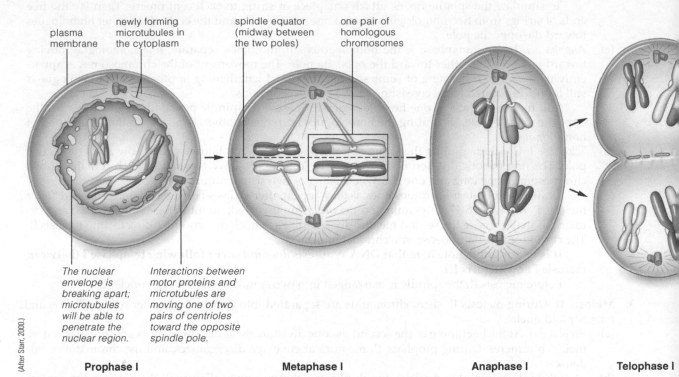

(After Starr, 2000.)

plasma
membrane

newly forming
microtubules in
the cytoplasm

spindle equator
(midway between
the two poles)

one pair of
homologous
chromosomes

*The nuclear
envelope is
breaking apart;
microtubules
will be able to
penetrate the
nuclear region.*

*Interactions between
motor proteins and
microtubules are
moving one of two
pairs of centrioles
toward the opposite
spindle pole.*

Prophase I   **Metaphase I**   **Anaphase I**   **Telophase I**

Figure 8-18 Meiosis in a generalized animal germ cell. Two pairs of chromosomes are shown. Maternal chromosomes are shaded purple. Paternal chromosomes are shaded light blue.

If gametes contained two chromatids, the zygote would have four, and duplication prior to zygote division would produce eight, twice as many as the organism should have. If DNA duplication within the zygote were not necessary for the onset of mitosis, this problem would not exist. Alas, DNA synthesis apparently is a necessity to initiate mitosis.

You can discover the second answer for yourself by continuing with the exercise, for although you have simulated meiosis, you have done so without showing what happens in *real* life. That's the next step . . .

B. Meiosis with Crossing Over

A very important event that results in a reshuffling of alleles on the chromatids occurs during prophase I. Recall that synapsis results in pairing of the homologues. During synapsis, the chromatids break, and portions of chromatids bearing genes for the same characteristic (but perhaps *different* alleles) are exchanged between *nonsister* chromatids. This event is called **crossing over,** and it results in recombination (shuffling) of alleles.

1. Look again at Figure 8-16. Distinguish between sister and nonsister chromatids. Now look at Figure 8-19, which demonstrates crossing over in one pair of homologues.
2. Return your chromosome models to the nucleus format with two pairs of homologues entering prophase I.
3. To simulate crossing over, break four beads from the arms of two nonsister chromatids in the long homologue pair, exchanging bead color between the two arms. During actual crossing over, the chromosomes may break anywhere within the arms.

   Crossing over is virtually a universal event in meiosis. Each pair of homologues may cross over in several places simultaneously during prophase I.
4. Manipulate your model chromosomes through meiosis I and II again and watch what happens to the distribution of the alleles as a consequence of the crossing over. Fill in Figure 8-20 as you did before, but this time show the effects of crossing over. Again, use different colors in your sketches.

   What are the genotypes of the gamete nuclei?

   _____

   Is the distribution of alleles present in the gamete nuclei after crossing over the same as that which was present without crossing over?

   _____

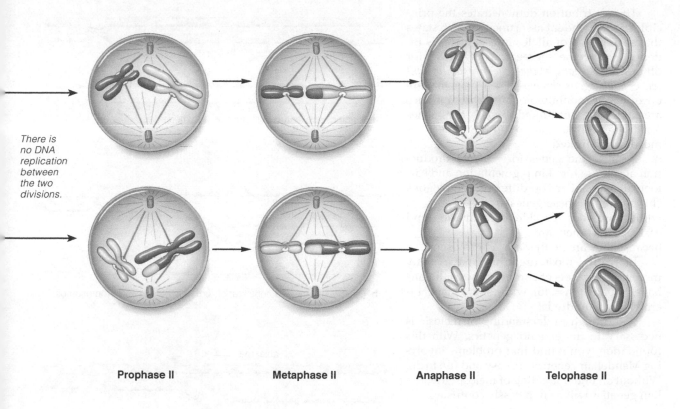

There is no DNA replication between the two divisions.

Prophase II   **Metaphase II**   **Anaphase II**   **Telophase II**

Is the distribution of alleles present in the gamete nuclei after crossing over the same as that in the nuclei after the first meiotic division?

_____

Crossing over provides for genetic recombination, resulting in increased variety. How many different genetic *types* of daughter chromosomes are present in the gamete nuclei without crossing over (Figure 8-17)?

_____

How many different types are present with crossing over (Figure 8-20)?

_____

We think you would agree that a greater number of *types* of daughter chromosomes indicates greater *variety*.

Recall that the parental nucleus contained a pair of homologues, each homologue consisting of two sister chromatids. Because sister chromatids are identical in all respects, they have the same alleles of a gene (see Figure 8-16). As your models showed, the alleles on nonsister chromatids may not (or may) be identical; they bear the same genes but may have different alleles, different forms of some genes.

What is the difference between a gene and an allele?

_____

_____

Let's look at a single set of alleles on your model chromosomes that are, say, the alleles for pigmentation, A and a. Both alleles were present in the parental nucleus. How many are present in the gametes?

_____

This illustrates Mendel's first principle, segregation. Segregation means that during gamete formation, pairs of alleles are separated (segregated) from each other and end up in different gametes.

C. Demonstrating Independent Assortment

Manipulate your model chromosomes again through meiosis with crossing over (Figure 8-20), searching for different possibilities in chromosome distribution that would make the gametes genetically different.

Does the distribution of the alleles for enzyme production to different gametes on the second set of homologues have any bearing on the distribution of the alleles on the first set (alleles for skin pigmentation and earlobe condition)? _____

This distribution demonstrates the principle of independent assortment, which states that segregation of alleles into gametes is independent of the segregation of alleles for other traits, *as long as the genes are on different sets of homologous chromosomes.* Genes that are on different (nonhomologous) chromosomes are said to be **nonlinked.** By contrast, genes for different traits that are on the same chromosome are **linked.**

Because the genes for enzyme production and those for skin pigmentation and earlobe attachment are on different homologous chromosomes, these genes are _____, while the genes for skin pigmentation and earlobe attachment are _____ because they are on the same chromosome.

In reality, most organisms have many more than two sets of chromosomes. Humans have 23 pairs (2n = 46), while some plants literally have hundreds!

A thorough understanding of meiosis is necessary to understand genetics. With this foundation, you'll find that problems involving Mendelian genetics are easy and fun to do. Without an understanding of meiosis, Mendelian genetics will be hopelessly confusing.

D. Nondisjunction and the Production of Gametes with Abnormal Chromosome Number

Errors in the process of meiosis can occur in many ways. Perhaps the best understood error process is that of **nondisjunction,** when one or more pairs of chromosomes fail to separate in anaphase. The result is gamete nuclei with too few or too many chromosomes.

1. Begin to manipulate your model chromosomes to show meiosis without crossing over. In modeling events at metaphase I, however, arrange the spindle fiber threads for the long pair of homologues so that they all extend to the same pole.

2. Model anaphase I, pulling the chromosomes toward their respective poles. Nondisjunction occurs in the long pair of homologues, with both duplicated chromosomes being pulled to the same pole. See Figure 8-21.

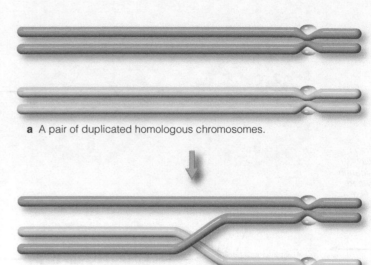

a A pair of duplicated homologous chromosomes.

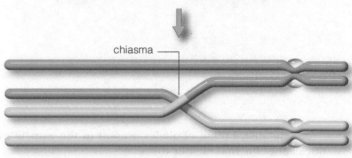

b Crossover between nonsister chromatids of the two chromosomes.

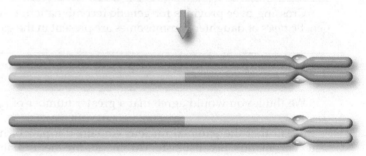

chiasma

c Nonsister chromatids exchange segments.

d Homologues have new combinations of alleles.

(After Starr, 2000.)

Figure 8-19 Crossing over in one pair of homologues. Maternal chromosomes are shaded purple. Paternal chromosomes are shaded light blue.

3. Continue to manipulate the model chromosomes through the remainder of the meiotic process.

How many chromosomes are found in gamete nuclei? _____

How does this compare to the chromosome number in normal gametes? _____

Recall that each chromosome bears a unique set of genes and speculate about the effect of nondisjunction on the resulting zygotes formed from fertilization with such a gamete.

_____

_____

_____

One of the most common human genetic disorders arises from nondisjunction during gamete (usually ovum) formation. Down syndrome results from nondisjunction in one of the 23 pairs of human

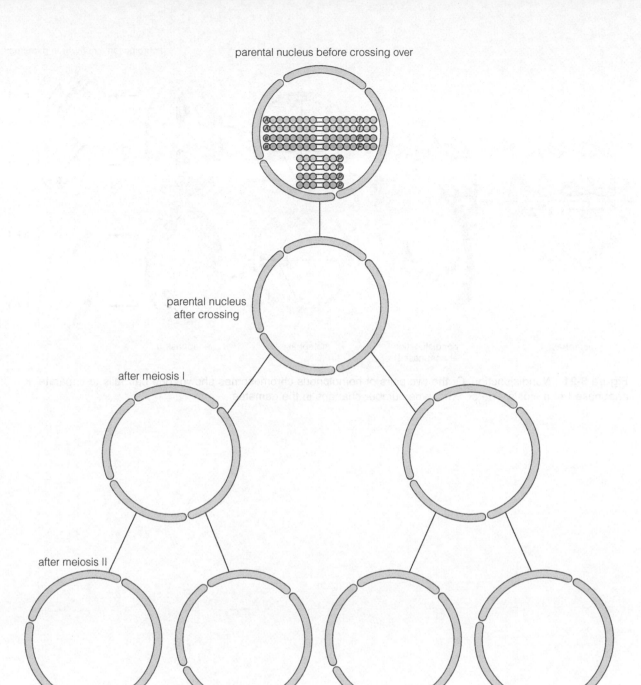

parental nucleus before crossing over

parental nucleus
after crossing

after meiosis I

after meiosis II

gamete nuclei

Figure 8-20 Meiosis with crossing over.

chromosomes, chromosome 21. An individual with Down syndrome has three copies of chromosome 21 instead of the normal two copies. While symptoms of this genetic disorder vary greatly, most individuals show moderate to severe mental impairment and a host of associated physical defects. Relatively few other human genetic disorders arise from nondisjunction, probably because the consequences of abnormal chromosome number are often lethal.

Note: **Remove marking ink from pop beads with 95% ethanol and tissues.**

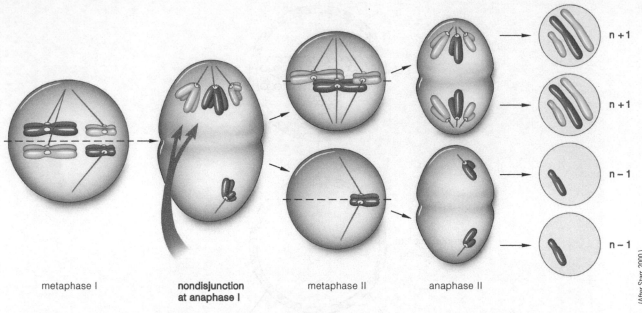

chromosome number in gametes:

n + 1

n + 1

n − 1

n − 1

(After Starr, 2000.)

metaphase I   **nondisjunction at anaphase I**   metaphase II   anaphase II

Figure 8-21 Nondisjunction. Of the two pairs of homologous chromosomes shown, one pair fails to separate at anaphase I of meiosis. The chromosome number changes in the gametes.

_____ 1. Reproduction in prokaryotes occurs primarily through the process known as
   (a) mitosis
   (b) cytokinesis
   (c) furrowing
   (d) fission

_____ 2. The genetic material (DNA) of eukaryotes is organized into
   (a) centrioles
   (b) spindles
   (c) chromosomes
   (d) microtubules

_____ 3. The process of cytoplasmic division is known as
   (a) meiosis
   (b) cytokinesis
   (c) mitosis
   (d) fission

_____ 4. The product of chromosome duplication is
   (a) two chromatids
   (b) two nuclei
   (c) two daughter cells
   (d) two spindles

_____ 5. The correct sequence of stages in _mitosis_ is
   (a) interphase, prophase, metaphase, anaphase, telophase
   (b) prophase, metaphase, anaphase, telophase
   (c) metaphase, anaphase, prophase, telophase
   (d) prophase, telophase, anaphase, interphase

_____ 6. During prophase, duplicated chromosomes
   (a) consist of chromatids
   (b) contain centromeres
   (c) consist of nucleoproteins
   (d) contain all of the above

_____ 7. During the S period of interphase
   (a) cell growth takes place
   (b) nothing occurs because this is a resting period
   (c) chromosomes divide
   (d) synthesis (or replication) of the nucleoproteins takes place

_____ 8. Chromatids separate during
   (a) prophase
   (b) telophase
   (c) cytokinesis
   (d) anaphase

_____ 9. Cell plate formation
   (a) occurs in plant cells but not in animal cells
   (b) usually begins during telophase
   (c) is a result of fusion of Golgi vesicles
   (d) is all of the above

_____ 10. Centrioles and a starburst cluster of spindle fibers would be found in
   (a) both plant and animal cells
   (b) only plant cells
   (c) only animal cells
   (d) none of the above

_____ 11. In meiosis, the number of chromosomes _____, while in mitosis, it _____.
   (a) is halved/is doubled
   (b) is halved/remains the same
   (c) is doubled/is halved
   (d) remains the same/is halved

_____ 12. The term "2n" means
   (a) the diploid chromosome number is present
   (b) the haploid chromosome number is present
   (c) chromosomes within a single nucleus exist in homologous pairs
   (d) both a and c

_____ 13. In higher animals, meiosis results in the production of
   (a) egg cells (ova)
   (b) gametes
   (c) sperm cells
   (d) all of the above

_____ 14. Recombination of alleles on nonsister chromatids occurs during
   (a) anaphase I
   (b) meiosis II
   (c) telophase II
   (d) crossing over

_____ 15. Alternative forms of genes are called
   (a) homologues
   (b) locus
   (c) loci
   (d) alleles

_____ 16. If both homologous chromosomes of each pair exist in the same nucleus, that nucleus is
   (a) diploid
   (b) unable to undergo meiosis
   (c) haploid
   (d) none of the above

_____ 17. DNA duplication occurs during
   (a) interphase
   (b) prophase I
   (c) prophase II
   (d) interkinesis

_____ 18. Nondisjunction
   (a) results in gametes with abnormal chromosome numbers
   (b) occurs at anaphase
   (c) results when homologues fail to separate properly in meiosis
   (d) is all of the above

_____ 19. Humans
   (a) don't undergo meiosis
   (b) have 46 chromosomes
   (c) produce gametes by mitosis
   (d) have all of the above characteristics

_____ 20. Gametogenesis in male animals results in
   (a) four sperm
   (b) one gamete and three polar bodies
   (c) four functional ova
   (d) a haploid ovum and three diploid polar bodies

EXERCISE 8

Mitosis, Meiosis, and Cytokinesis

POST-LAB QUESTIONS

Introduction

1. Distinguish among interphase, mitosis, and cytokinesis.

2. If a cell of an organism has 46 chromosomes before meiosis, how many chromosomes will exist in each nucleus after meiosis?

3. What basic difference exists between the life cycles of higher plants and higher animals?

4. In animals, meiosis results directly in gamete production, while in plants meiospores are produced. Where do the gametes come from in the life cycle of a plant?

5. How would you argue that meiosis is the basis for sexual reproduction in plants, even though the *direct* result is a spore rather than a gamete?

8.1 The Cell Cycle in Plant Cells: Onion Roots

6. If the chromosome number of a typical onion root tip cell is 16 before mitosis, what is the chromosome number of each newly formed nucleus after nuclear division has taken place?

7. In plants, what name is given to a region where mitosis occurs most frequently?

8. The cells in the following photomicrographs have been stained to show microtubules comprising the spindle apparatus. Identify the stage of mitosis in each and label the region indicated on (b). (Photos by Andrew S. Bajer.)

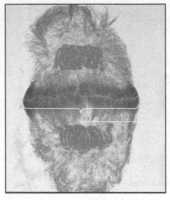

region? _____

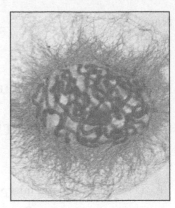

a stage? _____   **b** stage? _____   **c** stage? _____

8.2 The Cell Cycle in Animal Cells: Whitefish Blastula

9. Name two features of animal cell mitosis and cytokinesis you can use to distinguish these processes from those occurring in plant cells.

   a.

   b.

Food for Thought

10. Observe photomicrographs (**a**) and (**b**) below. Is (**a**) from a plant or an animal?

    Note the double nature of the blue "threads." Each individual component of the doublet is called a _____ _____. Is (**b**) from a plant or an animal? (Photos by Andrew S. Bajer.)

a plant or animal?

_____

structure?

_____

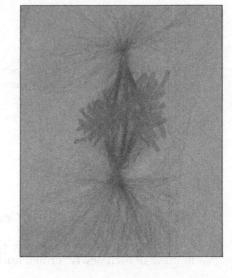

b plant or animal?

_____

11. Why do you suppose cytokinesis generally occurs in the cell's midplane?

12. Why must the DNA be duplicated during the S phase of the cell cycle, prior to mitosis?

13. What would happen if a cell underwent mitosis but not cytokinesis?

8.4 *Demonstrations of Meiosis Using Pop Beads*

14. Suppose one sister chromatid of a chromosome has the allele *H*. What allele will the other sister chromatid have? (Assume crossing over has not taken place.) _____

15. Suppose that two alleles on one homologous chromosome are *A* and *B*, and the other homologous chromosome's alleles are *a* and *b*.

    a. How many different genetic types of gametes would be produced *without* crossing over? _____

    b. What are the genotypes of the gametes? _____

    c. If crossing over were to occur, how many different genetic types of gametes could occur? _____

    d. List them. _____

16. Assume that you have built a homologous pair of *duplicated* chromosomes, one chromosome red and the other yellow. Describe or draw the appearance of two nonsister chromatids after crossing over.

17. Examine the meiotic diagram at right. Describe in detail what's wrong with it.

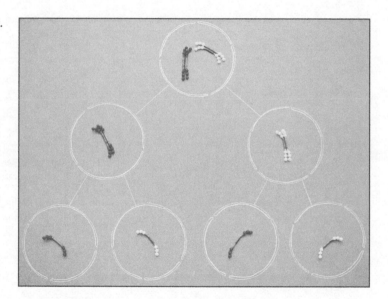

Food for Thought

18. From a genetic viewpoint, of what significance is fertilization?

Heredity, Nucleic Acids, & Biotechnology

OBJECTIVES

After completing this exercise, you will be able to

1. define *true-breeding, hybrid, monohybir cross, law of segregation, diploid, haploid, genotype, phenotype, dominant, recessive, complete dominance, homozygous, heterozygous, dihybrid cross, probability, chi-square test, DNA, RNA, purine, pyrimidine, principle of base pairing, replication, transcription, translation, codon, anticodon, peptide bond, gene, genetic engineering, recombinant DNA, plasmid, bacterial conjugation, bacterial transformation, plasmids, vectors, recombinant plasmids, genomic libraries, gene therapy, and genetic engineering;*

2. use sampling to determine phenotypic ratios of visible trait in the gametophytes of an F_1, C-fern hybrid;

3. observe sperm release and fertilization events that lead to an F_2 C-fern sporophyte generation;

4. form hypotheses about genotypic and phenotypic ratios in the F_2 C-fern sporophyte generation;

5. use a chi-square test to determine whether observed results are consistent with expected results;

6. identify the components of deoxyribonucleotides and ribonucleotides;

7. distinguish between DNA and RNA according to their structure and function;

8. describe DNA replication, transcription, and translation;

9. give the base sequence of DNA or RNA when presented with complementary strand;

10. identify a codon and anticodon on RNA models and describe the location and function of each;

11. give the base sequence of an anticodon when presented with that of a codon, and vice versa;

12. describe what is meant by the *one-gene, one-polypeptide hypothesis;*

13. describe the process of DNA recombination by bacterial conjugation;

14. explain the difference between DNA recombination by bacterial conjugation and the technique by which eukaryotic gene products are produced by bacteria;

15. explain how restriction enzymes and DNA ligase are used to insert a foreign gene into a plasmid;

16. diagram a typical plasmid used by scientists to transform bacteria;

17. state what can be harvested from cultures of transformed bacteria;

18. explain how bacterial transformation and similar procedures are used in health, and agriculture, and other industries.

INTRODUCTION

In 1866, an Austrian monk, Gregor Mendel, presented the results of painstaking experiments on the inheritance of the garden pea, but the scientific community ignored them, possibly because they didn't understand their significance. Now, more than a century later, Mendel's work seems elementary to modern-day geneticists, but its importance cannot be overstated. The principles generated by Mendel's pioneering experimentation are the foundation for the genetic counseling so important today to families with genetically based health disorders. They are also the framework for the modern research that is making inroads into treating diseases previously believed to be incurable. In this era of genetic engineering—the incorporation of foreign DNA into chromosomes of other species—it's easy to lose sight of the concepts underlying the processes that make it all possible. The initial experiments in this exercise take you through the basic processes fundamental to inheritance.

By 1900, Gregor Mendel had demonstrated patterns of inheritance, based solely on careful experimentation and observation. Mendel had no clear idea how the traits he observed were passed from generation to generation, although the seeds of that knowledge had been sown as early as 1869, when the physician-chemist Friedrich Miescher isolated the chemical substance of the nucleus. Miescher found the substance to be an acid with a large phosphorus content and named it "nuclein." Subsequently, nuclein was identified as **DNA,** short for **deoxyribonucleic acid.** Some 75 years would pass before the significance of DNA would be revealed.

Few would argue that the demonstration of DNA as the genetic material and the subsequent determination of its molecular structure are among the most significant discoveries of the twentieth century. Since the early 1950s, when James Watson and Francis Crick built on the discoveries of others before them to construct their first model of DNA, tremendous advances in molecular biology have occurred, many of them based on the structure of DNA. Today we speak of gene therapy and genetic engineering in household conversations. In the minds of some, these topics raise hopes for curing or preventing many of the diseases plaguing humanity. For others, thoughts turn to "playing with nature," undoing the deeds of God, or creating monstrosities that will wipe humanity off the face of the earth.

The final activities in this exercise will familiarize you with the basic structure of nucleic acids, their role in the cell, and the application of this knowledge in the science of biotechnology. Understanding the function of nucleic acids—both DNA and **RNA (ribonucleic acid)**—is central to understanding life itself. We hope you will gain an understanding that will allow you to form educated opinions concerning what science should do with its newfound technology.

9.1 Monohybrid Crosses

Garden peas have both male and female parts in the same flower and are able to self-fertilize. For his experiments, Mendel chose parental plants that were **true-breeding,** meaning that all self-fertilized offspring displayed the same form of a trait as their parent. For example, if a true-breeding purple-flowered plant self-fertilizes, all of its offspring will have purple flowers.

When parents that are true-breeding for *different* forms of a trait are crossed—for example, purple flowers and white flowers—the offspring are called **hybrids.** When only one trait is being studied, the cross is a **monohybrid cross.** We'll look first at monohybrid problems and crosses.

A. Experiment: Monohybrid Heredity in a Fern

A significant limitation of carrying out genetics experiments in the biology lab is that most take several months or years to collect relevant data. Even though Mendel could raise two generations of peas in a growing season, the experiments conducted in his garden plot often lasted several years. Two growing seasons were required to produce the corn monohybrid cross studied above. Fortunately, we can now look at inheritance in organisms with a much shorter life cycle. We'll use C-ferns to investigate a monohybrid cross.

Like all ferns, C-ferns have two independent life cycle phases: a structurally simple, haploid gametophyte and a more complex diploid sporophyte. A mature C-fern plant produces haploid spores via the process of meiosis. The spores germinate under suitable environmental conditions, and begin to divide mitotically to produce the gametophyte phase (Figure 9-1). This haploid phase develops very rapidly, with gametophytes maturing within 2 weeks.

At maturity, the gametophyte consists of a small (2 mm), simple, photoautrophic flattened structure with sex organs that produce *by mitosis* eggs in structures called archegonia and/or sperm in structures called antheridia. In the presence of water, flagellated sperm are discharged. The sperm are attracted to substances produced by the archegonia and swim toward the egg. Eventually one sperm fertilizes the egg, producing the first cell of the diploid sporophyte generation, the zygote.

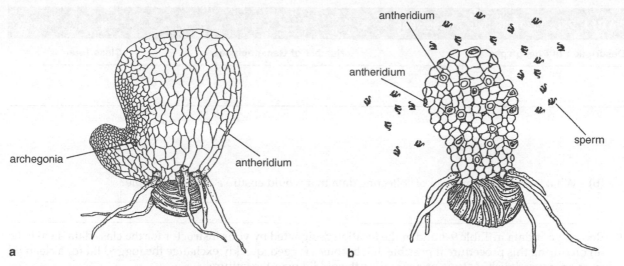

Figure 9-1 Mature C-fern gametophytes and gametes. (**a**) Hermaphroditic gametophytes produce both eggs and sperm and are somewhat heart-shaped. (**b**) Male gametophytes produce only sperm and appear tongue-shaped. (After University of Tennessee Research Foundation, 1998.)

The photoautotrophic sporophyte also develops rapidly, with roots and leaves visible within 1–2 weeks. The C-fern sporophytes reaches an ultimate height of 10–40+ cm. Spores are produced by meiosis in structures on the leaves, completing the life cycle.

A.1. Week 1—Observation of F_1 Hybrid Gametophytes *(About 30 min.)*

MATERIALS

Per student:

- 2-week-old C-fern culture in petri dish
- dissecting microscope
- sterile dH$_2$O
- sterile pipet
- marking pen
- calculator (optional)

Prior to this class, petri dishes with nutrient medium were inoculated with spores from an F_1 hybrid C-fern plant. The spores germinated and have grown into mature haploid gametophytes.

PROCEDURE

1. Observe the cultures under the dissecting microscope with the highest magnification possible and transmitted light (from below.) You can prevent your cultures from drying out from the heat of the microscope by leaving the lid on the culture as much as possible and by turning the light off or removing the culture to the lab bench when it is not being observed.

2. While you are observing the culture, tilt the lid up and use a sterile pipet to add 1–2 mL sterile distilled water. Lower the lid and tilt the plate back and forth to cover all of the gametophytes with water. Observe the release of swimming sperm from antheridia and their attempts to find and fertilize mature eggs within archegonia.

   Do all the gametophytes have the same phenotype? Describe any differences you observe.

   _____

   _____

   _____

   For this experiment, we will focus only on the larger, heart-shaped hermaphroditic gametophytes. Which of the phenotypes would you designate as a mutant? Why?

   _____

   _____

3. Take a random sample of the hermaphroditic gametophyte population by counting up to 50 individuals and tallying their phenotype.
   (a) Why is it important to take a random sample from the cultures?

   _____

TABLE 9-1 Gametophyte Phenotypes

Description of Phenotypes	Number of Gametophytes	Class Total

(b) What is a suitable method of collecting data that would ensure a random sample?

_____

_____

4. Record your data in Table 9-1 and in the location designated by your instructor for the class data. Leave the lid on during this procedure if possible. If it becomes fogged, quickly exchange the fogged lid for a clean lid from an unused dish. After scoring, replace the old lid over the culture.

5. When you've finished your observations, remove any excess water from the culture by lifting the lid slightly and pouring off the excess. Place the culture in the location designated by your instructor. Be sure the petri dish lid is in place.

(a) Are the plants in the culture dish haploid or diploid? _____

(b) What products will result from the fertilization events in the culture? _____
Will they be haploid or diploid? _____

(c) If the F_1 sporophyte is heterozygous for a single mutant trait, what genotypes will be present in the spores? _____
What genotypes will be present in the gametophytes? _____

(d) What is the expected ratio of genotypes? _____

(e) What is the approximate phenotypic ratio of the gametophytes? _____

(f) Can you determine the dominance relationships from the data in Table 9-1?

_____

_____

Biologists have assigned the designation *CP* to the single gene responsible for the two different phenotypes seen in this experiment. The dominant allele is thus designated *CP*, and the recessive allele *cp*.

(g) Predict the genetic outcome of the fertilizations taking place in the cultures by formulating a hypothesis to explain the inheritance of the trait. Indicate expected ratios of both the gametophyte and the F_2 generations.

_____

_____

A.2. Data Analysis (About 15 min.)

Gametes from true-breeding gametophyte parents (the P generation) combine to produce hybrid F_1 sporophyte fern plants. Meiosis within the F_1 plants produces spores. Each gametophyte you observe resulted from mitotic divisions of one of those spores. Their gametes will combine to produce the F_2 generation sporophytes.

In the space below, diagram the crosses involved in the F_1 and F_2 generations, indicating which generations/structures are haploid, and which are diploid.

You can now test your hypothesis concerning the method of trait inheritance to determine whether the data you collected support or do not support your model. Geneticists typically use the chi-square (χ^2) statistical test to determine whether experimentally obtained data are a satisfactory approximation of the expected data. In short, this test expresses the difference between expected (hypothetical) and observed (collected) numbers as a single value, χ^2. If the difference between observed and expected results is large, a large χ^2 results, while a small difference results in a small χ^2. Chi-square values are calculated according to the formula

$$\chi^2 = \sum \frac{(O - E)^2}{E}$$

where O = *observed* number of individuals
E = *expected* number of individuals
Σ = the sum of all values of $(O - E)^2/E$ for the various categories of phenotypes
Suppose 81 flowers are counted in a cross. Our hypothesis (expectation) is that three-fourths of them will be purple:

$$\frac{3}{4} \times 81 = 60.75$$

Similarly, we expect one-fourth to be white:

$$\frac{1}{4} \times 81 = 20.25$$

Suppose we actually count 64 purple flowers and 17 white flowers. Examine Table 9-2, noting how these values are used.

$$\chi^2 = \sum (0.174 + 0.522) = 0.696$$

TABLE 9-2 Calculations of Chi Square for Garden-Pea Monohybrid Cross

Phenotype	Genotype	O	E	$(O - E)$	$(O - E)^2$	$(O - E)^{2/E}$
Purple	$P\_$	64	60.75	3.25	10.56	0.174
White	pp	17	20.25	−3.25	10.56	0.522
Total		81	81	0		0.696

Now, how do we interpret the χ^2 value we found? Suppose the expected and observed values were identical. Then $\chi^2 = 0$. You might guess that a number very close to zero indicates close agreement between observed and expected and a large χ^2 value suggests that "something unusual" is taking place. The problem is that chance alone almost always causes small deviations between observed and expected results, *even when the hypothesis being tested is correct.*

When does the χ^2 value indicate that chance alone cannot explain the deviation? Geneticists generally agree on a probability value of 1 in 20 (or 5% = .05) as the lowest acceptable value derived from the χ^2 test. This number indicates that if the experiment is repeated many times, the deviations expected due to chance alone will be as large as or larger than those observed only about 5% or less of the time. Probabilities equal to or greater than .05 are considered to support the hypothesis, while probabilities lower than .05 do not support the hypothesis. Here we must consult a table of χ^2 values to make our decision (Table 9-3).

In our example, the χ^2 value is 0.696. Since this is a monohybrid problem with only two categories of possible outcomes (purple or white flowers), the number of degrees of freedom (n in the left-hand column of Table 9-3) is 1. Read across the table until you come to .05 and find the χ^2 value 3.84. Because 0.696, our calculated χ^2 value, is less than 3.841, it is likely that the variation in the observed and expected is the result of chance, and that our hypothesized outcome is correct. A value *greater than* 3.841, however, would indicate that chance alone cannot explain the deviation between observed and expected, and we would reject our hypothesis.

The term *degrees of freedom* requires further explanation. The number of degrees of freedom is always 1 *less* than the number of categories of possible outcomes. Thus, if you are dealing with a dihybrid problem with a ratio of 9:3:3:1 (four possible phenotypes), $n = 3$.

TABLE 9-3 Distribution of χ^2

| Degrees of Freedom, n | Probability of Obtaining a χ^2 Value as Large or Larger | | | |
	.10	.05	.01	.001
1	2.71	3.84	6.63	10.83
2	4.61	5.99	9.21	13.82
3	6.25	7.82	11.35	16.27
4	7.78	9.49	13.28	18.47

1. Transfer your individual or group data from the Totals columns in Table 9-1 to Table 9-4 and calculate χ^2.

TABLE 9-4 χ^2 Calculation from Gametophyte Data

Phenotype	Observed (*O*)	Expected (*E*)	(*O* – *E*)	(*O* – *E*)^2	(*O* – *E*)$^{2/E}$
Totals					$\chi^2 =$

2. Use Table 9-3 to determine the probability of obtaining this χ^2 value for the gametophyte data in Table 9-4. How many degrees of freedom are there? _____

Is your hypothesis supported or not supported? _____ If not, what might be changed in your hypothesis or in the experimental design?

_____

_____

A.3. Week 3—Observation of F$_2$ Sporophytes *(About 45 min.)*

MATERIALS

Per student:

- 4-week-old C-fern culture in petri dish
- dissecting microscope
- dissecting needle or toothpicks
- calculator (optional)

PROCEDURE

1. Examine your cultures with the dissecting microscope. Mutant and wild-type phenotypes are best observed using reflected light from the top or the side. Carefully observe the oldest leaves. Can you see mutant and wild-type phenotypes? _____

    Are the young sporophytes haploid or diploid? Why?

    _____

    _____

2. Sketch what you are observing, and label it with the following terms: gametophyte, sporophyte leaf, sporophyte root.

3. Take a random sample of the sporophyte population in a dish by counting up to 50 individuals and identifying their phenotype. You can remove the lid from the culture to do this. It may be easier to score the phenotype after gently and randomly pulling up individual sporophytes with a dissecting needle or toothpick and laying them out in a row on empty areas of the culture plate. Observe the largest leaf on each sporophyte and examine the differences carefully before recording data in Table 9-5 and in the location designated for class data.

4. Following scoring of phenotypes, place the lid back on the plate and return the culture to the designated location, or take it home so that you can observe it over the next several weeks to determine whether the phenotype of older sporophytes is apparent without use of a microscope.

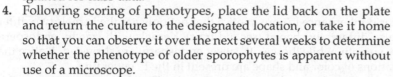

Description of Phenotypes	Number of Sporophytes	Class Total

5. Restate your hypothesis regarding the inheritance of the mutant and wild-type alleles, and your prediction of the genetic outcome in the F_2 sporophytes.

_____

_____

_____

6. Transfer your individual or class total data to Table 9-6 to calculate χ^2.

TABLE 9-6 χ^2 Calculation from Sporophyte Data

Phenotype	Observed (O)	Expected (E)[a]	(O − E)	(O − E)²	(O − E)²/E
Totals					$\chi^2=$

[a]This number should be based on the hypothesis you developed in this week's observations.

7. Use Table 9-3 to determine the probability of obtaining this χ^2 value for the sporophyte data in Table 9-6.
   How many degrees of freedom are there? _____
   What is the approximate probability? _____
   Is your hypothesis supported or not supported? _____
   Which allele is the dominant allele? _____ Which is recessive? _____
   If gametophytes had not expressed the phenotype, would you be able to form a hypothesis from observations of the gametophyte generation? _____ Why or why not? _____

_____

_____

Dihybrid Inheritance *(About 20 min.)*

In this section we'll examine cases in which two traits are involved: **dihybrid problems.**

MATERIALS

Per student group (table):

■ genetic corn ears illustrating a dihybrid cross

PROCEDURE

1. Examine the demonstration of dihybrid inheritance in corn. Notice that not only are the kernels two different colors (one trait), but they are also differently shaped (second trait). Kernels with starchy endosperm (the carbohydrate-storing tissue) are smooth, while those with sweet endosperm are shriveled. Notice that all *four* possible phenotypic combinations of color and shape are present in the F_2 generation.

   The P gene is involved in pigment production, with two alleles P and p. The S gene determines carbohydrate (sugar) storage, with two alleles S and s.

   Which genotypes of the parents produced the F_2 generation kernels? _____

2. Set up a Punnett square of this dihybrid cross:

   What is the predicted phenotypic ratio? _____

3. Count the number of kernels of each possible phenotype and record in Table 9-7. To increase your sample size, count three ears.

   Which traits seem dominant?

   _____

   Which traits seem recessive?

   _____

4. Calculate the actual phenotypic ratio you observed:

   _____

   Do your observed results differ from the expected results? _____

5. Use the chi-square test to determine if the deviation from the expected results can be accounted for by chance alone.

   Chi-square test results:_____

   _____

gametes of one parent

gametes of other parent

TABLE 9-7   Phenotypes in Dihybrid Corn Cross

	Number of Kernels with Phenotypes			
Ear	Yellow Smooth	Yellow Shriveled	Purple Smooth	Purple Shriveled
1				
2				
3				
Totals				

Genetics Problems. One of the best ways to solidify your understanding of different patterns of inheritance is to work genetics problems. Your instructor may assign you portions of Appendix 2, which is a collection of these useful and interesting problems.

Modeling the Structure and Function of Nucleic Acids and Their Products *(About 90 min.)*

MATERIALS

Per student pair or group:

■ DNA puzzle kit

Per lab room:

■ DNA model

PROCEDURE

Work in pairs or groups.

Note: Clear your work surface of everything except your lab manual and the DNA puzzle kit.

In this section, we are concerned with three processes: *replication, transcription,* and *translation.* But before we study these three *per se,* let's formulate an idea of the structure of DNA itself.

A. Nucleic Acid Structure

1. Obtain a DNA puzzle kit. It should contain the following parts:

    ■ 18 deoxyribose sugars
    ■ 9 ribose sugars
    ■ 18 phosphate groups
    ■ 4 adenine bases
    ■ 6 guanine bases
    ■ 6 cytosine bases

    ■ 4 thymine bases
    ■ 2 uracil bases
    ■ 3 transfer RNA (tRNA)
    ■ 3 amino acids
    ■ 3 activating units
    ■ ribosome template sheet

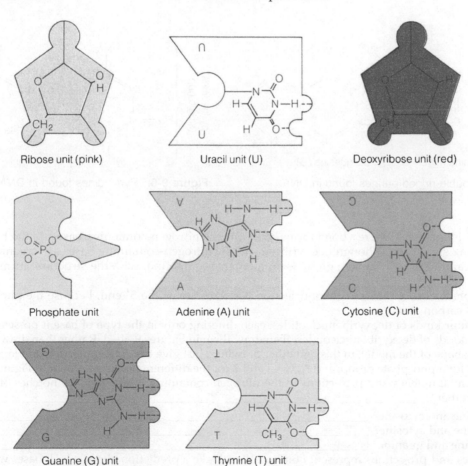

Ribose unit (pink)   Uracil unit (U)   Deoxyribose unit (red)

Phosphate unit   Adenine (A) unit   Cytosine (C) unit

Guanine (G) unit   Thymine (T) unit

2. Group the components into separate stacks. Select a single deoxyribose sugar, an adenine base (labeled A), and a phosphate, fitting them together as shown in Figure 9-2. This is a single nucleotide (specifically a *deoxy*ribonucleotide), a unit consisting of a sugar (deoxyribose), a phosphate group, and a nitrogen-containing base (adenine).

Let's examine each component of the nucleotide.

Deoxyribose (Figure 9-3) is a sugar compound containing five carbon atoms. Four of the five are joined by covalent bonds into a ring. Each carbon is given a number, indicating its position in the ring. (These numbers are read "1-prime, 2-prime," and so on. "Prime" is used to distinguish the carbon atoms from the position of atoms that are sometimes numbered in the nitrogen-containing bases.) This structure is usually drawn in a simplified manner, without actually showing the carbon atoms within the ring (Figure 9-4).

There are four kinds of nitrogen-containing bases in DNA. Two are **purines** and are double-ring structures. Specifically, the two purines are *adenine* and *guanine* (abbreviated A and G, respectively; Figure 9-5).

The other two nitrogen-containing bases are **pyrimidines,** specifically *cytosine* and *thymine* (abbreviated C and T, respectively). Pyrimidines are single-ring compounds, as shown in Figure 9-6.

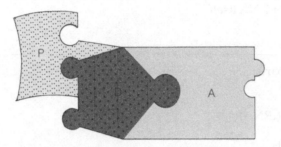

Figure 9-2 One deoxyribonucleotide.

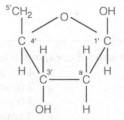

Figure 9-3 Deoxyribose.

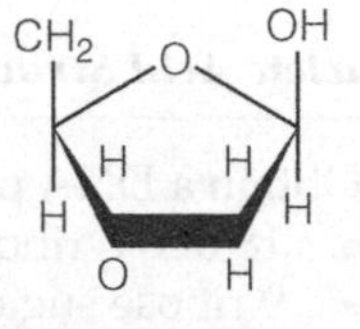

Figure 9-4 Simplified representation of deoxyribose.

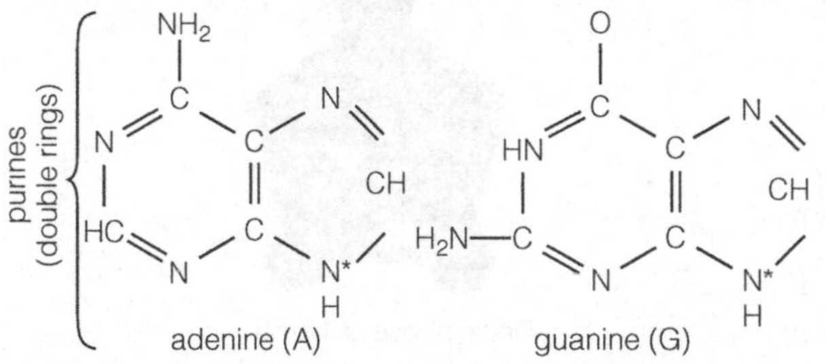

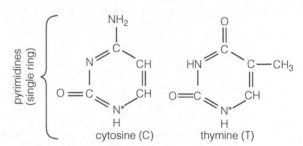

Figure 9-5 Double-ringed purines found in DNA.

Figure 9-6 Pyrimidines found in DNA.

The symbol * indicates where a bond forms between each nitrogen-containing base and the 1' carbon atom of the sugar ring structure. Although deoxyribose and the nitrogen-containing bases are organic compounds (they contain carbon), the phosphate group is an inorganic compound, with the structural formula shown in Figure 9-7.

The phosphate end of the deoxyribonucleotide is referred to as the 5' end, because the phosphate group bonds to the 5' carbon atom.

There are four kinds of deoxyribonucleotides, each differing only in the type of base it possesses. Construct the other three kinds of deoxyribonucleotides, then draw them in Figures 9-8b–d. Rather than drawing the somewhat complex shape of the model, in this and other drawings, just give the correct position and letters. Use D for deoxyribose, P for a phosphate group, and A, C, G, and T for the different bases (as shown in Figure 9-8a).

Note the small notches and projections in the nitrogen-containing bases. Will the notches of adenine and thymine fit together? _____

Will guanine and cytosine? _____
Will adenine and cytosine? _____
Will thymine and guanine? _____

The notches and projections represent bonding sites. Make a prediction about which bases will bond with one another. _____

Will a purine base bond with another purine? _____
Will a purine base bond with both types of pyrimidines? _____

3. Assemble the three additional deoxyribonucleotides, linking them with the adenine-containing unit, to form a nucleotide strand of DNA. Note that the sugar backbone is bonded together by phosphate groups. Your strand should appear like that shown in Figure 9-9.

4. Now assemble a second four-nucleotide strand, similar to that of Figure 9-9. However, this time make the base sequence T-A-C-G, from bottom to top. DNA molecules consist of *two* strands of nucleotides, each strand the *complement* of the other.

5. Assemble the two strands by attaching (bonding) the nitrogen bases of complementary strands. Note that the adenine of one nucleotide always pairs with the thymine of its complement; similarly, guanine always pairs with cytosine. This phenomenon is called the **principle of base pairing**. On Figure 9-10, attach letters to the model pieces indicating the composition of your double-stranded DNA model.

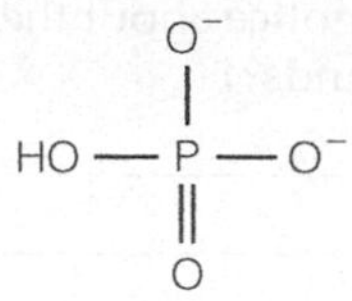

Figure 9-7 Phosphate group found in nucleic acids.

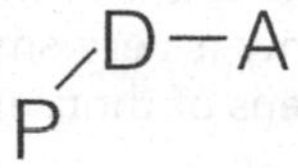

Deoxyribonucleotide containing adenine
a

Deoxyribonucleotide containing guanine
b

Deoxyribonucleotide containing cytosine
c

Deoxyribonucleotide containing thymine
d

Figure 9-8 Drawings of deoxyribonucleotides containing guanine, cytosine, and thymine.

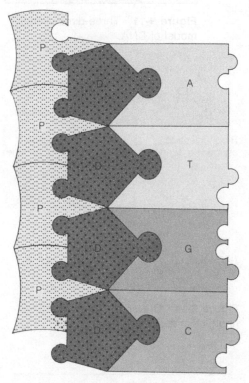

Figure 9-9 Four-nucleotide strand of DNA.

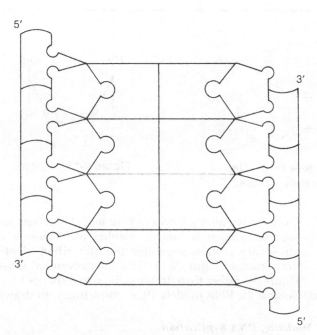

Figure 9-10 Drawing of a double strand of DNA.
Labels: A, T, G, C, D, P (all used more than once)

What do you notice about the *direction* in which each strand is running? (That is, are both 5′ carbons at the same end of the strands?)

_____

_____

_____

(Does the second strand of your drawing show this? It should.)

In life, the purines and pyrimidines are joined together by hydrogen bonds. Note again that the sugar backbone is linked by phosphate groups. Your model illustrates only a very small portion of a DNA molecule. The entire molecule may be tens of thousands of nucleotides in length!

6. Slide your DNA segment aside for the moment.

7. Examine the three-dimensional model of DNA on display in the laboratory (Figure 9-11). Notice that the two strands of DNA are twisted into a spiral-staircaselike pattern. This is why DNA is known as a *double helix*. Identify the deoxyribose sugar, nitrogen-containing bases, hydrogen bonds linking the bases, and the phosphate groups.

The second type of nucleic acid is RNA, short for ribonucleic acid. There are three important differences between DNA and RNA:

(a) RNA is a *single strand* of nucleotides.

(b) The sugar of RNA is **ribose.**

(c) RNA lacks the nucleotide that contains thymine. Instead, it has one containing the pyrimidine uracil (U) (see Figure 9-12).

Compare the structural formulas of ribose (Figure 9-13) and deoxyribose (Figure 9-4). How do they differ?

_____

Why is the sugar of DNA called *deoxy*ribose?

_____

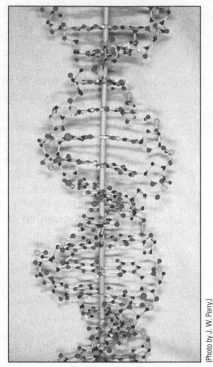

Figure 9-11 Three-dimensional model of DNA.

(Photo by J. W. Perry.)

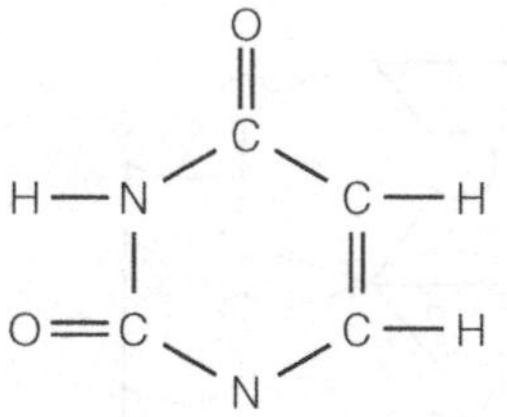

Figure 9-12 The pyrimidine uracil.

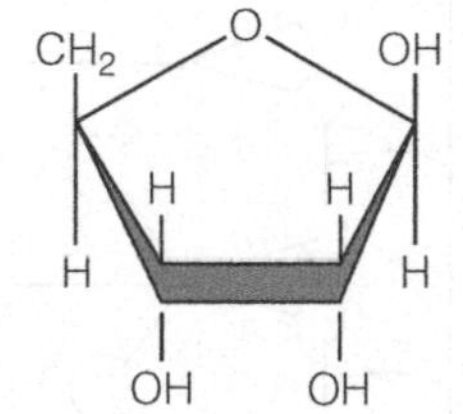

Figure 9-13 Ribose.

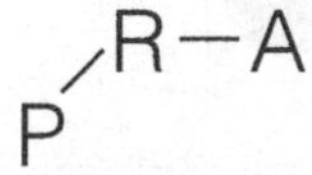

R—A
P
Ribonucleotide containing adenine
a

Ribonucleotide containing guanine
b

8. From the remaining pieces of your model kit, select four ribose sugars, an adenine, uracil, guanine, and cytosine, and four phosphate groups. Assemble the four **ribonucleotides** and draw each in Figure 9-14. (Use the convention illustrated in Figure 9-8 rather than drawing the actual shapes.)

Disassemble the RNA models after completing your drawing.

Ribonucleotide containing cytosine
c

Ribonucleotide containing uracil
d

Figure 9-14 Drawings of four possible ribonucleotides.

B. Modeling DNA Replication

DNA **replication** takes place during the S-stage of interphase of the cell cycle. Recall that the DNA is aggregated into chromosomes. Before mitosis, the chromosomes duplicate themselves so that the daughter nuclei formed by mitosis will have the same number of chromosomes (and hence the same amount of DNA) as did the parent cell.

Replication begins when hydrogen bonds between nitrogen bases break and the two DNA strands "unzip." Free nucleotides within the nucleus bond to the exposed bases, thus creating *two* new strands of DNA (as described below). The process of replication is controlled by enzymes called **DNA polymerases.**

1. Construct eight more deoxyribonucleotides (two of each kind) but don't link them into strands.
2. Now return to the double-stranded DNA segment you constructed earlier. Separate the two strands, imagining the zipperlike fashion in which this occurs within the nucleus.
3. Link the free deoxyribonucleotides to each of the "old" strands. When you are finished, you should have two double-stranded segments.

Note that one strand of each is the parental ("old") strand and the other is newly synthesized from free nucleotides. This illustrates the *semiconservative* nature of DNA replication. Each of the parent strands remains intact—it is *conserved*—and a new complementary strand is formed on it. Two "half-old, half-new" DNA molecules result.

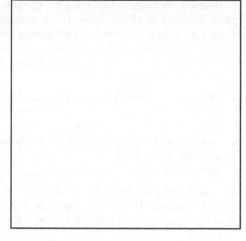

Figure 9-15 Drawing of two replicated DNA segments, illustrating their semiconservative nature.

4. Draw the two replicated DNA molecules in Figure 9-15, labeling the old and new strands. (Once again, use the convention shown in Figure 9-8.)

C. Transcription: DNA to RNA

DNA is an "information molecule" residing *within* the nucleus. The information it provides is for assembling proteins *outside* the nucleus, within the cytoplasm. The information does not go directly from the DNA to the cytoplasm. Instead, RNA serves as an intermediary, carrying the information from DNA to the cytoplasm.

Synthesis of RNA takes place within the nucleus by **transcription.** During transcription, the DNA double helix unwinds and unzips, and a single strand of RNA, designated **messenger RNA (mRNA),** is assembled using the nucleotide sequence of *one* of the DNA strands as a pattern (template). Let's see how this happens.

1. Disassemble the replicated DNA strands into their component deoxyribonucleotides.
2. Construct a new DNA strand consisting of nine deoxyribonucleotides. With the purines and pyrimidines pointing away from you, lay the strand out horizontally in the following base sequence: T-G-C-A-C-C-T-G-C
3. Now assemble RNA ribonucleotides complementary to the exposed nitrogen bases of the DNA strand. Don't forget to substitute the pyrimidine uracil for thymine.

What is the sequence from left to right of nitrogen bases on the mRNA strand?

_____

After the mRNA is synthesized within the nucleus, the hydrogen bonds between the nitrogen bases of the deoxyribonucleotides and ribonucleotides break.

4. Separate your mRNA strand from the DNA strand. (You can disassemble the deoxyribonucleotides now.) At this point, the mRNA moves out of the nucleus and into the cytoplasm.

By what avenue do you suppose the mRNA exits the nucleus? (*Hint:* Reexamine the structure of the nuclear membrane, as described in Exercise 3.)

_____

To *transcribe* means to "make a copy of." Is transcription of RNA from DNA the formation of an *exact* copy? _____ Explain.

_____

_____

You will use this strand of mRNA in the next section. Keep it close at hand.

D. Translation—RNA to Polypeptides

Once in the cytoplasm, mRNA strands attach to *ribosomes,* on which translation occurs. To *translate* means to change from one language to another. In the biological sense, **translation** is the conversion of the linear message encoded on mRNA to a linear strand of amino acids to form a polypeptide. (A *peptide* is two or more amino acids linked by a peptide bond.)

Translation is accomplished by the interaction of mRNA, ribosomes, and **transfer RNA (tRNA),** another type of RNA. The tRNA molecule is formed into a four-cornered loop. You can think of tRNA as a baggage-carrying molecule. Within the cytoplasm, tRNA attaches to specific free amino acids. This occurs with the aid of activating enzymes, represented in your model kit by the pieces labeled "glycine activating" or "alanine activating." The amino acid–carrying tRNA then positions itself on ribosomes where the amino acids become linked together to form polypeptides.

1. Obtain three tRNA pieces, three amino acid units, and three activating units.
2. Join the amino acids first to the activating units and then to the tRNA. Will a particular tRNA bond with *any* amino acid, or is each tRNA specific? _____
3. Now let's do some translating. In the space below, list the sequence of bases on the *messenger* RNA strand, starting at the left.
   (left, 3' end)_____(right, 5' end)

Translation occurs when a *three*-base sequence on mRNA is "read" by tRNA. This three-base sequence on mRNA is called a **codon.** Think of a codon as a three-letter word, read right (5') end to left (3') end. What is the order of the rightmost (first) mRNA codon? (Remember to list the letters in the *reverse* order of that in the mRNA sequence.)

   The first codon on the mRNA model is (5' end)_____ (3' end)

4. Slide the mRNA strand onto the ribosome template sheet, with the first codon at the 5' end.
5. Find the tRNA–amino acid complex that complements (will fit with) the first codon. The complementary three-base sequence on the tRNA is the **anticodon.** Binding between codons and anticodons begins at the P site of the 40s subunit (the smaller subunit) of the ribosome. The tRNA–amino acid complex with the correct anticodon positions itself on the P site.
6. Move the tRNA–amino acid complex onto the P site on the ribosome template sheet and fit the codon and anticodon together. In the boxes below, indicate the codon, anticodon, and the specific amino acid attached to the tRNA.

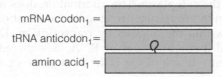

7. Now identify the second mRNA codon and fill in the boxes.

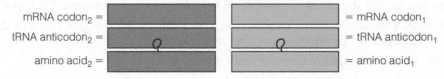

8. The second tRNA–amino acid complex moves onto the A site of the 40s subunit. Position this complex on the A site. An enzyme now catalyzes a condensation reaction, forming a **peptide bond** and linking the two amino acids into a dipeptide. (Water, HOH, is released by this condensation reaction.)
9. Separate amino acid₁ from its tRNA and link it to amino acid₂. (In reality, separation occurs somewhat later, but the puzzle doesn't allow this to be shown accurately; see below for correct timing.)

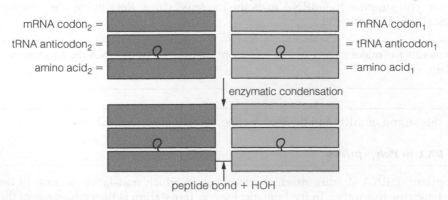

One tRNA–amino acid complex remains. It must occupy the A site of the ribosome in order to bind with its codon. Consequently, the dipeptide must move to the right.

10. Slide the mRNA to the right (so that tRNA$_2$ is on the P site) and fit the third mRNA codon and tRNA anti-codon to form a peptide bond, creating a model of a tripeptide. At about the same time that the second pep-tide bond is forming, the first tRNA is released from both the mRNA and the first amino acid. Eventually, it will pick up another specific amino acid.

What amino acid will tRNA$_1$ pick up? _____.

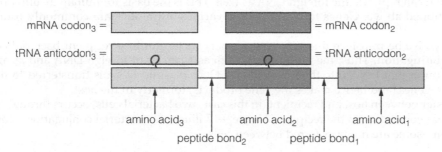

Record the tripeptide that you have just modeled. _____
You have created a short polypeptide. Polypeptides may be thousands of amino acids in length. As you see, the amino acid sequence is ultimately determined by DNA, because it was the original source of information.

Finally, let's turn our attention to the concept of a gene. A **gene** is a unit of inheritance. Our current understanding of a gene is that a gene codes for one polypeptide. This is appropriately called the **one-gene, one-polypeptide hypothesis.** Given this concept, do you think a gene consists of one, several, or many deoxyribonucleotides?

A gene probably consists of _____ deoxyribonucleotides.

Note: **Please disassemble your models and return them to the proper location.**

<table>
<tr><td>**9.4**</td><td>**Principles of Genetic Engineering: Recombination of DNA** *(About 15 min.)*</td></tr>
</table>

People suffering from Type 1 diabetes are unable to produce enough insulin, a hormone that is synthesized by the pancreas and that is instrumental in regulating the amount of blood sugar. Therapy for severe diabetes includes daily injections of insulin. Until recently, that insulin was extracted from the pancreas of slaughtered pigs and cows. With the advent of techniques commonly referred to as genetic engineering, human insulin is now pro-duced by bacteria. These organisms grow and reproduce rapidly, hence producing quantities of insulin en masse.

Genetic engineering is a convenient phrase to describe what is more properly called methods in recombinant DNA. **Recombinant DNA** is DNA into which a set of "foreign" nucleotides has been inserted. In the case of insulin production, researchers first located on human chromosomes the gene (set of nucleotides) that codes for insulin production. Once identified, the nucleo-tides were removed from the human DNA and inserted into the DNA within a bacterium. As this bacterial cell reproduces, each new generation contains the gene coding for human insulin. The bacteria produce the hormone, which is harvested and purified. Thus these recombinant bacteria are "insulin factories."

Bacteria have been exchanging genes with each other for millennia. In the process, new genetic strains of bacteria may be produced. The following demonstration will familiarize you with genetic recombination in bacteria; these principles are the basis for genetic engineering.

Two strains of the bacterium *Escherichia coli* will be used in this experiment:

■ Strain 1 carries a chromosomal gene that causes it to be resistant to the antibiotic drug streptomycin; it is susceptible to (killed by) another antibiotic, ampicillin. (See Figure 9-16a.)

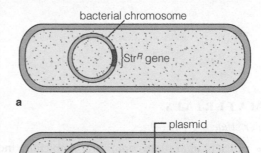

Figure 9-16 Genetic components of two strains of *Escherichia coli* bacteria.
(a) Strain 1 has a gene for resistance to the antibiotic streptomycin (designated Str^R) on its chromosome. **(b)** Strain 2 has a gene for resistance to the antibiotic ampicillin (designated Amp^R) on a plasmid within the cell.

■ Strain 2 is resistant to the antibiotic drug ampicillin but susceptible to streptomycin; the gene for resistance to ampicillin is located on a small extrachromosomal (that is, not on its chromosome) loop of DNA called a **plasmid.** (See Figure 9-16b.)

Plasmids contain relatively few genes compared to the bacterial chromosome. Like chromosomal DNA, plasmids can replicate. Insertion of a nonbacterial DNA segment (set of nucleotides) results in the formation of a hybrid plasmid that can replicate the foreign DNA as well. This is the basis for human insulin production within bacteria, as mentioned above. Genes for resistance to various antibiotics are commonly found on plasmids as well.

Plasmids may also be transferred from a host (donor) bacterium to a recipient bacterial cell by a process called **bacterial conjugation.** Thus, the plasmid acts both as a carrier of foreign DNA and as an agent (vector) for the introduction of that DNA into the recipient cell. Once plasmid DNA is transferred to the recipient, the recipient bears the genes (and hence makes the gene products) formerly in the host.

Plasmid transfer between host and recipient (in this case, two bacterial cells) occurs through a bridge formed by the host cell that connects it to the recipient. Figure 9-17 illustrates bacterial conjugation. Note that genes on the bacterial chromosome are not transferred between cells.

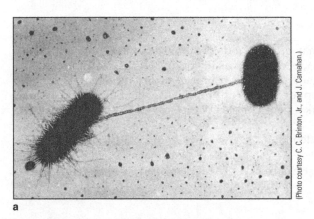

(Photo courtesy C. C. Brinton, Jr. and J. Carnahan.)

a

Figure 9-17 (a) Conjugation between two bacteria. (b) Plasmid gene transfer during bacterial conjugation. The bacterial chromosome is not shown.

nicked plasmid   conjugation tube

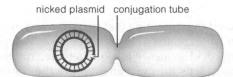

1 A conjugation tube has already formed between a donor and a recipient cell. An enzyme has nicked the donor's plasmid.

2 DNA replication starts on the nicked plasmid. The displaced DNA strand moves through the tube and enters the recipient cell.

3 In the recipient cell, replication starts on the transferred DNA.

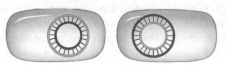

4 The cells separate from each other; the plasmids circularize.

b

(After Starr, 2000.)

MATERIALS

Per student group (4):

■ nutrient agar plate containing ampicillin, incubated with *E. coli* Strain 1 on one half, and *E. coli* Strain 2 on the other half
■ nutrient agar plate containing streptomycin, incubated with *E. coli* Strain 1 on one half, and *E. coli* Strain 2 on the other half

■ nutrient agar plate containing *both* ampicillin and streptomycin, incubated with a mating solution of both *E. coli* Strain 1 and *E. coli* Strain 2 spread across the plate (The mating solution is designed to allow bacterial conjugation to occur.)
■ demonstration nutrient agar plates showing growth of *E. coli* Strain 1 and *E. coli* Strain 2

PROCEDURE

1. Examine the plates for growth. (Your instructor will provide demonstration plates for you to examine so that you can recognize bacterial growth.)

2. Record your observations in Table 9-8, using a + to indicate the growth of bacteria, a – to indicate absence of growth. A bacterium that is sensitive to (killed by) an antibiotic will be unable to grow on nutrient media containing that specific antibiotic.

TABLE 9-8 Bacterial Growth on Antibiotic-Containing Plates

Growth of Bacteria on Nutrient Agar Containing

Strain of *E. coli*	Ampicillin	Streptomycin	Ampicillin plus Streptomycin
Strain 2 (donor)			
Strain 1 (recipient)			
"Mating mixture"			

3. Discard your plates in the designated location.

Was Strain 2 susceptible *or* resistant to ampicillin? _____

To streptomycin? _____

Was Strain 1 susceptible *or* resistant to ampicillin? _____

To streptomycin? _____

Make a conclusion about the presence and location of a gene in each of the two strains of *E. coli* for resistance to each of the antibiotics.

Strain 2

_____

_____

Strain 1

_____

_____

Make a conclusion about what happened when the two strains were mixed together. Incorporate your observations concerning antibiotic resistance into your conclusion.

_____

_____

_____

9.5   Bacterial Transformation on the Internet

Bacterial transformation occurs when bacterial cells pick up **plasmids** (tiny circles of double-stranded DNA) that carry additional genes. They occur naturally and are the smallest gene-carrying vehicles (**vectors**) that can enter, replicate, and express themselves within bacteria. Scientists use manipulated plasmids to introduce foreign genes into bacteria. They cut the plasmids open with **restriction enzymes,** allow foreign DNA fragments to fill the breach, and seal the cut ends with another enzyme, **DNA ligase** (Figure 9-18).

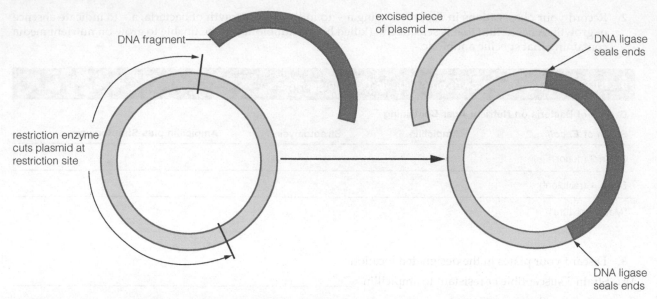

restriction enzyme cuts plasmid at restriction site

DNA fragment

excised piece of plasmid

DNA ligase seals ends

DNA ligase seals ends

Figure 9-18 The insertion of a DNA fragment into a plasmid.

The new plasmids (**recombinant plasmids**) and similar vectors are used in biotechnology to carry foreign genes into bacteria where (1) replication of the plasmid makes many copies of the foreign gene or (2) expression of the duplicated genes produces a high concentration of gene product. In either case, the plasmids or the gene products can be harvested later. A typical plasmid ring used for making additional recombinant plasmids contains a gene that confers antibiotic resistance and one or more **restriction sites** where one or more foreign genes can be inserted into the plasmid (Figure 9-19).

When recombinant plasmids are made, plasmids and DNA fragments are mixed in a solution that contains all the substances needed to open the plasmid and splice in a DNA fragment. However, not all plasmids are successfully cut and spliced back together with a DNA fragment. Similarly, when the recombinant plasmids are mixed with bacteria, typically only 1 in 1000 bacteria is transformed—that is, gets a plasmid. Thus as part of any procedure involving transformation, there must be ways to identify a transformed bacterium and to determine whether its plasmid has the inserted gene of interest.

Bacterial transformation plays a central role in many areas of biotechnology, including basic research, health, agriculture, and industry. One use for plasmids is to store cloned DNA fragments in **genomic libraries.** Like words and sentences in library books, such plasmids contain DNA sequences and genes. Now that the human genome is known, other vectors such as viruses may be commonly used to transform human cells to counter diseases caused by defective genes (**gene therapy**). Biotechnology industry harvests proteins produced by transformed bacteria cultured in large steady-state incubators for use in many fields. **Genetic engineering** (tinkering with the genes) of naturally and artificially selected plants has already produced enhanced cereals, fruits, and vegetables, which among other things offer better nutrition, longer storage lives, and resistance to disease and insects. As you probably know, all of this is, and will continue to be, increasingly controversial.

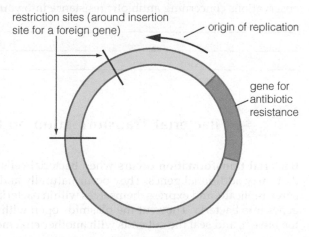

restriction sites (around insertion site for a foreign gene)

origin of replication

gene for antibiotic resistance

Figure 9-19 Typical plasmid.

MATERIALS

Per student:

- Internet access

PROCEDURE

1. Visit the National Center for Biotechnology Information. The current address is
   http://www.ncbi.nlm.nih.gov/
2. Type "plasmid" into the site's search engine and click the GO button. What kind of databases are checked by *Entrez*, The Life Sciences Search Engine?

   _____

   _____

   _____

Scientists from all over the world use these databases to report, share, and gather genetic information.

3. Go to several of the databases and open some of the files. Don't try to understand anything in particular; scan through the material instead. You should be able to find uses and gene maps of plasmids.
4. Now search for the plasmid pUC19, which is often used to store DNA fragments for clonal and subclonal surveys.
5. Visit the Human Genome Project. The current address is
   http://www.ornl.gov/TechResources/Human_Genome/home.html

What is the Human Genome Project?

   _____

   _____

   _____

You could literally take months to explore this site. Be sure to visit the Education and Ethical, Legal, Social Issues (ELSI) sections.

6. Another great site is managed by the Cold Spring Harbor Laboratory:
   http://www.cshl.org/

Be sure to search for plasmids and visit the Dolan DNA Learning Center. What is Barbara McClintock famous for in the field of plant genetics?

   _____

   _____

   _____

7. Since the preceding experiment involves bacterial growth, use your favorite search engine to find a video or cell cam of bacterial growth. The following site typically has excellent examples:
   http://www.cellsalive.com/
8. Use your favorite search engine to find five sites that have something to do with plasmids or bacterial transformation. List them below with a brief summary of their contents.

   http:// _____

   http:// _____

   http:// _____

   http:// _____

   http:// _____

_____ 1. In a monohybrid cross
(a) only one trait is being considered
(b) the parents are always dominant
(c) the parents are always heterozygous
(d) no hybrid is produced

_____ 2. The genetic makeup of an organism is its
(a) phenotype
(b) genotype
(c) locus
(d) gamete

_____ 3. An allele whose expression is completely masked by the expression or effect of its allelic partner is
(a) homologous
(b) homozygous
(c) dominant
(d) recessive

_____ 4. The physical appearance and physiology of an organism, resulting from interactions of its genetic makeup and its environment, is its
(a) phenotype
(b) hybrid vigor
(c) dominance
(d) genotype

_____ 5. When both dominant and recessive alleles are present within a single nucleus, the organism is ____ for the trait.
(a) diploid
(b) haploid
(c) homozygous
(d) heterozygous

_____ 6. A Punnett square is used to determine
(a) probable gamete genotypes
(b) possible parental phenotypes
(c) possible parental genotypes
(d) possible genetic outcomes of a cross

_____ 7. The gametophyte of a fern is
(a) haploid
(b) photoautotrophic
(c) a structure that produces eggs and/ or sperm
(d) all of the above

_____ 8. A chi-square test is used to
(a) determine if experimental data adequately matches what was expected
(b) analyze a Punnett square
(c) determine parental genotypes producing a given offspring genotype
(d) determine if a trait is dominant or recessive

_____ 9. Possible gamete genotypes produced by an individual of genotype $PpDd$ are
(a) Pp and Dd
(b) all $PpDd$
(c) PD and pd
(d) PD, Pd, pD, and pd

_____ 10. If you can roll your tongue.
(a) you have at least one copy of the dominant allele T
(b) you have two copies of the recessive allele t
(c) you must be male
(d) you are haploid

_____ 11. The individuals responsible for constructing the first model of DNA structure were
(a) Wallace and Watson
(b) Lamarck and Darwin
(c) Mendel and Meischer
(d) Crick and Watson

_____ 12. Deoxyribose is
(a) a five-carbon sugar
(b) present in RNA
(c) a nitrogen-containing base
(d) one type of purine

_____ 13. A nucleotide may consist of
(a) deoxyribose or ribose
(b) purines or pyrimidines
(c) phosphate groups
(d) all of the above

_____ 14. Which of the following is consistent with the principle of base pairing?
(a) purine-purine
(b) pyrimidine-pyrimidine
(c) adenine-thymine
(d) guanine-thymine

_____ 15. Nitrogen-containing bases between two complementary DNA strands are joined by
(a) polar covalent bonds
(b) hydrogen bonds
(c) phosphate groups
(d) deoxyribose sugars

_____ 16. The difference between deoxyribose and ribose is that ribose
(a) is a six-carbon sugar
(b) bonds only to thymine, not uracil
(c) has one more oxygen atom than deoxyribose has
(d) is all of the above

_____ 17. Replication of DNA
   (a) takes place during interphase
   (b) results in two double helices from one
   (c) is semiconservative
   (d) is all of the above

_____ 18. Transcription of DNA
   (a) results in formation of a complementary strand of RNA
   (b) produces two new strands of DNA
   (c) occurs on the surface of the ribosome
   (d) is semiconservative

_____ 19. An anticodon
   (a) is a three-base sequence of nucleotides on tRNA
   (b) is produced by translation of RNA
   (c) has the same base sequence as does the codon
   (d) is the same as a gene

_____ 20. Bacterial plasmids
   (a) are the only genetic material in bacteria
   (b) may carry genes for antibiotic resistance
   (c) may be transferred between bacteria during the process of conjugation
   (d) are both b and c

_____ 21. Transformed cells contain
   (a) transformants
   (b) new genes
   (c) bacteria
   (d) none of these choices

_____ 22. A tiny loop of double-stranded DNA describes a
   (a) bacterium
   (b) DNA ligase
   (c) plasmid
   (d) antibiotic

_____ 23. Enzymes used to insert a foreign gene into a plasmid include
   (a) DNA ligase
   (b) vectors
   (c) restriction enzymes
   (d) both a and c

_____ 24. A molecule or anything else that transports new genes into a cell is called a
   (a) vector
   (b) plasmid
   (c) restriction enzyme
   (d) bacterium

_____ 25. In genomic libraries:
   (a) naturally and artificially selected plants are genetically engineered
   (b) DNA fragments are stored in plasmids
   (c) a virus is used to transform human cells
   (d) none of the above are done

_____ 26. Characteristics of genetically engineered plants and plant products include:
   (a) better nutrition
   (b) longer storage lives
   (c) resistance to disease and insects
   (d) all of the above

EXERCISE 9

Heredity, Nucleic Acids, & Biotechnology

POST-LAB QUESTIONS

9.1 Monohybrid Crosses

1. Explain the implications of Mendel's law of segregation as it applies to the distribution of alleles in gametes.

2. Assume that production of hairs on a plant's leaves is controlled by a single gene with two alleles H (dominant) and h (recessive). Hairy leaves are dominant to smooth (nonhairy) leaves.
   a. Name the genotype(s) of a smooth-leaved plant. _____
   b. Name the genotype(s) of a hairy-leaved plant. _____
   c. What are the possible genotypes of gametes produced by the smooth-leaved plant? _____
   d. What are the possible genotypes of gametes produced by the hairy-leaved plant? _____

3. Non-true-breeding hairy-leaved plants are crossed with smooth-leaved plants.
   a. What genotypic and phenotypic ratios would you expect for the potential offspring? _____
   b. Suppose you perform such a cross, collect data, and do a chi-square test to aid in data analysis. How many degrees of freedom would there be? _____
   c. Suppose your chi-square value is very large (>25). What does this indicate about your experiment and/or hypothesis?

4. What genotypic ratio would you expect in the gametophyte generation of C-ferns produced by F_1 spores if two traits on separate chromosomes were being followed?

5. Were dominant and recessive traits observed equally in both gametophytes and sporophytes of C-ferns? How did you determine which character was dominant and which was recessive?

9.2 Dihybrid Inheritance

6. Suppose you have two traits controlled by genes on separate chromosomes. If sexual reproduction occurs between two heterozygous parents, what is the genotypic ratio of all possible gametes?

7. Explain the usefulness of the chi-square test.

8. Suppose students in previous semesters had removed some of the corn kernels from the genetic corn ears before you counted them. What effect would this have on your results?

9. Assume that one allele is completely dominant over the other for the following questions.
    a. Two individuals heterozygous for a *single* trait have children. What is the expected phenotypic ratio of the possible offspring? _____
    b. Two individuals heterozygous for *two* traits have children. What would be the expected phenotypic ratio of the possible offspring? _____
    c. Crossing two individuals heterozygous for two traits results in the same phenotypic ratio as for a single trait. Are the genes for these two traits on separate chromosomes or on the same chromosome? Explain your answer. (Remember that the gene for each trait is located at a locus, a physical region on the chromosome.)

10. How does probability differ from actuality?

9.3 *Modeling the Structure and Function of Nucleic Acids and Their Products*

11. The following diagram represents some of the puzzle pieces used in this section.
    a. Assembled in this form, do they represent a(an) amino acid, base, portion of messenger RNA, or deoxyribonucleotide?

    b. Justify your answer.

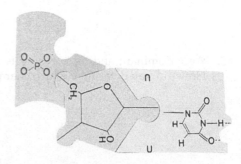

12. Why is DNA often called a *double helix*?

13. State the following ratios.
    a. Guanine to cytosine in a double-stranded DNA molecule: _____
    b. Adenine to thymine: _____

14. Define the following terms.
    a. replication

    b. transcription

    c. translation

    d. codon

    e. anticodon

15. What does it mean to say that DNA replication is *semiconservative*?

16. a. If the base sequence on one DNA strand is ATGGCCTAG, what is the sequence on the other strand of the helix?

    b. If the original strand serves as the template for transcription, what is the sequence on the newly formed RNA strand?

17. a. What amino acid would be produced if *transcription* takes place from a nucleotide with the three-base sequence ATA? _____

    b. A genetic mistake takes place during *replication* and the new DNA strand has the sequence ATG. What is the three-base sequence on an RNA strand transcribed from this series of nucleotides? _____

    c. Which amino acid results from this codon?

9.4 Principles of Genetic Engineering: Recombination of DNA

18. a. What is a plasmid?

    b. How are plasmids used in genetic engineering?

19. How does bacterial conjugation differ from the process by which eukaryotic gene products are produced by bacteria?

Food for Thought

20. What kinds of molecules are present in bacterial cells in addition to the DNA you might have isolated in Section 14.1?

9.5 Bacterial Transformation on the Internet

21. Define bacterial transformation.

22. What does the term *vector* mean in the field of biotechnology?

23. Draw and label a typical plasmid used by scientists to transform bacteria.

24. How are restriction and DNA ligase enzymes used to insert a foreign gene into a plasmid?

25. When is a plasmid considered a recombinant plasmid?

26. What two general types of products can be harvested from cultures of transformed bacteria?

27. Name the two elements needed for procedures similar to bacterial transformation to be useful in human gene therapy.

Food for Thought

28. When this exercise was written, the media campaign in support of golden rice was being launched. Search the Internet for the roots of the controversy over this genetically enhanced plant. Briefly describe the benefits of, and the concerns over, its introduction to the rice-growing areas of the world.

Evolutionary Agents

After completing this exercise, you will be able to

1. define *evolutionary agent, natural selection, fitness, directional selection, stabilizing selection, disruptive selection, gene flow, divergence, speciation, mutation, genetic drift, bottleneck effect, founder effect;*

2. determine the allele frequencies for a gene in a model population;

3. calculate expected ratios of phenotypes based on Hardy–Weinberg proportions;

4. describe the effects of nonrandom mating, natural selection, migration, genetic drift, and mutation on a model population;

5. describe the effects of different selection pressures on identical model populations;

6. identify the level at which selection operates in a population;

7. describe the impact of the founder effect on the genetic structure of populations.

INTRODUCTION

Heredity itself cannot cause changes in the frequencies of alternate forms of the same gene (alleles). If certain conditions are met, then the proportions of genotypes that make up a population of organisms should remain constant generation after generation according to the equation that describes the Hardy–Weinberg equilibrium:

$$p^2 + 2pq + q^2 = 1.0 \text{ (for two alleles)}$$

If p is the frequency of one allele, and q is the frequency of the other allele, then

$$p + q = 1.0$$

For example, if two alleles for coat color exist in a population of mice and the allele for white coats is present 70% of the time, then the alternate allele (black) must be present 30% of the time.

The Hardy–Weinberg equation describes the proportions of phenotypes present in succeeding generations, as long as conditions don't change. In our example, since $p = .7$, we would expect 49% (p^2) of the mice in our population to be homozygous for white coats. Thus, 42% ($2pq$) would have one of each allele and would appear gray if neither allele is dominant (that is, both alleles would have equal expression in the phenotype). What percentage of our population is homozygous for black coats? _____ %

In nature, however, the frequencies of genes in populations change over time. Natural populations never meet all of the conditions assumed for Hardy–Weinberg equilibrium. *Evolution is a process resulting in changes in the genetic makeup of populations through time;* therefore, factors that disrupt Hardy–Weinberg equilibrium are referred to as **evolutionary agents.** This exercise demonstrates the effect of these agents on the genetic structure of a simplified model population.

10.1   Natural Selection *(75 min.)*

The populations you will work with are composed of colored beads. White beads in our model represent individuals that are homozygous for the white allele ($C^W C^W$). Red beads are individuals homozygous for the red allele ($C^R C^R$), and pink beads are heterozygotes ($C^W C^R$). These beads exist in "ponds"—plastic dishpans filled with smaller beads. Counts of white, pink, and red individuals in the pond are made after straining the beads through a sieve. The smaller beads pass through the mesh, which retains the larger beads. Figure 10-1 shows the initial experimental setup.

When the individuals are recovered, the frequencies of the color alleles are determined using the Hardy–Weinberg equation. The alleles in our population are codominant. Thus, each white bead contains two white alleles; each pink bead, one white and one red allele; and each red bead, two red alleles. The total number of color alleles in a population of 40 individuals is 80. If such a population contains 10 white beads, 20 pink beads, and 10 red beads the frequency of the white allele is

$$p = \frac{(2 \times 10) + 20}{80} = .5$$

Because $p + q = 1.0$, the frequency of the red allele (q) must also be .5 if there are only two color alleles in this population.

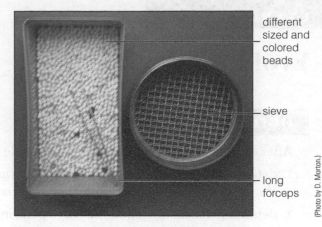

different sized and colored beads

sieve

long forceps

(Photo by D. Morton.)

Figure 10-1 Experimental setup used to demonstrate natural selection.

MATERIALS

Per student group (4):

- plastic dishpan (12″ × 7″ × 2″)
- 50 large (10-mm-diameter) white beads
- 50 large red beads
- 50 large pink beads
- 4000 small (8-mm diameter) white beads
- 4000 small red beads (optional)

- ruler with a cm scale
- pair of long forceps
- coarse sieve (9.5 mm)
- scientific calculator

Per lab room:

- clock with a second hand

PROCEDURE

A. Experiment: Natural Selection Acting Alone

Natural selection disturbs the Hardy–Weinberg equilibrium by discriminating between individuals with respect to their ability to produce young. Those individuals that survive to reproduce perpetuate more of their genes in the population. These individuals exhibit greater **fitness** than do those who leave no offspring or fewer offspring.

This experiment addresses the hypothesis that *individuals are more likely to survive and reproduce when their coloration makes it easier to hide from predators in the environment.*

1. Work in groups of four with each group member assuming one of the following roles: Predator, Data Recorder/Timer, Calculator, or Caretaker. Predators search for prey. Data Recorders/Timers record numerical results and time events. Calculators use a scientific calculator to crunch numbers as needed. Caretakers look after and manipulate the experimental setup.
2. Create a white pond by filling a dishpan with small white beads to a depth of about 5 cm and establish an initial population by mixing into the pond 10 large white beads, 10 large red beads, and 20 large pink beads. The Predator will prey on the large beads, removing as many as possible in a limited amount of time. The survivors will then reproduce the next generation and predation will begin again. This cycle will be repeated several times. Make a prediction as to the changing frequency in the population of the red allele over time and write it in the Prediction (Selection Alone) row of Table 10-3.
3. Search the pond for prey (large beads) and, using a pair of long forceps, remove as many of them as possible in 30 seconds.
4. Strain the pond with the sieve, and count the number of large white, pink, and red beads. Record the totals in the After row under Initial in Table 10-1.
5. Calculate the frequencies of the white (p) and red (q) alleles remaining in the population after selection and record them in First Generation after selection of Table 10-2. For example, if 6 white, 8 pink, and 8 red beads remain, the frequency of the white allele is

$$p = \frac{(2 \times 6) + 8}{44} = .45$$

TABLE 10-1 Large-Bead Counts Before and After Four Rounds of Simulated Predation

Population	White Beads	Pink Beads	Red Beads	Total Beads
Initial				
Before	10	20	10	40
After	_____	_____	_____	_____
Second Generation				
Before	_____	_____	_____	50
After	_____	_____	_____	_____
Third Generation				
Before	_____	_____	_____	50
After	_____	_____	_____	_____
Fourth Generation				
Before	_____	_____	_____	50
After	_____	_____	_____	_____

TABLE 10-2 Allele and Genotype Frequencies Due to Selection by Simulated Predation

Population	p	q	p^2	$2pq$	q^2
Initial	.5	.5	.25	.5	.25
First generation after selection					
Second generation after selection					
Third generation after selection					
Fourth generation after selection					

6. Using the new values for allele frequencies, calculate genotype frequencies for homozygous white (p^2), heterozygous pink ($2pq$), and homozygous red (q^2) individuals, and record them in Table 10-2. For example, if p now equals .45, the frequency of homozygous white individuals is

$$p^2 = (.45)^2 = .20$$

Assuming that 50 individuals comprise the next and succeeding generations (maximum number of individuals the pond can sustain), calculate the number of white, pink, and red individuals needed to create the population of a new pond and record these numbers in the Before row under Second Generation in Table 10-1. Here and in future calculations to generate new numbers of individuals, round up or down to the nearest whole number. For example, if $p^2 = .20$, the number of white beads needed is

$$p^2 \times 50 = .20 \times 50 = 10 \text{ white beads}$$

Using these numbers, construct a new pond.

7. Repeat steps 2–6 for three more rounds, filling in the remaining rows in Tables 10-1 and 10-2. When you are finished, copy the frequency of the red allele from Table 10-1 to Table 10-3 (Selection Alone column) and plot this data in Figure 10-2.

TABLE 10-3 Frequency of Red Allele (q) Due to Selection and Migration

Prediction (Selection Alone):

Prediction (Selection and Migration):

Generation	Selection Alone	Selection and Migration
1		
2		
3		
4		

Conclusion (Selection Alone):

Conclusion (Selection and Migration):

8. Write your conclusion as to your prediction in the Conclusion (Selection Alone) row of Table 10-3.
9. If you had started with a pond filled with small red beads as a background, how would the frequency of the red allele change?

_____

_____

_____

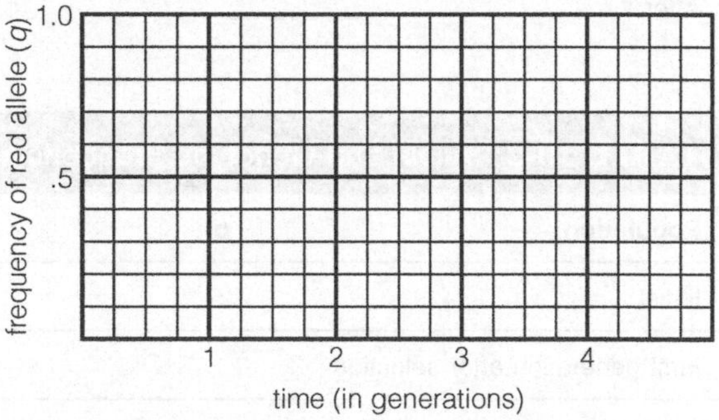

Figure 10-2 Effects of predation on allele frequencies.

10. Selection that favors one extreme phenotype over the other and causes allele frequencies to change in a predictable direction is known as **directional selection.** When selection favors an intermediate phenotype rather than one at the extremes, it's known as **stabilizing selection.** Selection that operates against the intermediate phenotype and favors the extreme ones is called **disruptive selection.** Which kind of selection is illustrated by simulated predation of white, pink, and red beads in a white pond? Explain why you made this choice.

_____

_____

_____

It is important to realize that selection operates on the entire phenotype so that the overall fitness of an organism is based on the result of interactions of thousands of genes.

11. If two identical populations (such as the mix of beads described in step 2) inhabited different environments (such as red and white ponds), how would the frequency of the color genes in each pond compare after a large number of generations?

_____

_____

_____

As two populations become genetically different through time (**divergence**), individuals from these populations can lose the ability to interbreed. If this happens, two species form from one ancestral species. This process is called **speciation.**

The frequencies of alleles in a population also change if new organisms immigrate and interbreed, or when old breeding members emigrate. **Gene flow** due to *migration* may be a powerful force in evolution. This activity demonstrates its effect.

1. Establish an initial population as in the previous section, step 2.
2. Begin selection as before, *except* add five new red beads to each generation before the new allele frequencies are determined. These beads represent migrants from a population where the red allele confers greater fitness. This experiment addresses the hypothesis that *gene flow resulting from the migration of a significant number of individuals into a population undergoing predation affects the change in allele frequencies expected from selection alone.* Write your prediction as to how the change in frequency of the red allele will be affected in the Conclusion (Selection and Migration) row of Table 10-3.
3. For each generation, record the frequencies of the red allele obtained with both selection and migration in Table 10-3.
4. How does migration influence the effectiveness of selection in this example?

   _____

   _____

   _____

5. Write your conclusion as to your prediction in the Conclusion (Selection and Migration) row of Table 10-3.
6. How would migration have influenced the change in gene frequencies if white instead of red individuals had entered the population?

   _____

   _____

   _____

Gene flow keeps local populations of the same species from becoming more and more different from each other. Things that serve as barriers to gene flow can accelerate the production of new species. Migration can also introduce new genes into a population and produce new genetic combinations. Imagine the result of a black allele being introduced into our model population and the new heterozygotes (perhaps gray and dark red) it would produce.

Note: If time is short, any or all of the remaining sections may be done as thought experiments, i.e., doing it in your head rather than actually setting it up.

10.2   Mutation *(About 15 min.)*

Another way new genetic information enters a population is through **mutation.** This usually represents an actual change in the information encoded by the DNA of an organism. As such, most mutations are harmful and will be eliminated by natural selection. Nevertheless, mutations do provide the raw material for evolution.

MATERIALS

Per student group (4):

- small bowl
- 10 large white beads
- 10 large red beads

- 20 large pink beads
- 1 large gray bead
- scientific calculator

PROCEDURE

1. Establish an initial population by placing 10 large white beads, 10 large red beads, and 20 large pink beads in a small bowl (without the small beads).
2. For the sake of expediency, establish a new generation by one group member picking, without looking, 20 large beads from the bowl. Replace one white bead with a gray bead. This represents a mutation in a gamete that one parent contributed to this generation.
3. Calculate the allele frequencies of the new generation, including the frequency of the new color allele (*r*). Record them in Table 10-4.

TABLE 10-4 Change in Allele Frequencies Due to Mutation

Population	p	q	r
Initial	.5	.5	
New generation with mutation			

4. Three alleles are present ($p + q + r = 1.0$), so the Hardy–Weinberg equation is expanded to $p^2 + 2pq + q^2 + 2pr + 2qr + r^2 = 1.0$, and in addition to white, pink, and red phenotypes, we now have gray, dark red, and potentially, in subsequent generations, black. If the next generation contains 50 individuals, how many offspring of each phenotype would you expect? Use Table 10-5 to calculate these numbers.

TABLE 10-5 Numbers of Each Phenotype Two Generations After a Single Mutation

Color	Genotype	Frequency	× 50	Number of Individuals
White	p^2			
Pink	$2pq$			
Red	q^2			
Gray	$2pr$			
Dark red	$2qr$			
Black	r^2			

Imagine a population made up of individuals in these proportions. What effect will natural selection have on these phenotypes in a white pond?

_____

_____

_____

How could conditions change to favor the selection of the rare black allele?

_____

_____

10.3 Genetic Drift (About 30 min.)

Chance is also a factor that results in shifts in gene frequencies over several generations (**genetic drift**). This is primarily due to the random aspects of reproduction and fertilization. Genetic drift is often a problem for small populations in that they can lose much of their genetic variability. In very small populations, chance can eliminate an allele from a population, such that p becomes 0 and the other allele becomes fixed ($q = 1.0$). This loss of genetic variation due to a small population size is known as the **bottleneck effect.**

MATERIALS

Per student group (4):

- small bowl
- 10 large white beads
- 10 large red beads

- 20 large pink beads
- scientific calculator

PROCEDURE

1. As in the previous section, place 10 large white beads, 10 large red beads, and 20 large pink beads in a small bowl. Listed in Table 10-6 are the expected allele frequencies for color in this population given all individuals participate in reproduction.

TABLE 10-6 Allele Frequencies Produced by Genetic Drift

| | | Actual Frequency in | |
	Expected Frequency	Small Cluster	Large Cluster
n			
p	.5		
q	.5		

2. Establish a cluster of reproductively lucky individuals by a group member choosing, without looking, 10 beads from the bowl.
3. In the second column of Table 10-6, record the allele frequencies present in this cluster.
4. Now replace the 10 beads you removed in step 2. Select beads at random again, but this time select 30 beads representing a larger cluster of reproductively lucky individuals.
5. Calculate the allele frequencies for this larger cluster and record them in the third column of Table 10-6.
6. Compare the allele frequencies in the three columns of Table 10-6. Sometimes chance determines whose gametes contribute to the next generation. What effect does the size of the number of individuals participating in reproduction have on gene flow to the next generation?

_____

_____

_____

7. Another way in which chance affects allele frequencies in a population is when migrants from old populations establish new populations. To model this effect, choose at random six individuals from an initial population to represent the migrants.
8. Move these individuals to a new unoccupied pond. (It is not necessary to actually set up a new pond for this demonstration. Use your imagination.)
9. Now calculate the allele frequencies in the new pond and record them in Table 10-7. How do they compare with the frequencies that characterized the pond from which these migrants came?

_____

_____

_____

The genetic makeup in future generations in the new population will more closely resemble the six migrants than the population from which the migrants came. This effect is known as the **founder effect.** The founder effect may not be an entirely random process because organisms that migrate from a population may be genetically different from the rest of the population to begin with. For example, if wing length in a population of insects is variable, we might expect insects with longer wings to be better at founding new populations because they can be carried farther by winds.

TABLE 10-7 Allele Frequencies in a Founder Population

	p	*q*
Initial Population	.5	.5
Founder Population		

10.4 Nonrandom Mating *(About 15 min.)*

Hardy–Weinberg equilibrium is also disturbed if individuals in a population don't choose mates randomly. Some members of a population may show a strong preference for mates with similar genetic makeups. This activity models this effect.

MATERIALS

Per student group (4):

- small bowl
- 10 large white beads
- 10 large red beads

- 20 large pink beads
- scientific calculator

PROCEDURE

1. Establish an initial population as in Section 10.3.
2. Assume that individuals will mate only with individuals of the same color. Arbitrarily assign sex to every bead so there are equal numbers of males and females in each color group.
3. If each pair of beads produces four offspring, record the number individuals with the same phenotype present in the next generation in Table 10-8. Remember that the pink pairs will produce one red, one white, and two pink individuals on average.

TABLE 10-8   Phenotype Changes Due to Nonrandom Mating Color

	Number in	
Color	Initial Generation	Next Generation
White	10	
Pink	20	
Red	10	

4. Calculate the genotype frequencies in this generation, record them in Table 10-9, and compare these with the frequencies in the initial generation.

TABLE 10-9   Genotype Frequency Changes Due to Nonrandom Mating

Genotype Frequency	Initial Generation	Next Generation
p^2		
$2pq$		
q^2		

5. What happens to the frequency of the heterozygote genotype in subsequent generations?

_____

_____

_____

_____ 1. If all conditions of Hardy–Weinberg equilibrium are met,
   (a) allele frequencies move closer to .5 each generation
   (b) allele frequencies change in the direction predicted by natural selection
   (c) allele frequencies stay the same
   (d) all allele frequencies increase

_____ 2. If a population is in Hardy–Weinberg equilibrium and $p = .6$,
   (a) $q = .5$
   (b) $q = .4$
   (c) $q = .3$
   (d) $q = .16$

_____ 3. Natural selection operates directly on
   (a) the genotype
   (b) individual alleles
   (c) the phenotype
   (d) color only

_____ 4. The process that discriminates between phenotypes with respect to their ability to produce offspring is known as
   (a) natural selection
   (b) gene flow
   (c) genetic drift
   (d) migration

_____ 5. Two populations that have no gene flow between them are likely to
   (a) become more different with time
   (b) become more alike with time
   (c) become more alike if the directional selection pressures are different
   (d) stay the same unless mutations occur

_____ 6. A process that results in individuals of two populations losing the ability to interbreed is referred to as
   (a) stabilizing selection
   (b) fusion
   (c) speciation
   (d) differential migration

_____ 7. Two ways in which new alleles can become incorporated in a population are
   (a) mutation and genetic drift
   (b) selection and genetic drift
   (c) selection and mutation
   (d) mutation and gene flow

_____ 8. If a new allele appears in a population, the Hardy–Weinberg formula
   (a) cannot be used because no equilibrium exists
   (b) can be used but only for two alleles at a time
   (c) can be used by lumping all but two phenotypes in one class
   (d) can be expanded by adding more terms

_____ 9. A shift from expected allele frequencies, resulting from chance, is known as
   (a) natural selection
   (b) genetic drift
   (c) mutation
   (d) gene flow

_____ 10. Genetic drift is a process that has a greater effect on populations that
   (a) are large
   (b) are small
   (c) are not affected by mutation
   (d) do not go through bottlenecks

EXERCISE 10

Evolutionary Agents

POST-LAB QUESTIONS

10.1 Natural Selection

1. What effect does increasing gene flow between two populations have on their genetic makeup?

2. How can selection cause two populations to become different with time?

3. Describe how the effects of directional selection can be offset by gene flow.

10.2 Mutation

4. In addition to mutation, what other mechanism allows for new genetic information to be introduced into a population? Explain your answer.

5. What is the fate of most new mutations?

6. If a population has three codominant color alleles, how many phenotypes are possible?

10.4 Nonrandom Mating

7. What effects can nonrandom mating exert on a population?

Food for Thought

8. What two evolutionary agents are most responsible for decreases in genetic variation in a population?

9. If a population has three color alleles and one is dominant over the other two, how many phenotypes are possible?

10. In humans, birth weight is an example of a characteristic affected by stabilizing selection. What does this mean to the long-term average birth weight of human babies? How might the increasing number of Caesarean sections be affecting this characteristic?

Evidences of Evolution

After completing this exercise, you will be able to

1. define *evolution, fossil, natural selection, population, species, fitness, mass extinction, adaptive radiation, anthropologist, hominid;*

2. explain how natural selection operates to alter the genetic makeup of a population;

3. describe the general sequence of evolution of life forms over geologic time;

4. recognize primitive and advanced characteristics of skull structure of human ancestors and relatives;

5. describe the evolutionary relationships among *Dryopithecus, Australopithecus afarensis,* and other australopith species, and the various *Homo* species.

INTRODUCTION

Have you ever marveled over the diversity of animals in a zoo? Lions, antelopes, zebras, giraffes, elephants, and chimpanzees are all mammals, as are we humans. All mammals living today, and many forms that are extinct, descended from a mammalian ancestor who lived more than 200 million years ago. The process by which this incredible diversity of mammals (and all the other living and extinct species) came to exist is called **evolution,** the focus of this exercise.

Evolution, the process that results in changes in the genetic makeup of populations of organisms through time, is the unifying framework for the whole of biology. Less than 200 years ago, it seemed obvious to most people that living organisms had not changed over time—that dandelions looked like dandelions and humans looked like humans, year after year, generation after generation, without change. As scientists studied the natural world more closely, however, evidence of change and the relatedness of all living organisms emerged from geology and the fossil record, as well as from comparative morphology, developmental patterns, and biochemistry.

Charles Darwin (and, independently, Alfred Russel Wallace) postulated the major mechanism of evolution to be **natural selection**—the difference in survival and reproduction that occurs among individuals of a population that differ in one or more alleles. (See Exercise 8 for the definition of *allele*.) A **population** is a group of individuals of the same species occupying a given area. A **species** is one or more populations that closely resemble each other, interbreed under natural conditions, and produce fertile offspring.

Individuals in a population that have particular combinations of alleles better adapted to their environment survive and produce more offspring than individuals with different genetic makeups. The offspring have received alleles that also make their chances for survival and reproduction greater, and so it goes, generation after generation. The population evolves as some traits become more common and others decrease or disappear over time.

Many people think of evolution in historic terms—as something that produced the dinosaurs or the Galapagos finches Darwin studied but that no longer operates in today's world. Remember, though, that *the process of genetic change over time continues today* and is a dominant force shaping the living organisms of our planet.

The scientific evidence for evolution is overwhelming, although scientists continue to debate the exact mechanisms by which natural selection and other evolutionary agents change allelic frequencies. In this exercise, you will observe the effects of natural selection in a living plant population; consider the time over which evolution has occurred; and examine the fossil record for evidence of large-scale trends and change among our human ancestors and relatives.

Natural selection discriminates among individuals with respect to their ability to produce offspring. Those individuals that survive and reproduce will perpetuate more of their alleles in the population. These individuals are said to exhibit greater **fitness** than those who leave no or fewer offspring.

You will examine individuals of two different populations of dandelions. Dandelions are interesting because, despite their showy yellow flowers, they reproduce primarily asexually. Instead of the sexual reproduction that typically occurs in flowers, dandelions form seeds without fertilization. Thus, each dandelion is essentially genetically (and physically) identical to its parent plant.

One set of dandelion seedlings in your laboratory was grown from seeds collected from an area that has endured frequent, close mowing for many years. The second group of seedlings was grown from seeds collected from plants mowed infrequently, if at all. Your instructor will describe in more detail the conditions under which the parent plants of your seedlings have grown.

This experiment tests the hypothesis that *dandelion populations grown for generations under different mowing regimes will have different growth forms*. Write a prediction regarding this experiment in Table 11-2. What would you predict the dandelion plants grown from seeds produced in mowed and unmowed areas to look like?

MATERIALS

Per student group (4):

- flat of dandelion seedlings, labeled Mowed or Unmowed
- trowel or large spoon
- metric ruler
- knife or single-edged razor blade
- calculator

Per lab room:

- several dishpans half-filled with water
- paper towels

PROCEDURE

1. Observe the flats of young dandelion plants. Note and describe any general differences in appearance between those grown from seeds of unmowed plants versus those grown from seeds of mowed plants.

   _____

   _____

   _____

2. Take one flat of plants to your lab bench. Remove the plants carefully from the flat with a trowel or spoon. Be careful to remove all the root system but do not damage or lose root material. Wash each plant gently in the dishpan provided, *not in the sink,* to remove the residual soil. Blot the plants dry with the paper towels.

3. Cut each plant at the region separating the root from the leafy shoot, keeping each root portion with its corresponding shoot. See Figure 11-1.

4. Measure shoot length to the nearest millimeter as the length between the cut surface and the tip of the longest leaf. Record this measurement in Table 11-1.

5. Measure root length to the nearest millimeter as the length between the cut surface and the tip of the longest tap root. Sometimes it's hard to tell where the tap root ends, but try to avoid measuring branch roots. See Figure 11-1. Record this data in Table 11-1.

6. For each plant, divide the shoot length by the root length to calculate the shoot-to-root ratio. Record the ratio in Table 11-1.

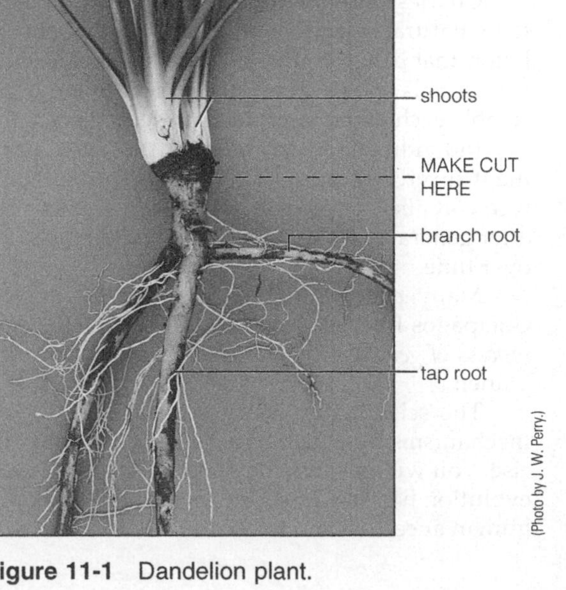

Figure 11-1   Dandelion plant.

---

*Adapted from an exercise by Thomas Hilbish and Minnie Goodwin, University of South Carolina.

TABLE 11-1	Measurements of Dandelions (Mowed or Unmowed? _____)		
Plant	Shoot Length (mm)	Root Length (mm)	Shoot-to-Root Ratio
1			
2			
3			
4			
5			
Total number of plants =			Average =

7. Graph the distribution of the shoot-to-root ratio of each plant in your flat in Figure 11-2.

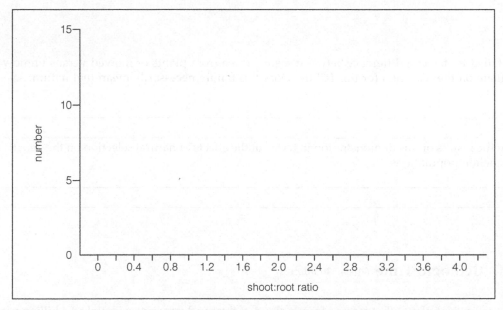

Figure 11-2 Distribution of shoot-to-root ratios of length measurements of dandelions.

8. Calculate the average shoot-to-root ratio of your plants by adding the shoot-to-root ratios of all the individual plants and dividing that number by the total number of plants.

9. Your instructor has drawn two graphs like that in Figure 11-2, one for mowed plants and one for unmowed plants. Pool your data with those of your classmates who have measured plants of the same type and enter that pooled data in Table 11-2 and on the corresponding graph. Calculate the average shoot-to-root ratio of the pooled data for each set of plants.

Are there noticeable differences in the distribution of shoot-to-root ratios between descendants of dandelions from intensely mowed versus unmowed areas? _____

If so, what are those differences?

_____

_____

_____

Do the results support your hypothesis? Write a conclusion accepting or rejecting the hypothesis in Table 11-2.

TABLE 11-2 Class Shoot-to-Root Ratios for Mowed and Unmowed Plants		
Prediction:		
Group	**Mowed**	**Unmowed**
1		
2		
3		
4		
5		
6		
7		
8		
Average		
Conclusion:		

If you failed to detect a difference between seeds grown from plants of mowed versus unmowed populations, speculate on the reason(s) for this failure. Does this failure necessarily mean that natural selection is not occurring?

_____

_____

Explain the results of this demonstration in terms of the effects of natural selection on the genetic makeup of the two dandelion populations.

_____

_____

11.2 Geologic Time *(About 30 min.)*

Our earth is an ancient planet that formed from a cloud of dust and gas approximately 4.6 billion years ago.

Life formed relatively quickly on the young planet, and the first cells emerged in the seas by 3.8 billion years ago. For more than a billion years, the only living organisms on earth were prokaryotic bacteria and bacterialike cells. Eventually, though, more complex, eukaryotic organisms evolved. The seas were colonized by single-celled organisms first, followed by multicelled and colonial plants and animals (Figure 11-3). Only much later did life move onto the land.

Throughout the history of life on earth, there have been many episodes of **mass extinctions** (catastrophic global events in which major groups of species are wiped out) followed by **adaptive radiation** (in which a lineage fills a wide range of habitats in a burst of evolutionary activity). For example, scientists postulate that a huge asteroid impacted the earth about 65 million years ago. The resulting environmental destruction caused a mass extinction in which nearly all the dinosaurs and marine organisms became extinct. After this extinction, the early mammals underwent an adaptive radiation and filled the habitats that had previously been occupied by dinosaurs.

Geologic time, "deep time," is difficult for humans, with our recent evolution and 75-year life spans to grasp. In this section, you will construct a geologic time line to help you gain perspective on the almost incomprehensible sweep of time over which evolution has operated.

MATERIALS

Per student pair:

- one 4.6-m rope or string
- meter stick and/or metric ruler
- masking tape
- calculator

PROCEDURE

1. Obtain a 4.6-m length of rope or string. The string represents the entire length of time (4.6 billion years) since the earth was formed. On this time line, you will attach masking-tape labels to mark the points when the events below occurred. Measure from the starting point with the meter stick or metric ruler.
2. Locate and mark with tape on the time line the following events (bya = billion years ago, mya = million years ago, ya = years ago):

4.6 bya	Formation of earth
3.8 bya	First living prokaryotic cells
2.5 bya	Oxygen-releasing photosynthetic pathway
1.2 bya	Origin of eukaryotic cells
580 mya	Origin of multicellular marine plants and animals
500 mya	First vertebrate animals (marine fishes)
435 mya	First vascular land plants and land animals
345 mya	Great coal forests; amphibians and insects undergo great adaptive radiation
240 mya	Mass extinction of nearly all living organisms
225 mya	Origin of mammals, dinosaurs; seed plants dominate
135 mya	Dinosaurs reach peak; flowering plants arise
65 mya	Mass extinction of dinosaurs and most marine organisms
63 mya	Flowering plants, mammals, birds, insects dominate
6 mya	First hominid (upright walking) human ancestors
2.5 mya	Human ancestors (*Homo*) using stone tools, forming social life
200,000 ya	Evolution of *Homo sapiens*
6400 ya	Egyptian pyramids constructed
2000 ya	Peak of Roman Empire
225 ya	Signing of the Declaration of Independence
In A.D. 1969	Humans step on earth's moon

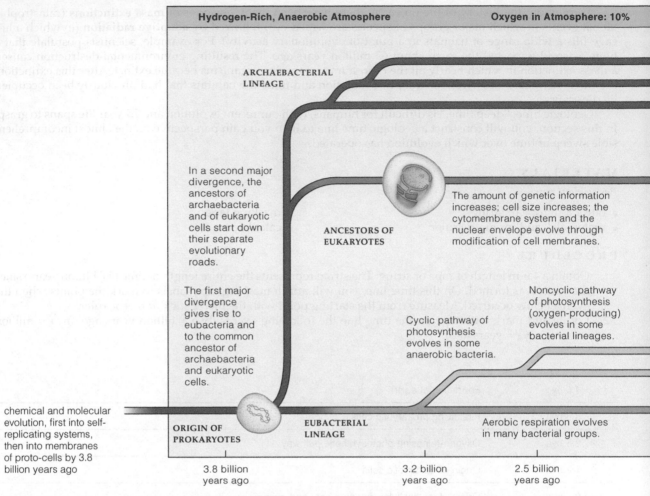

Figure 11-3 An evolutionary tree of life that reflects mainstream thinking about the connections among major lineages and origins of some eukaryotic organelles. (After Starr, 2000.)

Over what proportion of the earth's history were there only single-celled living organisms? _____
Over what proportion of the earth's history have multicelled organisms existed? _____
Over what proportion of the earth's history have mammals been a dominant part of the fauna? _____
Over what proportion of the earth's history have modern humans existed? _____

11.3   The Fossil Record and Human Evolution *(About 30 min.)*

The fossil record provides us with compelling evidence of evolution. Most fossils are parts of organisms, such as shells, teeth, and bones of animals, or stems and seeds of plants. These parts are replaced by minerals to form stone or are surrounded by hardened material that preserves the external form of the organism.

Anthropologists, scientists who study human origin and cultures, are piecing together the story of human evolution through the study of fossils, as well as through comparative biochemistry and anatomy. The fossil record of our human lineage is fragmentary at times and generates much discussion and differing interpretations. However, scientists agree that **hominids,** all species on the evolutionary branch leading to modern humans, arose in Africa between 10 and 5 million years ago from the same genetic line that also produced gorillas and chimpanzees.

Many anatomical changes occurred in the course of evolution from hominid ancestor to modern humans. Arms became shorter, feet flattened and then developed arches, and the big toe moved in line with the other toes. The legs moved more directly under the pelvis. These features allowed upright posture and bipedalism (walking on two legs), and they also allowed the use of hands for tasks other than locomotion.

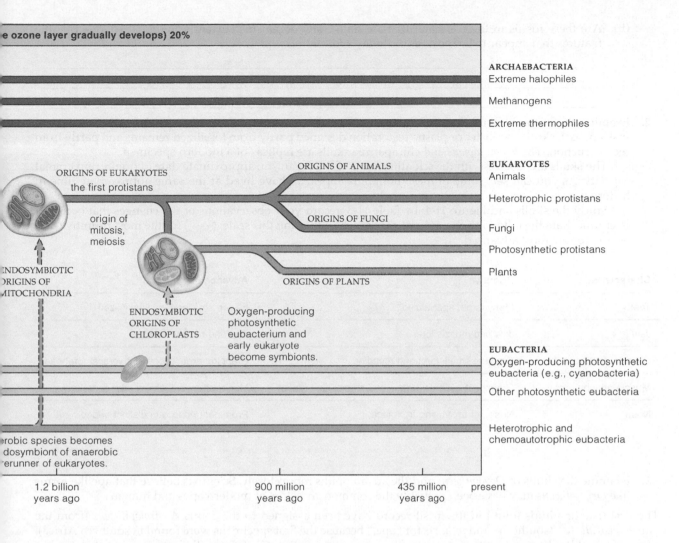

e ozone layer gradually develops) 20%

ARCHAEBACTERIA
Extreme halophiles

Methanogens

Extreme thermophiles

EUKARYOTES
Animals

Heterotrophic protistans

Fungi

Photosynthetic protistans

Plants

EUBACTERIA
Oxygen-producing photosynthetic
eubacteria (e.g., cyanobacteria)

Other photosynthetic eubacteria

Heterotrophic and
chemoautotrophic eubacteria

ORIGINS OF EUKARYOTES
the first protistans

ORIGINS OF ANIMALS

origin of
mitosis,
meiosis

ORIGINS OF FUNGI

ENDOSYMBIOTIC
ORIGINS OF
MITOCHONDRIA

ENDOSYMBIOTIC
ORIGINS OF
CHLOROPLASTS

ORIGINS OF PLANTS

Oxygen-producing
photosynthetic
eubacterium and
early eukaryote
become symbionts.

erobic species becomes
dosymbiont of anaerobic
erunner of eukaryotes.

1.2 billion
years ago

900 million
years ago

435 million
years ago

present

Still other anatomical changes were associated with the head. In this section, you will observe some of the evolutionary trends associated with skull structure by studying reproductions of the skulls of several ancestors and relatives of modern humans.

MATERIALS

Per laboratory room:

- 1 set of skull reproductions and replicas, including *Dryopithecus africanus*, *Australopithecus afarensis*, *Australopithecus africanus*, *Paranthropus robustus*, *Homo habilis*, *Homo erectus*, *Homo neandertalensis*, *Homo sapiens*, and chimpanzee (*Pan troglodytes*)

- collection of fossil plants and animals (optional)
- guide to fossils (optional)
- wall chart of human evolution (optional)

PROCEDURE

1. If they are available, examine the fossil collection and field guide to fossils. Note which fossils are forms that lived in the seas and which lived on land.
    (a) Do any of the fossils resemble organisms now living? If so, describe the similarities between fossil and current forms.

    _____

    _____

    _____

(b) Are there fossils in the collection that are unlike any organisms currently living? If so, describe the features that appear to be most unlike today's forms.

_____

_____

_____

2. Examine the hominid skulls in the laboratory. Note that these skulls are not actual fossils themselves. Instead, each skull is a plaster or plastic restoration designed partly from fossilized remains and partly from reconstruction. The *Homo sapiens* and chimpanzee skulls are replicas of a modern specimen.

   The skulls are labeled with the scientific name of the organism, approximate date of origin, and cranial capacity. As you can see, some of these hominids appear to have lived at the same approximate time in history.

   Study the skulls and Figure 11-4. In Table 11-3, record your observations of the changes that occurred over time. Rate the following characteristics for each skull using this scale: (– – –) for the most primitive and (+ + +) for the most advanced.

Characteristic	Primitive	Advanced
Teeth	Many large, specialized	Fewer, smaller, less specialized
Jaw	Jaw and muzzle long	Jaw and muzzle short
Cranium (braincase)	Cranium small, forehead receding	Cranium large, prominent vertical forehead
Muscle attachments	Prominent eyebrow ridges and cranial keel (ridge)	Much reduced eyebrow ridges, no keel
Nose	Nose not protruding from face	Prominent nose with distinct bridge

3. Examine the skulls of *Dryopithecus* and the australoaths more closely. Scientists believe that apelike forms like *Dryopithecus* may have been probably the common ancestors of modern apes and humans.

The first true hominids found in the fossil record have been assigned to the genus *Australopithecus* (from the Greek *australis* for "southern" and *pithecus* for "ape," because the first specimens were found in southern Africa). Some of the oldest known fossils of *Australopithecus* are of the species *A. afarensis* (the most famous skeleton has been named Lucy). Lucy and her conspecifics (members of the same species) lived approximately 3.5 million years ago and may have been the ancestors to the later forms of *Australopithecus*. (See Figure 11-5.)

(a) Study the skull of *Dryopithecus* and the skull of *A. afarensis*. What resemblances do you see between the two specimens?

_____

_____

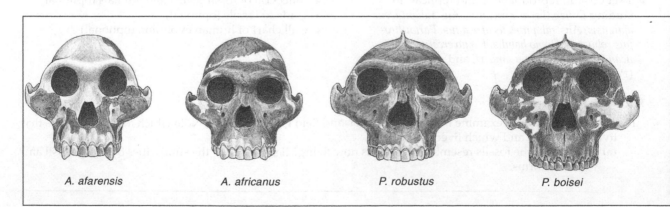

A. afarensis   A. africanus   P. robustus   P. boisei

Figure 11-4 Comparison of skull shapes of human ancestors and relatives. (After Starr and Taggart, 1992.)

TABLE 11-3 Skull Characteristics of Human Ancestors and Relatives, from (− − −) = Most Primitive to (+ + +) = Most Advanced

Genus and Species	Teeth	Jaw	Structure	Cranium	Muscle	Attachments	Nose
Dryopithecus							
Australopithecus afarensis							
Australopithecus africanus							
Paranthropus robustus							
Homo habilis							
Homo erectus							
Homo neandertalensis							
Homo sapiens							
Pan troglodytes							

(b) What structural differences do you see between the two skull specimens?

_____

_____

(c) What advances do you see in skull characteristics between *Dryopithecus* and *Australopithecus africanus*?

_____

_____

(d) Between *Dryopithecus* and *Paranthropus robustus*?

_____

_____

The genetic line(s) of *Australopithecus africanus* and *P. robustus* became extinct approximately 2–1.5 million years ago. There is clear evidence, though, that these hominids were fully bipedal, though not as upright in stance as modern humans. These australopiths also probably used unworked stones and pieces of wood as tools.

4. Now turn your attention to the skulls of the *Homo* lineage. There is debate also about the lineage of our own genus, *Homo* ("Man"). *Homo habilis*, often dubbed simply "early *Homo*," lived in eastern and southern Africa from about 2.5–1.6 million years ago. *Homo erectus* ("upright man") also arose in Africa. However, it was *Homo erectus* whose populations left Africa in waves between 2 million and 500,000 years ago. Our own species, *Homo sapiens* ("wise man"), arose from other *Homo* ancestors.

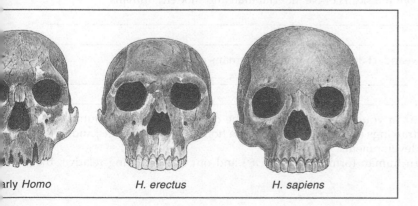

early *Homo*   *H. erectus*   *H. sapiens*

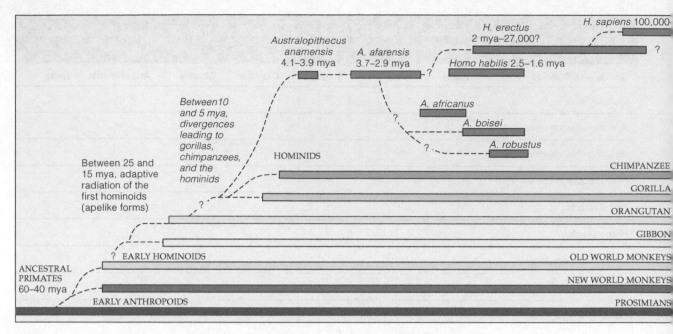

Figure 11-5 Timeline for appearance and extinction of major hominid species.

(a) What changes from primitive to advanced characteristics do you see in the evolutionary line from *A. afarensis* to *Homo sapiens*?

_____

_____

_____

(b) What is the trend in cranial capacity between the ancestral *A. afarensis* and *Homo erectus*?

_____

_____

(c) Between *Homo erectus* and *Homo sapiens*?

_____

_____

Fossils of the Neandertals (*Homo neandertalensis*) are known from Europe, the Near East, and China. These humans hunted and gathered food, made many kinds of tools, and buried their dead, which has been interpreted as showing their capability for abstract and religious thought. The Neandertals disappeared mysteriously 40,000–35,000 years ago, soon after the modern form of *Homo sapiens* arrived from Africa in their range.

(d) What differences do you see between the skulls of Neandertal man and modern humans?

_____

_____

(e) How do the cranial capacities of Neandertal man and modern humans compare?

_____

_____

5. If one is available, study the wall chart in your lab room showing the course of human evolution. Pay particular attention to the conceptual drawings of the appearance of the hominids and humans, and to the evolution of tool making and cultural development.
6. Finally, study the skulls of the modern human form (*Homo sapiens*) and our nearest living relative, the chimpanzee (*Pan troglodytes*).

Fossil evidence, biochemistry, and genetic analyses indicate that chimpanzees and humans are the most closely related of all living primates. In fact, comparisons of amino acid sequences of proteins and DNA sequences show that chimpanzees and humans share about 99% of their genes! Other evidence shows that the separation from the hominid line of the lineage that led to chimpanzees occurred no more than 4–6 million years ago.

(a) How does the chimpanzee skull compare to the human skull with respect to primitive and advanced anatomical features?

_____

_____

(b) How does the chimpanzee skull compare to the skull of *Dryopithecus*, the apelike hominid ancestor?

_____

_____

_____ 1. Evolution is
   (a) the difference in survival and reproduction of a population
   (b) the process that results in changes in the genetic makeup of a population over time
   (c) a group of individuals of the same species occupying a given area
   (d) change in an individual's genetic makeup over its lifetime

_____ 2. The major mechanism of evolution is
   (a) adaptive radiation
   (b) drift
   (c) fitness
   (d) natural selection

_____ 3. An organism whose genetic makeup allows it to produce more offspring than another of its species is said to have greater
   (a) fitness
   (b) evolution
   (c) adaptive radiation
   (d) selection

_____ 4. The earth was formed approximately
   (a) 4600 years ago
   (b) 1 million years ago
   (c) 1 billion years ago
   (d) 4.6 billion years ago

_____ 5. The first living organisms appeared on the earth approximately
   (a) 4.6 billion years ago
   (b) 3.8 billion years ago
   (c) 1 billion years ago
   (d) 6400 years ago

_____ 6. A catastrophic global event in which major groups of species disappear is called
   (a) adaptive radiation
   (b) natural selection
   (c) mass extinction
   (d) evolution

_____ 7. Fossils
   (a) are remains of organisms
   (b) are formed when organic materials are replaced with minerals
   (c) provide evidence of evolution
   (d) are all of the above

_____ 8. The organism believed to be the common ancestor of modern apes and humans is
   (a) *Australopithecus africanus*
   (b) *Homo habilis*
   (c) the chimpanzee
   (d) *Dryopithecus*

_____ 9. Hominids with evolutionarily advanced skull characteristics would have
   (a) many large, specialized teeth
   (b) a large cranium with prominent vertical forehead
   (c) prominent eyebrow ridges
   (d) a long jaw and muzzle

_____ 10. The human ancestor whose populations dispersed from Africa to other parts of the world was
   (a) *Homo sapiens*
   (b) *Australopithecus afarensis*
   (c) *Dryopithecus*
   (d) *Homo erectus*

EXERCISE 11

Evidences of Evolution

POST-LAB QUESTIONS

Introduction

1. Define the following terms.
   a. evolution

   b. natural selection

11.1 Experiment: Natural Selection

2. Describe how natural selection would operate to change the genetic makeup of a dandelion population that had been growing in an unmowed area if that area became subject to frequent low mowing.

11.2 Geologic Time

3. Which event occurred first in the earth's history?
   a. Photosynthesis or eukaryotic cells? _____
   b. Vertebrate animals or flowering plants? _____
   c. Extinction of dinosaurs or origin of hominids? _____

11.3 The Fossil Record and Human Evolution

4. What primitive characteristics are visible in the skull pictured here?

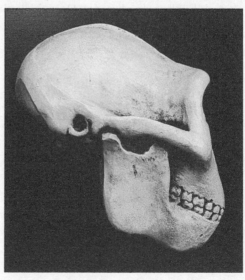

(Photo by J. W. Perry.)

5. Describe anatomical changes in the skull that occurred in human evolution between a *Dryopithecus*-like ancestor and *Homo sapiens*.

6. Compare the slope of the forehead of chimpanzees with that of modern humans.

7. Compare the teeth of *Australopithecus afarensis* and *Homo sapiens*.
   a. Describe similarities and differences between the two species.

   b. Write a hypothesis about dietary differences between the two species based on their teeth.

Food for Thought

8. Why was the development of bipedalism a major advancement in human evolution?

9. Do you believe humans are still evolving? If not, why not? If yes, explain in what ways humans may be evolving.

10. Humans have selectively bred many radically different domestic animals (for example, St. Bernard and chihuahua dog breeds). Does this activity result in evolution? Why or why not?

Taxonomy: Classifying and Naming Organisms

OBJECTIVES

After completing this exercise, you will be able to

1. define *common name, scientific name, binomial, genus, specific epithet, species, taxonomy, phylogenetic system, dichotomous key, herbarium;*

2. distinguish common names from scientific names;

3. explain why scientific names are preferred over common names in biology;

4. identify the genus and specific epithet in a scientific binomial;

5. write out scientific binomials in the form appropriate to the Linnean system;

6. construct a dichotomous key;

7. explain the usefulness of an herbarium;

8. use a dichotomous key to identify plants, animals, or other organisms as provided by your instructor.

INTRODUCTION

We are all great classifiers. Every day, we consciously or unconsciously classify and categorize the objects around us. We recognize an organism as a cat or a dog, a pine tree or an oak tree. But there are numerous kinds of oaks, so we refine our classification, giving the trees distinguishing names such as "red oak," "white oak," or "bur oak." These are examples of **common names,** names with which you are probably most familiar.

Scientists are continually exchanging information about living organisms. But not all scientists speak the same language. The common name "white oak," familiar to an American, is probably not familiar to a Spanish biologist, even though the tree we know as white oak may exist in Spain as well as in our own backyard. Moreover, even within our own language, the same organism can have several common names. For example, within North America a gopher is also called a ground squirrel, a pocket mole, and a groundhog. On the other hand, the same common name may describe many different organisms; there are more than 300 different trees called "mahogany"! To circumvent the problems associated with common names, biologists use **scientific names** that are unique to each kind of organism and that are used throughout the world.

A scientific name is two-parted, a binomial. The first word of the binomial designates the group to which the organism belongs; this is the **genus** name (the plural of genus is *genera*). All oak trees belong to the genus *Quercus,* a word derived from Latin. Each kind of organism within a genus is given a **specific epithet.** Thus, the scientific name for white oak is *Quercus alba* (specific epithet is *alba*), while that of bur oak is *Quercus macrocarpa* (specific epithet is *macrocarpa*).

Notice that the genus name is always capitalized; the specific epithet usually is not capitalized (although it can be if it is the proper name of a person or place). The binomial is written in *italics* (since these are Latin names); if italics are not available, the genus name and specific epithet are underlined.

You will hear discussion of "species" of organisms. For example, on a field trip, you may be asked "What species is this tree?" Assuming you are looking at a white oak, your reply would be *"Quercus alba."* The scientific name of the **species** includes *both* the genus name and specific epithet.

If a species is named more than once within textual material, it is accepted convention to write out the full genus name and specific epithet the first time and to abbreviate the genus name every time thereafter. For example, if white oak is being described, the first use is written *Quercus alba,* and each subsequent naming appears as *Q. alba.*

Similarly, when a number of species, all of the same genus, are being listed, the accepted convention is to write both the genus name and specific epithet for the first species and to abbreviate the genus name for each species listed thereafter. Thus, it is acceptable to list the scientific names for white oak and bur oak as *Quercus alba* and *Q. macrocarpa,* respectively.

Taxonomy is the science of classification (categorizing) and nomenclature (naming). Biologists prefer a system that indicates the evolutionary relationships among organisms. To this end, classification became a **phylogenetic system;** that is, one indicating the presumed evolutionary ancestry among organisms.

Current taxonomic thought separates all living organisms into six kingdoms:

- Kingdom Bacteria (prokaryotic cells that include pathogens)
- Kingdom Archaea (prokaryotic organisms that are evolutionarily closer to eukaryotes than bacteria)
- Kingdom Protista (euglenids, chrysophytes, diatoms, dinoflagellates, slime molds, and protozoans)
- Kingdom Fungi (fungi)
- Kingdom Plantae (plants)
- Kingdom Animalia (animals)

Let's consider the scientific system of classification, using ourselves as examples. All members of our species belong to

- Kingdom Animalia (animals)
- Phylum Chordata (animals with a notochord)
- Class Mammalia (animals with mammary glands)
- Order Primates (mammals that walk upright on two legs)
- Family Hominidae (human forms)
- Genus *Homo* (mankind)
- Specific epithet *sapiens* (wise)
- Species: *Homo sapiens*

The more closely related evolutionarily two organisms are, the more categories they share. You and I are different individuals of the same species. We share the same genus and specific epithet, *Homo* and *sapiens*. A creature believed to be our closest extinct ancestor walked the earth 1.5 million years ago. That creature shared our genus name but had a different specific epithet, *erectus*. Thus, *Homo sapiens* and *H. erectus* are *different* species.

Like all science, taxonomy is subject to change as new information becomes available. Modifications are made to reflect revised interpretations.

12.1   Constructing a Dichotomous Key *(About 45 min.)*

To classify organisms, you must first identify them. A *taxonomic key* is a device for identifying an object unknown to you but that someone else has described. The user chooses between alternative characteristics of the unknown object and, by making the correct choices, arrives at the name of the object.

Keys that are based on successive choices between two alternatives are known as **dichotomous keys** (*dichotomous* means "to fork into two equal parts"). When using a key, always read both choices, even though the first appears to describe the subject. Don't guess at measurements; use a ruler. Since living organisms vary in their characteristics, don't base your conclusion on a single specimen if more are available.

MATERIALS

Per lab room:

- several meter sticks or metric height charts taped to a wall

PROCEDURE

1. Suppose the geometric shapes below have unfamiliar names. Look at the dichotomous key following the figures. Notice there is a 1a and a 1b. Start with 1a. If the description in 1a fits the figure you are observing better than description 1b, then proceed to the choices listed under 2, as shown at the end of line 1a. If 1a does *not* describe the figure in question, 1b does. Looking at the end of line 1b, you see that the figure would be called an Elcric.
2. Using the key provided, determine the hypothetical name for each object. Write the name beneath the object and then check with your instructor to see if you have made the correct choices.

   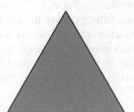 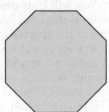

_____   _____   _____   _____   _____

Key		
1a.	Figure with distinct corners	2
1b.	Figure without distinct corners	Elcric
2a.	Figure with 3 sides	3
2b.	Figure with 4 or more sides	4
3a.	All sides of equal length	Legnairt
3b.	Only 2 sides equal	Legnairtosi
4a.	Figure with only right angles	Eraqus
4b.	Figure with other than right angles	Nogatco

3. Now you will construct a dichotomous key, using your classmates as subjects. The class should divide up into groups of eight (or as evenly as the class size will allow). Working with the individuals in your group, fill in Table 12-1, measuring height with a metric ruler or the scale attached to the wall.

4. To see how you might plan a dichotomous key, examine the following branch diagram. If there are both men and women in a group, the most obvious first split is male/female (although other possibilities for the split could be chosen as well). Follow the course of splits for two of the men in the group.

   Note that each choice has *only* two alternatives. Thus, we split into "under 1.75 m" and "1.75 m or taller." Likewise, our next split is into "blue eyes" and "nonblue eyes" rather than all the possibilities.

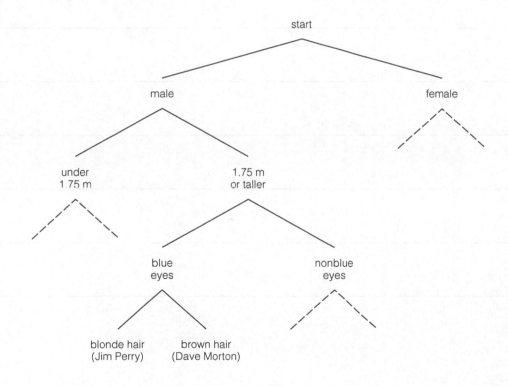

5. On a separate sheet of paper, construct a branch diagram for your group using the characteristics in Table 12-1 and then condense it into the dichotomous key that follows. When you have finished, exchange your key with that of an individual in another group. Key out the individuals in the other group without speaking until you believe you know the name of the individual you are examining. Ask that individual if you are correct. If not, go back to find out where you made a mistake, or possibly where the key was misleading. (Depending on how you construct your key, you may need more or fewer lines than have been provided.)

TABLE 12-1	Characteristics of Students				
Student (name)	Sex (m/f)	Height (m)	Eye Color	Hair Color	Shoe Size
1.					
2.					
3.					
4.					
5.					
6.					
7.					
8.					

Key to Students in Group _____

1a.	
1b.	
2a.	
2b.	
3a.	
3b.	
4a.	
4b.	
5a.	
5b.	
6a.	
6b.	
7a.	
7b.	
8a.	
8b.	

12.2 Using a Taxonomic Key

A. Some Microscopic Members of the Freshwater Environment (About 30 min.)

Suppose you want to identify the specimens in some pond water. The easiest way is to key them out with a dichotomous key, now that you know how to use one. In this section, you will do just that.

MATERIALS

Per student:

- compound microscope
- microscope slide
- coverslip
- dissecting needle

Per student group (table):

- cultures of freshwater organisms
- 1 disposable plastic pipet per culture
- methylcellulose in dropping bottle

PROCEDURE

1. Obtain a clean glass microscope slide and clean coverslip.
2. Using a disposable plastic pipet or dissecting needle, withdraw a small amount of the culture provided.
3. Place *one* drop of the culture on the center of the slide.
4. Gently lower the coverslip onto the liquid.
5. Using your compound light microscope, observe your wet mount. Focus first with the low-power objective and then with the medium or high-dry objective, depending on the size of the organism in the field of view.
6. Concentrate your observation on a single specimen, keying out the specimen using the Key to Selected Freshwater Inhabitants that follows.
7. In the space provided, write the scientific name of each organism you identify. After each identification, have your instructor verify your conclusion.
8. Clean and reuse your slide and coverslip after each identification.

Key to Selected Freshwater Inhabitants		
1a.	Filamentous organism consisting of green, chloroplast-bearing threads	2
1b.	Organism consisting of a single cell or nonfilamentous colony	4
2a.	Filament branched, each cell mostly filled with green chloroplast	*Cladophora*
2b.	Filament unbranched	3
3a.	Each cell of filament containing 1 or 2 spiral-shaped green chloroplasts	*Spirogyra*
3b.	Each cell of filament containing 2 star-shaped green chloroplasts	*Zygnema*
4a.	Organism consisting of a single cell	5
4b.	Organism composed of many cells aggregated into a colony	6
5a.	Motile, teardrop-shaped or spherical organism	*Chlamydomonas*
5b.	Nonmotile, elongate cell on either end; clear, granule-containing regions at ends	*Closterium*
6a.	Colony a hollow round ball of more than 500 cells; new colonies may be present inside larger colony	*Volvox*
6b.	Colony consisting of less than 50 cells	7
7a.	Organism composed of a number of tooth-shaped cells	*Pediastrum*
7b.	Colony a loose square or rectangle of 4–32 spherical cells	*Gonium*

Organism 1 is _____

Organism 2 is _____

Organism 3 is _____

Organism 4 is _____

Organism 5 is _____

Organism 6 is _____

Organism 7 is _____

Organism 8 is _____

Suppose you want to identify the trees growing on your campus or in your yard at home. Without having an expert present, you can now do that, because you know how to use a taxonomic key. But how can you be certain that you have keyed your specimen correctly?

Typically, scientists compare their tentative identifications against *reference specimens*—that is, preserved organisms that have been identified by an expert *taxonomist* (a person who names and classifies organisms).

If you are identifying fishes or birds, the reference specimen might be a bottled or mounted specimen with the name on it. In the case of plants, reference specimens most frequently take the form of *herbarium mounts* (Figure 12-1) of the plants. An **herbarium** (plural, *herbaria*) is a repository, a museum of sorts, of preserved plants. The taxonomist flattens freshly collected specimens in a plant press. They are then dried and mounted on sheets of paper. Herbarium labels are affixed to the sheets, indicating the scientific name of the plant, the person who collected it, the location and date of collection, and often pertinent information about the habitat in which the plant was found.

It is likely that your school has an herbarium. If so, your instructor may show you the collection. To some, this endeavor may seem boring, but herbaria serve a critical function. The appearance or disappearance of plants from the landscape often gives a very good indication of environmental change. An herbarium records the diversity of plants in the area, at any point in history since the start of the collection.

Label indicates name of specimen, site and date of collection, associated species at same site, name(s) of collector(s)

(Photo by J. W. Perry)

Figure 12-1 A typical herbarium mount.

MATERIALS

Per student group (table):

- set of 8 tree twigs with leaves (fresh or herbarium specimens) *or*
- trees and shrubs in leafy condition (for an outdoor lab)

PROCEDURE

Use the appropriate following key to identify the tree and shrub specimens that have been provided in the lab or that you find on your campus. Refer to the *Glossary to Tree Key* and Figures 12-2 through 12-9 when you encounter an unfamiliar term. When you have finished keying a specimen, confirm your identification by checking the herbarium mounts or asking your instructor.

Note: **Some descriptions within the key have more characteristics than your specimen will exhibit. For example, the key may describe a fruit type when the specimen doesn't have a fruit on it. However, other specimen characteristics are described, and these should allow you to identify the specimen.**

Note: **The keys provided are for *selected* trees of your area. In nature, you will find many more genera than can be identified by these keys.**

Common names within parentheses follow the scientific name. A metric ruler is provided on page 204 for use where measurements are required.

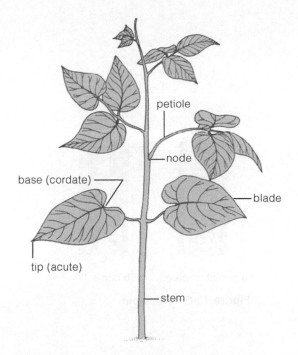

Figure 12-2 Structure of a typical plant (bean).

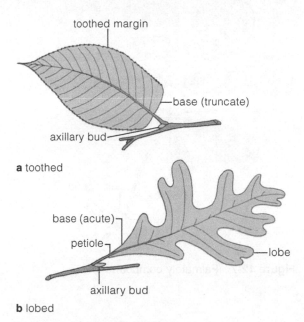

a toothed

b lobed

Figure 12-3 Simple leaves.

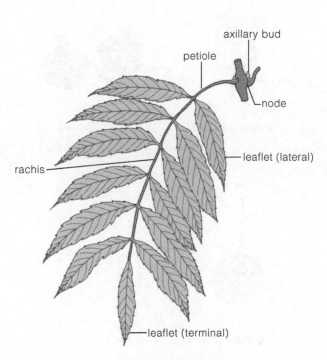

Figure 12-4 Pinnately compound leaf.

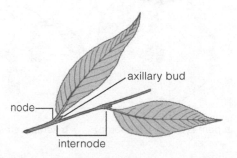

Figure 12-5 Simple leaves—alternating.

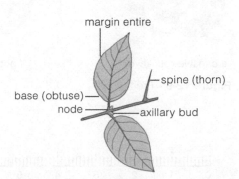

Figure 12-6 Simple leaves—opposite.

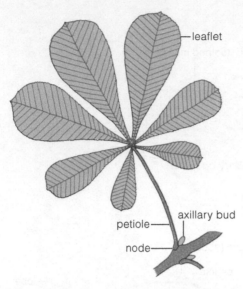

Figure 12-7 Palmately compound leaf.

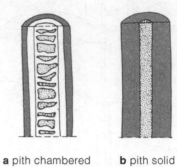

a pith chambered   **b** pith solid

Figure 12-8 Pith types.

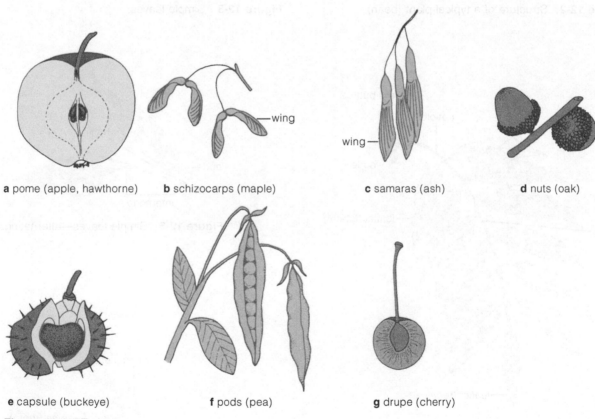

a pome (apple, hawthorne)   **b** schizocarps (maple)   **c** samaras (ash)   **d** nuts (oak)

e capsule (buckeye)   **f** pods (pea)   **g** drupe (cherry)

Figure 12-9 Fruit types.

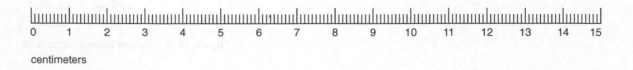

centimeters

Key to Some Common Genera of Trees of the Midwestern and Eastern United States and Canada

1a.	Leaves broad and flat; plants producing flowers and fruits (angiosperms)	2
1b.	Leaves needlelike or scalelike; plants producing cones, but no flowers or fruits (gymnosperms)	22
2a.	Leaves compound	3
2b.	Leaves simple	9
3a.	Leaves alternate	4
3b.	Leaves opposite	7
4a.	Leaflets short and stubby, less than twice as long as broad; branches armed with spines or thorns; fruit a beanlike pod	5
4b.	Leaflets long and narrow, more than twice as long as broad; trunk and branches unarmed; fruit a nut	6
5a.	Leaflet margin without teeth; terminal leaflet present; small deciduous spines at leaf base	*Robinia* (black locust)
5b.	Leaflet margin with fine teeth; terminal leaflet absent; large permanent thorns on trunk and branches	*Gleditsia* (honey locust)
6a.	Leaflets usually numbering less than 11; pith of twigs solid	*Carya* (hickory)
6b.	Leaflets numbering 1 or more, pith of twigs divided into chambers	*Juglans* (walnut, butternut)
7a.	Leaflets pinnately arranged; fruit a light-winged samara	8
7b.	Leaflets palmately arranged; fruit a heavy leathery spherical capsule	*Aesculus* (buckeye)
8a.	Leaflets numbering mostly 3–5; fruit a schizocarp with curved wings	*Acer* (box elder)
8b.	Leaflets numbering mostly more than 5; samaras borne singly, with straight wings	*Fraxinus* (ash)
9a.	Leaves alternate	10
9b.	Leaves opposite	21
10a.	Leaves very narrow, at least 3 times as long as broad; axillary buds flattened against stem	*Salix* (willow)
10b.	Leaves broader, less than 3 times as long as broad	11
11a.	Leaf margin without small, regular teeth	12
11b.	Leaf margin with small, regular teeth	13
12a.	Fruit a pod with downy seeds; leaf blade obtuse at base; petioles flattened, or if rounded, bark smooth	*Populus* (poplar, popple, aspen)
12b.	Fruit an acorn; leaf blade acute at the base; petioles rounded; bark rough	*Quercus* (oaks)
13a.	Leaves (at least some of them) with lobes or other indentations in addition to small, regular teeth	14
13b.	Leaves without lobes or other indentations except for small, regular teeth	16
14a.	Lobes asymmetrical, leaves often mitten-shaped	*Morus* (mulberry)
14b.	Lobes or other indentations fairly symmetrical	15

15a.	Branches thorny (armed); fruit a small applelike pome	*Crataegus* (hawthorne)
15b.	Branches unarmed	17
16a.	Bark smooth and waxy, often separating into thin layers; leaf base symmetrical	*Betula* (birch)
16b.	Bark rough and furrowed, leaf base asymmetrical	*Ulmus* (elm)
17a.	Leaf base asymmetrical, strongly heart-shaped, at least on one side	*Tilia* (basswood or linden)
17b.	Leaf base acute, truncate, or slightly cordate	18
18a.	Leaf base asymmetrical; bark on older stems (trunk) often warty	*Celtis* (hackberry)
18b.	Leaf base symmetrical	19
19a.	Leaf blade usually about twice as long as broad, generally acute at the base; fruit fleshy	20
19b.	Leaf not much longer than broad, generally truncate at base; fruit a dry pod	*Populus* (poplar, popple, aspen)
20a.	Leaf tapering to a pointed tip, glandular at base	*Prunus* (cherry)
20b.	Leaf spoon-shaped with a rounded tip, no glands at base	*Crataegus* (hawthorne)
21a.	Leaf margins with lobes and points, fruit a schizocarp	*Acer* (maple)
21b.	Leaf margins without lobes or points; fruit a long capsule	*Catalpa* (catalpa)
22a.	Leaves needlelike, with 2 or more needles in a cluster	23
22b.	Leaves needlelike or scalelike, occurring singly	24
23a.	Leaves more than 5 in a cluster, soft, deciduous, borne at the ends of conspicuous stubby branches	*Larix* (larch, tamarack)
23b.	Leaves 2–5 in a cluster	*Pinus* (pines)
24a.	Leaves soft, not sharp to the touch	25
24b.	Leaves stiff or sharp and often unpleasant to touch	27
25a.	Leaves about 0.2 cm and scalelike, overlapping	*Thuja* (white cedar, arbor vitae)
25b.	Leaves needlelike, appear to form two ranks on twig	26
26a.	Leaves with distinct petioles, 0.8–1.5 cm long; twigs rough; female cones drooping from branches	*Tsuga* (hemlock)
26b.	Leaves without distinct petioles, 1–3 cm long; twigs smooth; female cones erect on branches	*Abies* (firs)
27a.	Leaves appear triangular-shaped, about 0.5 cm, and tightly pressed to twig; cone blue, berrylike	*Juniperus* (juniper, Eastern red cedar)
27b.	Leaves elongated and needlelike	28
28a.	Tree; leaves 4-sided, protrude stiffly from twig; female cones droop from branch	*Picea* (spruces)
28b.	Shrub; leaves flattened, pressed close to twig at base; seed partially covered by a fleshy coat, usually red	*Taxus* (yew)

Key to Some Common Genera of Trees of the Pacific Region of the United States and Canada

1a.	Leaves broad and flat; plants producing flowers and fruits (angiosperms)	2
1b.	Leaves needlelike or scalelike; plants producing cones, but no flowers or fruits (gymnosperms)	15
2a.	Leaves compound	3
2b.	Leaves simple	6
3a.	Leaves pinnately arranged	4
3b.	Leaves palmately arranged	5
4a.	Leaflets number 7, fruit a samara	*Fraxinus* (ash)
4b.	Leaflets 15–17, fruit a nut	*Juglans* (walnut)
5a.	Leaflets numbering 3, lobed; fruit a schizocarp	*Acer* (box elder)
5b.	Leaflets numbering more than 3; fruit a smooth or spiny capsule	*Aesculus* (buckeye)
6a.	Three or more equal-sized veins branching from leaf base	7
6b.	Leaf with single large central vein with other main veins branching from the central vein	9
7a.	Leaves opposite; fruit a schizocarp	*Acer* (maple)
7b.	Leaves alternate	8
8a.	Leaves nearly round in outline; fruit a pod	*Cercis* (redbud)
8b.	Leaves deeply lobed, very hairy beneath; fruit consisting of an aggregation of many 1-seeded nutlets surrounded by long hairs	*Platanus* (sycamore)
9a.	Leaf lobed; fruit a nut	*Quercus* (oak)
9b.	Leaf not lobed	10
10a.	Leaves opposite; fruit a drupe; halves of leaf remain attached by "threads" after blade has been creased and broken	*Cornus* (dogwood)
10b.	Leaves alternate	11
11a.	Upon crushing, blade gives off strong, penetrating odor	*Eucalyptus* (eucalyptus)
11b.	Blade not strongly odiferous upon crushing	12
12a.	Branch bark smooth, conspicuously red-brown; fruit a red ororange berry	*Arbutus* (madrone)
12b.	Branch rough, not colored red-brown	13
13a.	Undersurface of leaves golden-yellow; fruit a spiny, husked nut	*Catanopsis* (golden chinquapin)
13b.	Leaves green beneath	14
14a.	Petiole hairy; leathery blade with a stubby spine at end of each main vein; fruit a nut	*Lithocarpus* (tanoak)
14b.	Petiole and leaf lacking numerous hairs, leaves long and narrow, more than twice as long as wide	*Salix* (willow)

15a.	Leaves needlelike	16
15b.	Leaves scalelike	23
16a.	Leaves needlelike, with 2 or more needles in a cluster	17
16b.	Leaves needlelike, occurring singly	18
17a.	Needles 2–5 in a cluster	*Pinus* (pine)
17b.	Needles 6 or more per cluster, soft, deciduous, borne at the ends of stubby, conspicuous branches	*Larix* (larch)
18a.	Round scars on twigs where old needles have fallen off; twigs smooth; needles soft to the grasp; cones pointing upward with reference to stem	*Abies* (fir)
18b.	Twigs rough, with old needle petioles remaining	19
19a.	Needles angled, stiff, sharp, pointed, unpleasant to grasp; cones hanging downward from branch	*Picea* (spruce)
19b.	Needles soft, not sharp when grasped	20
20a.	Needles round in cross section, can be rolled easily between thumb and index finger; needles less than 1.3 cm long; cones small, less than 1.5 cm	*Tsuga* (hemlock)
20b.	Needles too flat to be rolled easily	21
21a.	Tips of needles blunt or rounded, undersurface with 2 white bands; cones with long, conspicuous, 3-lobed bracts	*Pseudotsuga* (Douglas fir)
21b.	Tips of needles pointed	22
22a.	Tops of needles grooved; woody seed cones broadly oblong in outline	*Sequoia* (redwood)
22b.	Tops of needles with ridges; lacking in cones, instead having a red, fleshy, cuplike seed covering	*Taxus* (yew)
23a.	Twig ends appear as if jointed	*Calocedrus* (incense cedar)
23b.	Tips of branches flattened, not jointed in appearance	24
24a.	Leaves glossy and fragrant	*Thuja* (Western red cedar)
24b.	Leaves awl-shaped, arranged spirally on twig	*Sequoiadendron* (giant sequoia)

Glossary to Tree Key

- *Acorn*—The fruit of an oak, consisting of a nut and its basally attached cup (Fig. 12-9d)
- *Acute*—Sharp-pointed (Fig. 12-2)
- *Alternate*—Describing the arrangement of leaves or other structures that occur singly at successive nodes or levels; not opposite or whorled (Fig. 12-5)
- *Angiosperm*—A flowering seed plant (e.g., bean plant, maple tree, grass)
- *Armed*—Possessing thorns or spines
- *Asymmetrical*—Not symmetrical
- *Axil*—The upper angle between a branch or leaf and the stem from which it grows
- *Axillary bud*—A bud occurring in the axil of a leaf (Figs. 12-3 through 12-7)
- *Basal*—At the base
- *Blade*—The expanded, more or less flat portion of a leaf (Fig. 12-2)
- *Bract*—A much reduced leaf
- *Capsule*—A dry fruit that splits open at maturity (e.g., buckeye; Fig. 12-9e)

- *Compound leaf*—Blade composed of 2 or more separate parts (leaflets) (Figs. 12-4, 12-7)
- *Cordate*—Heart-shaped (Fig. 12-2)
- *Deciduous*—Falling off at the end of a functional period (such as a growing season)
- *Drupe*—Fleshy fruit containing a single hard stone that encloses the seed (e.g., cherry, peach, or dogwood; Fig. 12-9g)
- *Fruit*—A ripened ovary, in some cases with associated floral parts (Figs. 12-9a–g)
- *Glandular*—Bearing secretory structures (glands)
- *Gymnosperm*—Seed plant lacking flowers and fruits (e.g., pine tree)
- *Lateral*—On or at the side (Fig. 12-4)
- *Leaflet*—One of the divisions of the blade of a compound leaf (Figs. 12-4, 12-7)
- *Lobed*—Separated by indentations (sinuses) into segments (lobes) larger than teeth (Fig. 12-3b)
- *Node*—Region on a stem where leaves or branches arise (Figs. 12-2 through 12-7)
- *Nut*—A hard, 1-seeded fruit that does not split open at maturity (e.g., acorn; Fig. 12-9d)
- *Obtuse*—Blunt (Fig. 12-6)
- *Opposite*—Describing the arrangement of leaves of other structures that occur 2 at a node, each separated from the other by half the circumference of the axis (Fig. 12-6)
- *Palmately compound*—With leaflets all arising at apex of petiole (Fig. 12-7)
- *Petiole*—Stalk of a leaf (Figs. 12-2, 12-3, 12-4, 12-7)
- *Pinnately compound*—A leaf constructed somewhat like a feather, with the leaflets arranged on both sides of the rachis (Fig. 12-4)
- *Pith*—Internally, the centermost region of a stem (Figs. 12-8a, b)
- *Pod*—A dehiscent, dry fruit; a rather general term sometimes used when no other more specific term is applicable (Fig. 12-9f)
- *Pome*—Fleshy fruit containing several seeds (e.g., apple or pear; Fig. 12-9a)
- *Rachis*—Central axis of a pinnately compound leaf (Fig. 12-4)
- *Samara*—Winged, 1-seeded, dry fruit (e.g., ash fruits; Fig. 12-9c)
- *Schizocarp*—Dry fruit that splits at maturity into two 1-seeded halves (Fig. 12-9b)
- *Simple leaf*—One with a single blade, not divided into leaflets (Figs. 12-3, 12-5, 12-6)
- *Spine*—Strong, stiff, sharp-pointed outgrowth on a stem or other organ (Fig. 12-6)
- *Symmetrical*—Capable of being divided longitudinally into similar halves
- *Terminal*—Last in a series (Fig. 12-4)
- *Thorn*—Sharp, woody, spinelike outgrowth from the wood of a stem; usually a reduced, modified branch
- *Tooth*—Small, sharp-pointed marginal lobe of a leaf (Fig. 12-3a)
- *Truncate*—Cut off squarely at end (Fig. 12-3a)
- *Unarmed*—Without thorns or spines
- *Whorl*—A group of 3 or more leaves or other structures at a node

12.3   What Species Is Your Christmas Tree? *(About 20 min.)*

Each year millions of "evergreen" trees become the center of attraction in human dwellings during the Christmas season. The process of selecting the all-important tree is the same whether you reside in the city where you buy your tree from a commercial grower, or whether you cut one off your "back forty." You ponder and evaluate each specimen until, with the utmost confidence, you bring home that perfect tree. Now that you have it, just what kind of tree stands in your home, looking somewhat like a cross between Old Glory and the Sistine Chapel? This key contains most of the trees that are used as Christmas trees; other gymnosperm trees are included, too. The common "Christmas trees" have an asterisk after their scientific name. Note that this key, unlike those in the preceding sections, indicates actual species designations.

1a.	Tree fragrant, boughs having supported (on clear moonlit nights) masses of glistening snow on their green needles; tree a product of nature	2
1b.	Tree not really a tree but rather a product of a cold and insensitive society; tree never giving life and never having life	17
2a.	Leaves persistent and green throughout the winter, needlelike, awl-shaped, or scalelike	3
2b.	Leaves deciduous; for this reason not a desirable Christmas tree	4
3a.	Leaves in clusters of 2–5, their bases within a sheath	5
3b.	Leaves borne singly, not in clusters	9
4a.	Cones 1.25–1.8 cm long, 12–15 scales making up cone	*Larix laricina* (tamarack)
4b.	Cones 1.8–3.5 cm long, 40–50 scales comprising cone	*Larix decidua* (larch)
5a.	Leaves 5 in a cluster, cones 10–25 cm long	*Pinus strobus** (white pine)
5b.	Leaves 2 in a cluster, cones less than 10 cm long	6
6a.	Leaves 2.5–7.5 cm long	7
6b.	Leaves 7.5–15 cm long	8
7a.	Leaves with a bluish cast; cones with a stout stalk, pointing away from the tip of the branch; bark orange in the upper part of the tree	*Pinus sylvestris** (scotch pine)
7b.	Leaves 1.25–3.75 cm long; cones stalkless, pointing forward toward the tip of branch	*Pinus banksiana* (jack pine)
8a.	Leaves slender, shiny; bark of trunk red-brown; cones 5–7.5 cm long; scales of cones without any spine at tip	*Pinus resinosa** (red pine)
8b.	Leaves thickened, dull; bark of trunk gray to nearly black; cones 5 to 7.5 cm long; scales of cone armed with short spine at tip	*Pinus nigra** (Austrian pine)
9a.	Leaves scalelike or awl-shaped	10
9b.	Leaves needlelike	11
10a.	Twigs flattened, leaves all of one kind, scalelike, extending down the twig below the point of attachment	*Thuja occidentalis* (white cedar, arbor vitae)
10b.	Twigs more or less circular in cross section; leaves of 2 kinds, either scalelike or awl-shaped, often both on same branch, not extending down the twig; coneless but may have a blue berrylike structure	*Juniperus virginiana* (red cedar)
11a.	Leaves with petioles	12
11b.	Leaves lacking petioles, leaf tip notched, needles longer than 1.25 cm	*Abies balsamea** (balsam fir)
12a.	Leaves angular, 4-sided in cross section, harsh to the touch; petiole adheres to twig	13
12b.	Leaves flattened	15
13a.	Leaves 3–10 mm long, blunt-pointed; twigs rusty and hairy	*Picea mariana** (black spruce)
13b.	Leaves 2 cm long, sharp-pointed; twigs smooth	14

14a.	Cones 2.5–5 cm long; leaves ill-scented when bruised or broken; smaller branches mostly horizontal	*Picea glauca** (white spruce)
14b.	Cones 7.5–15 cm long; scales comprising cone with finely toothed markings; leaves not ill-scented when bruised or broken; smaller branches drooping	*Picea abies** (Norway spruce)
15a.	Leaves pointed, over 1.25 cm long; red fleshy, berrylike structures present	16
15b.	Leaves rounded at tip, less than 1.25 cm long, with 2 white lines on underside	*Tsuga canadensis* (hemlock)
16a.	Leaves 2–2.5 cm long, dull dark green on top, with 2 broad yellow bands on undersurface; petiole yellowish	*Taxus cuspidata* (Japanese yew)
16b.	Leaves 1.25–2 cm long, without yellow bands on underside	*Taxus canadensis* (American yew)
17a.	Tree a glittering mass of structural aluminum, sometimes illuminated by multicolored floodlights	*Aluminous ersatzenbaum** (aluminum substitute)
17b.	Tree green, produced with petroleum products, increasing our dependence upon oil; used year after year; exactly like all others of its manufacture	*Plasticus perfectus** (plastic substitute)

_____ 1. The name "human" is an example of a
  (a) common name
  (b) scientific name
  (c) binomial
  (d) polynomial

_____ 2. Current scientific thought places organisms in one of ___ kingdoms.
  (a) two
  (b) four
  (c) five
  (d) six

_____ 3. The scientific name for the ruffed grouse is *Bonasa umbellus. Bonasa* is
  (a) the family name
  (b) the genus
  (c) the specific epithet
  (d) all of the above

_____ 4. A binomial is always a
  (a) genus
  (b) specific epithet
  (c) scientific name
  (d) two-part name

_____ 5. The science of classifying and naming organisms is known as
  (a) taxonomy
  (b) phylogeny
  (c) morphology
  (d) physiology

_____ 6. Which scientific name for the wolf is presented correctly?
  (a) Canis lupus
  (b) canis lupus
  (c) *Canis lupus*
  (d) Canis Lupus

_____ 7. A road that dichotomizes is
  (a) an intersection of two crossroads
  (b) a road that forks into two roads
  (c) a road that has numerous entrances and exits
  (d) a road that leads nowhere

_____ 8. Most scientific names are derived from
  (a) English
  (b) Latin
  (c) Italian
  (d) French

_____ 9. One objection to common names is that
  (a) many organisms may have the same common name
  (b) many common names may exist for the same organism
  (c) the common name may not be familiar to an individual not speaking the language of the common name
  (d) all of the above are true

_____ 10. Phylogeny is the apparent
  (a) name of an organism
  (b) ancestry of an organism
  (c) nomenclature
  (d) dichotomy of a system of classification

Name _____ Section Number _____

Taxonomy: Classifying and Naming Organisms

POST-LAB QUESTIONS

Introduction

1. If you were to use a binomial system to identify the members of your family (mother, father, sisters, brothers), how would you write their names so that your system would most closely approximate that used to designate species?

2. Describe several advantages of using scientific names instead of common names.

3. Based on the following classification scheme, which two organisms are most closely phylogenetically related? Why?

	Organism 1	Organism 2	Organism 3	Organism 4
Kingdom	Animalia	Animalia	Animalia	Animalia
Phylum	Arthopoda	Arthropoda	Arthropoda	Arthropoda
Class	Insecta	Insecta	Insecta	Insecta
Order	Coleoptera	Coleoptera	Coleoptera	Coleoptera
Genus	*Caulophilus*	*Sitophilus*	*Latheticus*	*Sitophilus*
Specific epithet	*oryzae*	*oryzae*	*oryzae*	*zeamaize*
Common name	Broadnosed grain weevil	Rice weevil	Longheaded flour beetle	Maize weevil

12.2 Using a Taxonomic Key

Consider the drawing of plants A and B in answering questions 4–6.

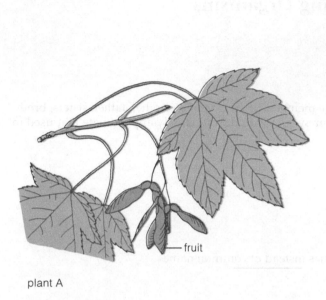

fruit

plant A

cone

plant B

4. Using the taxonomic key in the exercise, identify the two plants as either angiosperms or gymnosperms.

Plant A is a (an) _____.

Plant B is a (an) _____.

5. To what genus does plant A belong? What is its common name?

genus: _____

common name: _____

6. To what genus does plant B belong? What is its common name?

genus: _____

common name: _____

Consider the drawing of plants C and D in answering questions 7–9.

plant C

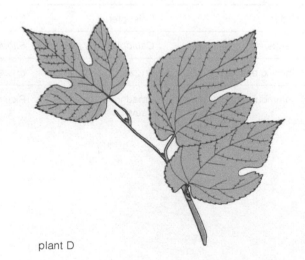

plant D

7. As completely as possible, describe the leaf of plant C.

8. To what genus does plant C belong? What is its common name?

        genus: _____

    common name: _____

9. Using the taxonomic key in the exercise, identify the genus of the organism below.

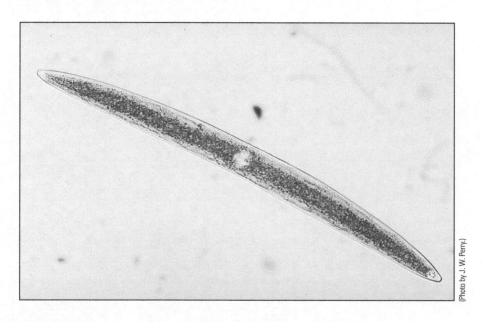

(Photo by J. W. Perry.)

        genus: _____

Food for Thought

10. If you owned a large, varied music collection, how might you devise a key to keep track of all your different kinds of music?

Bacteria and Protists

OBJECTIVES

After completing this exercise, you will be able to

1. define *prokaryotic, eukaryotic, pathogen, decomposer, coccus, bacillus, spirillum, Gram stain, antibiotic, symbiosis, parasitism, commensalism, mutualism, nitrogen fixation, monoecious, dioecious, plasmodium, obligate mutualism, gametangia, phagocytosis, vector; phytoplankton, phyt, phycobilin, agar, fucoxanthin, algin, kelp, gametangium, oogonia* and *antheridia;*

2. describe characteristics distinguishing bacteria from protists;

3. identify and classify the organisms studied in this exercise;

4. identify structures (those in **boldface** in the procedure sections) in the organisms studied;

5. distinguish Gram-positive and Gram-negative bacteria, indicating their susceptibility to certain antibiotics;

6. suggest measures that might be used to control malaria;

7. recognize and classify selected members of the seven phyla represented in this exercise;

8. distinguish among the structures associated with asexual and sexual reproduction described in this exercise;

9. identify the structures of the algae (those in **boldface** in the procedure sections).

INTRODUCTION

The bacteria (Domain Bacteria, Kingdom Bacteria) and archaea (Domain Archaea, Kingdom Archaea) and protists (Domain Eukarya, Kingdom Protista) are among the simplest of living organisms. These have unicellular organisms within them, but that's where the similarity ends.

Bacteria and archaea are **prokaryotic** organisms, meaning that their DNA is free in the cytoplasm, unbounded by a membrane. They lack organelles (cytoplasmic structures surrounded by membranes). By contrast, the protists are **eukaryotic** organisms: The genetic material contained within the nucleus and many of their cellular components are compartmentalized into membrane-bound organelles.

Both bacteria and archaea, and unicellular protists, are at the base of the food chain. From an ecological standpoint, these simple organisms are among the most important organisms on our planet. Ecologically, they're much more important than we are.

13.1   Domain Bacteria, Kingdom Bacteria *(About 40 min.)*

The Kingdom Bacteria consists of bacteria and cyanobacteria. Most bacteria are heterotrophic, dependent upon an outside source for nutrition, while cyanobacteria are autotrophic (photosynthetic), able to produce their own carbohydrates.

Some heterotrophic bacteria are **pathogens,** causing plant and animal diseases, but most are **decomposers,** breaking down and recycling the waste products of life. Others are nitrogen fixers, capturing the gaseous nitrogen in the atmosphere and making it available to plants via a symbiotic association with their roots.

MATERIALS

Per student:

- bacteria type slide
- microscope slide
- coverslip
- dissecting needle
- compound microscope

Per lab room:

- Gram-stained bacteria (3 demonstration slides)
- *Oscillatoria*—living culture; disposable pipet
- *Azolla*—living plants

A. Bacteria: Are Bacteria Present in the Lab? *(About 15 min.)*

MATERIALS

Per student group (4):

- dH$_2$O in dropping bottle
- 4 nutrient agar culture plates
- sterile cotton swabs
- china marker

- 2 bottles labeled A and B (A contains tap water, B contains 70% ethyl alcohol)
- transparent adhesive tape
- paper towels

PROCEDURE

Work in groups of four.

1. Obtain four petri dishes containing sterile nutrient agar. Using a china marker, label one dish "Dish 1: Control." Label the others "Dish 2: Dry Swab," "Dish 3: Treatment A," and "Dish 4: Treatment B." Also include the names of your group members.
2. Run a sterile cotton swab over a surface within the lab. Some examples of things you might wish to sample include the surface of your lab bench, the floor, and the sink. Be creative!
3. Lift the lid on Dish 2 as little as is necessary to run the swab over the surface of the agar.

Note: Be careful that you don't break the agar surface.

4. Tape the lid securely to the bottom half of the dish.
5. Soak one paper towel with liquid A and a second paper towel with liquid B.
6. Wipe down one-half of the surface you just sampled with liquid A (tap water), the other half with liquid B (70% ethyl alcohol). After the areas have dried, using dishes 3 and 4, repeat the procedures described in step 2. Place the cultures in a desk drawer to incubate until the next lab period. At that time, examine your culture for bacterial colonies, noting the color and texture of the bacterial growth.
7. Make a prediction of what you will find in the culture plates after the next lab period.

_____

_____

8. Describe what you see.
   Source of sample: _____

   Dish 1: Control. _____

   _____

   _____

   Dish 2: Dry Swab. _____

   _____

   _____

   Dish 3: Treatment A. _____

   _____

   Dish 4: Treatment B. _____

> **Caution**
>
> *Leave the lid on as you examine the cultures to prevent the spread of any potentially pathogenic (disease-causing) organisms. While the probability is small that pathogens are present, you should always err on the side of caution.*

9. Make a conclusion about the usefulness of tap water and 70% ethyl alcohol as disinfectants.

_____

_____

B. Bacteria: Bacterial Shape and Sensitivity to Antibiotics *(About 10 min.)*

MATERIALS

Per student:

- bacteria type slide
- compound microscope

Per lab room:

- Gram-stained bacteria (3 demonstration slides)

Work alone for this and the rest of these activities.

Bacteria exist in three shapes: **coccus** (plural, *cocci;* spherical), **bacillus** (plural, *bacilli;* rods), and **spirillum** (plural, *spirilla;* spirals).

PROCEDURE

1. Study a bacteria type slide illustrating these three shapes. You'll need to use the highest magnification available on your compound microscope. In Figure 13-1, draw the bacteria you are observing.

In addition to being differentiated by shape, bacteria can be separated according to how they react to a staining procedure called **Gram stain,** named in honor of a nineteenth-century microbiologist, Hans Gram. *Gram-positive* bacteria are purple after being stained by the Gram stain procedure, while *Gram-negative* bacteria appear pink. The Gram stain reaction is important to bacteriologists because it is one of the first steps in identifying an unknown bacterium. Furthermore, the Gram stain reaction indicates a bacterium's susceptibility or resistance to certain **antibiotics,** substances that inhibit the growth of bacteria.

2. Examine the *demonstration slides* illustrating Gram-stained bacteria. Gram-positive bacteria are susceptible to penicillin, while Gram-negative bacteria are not. In Table 13-1, list the species of bacteria that you have examined and their staining characteristics.

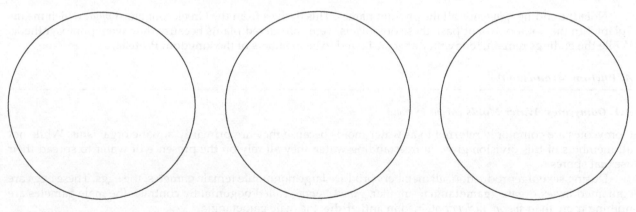

Figure 13-1 Drawings of the three bacterial shapes (_____×).

TABLE 13-1 Gram Stain Reaction of Various Bacteria	
Bacterial Species	**Gram Reactions (+ or −)**

13.2   Domain Eukarya, Kingdom Protista *(About 2 hr.)*

The protists are a diverse assemblage of organisms, both green (photoautotrophic) and nongreen (heterotrophic). They are so diverse that different protists have previously been classified as fungi, animals, and plants!

The protists make an enormous impact on the biosphere, both positively and negatively. Among their greatest contribution is the production of oxygen, because these are the photosynthetic protistans. Many are given the common name "algae."

Pond scum, frog spittle, seaweed, the stuff that clogs your aquarium if it's not cleaned routinely, the debris on an ocean beach after a storm at sea, the nuisance organisms of a lake—these are the images that pop into our mind when we first think about the organisms called algae. But many algae are also **phytoplankton,** the weakly swimming or floating algae, at the base of the aquatic food chain.

The following organisms are examined in this exercise:

Phylum	Common Name
Euglenozoa	Euglenoids
Chrysophyta	Yellow-green algae and diatoms
Dinoflagellata	Dinoflagellates
Rhodophyta	Red algae
Phaeophyta	Brown algae
Chlorophyta	Green algae
Oomycotes	Water molds
Amoebozoa	Slime molds
Alveolata	Apicomplexans

Note the ending *-phyta* for all the phylum names. This derives from the Greek root word *phyt*, which means "plant." In the not too distant past, these organisms were considered plants because they were photosynthetic. While the endings remain, today they are considered to be members of the kingdom Protista.

A. Phylum Stramenopila

A.1. Oomycotes: Water Molds *(About 15 min.)*

Oomycotes are commonly referred to as water molds because they are primarily aquatic organisms. While not all members of this division grow in freestanding water, they all rely on the presence of water to spread their asexual spores.

During sexual reproduction, all members produce large nonmotile female gametes, the eggs. These eggs are contained in sex organs (**gametangia;** singular, *gametangium*) called **oogonia.** By contrast, the male gametes are nothing more than *sperm nuclei* contained in **antheridia,** the male gametangia.

MATERIALS

Per student:

- culture of *Saprolegnia* or *Achlya*
- glass microscope slide
- compound microscope

A.1.a. Phytophthora

Some of the most notorious and historically important plant pathogens known to humans are water molds. Included in this group is *Phytophthora infestans,* the fungus that causes the disease known as late blight of potato. This disease spread through the potato fields of Ireland between 1845 and 1847. Most of the potato plants died, and 1 million Irish working-class citizens who had depended on potatoes as their primary food source starved to death. Another 2 million emigrated, many to the United States.

Perhaps no other plant pathogen so poignantly illustrates the importance of environmental factors in causing disease. While *Phytophthora infestans* had been present previous to 1845 in the potato-growing fields of Ireland, it was not until the region experienced several consecutive years of wet and especially cool growing seasons that late blight became a major problem. We can easily observe the importance of these climate-associated factors by studying another *Phytophthora* species, *P. cactorum.*

MATERIALS

Per student:

- culture of *Phytophthora cactorum*
- glass microscope slide
- dissecting needle
- compound microscope

Per lab room:

- refrigerator

or

Per student group (4):

- ice bath

PROCEDURE

1. Obtain a culture of *P. cactorum* that has been flooded with distilled water. Note that the agar has been removed from the edges of the petri dish and that the mycelium has grown from the agar edge into the water.
2. Place a glass slide on the stage of your compound microscope. This slide will serve as a platform for the culture, allowing you to use the mechanical stage of the microscope to move the culture (if the microscope is so equipped).
3. Remove the lid from the culture, carefully place the culture dish on the platform, and examine the culture with the low-power objective.
4. Search the surface of the mycelium, especially at the edges, until you find the rather *pear-shaped* **zoosporangia**. Switch to the medium-power objective for closer observation and then draw in Figure 13-2 a single zoosporangium.
5. Return your microscope to the low-power objective, remove the culture, replace the cover, and place it in a refrigerator or on ice for 15–30 minutes.
6. After the incubation time, again observe the zoosporangia microscopically. Find one in which the **zoospores** are escaping from the zoosporangium. Each zoospore has the potential to grow into an entirely new mycelium! Draw the zoospores in Figure 13-2.

Figure 13-2 Drawing of zoosporangium and zoospores of *Phytophthora cactorum.*
(_____×).
Labels: zoosporangium, zoospores

Like the other water molds, *Phytophthora* reproduces sexually. Your cultures contain the sexual structures as well as asexual zoosporangia.

7. Using a dissecting needle, cut a section about 1 cm square from the agar colony and *invert* it on a glass slide (so that the bottom side of the agar is now uppermost).
8. Place a coverslip on the agar block and observe with your compound microscope, first with the low-power objective, then with the medium-power, and finally with the high-dry objective.
9. Identify the spherical **oogonia** that contain **eggs** or thick-walled **zygotes** (depending on the stage of development). If present, the **antheridia** are club shaped and plastered to the wall of the oogonium.
10. In Figure 13-3, draw an oogonium, eggs (zygotes), and antheridia.

B. Phylum Amoebozoa: Slime Molds

The slime molds have both plantlike and animal-like characteristics. Because they engulf their food and lack a cell wall in their vegetative (nonreproductive) state, they are placed in the kingdom Amoebozoa. However, when they reproduce, they produce spores with a rigid cell wall, similar to plants.

B.1. Amoeba *(About 20 min.)*

Amoebas continually change shape by forming projections called *pseudopodia* (singular, *pseudopodium,* "false foot"). One notorious human pathogen is *Entamoeba histolytica,* the cause of amoebic dysentery.

Figure 13-3 Drawing of the gametangia of *Phytophthora cactorum.*
(_____×).
Labels: oogonium, eggs (zygotes), antheridium

MATERIALS

Per student:

- depression slide
- coverslip
- dissecting needle
- compound microscope

Per group (table):

- carmine, in 2 screw-cap bottles
- tissue paper

Per lab room:

- *Amoeba*—living culture on demonstration at dissecting microscope; disposable pipet

PROCEDURE

1. Observe the *Amoeba*-containing culture on the stage of a demonstration dissecting microscope. (One species of amoeba has the scientific name *Amoeba*.) The microscope has been focused on the bottom of the culture dish, where the amoebas are located. Look for gray, irregularly shaped masses moving among the food particles in the culture.
2. Using a clean pipette, remove an amoeba and place it, along with some of the culture medium, in a depression slide.
3. Examine it with your compound microscope using the medium-power objective. You will need to adjust the diaphragm to increase the contrast (see Exercise 1) because *Amoeba* is nearly transparent.
4. Refer to Figure 13-4. Locate the **pseudopodia.** At the periphery of the cell, identify the **ectoplasm,** a thin, clear layer that surrounds the inner, granular **endoplasm.**

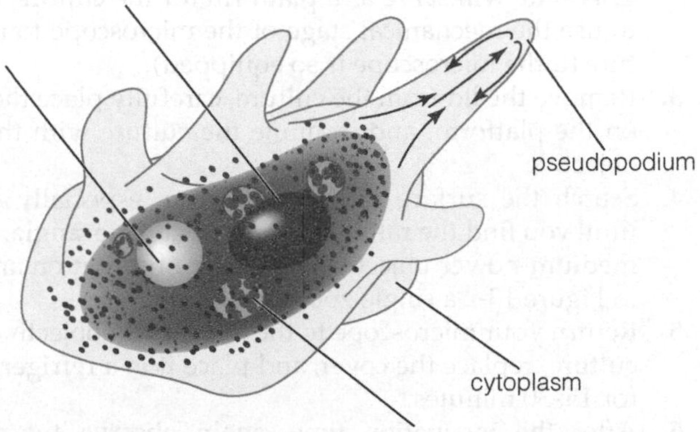

Figure 13-4 Artist's rendering of *Amoeba.*
Labels: ectoplasm, endoplasm, nucleus, contractile vacuole, food vacuole

Watch the organism as it changes shape. Which region of the endoplasm appears to stream, the outer or the inner? _____

This region, called the *plasmasol*, consists of a fluid matrix that can undergo phase changes with the semisolid *plasmagel*, the outer layer of the endoplasm. Pseudopodium formation occurs as the plasmasol flows into new environmental frontiers and then changes to plasmagel.

Numerous granules will be found within the endoplasm. Some of these are organelles; others are food granules.

5. Within the endoplasm, try to locate the **nucleus,** a densely granular, spherical structure around which the cytoplasm is streaming.
6. Find the clear, spherical **contractile vacuoles,** which regulate water balance within the cell. Watch for a minute or two to observe the action of contractile vacuoles. Label Figure 13-4.

Amoeba feeds by a process called **phagocytosis,** engulfing its food. Pseudopodia form around food particles, and then the pseudopodia fuse, creating a **food vacuole** within the cytoplasm. Enzymes are then emptied into the food vacuole, where the food particle is digested into a soluble form that can pass through the vacuolar membrane. You can stimulate feeding behavior by drawing carmine under the coverslip:

7. Place a drop of distilled water against one edge of the coverslip.
8. Pick up some carmine crystals by dipping a dissecting needle into the bottle; and deposit them into the water droplet.
9. Draw the suspension beneath the coverslip by holding a piece of absorbent tissue against the coverslip on the side *opposite* the carmine suspension.
10. Observe the *Amoeba* with your microscope again—you may catch it in the act of feeding.

C. Phylum Alveolata: Apicomplexans *(About 20 min.)*

All apicomplexans are parasites, infecting a wide range of animals, including humans. *Plasmodium vivax* causes one type of malaria in humans. In this section, you will study its life cycle with demonstration slides. Refer to Figure 13-3 as you proceed.

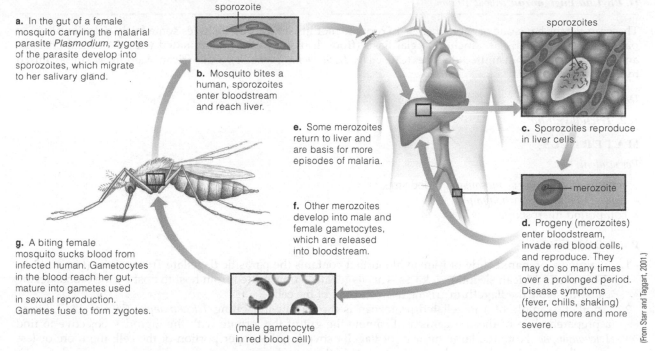

a. In the gut of a female mosquito carrying the malarial parasite *Plasmodium*, zygotes of the parasite develop into sporozoites, which migrate to her salivary gland.

sporozoite

b. Mosquito bites a human, sporozoites enter bloodstream and reach liver.

e. Some merozoites return to liver and are basis for more episodes of malaria.

sporozoites

c. Sporozoites reproduce in liver cells.

f. Other merozoites develop into male and female gametocytes, which are released into bloodstream.

merozoite

d. Progeny (merozoites) enter bloodstream, invade red blood cells, and reproduce. They may do so many times over a prolonged period. Disease symptoms (fever, chills, shaking) become more and more severe.

g. A biting female mosquito sucks blood from infected human. Gametocytes in the blood reach her gut, mature into gametes used in sexual reproduction. Gametes fuse to form zygotes.

(male gametocyte in red blood cell)

(From Starr and Taggart, 2001.)

Figure 13-5 Life cycle of *Plasmodium vivax*, causal agent of malaria.

Plasmodium vivax

MATERIALS

Per lab room:

- demonstration slide of *Plasmodium vivax*, sporozoites
- demonstration slide of *P. vivax*, merozoites
- demonstration slide of *P. vivax*, immature gametocytes

PROCEDURE

1. Examine Figure 13-5a. *P. vivax* is transmitted to humans through the bite of an infected female *Anopheles* mosquito. The mosquito serves as a **vector**, a means of transmitting the organism from one host to another. Male mosquitos cannot serve as vectors, because they lack the mouth parts for piercing skin and sucking blood. If the mosquito carries the pathogen, **sporozoites** enter the host's bloodstream with the saliva of the mosquito.
2. Examine the demonstration slide of sporozoites (Figure 13-5b).
3. The sporozoites travel through the bloodstream to the liver, where they penetrate certain cells, grow, and multiply. When released from the liver cells, the parasite is in the form of a **merozoite** and infects the red blood cells. Figures 13-5c and 13-5d show this process.

(There are two intervening stages between sporozoites and merozoites. These are the *trophozoites* and *schizonts*, both developmental stages in red blood cells.)

4. Now examine the demonstration slide illustrating merozoites in red blood cells (Figure 13-5d).
5. Within the red blood cells, merozoites divide, increasing the merozoite population. At intervals of 48 or 72 hours, the infected red blood cells break down, releasing the merozoites. At this time, the infected individual has disease symptoms, including fever, chills, and shaking caused by the release of merozoites and metabolic wastes from the red blood cells. Some of these merozoites return to the liver cells, where they repeat the cycle and are responsible for recurrent episodes of malaria.
6. Merozoites within the bloodstream can develop into **gametocytes.** For development of a gametocyte to be completed, the gametocyte must enter the gut of the mosquito. This occurs when a mosquito feeds upon an infected (diseased) human.
7. Observe the demonstration slide of an **immature gametocyte** in a red blood cell (Figure 13-5f).
8. Within the gut of the mosquito, the gametocyte matures into a gamete (Figure 13-5g). When gametes fuse, they form a zygote that matures into an **oocyst.** Within each oocyst, sporozoites form, completing the life cycle of *Plasmodium vivax*. These sporozoites migrate to the mosquito's salivary glands to be injected into a new host.

D. Phylum Euglenozoa *(About 10 min.)*

These protistans have one or more flagella to provide motility. The group includes some fairly notorious human parasites that cause disease, including giardiasis (from drinking water contaminated with the causal protozoans), some sexually transmitted diseases (caused by *Trichomonas vaginalis*), and African sleeping sickness, caused by *Trypanosoma brucei*.

D.1. Trypanosoma

D.2. Trichonympha

MATERIALS

Per student:

- prepared slide of *Trypanosoma* in blood smear
- prepared slide of *Trichonympha*
- compound microscope

PROCEDURE

1. Examine a prepared slide of human blood that contains the parasitic flagellate *Trypanosoma* (Figure 13-6), the cause of African sleeping sickness. This flagellate is transmitted from host to host by the bloodsucking tsetse fly. Note the **flagellum** arising from one end of the cell.
2. Another example of a flagellated protozoan is the termite-inhabiting *Trichonympha* (Figure 13-7). Study a prepared slide of these organisms. Examine the gut of the termite with the high-dry objective to find *Trichonympha*. Note the large number of **flagella** covering the upper portion of the cell, the more or less centrally located **nucleus** and wood fragments in the cytoplasm.

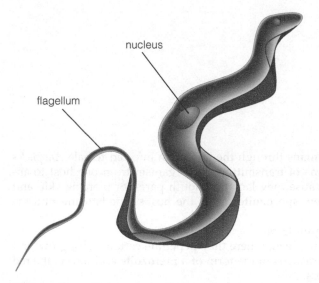

Figure 13-6 *Trypanosoma.*

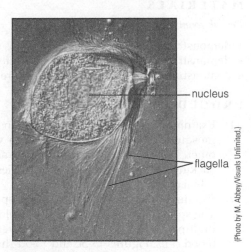

(Photo by M. Abbey/Visuals Unlimited.)

Figure 13-7 *Trichonympha* (900×).

The association of the termite and *Trichonympha* is an example of **obligate mutualism,** in which neither organism is capable of surviving without the other. Termites lack the enzymes to metabolize cellulose, a major component of wood. Wood particles ingested by termites are engulfed by *Trichonympha*, whose enzymes break the cellulose into soluble carbohydrates that are released for use by the termite.

13.3   Phylum Chrysophyta: Diatoms *(About 10 min.)*

The phylum Chrysophyta includes yellow green algae, and diatoms. You will examine only the diatoms here.

   Diatoms are called the organisms that live in glass houses because their cell walls are composed largely of opaline *silica* ($SiO_2 \cdot nH_2O$). Diatoms are important as primary producers in the food chain of aquatic environments, and their cell walls are used for a wide variety of industrial purposes, ranging from the polishing agent in toothpaste to a reflective roadway paint additive. Massive deposits of cell walls of long-dead diatoms make up diatomaceous earth (Figure 13-8).

MATERIALS

Per student:

- microscope slide
- coverslip
- prepared slide of freshwater diatom
- compound microscope

Per group (table):

- diatomaceous earth
- dH$_2$O in 2 dropping bottles

Per lab room:

- diatoms—living culture; disposable pipet

PROCEDURE

1. Prepare a wet mount of living diatoms. Use the high-dry objective to note the golden brown chloroplasts within the cytoplasm and the numerous holes in the cell walls (shells).
2. In Figure 13-9, sketch several of the diatoms you are observing.
3. Now obtain a prepared slide of diatoms. These cells have been "cleaned," making the perforations in the cell wall especially obvious if you close the iris diaphragm on your microscope's condenser to increase the contrast. Study with the high-dry objective.

The pattern of the holes in the walls are characteristic of a given species. Before the advent of electronic techniques, microscopists observed diatom walls to assess the quality of microscope lenses. The resolving power (see discussion of resolving power in Exercise 1) could be determined if one knew the diameter of the holes under observation.

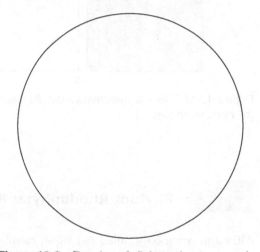

Figure 13-8 Diatomaceous earth quarry near Quincy, Washington.

(Photo courtesy Dan Williams.)

Figure 13-9 Drawing of diatoms (_____×).

13.4   Phylum Dinoflagellata: Dinoflagellates *(About 10 min.)*

Dinoflagellates spin as they move through the water due to the position of their flagellum. On occasion, populations of certain dinoflagellates may increase dramatically, causing the seas to turn red or brown. These are the **red tides,** which can devastate fish populations because neurotoxins produced by the dinoflagellates poison fish.

MATERIALS

Per student:

- prepared slide of a dinoflagellate (for example, *Gymnodinium, Ceratium,* or *Peridinium*)
- compound microscope

Per lab room:

- dinoflagellate—living culture; disposable pipet (optional)

PROCEDURE

1. With the high-dry objective of your compound microscope, examine a prepared slide or living representative.
2. Compare what you are observing with Figures 13-10a and b.
3. Attempt to identify the stiff cellulosic plates encasing the cytoplasm.
4. Locate the two grooves formed by the junction of the plates. This is where the flagella are located. If you are examining living specimens, chloroplasts may be visible beneath the cellulose plates.
5. Examine the scanning electron micrograph of a dinoflagellate in Figure 13-11.

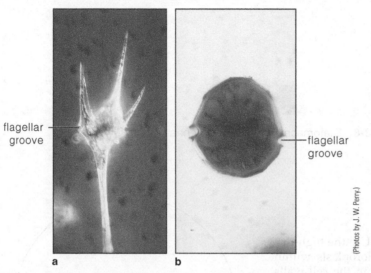

a   b

(Photos by J. W. Perry.)

Figure 13-10 Two representative dinoflagellates. **(a)** *Ceratium* (240×). **(b)** *Peridinium* (640×).

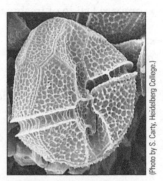

(Photo by S. Carty, Heidelberg College.)

Figure 13-11 Scanning electron micrograph of a dinoflagellate (1000×).

13.5   Phylum Rhodophyta: Red Algae *(About 10 min.)*

Although commonly called red algae, members of the Rhodophyta vary in color from red to green to purple to greenish-black. The color depends on the quantity of their photosynthetic accessory pigments, the **phycobilins,** which are blue and red. These accessory pigments allow capture of light energy across the entire visible spectrum. This energy is passed on to chlorophyll for photosynthesis. One phycobilin, the red phycoerythrin, allows some red algae to live at great depths where red wavelengths, those of primary importance for green and brown algae, fail to penetrate.

Which wavelengths (colors) would be absorbed by a red pigment?

_____

Most abundant in warm marine waters, red algae are the source of **agar** and carageenan substances extracted from their cell walls. Agar is the solidifying agent in microbiological media and carageenan is a thickener used in ice cream and bean.

MATERIALS

Per lab room:

- culture dish containing dried agar and petri dish of hydrated agar
- demonstration slide of *Porphyridium*
- demonstration specimen and slide of *Porphyra* (nori)

PROCEDURE

1. Observe the dried agar that is on demonstration.
2. Now observe the petri dish containing agar that has been hydrated, heated, and poured into the dish.

3. Examine these cells of the unicellular *Porphyridium* at the demonstration microscope, noting the reddish chloroplast.
4. In Figure 13-12, sketch a cell of *Porphyridium*.
5. Examine a portion of the multicellular membranous *Porphyra* specimen on demonstration. Draw what you see in Figure 13-13.
6. Compare the specimen with the photo of live *Porphyra* (Figure 13-14).
7. In the adjacent demonstration microscope, note the microscopic appearance of *Porphyra* (Figure 13-15). Note that the clear areas are actually the cell walls.

Porphyra is used extensively as a food substance in Asia, where it's commonly sold under the name *nori*. In Japan, nori production is valued at $20 million annually.

Figure 13-12 Drawing of *Porphyridium* (_____ ×).

Figure 13-13 Drawing of the macroscopic appearance of *Porphyra* (nori) (_____ ×).

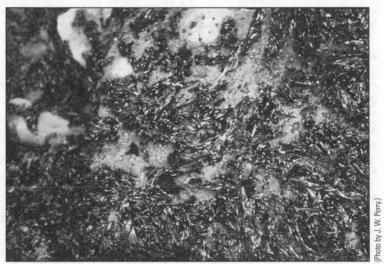

(Photo by J. W. Perry.)

Figure 13-14 Living *Porphyra* (0.5×).

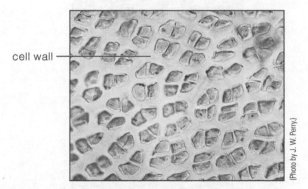

cell wall

(Photo by J. W. Perry.)

Figure 13-15 Microscopic appearance of *Porphyra* (nori) (278×).

The vast majority of the brown algae are found in cold, marine environments. All members are multicellular, and most are macroscopic. Their color is due to the accessory pigment fucoxanthin, which is so abundant that it masks the green chlorophylls. Some species are used as food, while others are harvested for fertilizers. Of primary economic importance is **algin,** a cell wall component of brown algae that makes ice cream smooth, cosmetics soft, and paint uniform in consistency, among other uses.

A. *Kelps*—Laminaria and Macrocystis

Kelps are large (up to 100 m long), complex brown algae. They are common along the seashores in cold waters.

MATERIALS

Per lab room:

- demonstration specimen of *Laminaria*
- demonstration specimen of *Macrocystis*

PROCEDURE

Examine specimens of *Laminaria* (Figure 13-16) and *Macrocystis* (Figure 13-17). On each, identify the rootlike **holdfast** that anchors the alga to the substrate; the **stipe,** a stemlike structure; and the leaflike **blades.**

B. *Rockweed*—Fucus

Fucus is a common brown alga of the coastal shore, especially abundant attached to rocks where the plants are periodically wetted by splashing waves and the tides.

MATERIALS

Per lab room:

- demonstration specimen of *Fucus*

PROCEDURE

1. Examine demonstration specimens of *Fucus* (Figure 13-18), noting the branching nature of the body.

2. Locate the short **stipe** and **blade.**

Figure 13-16 *Laminaria.*

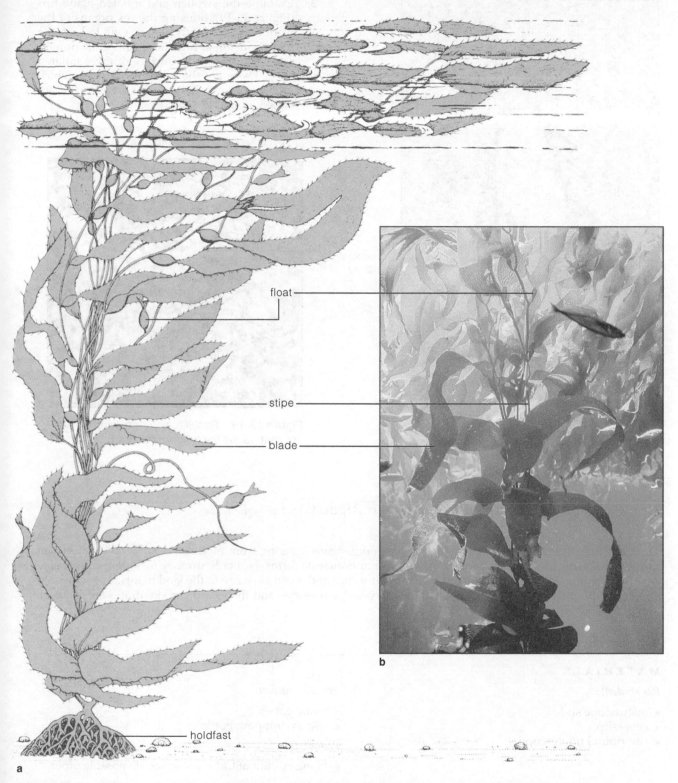

float

stipe

blade

holdfast

a

b

Figure 13-17 *Macrocystis.* (a) Artist conception. (After M. Neushul in Scagel et al., 1982.) (b) Photograph taken at Monterey Bay Aquarium, Monterey, CA. (Photo by J. W. Perry.) (Both 0.01×)

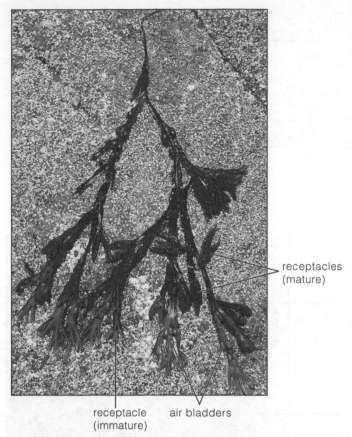

Figure 13-18 The rockweed, *Fucus* (0.60×). (Photo by J. W. Perry.)

receptacles (mature)

receptacle (immature)   air bladders

3. Examine the swollen and inflated blade tips (Figure 13-19), housing the sex organs of the plant and apparently also serving to keep the plant buoyant at high tide. Notice the numerous tiny dots on the surface of these inflated ends. These are the openings through which motile sperm cells swim to fertilize the enclosed, nonmotile egg cells.

receptacles   blade

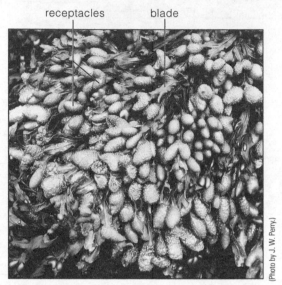

Figure 13-19 *Fucus* (colony, showing swollen tips that house the sex organs (0.35×).

(Photo by J. W. Perry.)

13.7   Phylum Chlorophyta: Green Algae *(About 15 min.)*

The Chlorophyta is a diverse assemblage of green organisms, ranging from motile and nonmotile unicellular forms to colonial, filamentous, membranous, and multinucleate forms. Not only do they include the most species, they are also important phylogenetically because ancestral green gave rise to the land plants. As you would expect, the photosynthetic pigments—both primary and accessory—and the stored carbohydrates are identical in the green algae and land plants.

A. Chlamydomonas—A Motile Unicell

MATERIALS

Per student:

- microscope slide
- coverslip
- compound microscope

Per student pair:

- tissue paper
- I₂KI in dropping bottle

Per lab room:

- living culture of *Chlamydomonas*, disposable pipet

PROCEDURE

1. Prepare a wet mount of *Chlamydomonas* cells from the culture provided.
2. First examine with the low-power objective of your compound microscope. Notice the numerous small cells swimming across the field of view.
3. It will be difficult to study the fast-swimming cells, so kill the cells by adding a drop of I₂KI to the edge of the coverslip; draw the I₂KI under the coverslip by touching a folded tissue to the opposite edge. To observe the cells, switch to the highest power objective available.
4. Identify the large, green cup-shaped **chloroplast** filling most of the cytoplasm.

5. *Chlamydomonas* stores its excess photosynthate as *starch grains*, which appear dark blue or black when stained with I₂KI. Locate the starch grains.

6. If the orientation of the cell is just right, you may be able to detect an orange **stigma** (eyespot) that serves as a light receptor.

7. Close the iris diaphragm to increase the contrast and find the two **flagella** at the anterior end of the cell.

8. Examine Figure 13-20, a transmission electron micrograph of *Chlamydomonas*. The electron microscope makes much more obvious the structures you can barely see with your light microscope. Notice the magnification.

9. The green algae have a specialized center for starch synthesis located within their chloroplast, the **pyrenoid**. Find the pyrenoid.

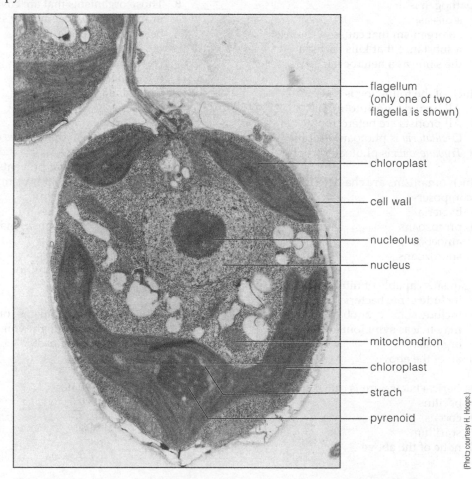

flagellum
(only one of two
flagella is shown)

chloroplast

cell wall

nucleolus

nucleus

mitochondrion

chloroplast

strach

pyrenoid

(Photo courtesy H. Hoops.)

Figure 13-20 Transmission electron micrograph of *Chlamydomonas* (9750×).

B. Ulva—A Membranous Form

The second representative of the green algae illustrates the morphological form of a membranous (tissuelike) body.

MATERIALS

Per lab room:

■ demonstration specimen of *Ulva*

PROCEDURE

Examine living or preserved specimens of *Ulva*, commonly known as sea lettuce (Figure 13-21). The broad, leaflike body is called a *thallus*, a general term describing a vegetative body with relatively little cell differentiation. The *Ulva* body is only two cells thick!

(Photo by J. W. Perry.)

Figure 13-21 The sea lettuce, *Ulva* (0.5×).

_____ 1. Some members of the Domain Bacteria have
(a) a nucleus
(b) membrane-bound organelles
(c) chloroplasts
(d) photosynthetic ability

_____ 2. A pathogen is
(a) a disease
(b) an organism that causes a disease
(c) a substance that kills bacteria
(d) the same as a heterocyst

_____ 3. Which of the following is true?
(a) All bacteria are autotrophic
(b) All protists are heterotrophic
(c) *Oscillatoria* is photoautotrophic
(d) *Trypanosoma* is photoautotrophic

_____ 4. Which organisms are characterized as decomposers?
(a) bacteria
(b) protozoans
(c) amoebas
(d) sporozoans

_____ 5. Organisms capable of nitrogen fixation
(a) include some bacteria
(b) include some cyanobacteria
(c) may live as symbionts with other organisms
(d) all of the above

_____ 6. A spherical bacterium is called a
(a) bacillus
(b) coccus
(c) spirillum
(d) none of the above

_____ 7. Gram stain is used to distinguish between different
(a) bacteria
(b) protistans
(c) dinoflagellates
(d) all of the above

_____ 8. Those organisms that are covered by numerous, tiny locomotory structures belong to the phylum
(a) Euglenozoa
(b) Stramenopila
(c) Amoebozoa
(d) Ciliata

_____ 9. A vector is
(a) an organism that causes disease
(b) a disease
(c) a substance that prevents disease
(d) an organism that transmits a disease causing organism

_____ 10. The organism that causes malaria is
(a) a pathogen
(b) *Plasmodium vivax*
(c) carried by a mosquito
(d) all of the above

_____ 11. Which of these organisms (or parts of the organisms) might you find as an ingredient in toothpaste?
(a) euglenoids
(b) diatoms
(c) dinoflagellates
(d) stoneworts

12. Red tides are caused by
    (a) dinoflagellates
    (b) red algae
    (c) brown algae
    (d) diatoms

13. Agar is derived from
    (a) red algae
    (b) brown algae
    (c) green algae
    (d) all of the above

14. Which is the correct plural form of the word for the organisms studied in this exercise?
    (a) alga
    (b) algae
    (c) algas
    (d) algaes

15. Phycobilins are
    (a) photosynthetic pigments
    (b) found in the red algae
    (c) blue and red pigments
    (d) all of the above

16. The cell wall component algin is
    (a) found in the brown algae
    (b) used in the production of ice cream
    (c) used as a medium on which microorganisms are grown
    (d) both a and b

17. Specifically, female sex organs are known as
    (a) oogonia
    (b) gametangia
    (c) antheridia
    (d) zygotes

18. A reagent that stains the stored food of a green alga black is
    (a) India ink
    (b) I_2KI
    (c) methylene blue
    (d) both a and b

19. The starch production center within many algal cells is the
    (a) nucleus
    (b) cytoplasm
    (c) stipe
    (d) pyrenoid

20. The phylum of organisms *most* closely linked to the evolution of land plants is the
    (a) Chlorophyta
    (b) Charophyta
    (c) Phaeophyta
    (d) Euglenozoa

EXERCISE 13

Bacteria and Protists

POST-LAB QUESTIONS

Introduction

1. What major characteristic distinguishes bacteria from protists?

13.1A Bacteria

2. What form of bacterium is shown in this photomicrograph?

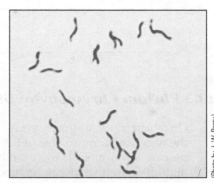

(1100×).

13.2A Phylum Stramenopila

3. While the organism that resulted in the Irish potato famine of 1845–1847 had long been present, environmental conditions that occurred during this period resulted in the destructive explosion of disease. Indicate what those environmental conditions were and why they resulted in a major disease outbreak.

13.2B Phylum Amoebozoa

4. This photomicrograph was taken from a prepared slide of a stained specimen. You observed unstained living specimens in lab. Describe the mechanism by which it moves from place to place.

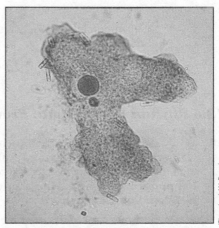

(425×).

5. What is phagocytosis? What function does it serve?

Food for Thought

6. If you were to travel to a region where rice is grown in paddies, you would see lots of the water fern *Azolla* growing in the water. A farmer would tell you this is done because *Azolla* is considered a "natural fertilizer." Explain why this is the case.

7. Based on your knowledge of the life history of *Plasmodium vivax*, suggest two methods for controlling malaria. Explain why each method would work.

    a. _____

    b. _____

13.3 Phylum Chrysophyta: Diatoms

8. On a field trip to a stream, you collect a leaf that has fallen into the water and scrape some of the material from its surface, and prepare a wet mount. You examine your preparation with the high-dry objective of your compound microscope, finding the organism pictured at the right. What substance makes up a significant portion of the cell wall?

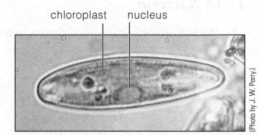

chloroplast   nucleus

(235×).

(Photo by J. W. Perry.)

13.5 Phylum Rhodophyta: Red Algae

9. While wading in the warm salt water off the beaches of the Florida Keys on spring break, you stoop down to look at the feathery alga shown here.

    a. What pigment gives this organism its coloration?

    b. What commercial products are derived from this phylum?

(0.25×).

(Photo by J. W. Perry.)

13.6 Phylum Phaeophyta: Brown Algae

10. While walking along the beach at Point Lobos, California, the fellow pictured walks up to you with alga in hand. Figuring you to be a college student who has probably had a good introductory biology course, he asks if you know what it is.

    a. What is the cell wall component of the organism that has commercial value?

    b. Name three uses for that cell wall component.

(Photo by J. W. Perry.)

Food for Thought

11. Some botanists consider the stoneworts to be a link between the higher plants and the algae. As you will learn in future exercises, higher plants, such as the mosses, have both haploid and diploid stages that are *multicellular*.

    a. Describe the multicellular organism in the charophytes.

    b. Is this organism haploid or diploid?

    c. Is the zygote haploid or diploid?

    d. Is the zygote unicellular or multicellular?

12. a. What color are the marker lights at the edge of an airport taxiway?

    _____

    b. Are the wavelengths of this color long or short, relative to the other visible wavelengths?

    _____

    c. Which wavelengths penetrate deepest into water, long or short?

    _____

    d. Make a statement regarding why phycobilin pigments are present in deep-growing red algae.

    e. What benefit is there to the color of airport taxiway lights for a pilot attempting to taxi during foggy weather?

13. a. How is an algal holdfast similar to a root?

    b. How is it different?

14. Why do you suppose the Swedish automobile manufacturer Volvo chose this company name?

15. List three reasons why algae are useful and important to life.

    a.

    b.

    c.

Fungi

OBJECTIVES

After completing this exercise, you will be able to

1. define *parasite, saprobe, mutualist, gametangium, hypha, mycelium, multinucleate, sporangium, rhizoid, zygosporangium, ascus, conidium, ascospore, ascocarp, basidium, basidiospore, basidiocarp, lichen, mycorrhiza;*

2. recognize representatives of the major phyla of fungi;

3. distinguish structures that are used to place various representatives of the fungi in their proper phyla;

4. list reasons why fungi are important;

5. distinguish between the structures associated with asexual and sexual reproduction described in this exercise;

6. identify the structures (in **boldface**) of the fungi examined;

7. determine the effect of light on certain species of fungi.

INTRODUCTION

As you walk in the woods following a warm rain, you are likely to be met by a vast assemblage of colorful fungi. Some of them grow on dead or diseased trees, some on the surface of the soil, others in pools of water. Some are edible, some deadly poisonous.

Fungi (kingdom Fungi) are *heterotrophic* organisms; that is, they are incapable of producing their own food material. They secrete enzymes from their bodies that digest their food externally. The digested materials are then absorbed into the body.

Depending on the relationship between the fungus and its food source, fungi can be characterized in one of three ways:

1. Parasitic fungi (**parasites**) obtain their nutrients from the organic material of another living organism, and in doing so adversely affect the food source, often causing death.
2. Saprotrophic fungi (**saprobes**) grow on nonliving organic (carbon-containing) matter. Some are even **mutualists.**
3. Mutualistic fungi (**mutualists**) form a partnership beneficial to both the fungus and its host.

Fungi, along with the bacteria, are essential components of the ecosystem as decomposers. These organisms recycle the products of life, making the products of death available so that life may continue. Without them we would be hopelessly lost in our own refuse. Fungi and fungal metabolism are responsible for some of the food products that enrich our lives—the mushrooms of the field, the blue cheese of the dairy case, even the citric acid used in making soft drinks.

The kingdom Fungi is divided into four separate phyla, based on structures formed during sexual reproduction (Table 14.1). Certain fungi (the "imperfect fungi") do not reproduce sexually; hence they are considered an informal "group."

TABLE 14-1	Classification of the Fungi	
Phylum	**Common Name**
Chytridiomycota	Chytrids
Zygomycota	Zygosporangium-forming fungi
Ascomycota	Sac fungi
Basidiomycota	Club fungi
"Imperfect Fungi"	Fungi without sexual reproduction

14.1   Phylum Chytridiomycota *(About 5 min.)*

Commonly known as chytrids, these fungi are among the most simple in body form. While most live on dead organic matter, some are parasitic on economically important plants, resulting in damage and/or death of the host plant. Parasitic organisms that cause disease are called **pathogens.**

MATERIALS

Per lab room:

■ preserved specimen of potato tuber with black wart disease

PROCEDURE

1. Examine the preserved specimen (if available) and Figure 14-1 of the potato tuber that exhibits the disease known as black wart. Notice the warty eruptions on the surface of the tuber. The warts are caused by the presence of numerous cells of a chytrid infecting the tuber.
2. Now look at Figure 14-2, a photomicrograph of a chytrid similar to that which causes black wart disease.

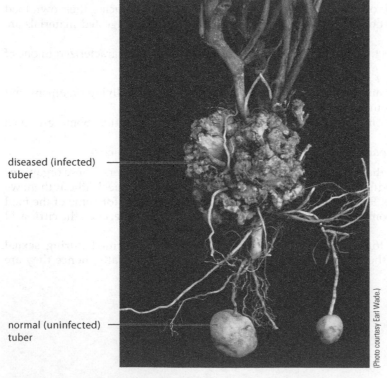

diseased (infected) tuber

normal (uninfected) tuber

(Photo courtesy Earl Wade.)

Figure 14-1   Potato tuber infected by black wart pathogen (0.5×).

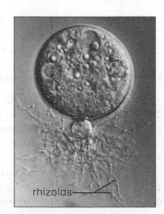

rhizoids

Figure 14-2   Chytrid. (71×). (Photo by M. S. Fuller, *Zoosporic Fungi in Teaching and Research*, Fuller and A. Jaworski (eds.), 1987, Southeastern Publishing Company, Athens, GA.)

Commonly called "zygomycetes," all members of this phylum produce a thick-walled zygote called a **zygosporangium.** Most zygomycetes are saprobes. The common black bread mold, *Rhizopus*, is a representative zygomycete. Before the introduction of chemical preservatives into bread, *Rhizopus* was an almost certain invader, especially in high humidity.

MATERIALS

Per student:

- culture of *Rhizopus*
- prepared slide of *Rhizopus*
- dissecting needle
- glass microscope slide
- coverslip
- compound microscope

Per student pair:

- dH$_2$O in dropping bottle

Per lab room:

- demonstration culture of *Rhizopus* zygosporangia, on dissecting microscope

PROCEDURE

Examine Figure 14-7, the life cycle of *Rhizopus*, as you study this organism.

1. Obtain a petri dish culture of *Rhizopus*.
2. Observe the culture, noting that the body of the organism consists of many fine strands. These are called **hyphae** (Figure 14-3). All of the hyphae together are called the **mycelium.** Biologists use the term *mycelium* instead of saying "the fungal body."
3. Now note that the culture contains numerous black "dots." These are the **sporangia** (singular, *sporangium*; Figures 14-7a and b). Sporangia contain **spores** (Figures 14-7c–d) by which *Rhizopus* reproduces asexually.
4. Using a dissecting needle, remove a small portion of the mycelium and prepare a wet mount. Examine your preparation with the high-dry objective of your compound microscope.
5. It's likely that when you added the coverslip, you crushed the sporangia, liberating the spores. Are there many or few spores within a single sporangium? _____
6. Identify the **rhizoids** (Figure 14-7a) at the base of a sporangium-bearing hypha. Rhizoids anchor the mycelium to the substrate.
7. Now observe the demonstration culture of sexual reproduction in *Rhizopus* (Figure 14-4). Two different mycelia must grow in close proximity before sexual reproduction can occur. (The difference in the mycelia is genetic rather than structural. Because they are impossible to distinguish, the mycelia are simply referred to as + and – mating types, as indicated in Figure 14-7.)
8. Note the black "line" running down the center of the culture plate. These are the numerous **zygosporangia,** the products of sexual reproduction. Here's how the zygosporangia are formed:

Figure 14-3 Bread mold culture (0.5×).

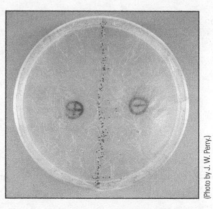

Figure 14-4 Sexual reproduction in the bread mold. Black line consists of zygosporangia (0.5×).

   (a) As the hyphae from each mating type grow close together, chemical messengers produced within the hyphae signal them to produce protuberances (Figure 14-7f).
   (b) When the protuberances make contact, gametangia (Figures 14-5, 14-7g) are produced at their tips. Each gametangium contains many haploid nuclei of a single mating type.
   (c) The wall between the two gametangia then dissolves, and the cytoplasms of the gametangia mix.
   (d) Eventually the many haploid nuclei from each gametangium fuse (Figure 14-7h). The resulting cell contains many diploid nuclei resulting from the fusion of gamete nuclei of opposite mating types. Each diploid nucleus is considered a *zygote*. This multinucleate cell is called a zygospore.

(e) Eventually a thick, bumpy wall forms around this zygospore. Because germination of the zygospore results in the production of a sporangium, the thick-walled zygospore-containing structure is called a zygosporangium. Meiosis takes place inside the zygopsporangium so that the spores formed are haploid, some of one mating type, some of the opposite type (Figures 14-6, 14-7i).

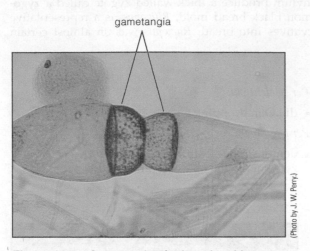

Figure 14-5 Gametangia of a bread mold (230×).

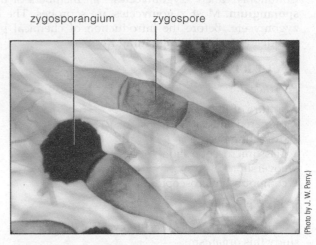

(Photo by J. W. Perry.)

Figure 14-6 Zygospore and zygosporangium of the bread mold (230×).

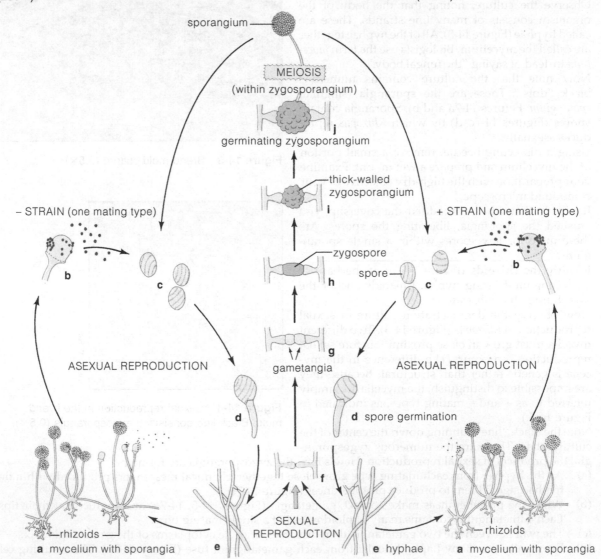

Figure 14-7 Life cycle of *Rhizopus*, black bread mold. (Green structures are 2n.)

9. Obtain a prepared slide of *Rhizopus*. Examine it with the medium- and high-power objective of your compound microscope.

10. Find the stages of sexual reproduction in *Rhizopus*, including gametangia, zygospores, and zygosporangia.

14.3   Experiment: Bread Mold and Food Preservatives *(Setup: about 10 min.)*

Knowing that fungal spores are all around us, why don't we have fungi covering everything we own? Well, given the proper environmental conditions just such a situation can occur. This is particularly true of food products. That's one reason we refrigerate many of our foodstuffs. Manufacturers often incorporate chemicals into some of our foods to retard spoilage, which includes preventing the growth of fungi.

This experiment addresses the hypothesis that *the chemicals placed in bread retard the growth of Rhizopus*.

MATERIALS

Per student group:

- one slice of bread without preservatives
- one slice of bread with preservatives
- suspension culture of bread mold fungus
- sterile dH$_2$O
- 2 large culture bowls

- plastic film
- pipet and bulb
- china marker
- metric ruler

PROCEDURE

1. With the china marker, label the side of one culture bowl "NO PRESERVATIVES," the other "WITH PRESERVATIVES." Place the appropriate slice of bread in its respective bowl.
2. Add a small amount of dH$_2$O to the bottom of each culture bowl. This provides humidity during incubation.
3. Add 5 drops of the spore suspension of the bread mold fungus to the center of each slice of bread.
4. Cover the bowls with plastic film and place them in a warm place designated by your instructor.
5. In Table 14-2, list the ingredients of both types of bread.
6. Make a prediction of what you think will be the outcome of the experiment, writing it in Table 14-2.
7. Check your cultures over the next three days, looking for evidence of fungal growth. Measure the diameter of the mycelium present in each culture, using a metric ruler.
8. Record your measurements and conclusion in Table 14-2, accepting or rejecting the hypothesis.

TABLE 14-2 Growth of Bread Mold			
Ingredients			
Bread without preservatives:			
Bread with preservatives:			
Prediction:			
		Mycelial Diameter (cm)	
Culture	Day 1	Day 2	Day 3
With preservatives			
No preservatives			
Conclusion:			

Design an experiment that would allow you to identify the substance that has the effect you observed.

_____

_____

| 14.4 | Phylum Ascomycota: Sac Fungi *(About 30 min.)* |

Members of the ascomycetes produce spores in a sac, the **ascus** (plural, *asci*), which develops as a result of sexual reproduction. Asexual reproduction takes place when the fungus produces asexual spores called **conidia** (singular, *conidium*). The phylum includes organisms of considerable importance, such as the yeasts crucial to the baking and brewing industries, as well as numerous plant pathogens. A few are highly prized as food, including morels and truffles. Truffles cost in excess of $400 per pound!

MATERIALS

Per student:

- glass microscope slide
- coverslip
- dissecting needle
- prepared slide of *Peziza*
- compound microscope

Per student pair:

- culture of *Eurotium*
- dH₂O in dropping bottle
- large preserved specimen of *Peziza* or another cup fungus

A. Eurotium: *A Blue Mold*

The blue mold *Eurotium* gets its common name from the production of blue-walled asexual conidia.

PROCEDURE

1. Use a dissecting needle to scrape some **conidia** from the agar surface of the culture provided, then prepare a wet mount. (Try to avoid the yellow bodies—more about them in a bit.)
2. Refer to Figure 14-8 as you observe your preparation with the high-dry objective of your compound microscope.
3. Note that the conidia are produced at the end of a specialized hypha that has a swollen tip.

This arrangement has been named *Aspergillus*. This structure, the *conidiophore*, somewhat resembles an aspergillum used in the Roman Catholic Church to sprinkle holy water, from which its name is derived.

These tiny conidia are carried by air currents to new environments, where they germinate to form new mycelia.

(You may be confused about why we italicize *Aspergillus*. The reason is that this is a scientific name. Fungi other than *Eurotium* produce the same type of asexual structure; hence the asexual structure itself is given a scientific name.)

4. Note the yellow bodies on the culture medium. These are the *fruiting bodies,* known as **ascocarps,** which are the products of sexual reproduction (Figure 14-9).
5. With your dissecting needle, remove an ascocarp from the culture and prepare a wet mount.

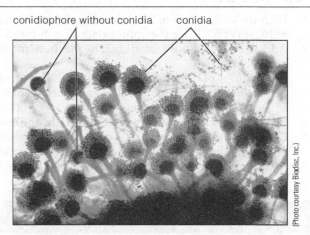

conidiophore without conidia   conidia

Figure 14-8 Conidiophore and conidia of the blue mold, *Eurotium* (138×).

(Photo courtesy Biodisc, Inc.)

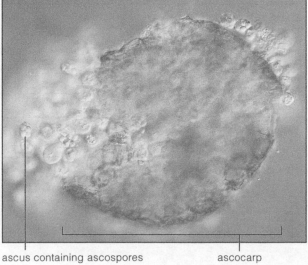

ascus containing ascospores   ascocarp

(Photo by J. W. Perry.)

Figure 14-9 Crushed ascocarp of *Eurotium* (287×).

6. Using your thumb, carefully press down on the coverslip to rupture the ascocarp.
7. Observe your preparation with the medium-power and high-dry objectives of your compound microscope. Identify the **asci,** which contain dark-colored, spherical **ascospores** (Figure 14-9).

The sexual cycle of the sac fungi is somewhat complex and is summarized in Figure 14-10.

The female gametangium, the **ascogonium** (Figure 14-10a; plural, *ascogonia*) is fertilized by male nuclei from antheridia (Figure 14-10a).

The male nuclei (darkened circles in Figure 14-10a) pair with the female nuclei (open circles) but do not fuse immediately.

Papillae grow from the ascogonium, and the paired nuclei flow into these papillae (Figure 14-10b).

Now cell walls form between each pair of sexually compatible nuclei (Figure 14-10c).

Subsequently, the two nuclei fuse; the resultant cell is the diploid ascus (Figure 14-10d). (Consequently, the ascus is a zygote.)

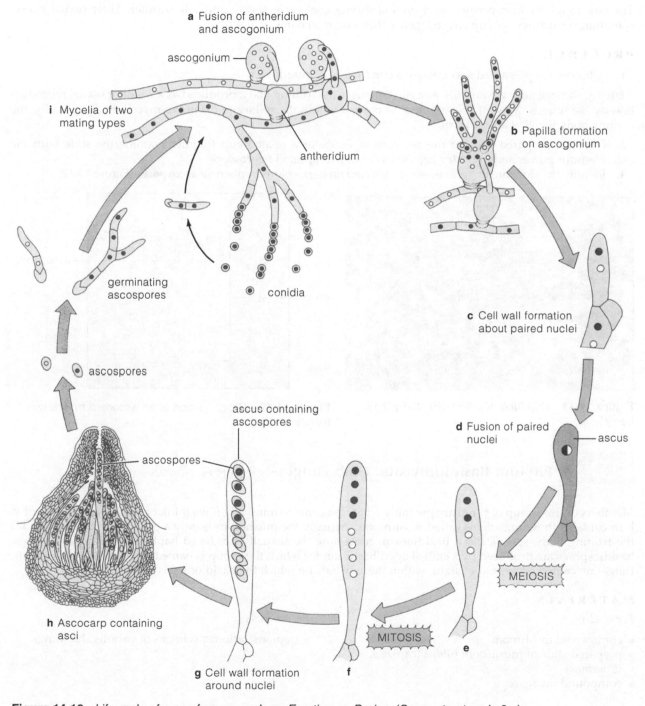

a Fusion of antheridium and ascogonium

ascogonium

i Mycelia of two mating types

antheridium

b Papilla formation on ascogonium

c Cell wall formation about paired nuclei

germinating ascospores

conidia

ascospores

ascus containing ascospores

ascospores

d Fusion of paired nuclei — ascus

MEIOSIS

h Ascocarp containing asci

MITOSIS

e

g Cell wall formation around nuclei

f

Figure 14-10 Life cycle of a sac fungus, such as *Eurotium* or *Peziza*. (Green structure is 2n.)

The nucleus of the ascus undergoes meiosis to form four nuclei (Figure 14-10e).

Mitosis then produces eight nuclei from these four (Figure 14-10f). (Notice that cytokinesis—cytoplasmic division—does not follow meiosis.)

Next, a cell wall forms about each nucleus, some of the cytoplasm of the ascus being included within the cell wall (Figure 14-10g).

Thus, eight haploid uninucleate ascospores (Figure 14-10g) have been formed.

While asci and ascospores are forming, surrounding hyphae proliferate to form the ascocarp (Figure 14-10h) around the asci.

As you see, two different types of ascospores are produced during meiosis. Each ascospore gives rise to a mycelium having only one mating type (Figure 14-10i).

B. Peziza: *A Cup Fungus*

The cup fungi are commonly found on soil during cool early spring and fall weather. Their sexual spore-containing structures are cup shaped, hence their common name.

PROCEDURE

1. Observe the preserved specimen of a cup fungus (Figure 14-11).

Actually, the structure we identify as a cup fungus is the fruiting body, produced as a result of sexual reproduction by the fungus. Most of the organism is present within the soil as an extensive mycelium. Specifically, the fruiting body is called an **ascocarp.**

2. Obtain a prepared slide of the ascocarp of *Peziza* or a related cup fungus. Examine the slide with the medium-power and high-dry objectives of your compound microscope.

3. Identify the elongate fingerlike **asci**, which contain dark-colored, spherical **ascospores** (Figure 14-12).

Figure 14-11   Cup fungi (0.25×). (Photo by J. W. Perry.)

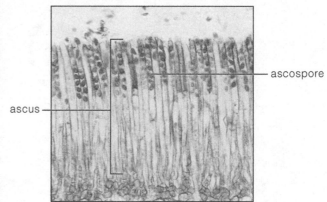

Figure 14-12   Cross section of an ascocarp from a cup fungus (186×).

14.5   Phylum Basidiomycota: Club Fungi *(About 30 min.)*

Members of this group of fungi are probably what first come to mind when we think of fungi, because this phylum contains those organisms called mushrooms. Actually the mushroom is only a portion of the fungus—it's the fruiting body, specifically a **basidiocarp,** containing the sexually produced haploid **basidiospores.** These basidiospores are produced by a club-shaped **basidium** for which the group is named. Much (if not most) of the fungal mycelium grows out of sight, within the substrate on which the basidiocarp is found.

MATERIALS

Per student:

- commercial mushroom
- prepared slide of mushroom pileus (cap), c.s. (*Coprinus*)
- compound microscope

Per lab room:

- demonstration specimens of various club fungi

A. Gill Fungi: The Mushrooms

Mushrooms called gill fungi have their sexual spores produced on sheets of hyphae that look like the gills of fish.

PROCEDURE

1. Obtain a fresh fruiting body, more properly called a **basidiocarp** (Figure 14-13). Identify the **stalk** and **cap.**

2. Look at the bottom surface of the cap, noting the numerous gills. It is on the surface of these gills that the haploid basidiospores are produced. Remember that all the structures you are examining are composed of aggregations of fungal hyphae.

3. Obtain a prepared slide of a cross section of the cap of a mushroom (Figure 14-14). Observe the slide first with the low-power objective of your compound microscope.

4. In the center of the cap, identify the **stalk.** The **gills** radiate from the stalk to the edge of the cap, much as spokes of a bicycle wheel radiate from the hub to the rim.

5. Switch to the high-dry objective to study a single gill (Figure 14-15). Note that the component hyphae produce club-shaped structures at the edge. These are the **basidia** (singular, *basidium*).

6. Each basidium produces four haploid **basidiospores.** Find them. (All four may not be in the same plane of section.)

Each basidiospore is attached to the basidium by a tiny hornlike projection. As the basidiospore matures, it is shot off the projection due to a buildup of turgor pressure within the basidium.

The life cycle of a typical mushroom is shown in Figure 14-16.

Because of genetic differences, basidiospores are of two different types. Mycelia produced by the two types of basidiospores are of two mating strains. Figure 14-16 shows these two different nuclei as open and closed (darkened) circles.

When a haploid basidiospore (Figure 14-16a) germinates, it produces a haploid *primary mycelium* (Figure 14-16b). The primary mycelium is incapable of producing a fruiting body.

Fusion between two sexually compatible mycelia (Figure 14-16b) must occur to continue the life cycle.

Surprisingly, the nuclei of the two mycelia don't fuse immediately; thus, each cell of this so-called *secondary mycelium* (Figure 14-16c) contains two genetically *different* nuclei. This condition is called "dikaryotic."

The secondary mycelium forms an extensive network within the substrate. An environmental or genetic trigger eventually stimulates the formation of the aerial basidiocarp (Figure 14-16d).

Each cell of the basidiocarp has two genetically different nuclei, including the basidia (Figure 14-16e) on the gills.

Within the basidia, the two nuclei fuse; the basidia are now diploid (Figure 14-16f).

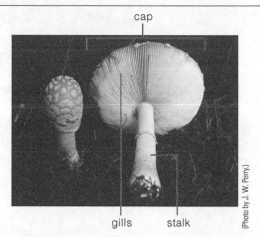

Figure 14-13 Mushroom basidiocarps. The one on the left is younger than that on the right (0.25×).

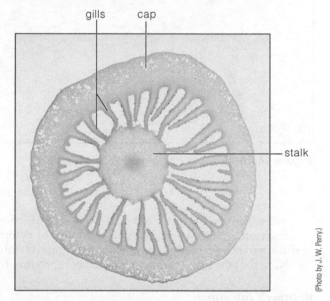

Figure 14-14 Cross section of mushroom cap (23×).

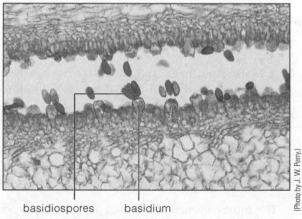

Figure 14-15 High magnification of a mushroom gill (287×).

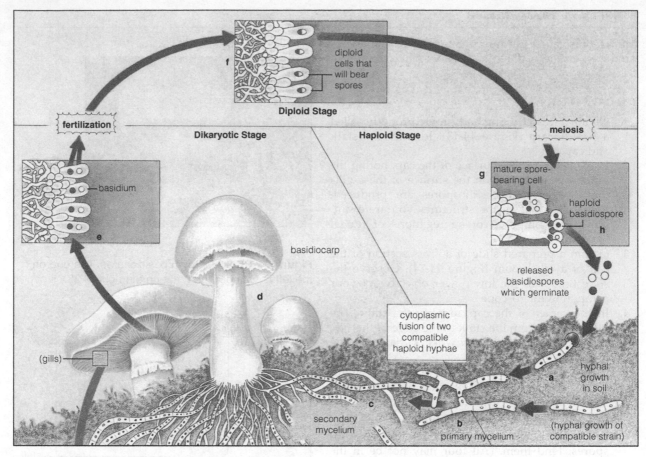

Figure 14-16   Life cycle of a mushroom.

Subsequently, these diploid nuclei undergo meiosis, forming genetically distinct nuclei (Figure 14-16g).

Each nucleus flows with a small amount of cytoplasm through the hornlike projections at the tip of the basidium to form a haploid basidiospore (Figure 14-16h).

Note that the gilled mushrooms do *not* reproduce by means of asexual conidia.

B. Other Club Fungi

A wide variety of basidiomycetes are not mushrooms. Let's examine a few representatives that you're likely to see in the field.

PROCEDURE

1. Examine the representatives of fruiting bodies of other members of the club fungi available in the laboratory.
2. Observe "puffballs" (Figure 14-17).The basidiospores of puffballs are contained within a spherical basidiocarp that develops a pore at the apex. Basidiospores are released when the puffball is crushed or hit by driving rain.
3. Examine "shelf fungi" (Figure 14-18). What you are looking at is actually the basidiocarp of the fungus that has been removed from the surface of a tree.

The presence of a basidiocarp of the familiar shelf fungi indicates that an extensive network of fungal hyphae is growing within a tree, digesting the cells of the wood. As a forester assesses a woodlot to determine the potential yield of usable wood, one of the things noted is the presence of shelf fungi, which indicates low-value (diseased) trees.

The most common shelf fungi are called polypores because of the numerous holes ("pores") on the lower surface of the fruiting body. These pores are lined with basidia bearing basidiospores.

4. Observe the undersurface of the basidiocarp that is on demonstration at a dissecting microscope and note the pores.

Figure 14-17 Puffballs. Note pore for spore escape (0.31×).

(Photo by J. W. Perry.)

Figure 14-18 Shelf fungus (0.12×).

(Photo by J. W. Perry.)

14.6 "Imperfect Fungi" *(About 15 min.)*

This group consists of fungi for which no sexual stage is known. Someone once decided that in order for a life to be complete—to be "perfect"—sex was necessary. Thus, these fungi are "imperfect." Reproduction takes place primarily by means of asexual conidia.

These fungi are among the most economically important, producing antibiotics (for example, one species of *Penicillium* produces penicillin). Others produce the citric acid used in the soft-drink industry; still others are used in the manufacture of cheese. Some are important pathogens of both plants and animals.

MATERIALS

Per student:

- dissecting needle
- glass microscope slide
- coverslip
- prepared slide of *Penicillium* conidia (optional)
- compound microscope

Per student pair:

- dH$_2$O in dropping bottle
- culture of *Alternaria*

Per lab room:

- demonstration of *Penicillium*-covered food *and/or* plate cultures

A. Penicillium*

The genus *Penicillium* has numerous species. The common feature of all species in this genus is the distinctive shape of the hypha that produce the conidia (Figure 14-19).

PROCEDURE

1. Examine demonstration specimens of moldy oranges or other foods. The blue color is attributable to a pigment in the numerous conidia produced by this fungus, *Penicillium*.
2. With a dissecting needle, scrape some of the conidia from the surface of the moldy specimen (or from a petri dish culture plate containing *Penicillium*) and

*Some species of Penicillium reproduce sexually, forming ascocarps. These species are classified in the phylum Ascomycota. However, not all fungi producing the conidiophore form called Penicillium reproduce sexually. Those that reproduce only by asexual means are considered "imperfect fungi."

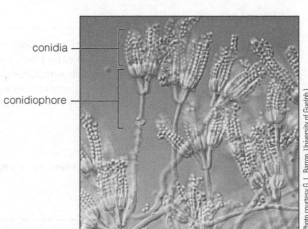

conidia

conidiophore

Figure 14-19 *Penicillium* (300×).

(Photo courtesy G. L. Barron, University of Guelph.)

prepare a wet mount. Observe your preparation using the high-dry objective. (Prepared slides may also be available.)

3. Identify the **conidiophore** and the numerous tiny, spherical **conidia** (Figure 14-19). The name *Penicillium* comes from the Latin word *penicillus*, meaning "a brush." (Appropriate, isn't it?)

B. Alternaria

Perhaps no other fungus causes more widespread human irritation than *Alternaria*, an allergy-causing organism. During the summer, many weather programs announce the daily pollen (from flowering plants, Exercise 17) and *Alternaria* spore counts as an index of air quality for allergy sufferers.

PROCEDURE

1. From the petri dish culture plate provided, remove a small portion of the mycelium and prepare a wet mount.
2. Examine with the high-dry objective of your compound microscope to find the **conidia** (Figure 14-20). Produced in chains, *Alternaria* conidia are multicellular, unlike those of *Penicillium* or *Aspergillus.* This makes identification very easy, since the conidia are quite distinct from all others produced by the fungi.

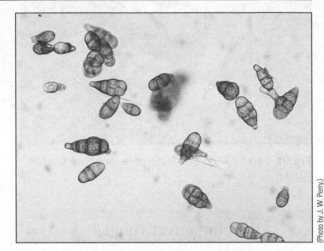
(Photo by J. W. Perry.)

Figure 14-20 Conidia of *Alternaria* (332×).

14.7 Mutualistic Fungi (About 20 min.)

Fungi team up with plants or members of the bacterial or protist kingdoms to produce some remarkable mutualistic relationships. The best way to envision such relationships is to think of them as good interpersonal relationships, in which both members benefit and are made richer than would be possible for each individual on its own.

MATERIALS

Per lab room:

- demonstration specimens of crustose, foliose, and fruticose lichens
- demonstration slide of lichen, c.s. on compound microscope
- demonstration slide of mycorrhizal root, c.s. on compound microscope

A. Lichens

Lichens are organisms made up of fungi and either green algae (kingdom Protista) or cyanobacteria (kingdom Eubacteria). The algal or cyanobacterial cells photosynthesize, and the fungus absorbs a portion of the carbohydrates produced. The fungal mycelium provides a protective, moist shelter for the photosynthesizing cells. Hence, both partners benefit from the relationship.

PROCEDURE

1. Examine the demonstration specimen of a **crustose lichen** (Figure 14-21).
2. What is it growing on?

_____

_____

3. Write a sentence describing the crustose lichen.

_____

_____

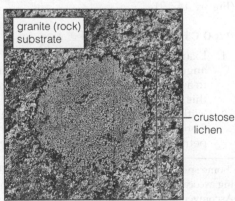
granite (rock) substrate

crustose lichen

(Photo by J. W. Perry.)

Figure 14-21 Crustose lichen (0.5×).

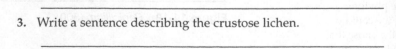

4. Observe the demonstration specimen of a **foliose lichen** (Figure 14-22).

5. What is it growing on? _____

6. Write a sentence describing the foliose lichen.

   _____

   _____

   _____

   _____

7. Examine the demonstration specimen of a **fruticose lichen** (Figure 14-23).

8. What is it growing on?

   _____

   _____

   _____

Figure 14-22 Foliose lichen (0.5×).

(Photo by J. W. Perry.)

9. Write a sentence describing the fruticose lichen.

   _____

   _____

   _____

apothecia (spore containers)

moss

Figure 14-23 Fruticose lichen (1×).

(Photo by J. W. Perry.)

10. Observe the cross section of a lichen at the demonstration microscope. This particular lichen is composed of fungal and green algae cells (Figure 14-24).
11. Locate the **fungal mycelium** and then the **algae cells.**
12. It's likely that this slide has a cup-shaped fruiting body. To which fungal phylum does this fungus belong?

_____

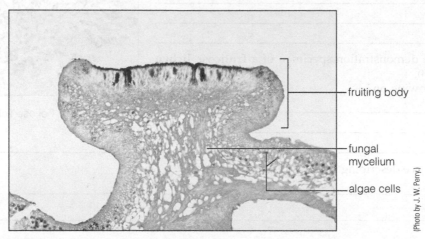

Figure 14-24 Section of lichen (84×).

B. Mycorrhizae

Mycorrhiza literally means "fungus root." A mycorrhiza is a mutualistic association between plant roots and certain species of fungi. In fact, many of the mushroom species (see Section 14-5A) you find in forests (especially those that appear to grow right from the ground) are part of a mycorrhizal association.

The fungus absorbs carbohydrates from the plant roots. The fungal hyphae penetrate much farther into the soil than plant root hairs can reach, absorbing water and dissolved mineral ions that are then released to the plant. Again, both partners benefit. In fact, many plants don't grow well in the absence of mycorrhizal associations.

There are two different types of mycorrhizal roots, **ectomycorrhizae** and **endomycorrhizae.** "Ectos" produce a fungal sheath around the root, and the hyphae penetrate between the cell walls of the root's cortex, but not into the root cells themselves. By contrast, "endos" do not form a covering sheath, and the fungi are found within the cells. We'll look at both types.

MATERIALS

Per student:

- ectomyccorhizal pine root, prep. slide, c.s.
- endomyccorhizal root, prep. slide, c.s.
- compound microscope

PROCEDURE

1. Obtain a prepared slide of an ectomycorrhizal pine root. Examine it with the medium- and high-power objective of your compound microscope.
2. Identify the sheath of fungal hyphae that surround the root.
3. Now look for hyphae between the cell walls within the cortex of the root.
4. Draw what you see in Figure 14-25.
5. Obtain a prepared slide of an endomycorrhizal root (Figure 14-26) and examine it as you did above.
6. Identify the hyphae within the plant root cells.

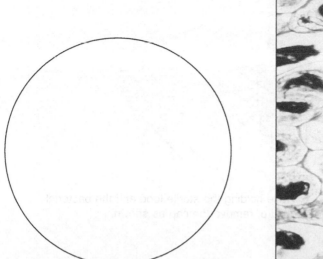

Figure 14-25 Drawing of ectomycorrhizal pine root (_____×).

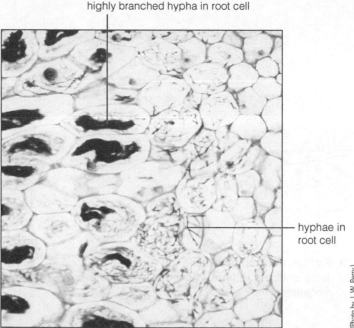

(Photo by J. W. Perry.)

highly branched hypha in root cell

hyphae in root cell

Figure 14-26 Cross section of endomycorrhizal root (110×).

14.8   Experiment: Environmental Factors and Fungal Growth
(About 45 min. to set up)

Like all living organisms, environmental cues are instrumental in causing growth responses in fungi. Intuitively, we know that light is important for plant growth. Light also affects animals in a myriad of ways, from triggering reproductive events in deer to affecting the complex psyche of humans. What effect might light have on fungi? These experiments enable you to determine whether this environmental factor has any effect on fungal growth.

A. Light and Darkness

This experiment addresses the hypothesis that *light triggers spore production of certain fungi.*

MATERIALS

Per experimental group (student pair):

- transfer loop
- bunsen burner (or alcohol lamp)
- matches or striker
- 3 petri plates containing potato dextrose agar (PDA)
- grease pencil or other marker

Per lab bench:

- test tubes with spore suspensions of *Trichoderma viride*, *Penicillium claviforme*, and *Aspergillus ornatus*

PROCEDURE

Work in pairs, and refer to Figure 14-27, which shows how to inoculate the cultures.

1. Receive instructions from your lab instructor as to which fungal culture you should use.

    Note: **Do not open the petri plates until step 3.**

2. Use a grease pencil or other lab marker to label each cover with your name and the name of the fungus you have been assigned. Write "CONTINUOUS LIGHT" on one petri plate, "CONTINUOUS DARKNESS" on a second plate, and "ALTERNATING LIGHT AND DARKNESS" on the third one.
3. Following the directions in Figure 14-27, inoculate each plate by touching the loop gently on the agar surface approximately in the middle of the plate.

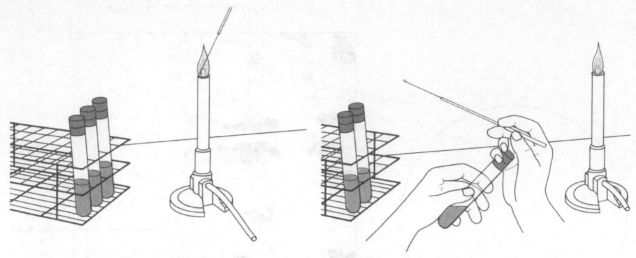

a Sterilize the loop by holding the wire in a flame until it is red hot. Allow it to cool before proceeding.

b While holding the sterile loop and the bacterial culture, remove the cap as shown.

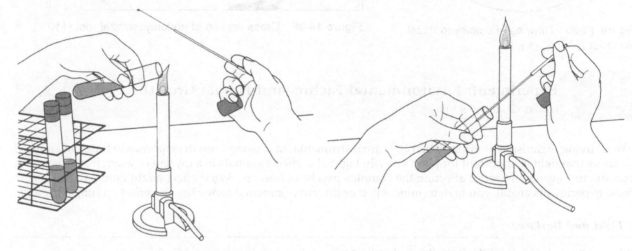

c Briefly heat the mouth of the tube in a burner flame before inserting the loop for an inoculum.

d Get a loopful of culture, withdraw the loop, heat the mouth of the tube, and replace the cap.

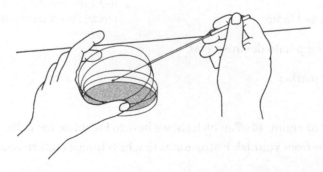

e To inoculate a solid medium in a petri plate, place the plate on a table and lift one edge of the cover.

Figure 14-27 Procedure for inoculating a culture plate. (After Case and Johnson, 1984.)

4. Place one plate in each condition:
   - continuous light
   - continuous darkness
   - alternating light and dark
5. In Table 14-3, write your prediction of what will occur in the different environmental conditions.
6. Check the results after 5–7 days and make conclusions concerning the effect of light on the growth of the cultures. Record your results in Table 14-3. Draw what you observe in Figure 14-28.
7. Compare the results you obtained with the results of other experimenters who used different species. Record your conclusions in Table 14-3.

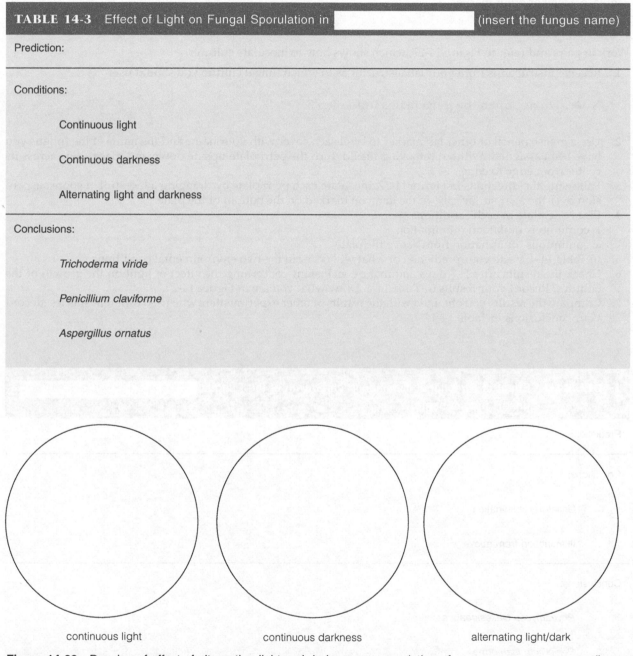

TABLE 14-3	Effect of Light on Fungal Sporulation in _____ (insert the fungus name)

Prediction:

Conditions:

     Continuous light

     Continuous darkness

     Alternating light and darkness

Conclusions:

     Trichoderma viride

     Penicillium claviforme

     Aspergillus ornatus

continuous light   continuous darkness   alternating light/dark

Figure 14-28 Drawing of effect of alternating light and darkness on sporulation of _____ (insert name of fungus).

B. Directional Illumination

This experiment addresses the hypothesis that *light from one direction causes the spore-bearing hyphae of certain fungi to grow in the direction of the light.*

MATERIALS

Per experimental group (student pair):

- transfer loop
- bunsen burner (or alcohol lamp)
- matches or striker
- 2 petri plates containing either potato dextrose agar (PDA) or cornmeal (CM) agar
- grease pencil or other marker

Per lab bench:

- test tubes with spore suspensions of
  Phycomyces blakesleeanus (inoculate onto PDA)
  Penicillium isariforme (inoculate onto PDA)
  Aspergillus giganteus (inoculate onto CM)

PROCEDURE

Work in pairs and refer to Figure 14-27, which shows how to inoculate cultures.

1. Receive instructions from your lab instructor as to which fungal culture you should use.

   Note: Do not open the petri plates until step 3.

2. Use a grease pencil or other lab marker to label each cover with your name and the name of the fungus you have been assigned. Without removing the lid, turn the petri plate upside down and place a line across its center from edge to edge.
3. Following the directions in Figure 14-27, inoculate each petri plate by dragging a loop full of spore suspension over the agar surface above the line you marked on the bottom of the plate.
4. Place one plate in each condition:
   - continuous unilateral illumination
   - continuous illumination from above the plate
5. In Table 14-4, write your prediction of what will occur in the two environmental conditions.
6. Check the results after 5–7 days and make conclusions concerning the effect of light on the growth of the cultures. Record your results in Table 14-4. Draw what you see in Figure 14-29.
7. Compare the results you obtained with the results of other experimenters who used different species. Record your conclusions in Table 14-4.

TABLE 14-4	Effect of Directional Illumination on Growth of the Fungus _____ (insert fungus name)
Prediction:	
Conditions:	
Unilateral illumination	
Illumination from above	
Conclusions:	
Phycomyces blakesleeanus	
Penicillium isariforme	
Aspergillus giganteus	

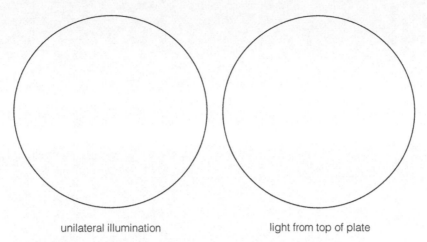

unilateral illumination   light from top of plate

Figure 14-29 Drawing of effect of unilateral illumination on growth of ____
_____ (insert name of fungus).

PRE-LAB QUESTIONS

_____ 1. An organism that grows specifically on nonliving organic material is called
(a) an autotroph
(b) a heterotroph
(c) a parasite
(d) a saprophyte

_____ 2. Taxonomic separation into fungal phyla is based on
(a) sexual reproduction, or lack thereof
(b) whether the fungus is a parasite or saprophyte
(c) the production of certain metabolites, like citric acid
(d) the edibility of the fungus

_____ 3. Fungi is to fungus as _____ is to

_____.

(a) mycelium, mycelia
(b) hypha, hyphae
(c) sporangia, sporangium
(d) ascus, asci

_____ 4. Which statement is not true of the zygospore-forming fungi?
(a) They are in the phylum Zygomycota
(b) Ascospores are found in an ascus
(c) *Rhizopus* is a representative genus
(d) A zygospore is formed after fertilization

_____ 5. Which structures would you find in a sac fungus?
(a) ascogonium, antheridium, zygospores
(b) ascospores, oogonia, asci, ascocarps
(c) basidia, basidiospores, basidiocarps
(d) ascogonia, asci, ascocarps, ascospores

_____ 6. The club fungi are placed in the phylum Basidiomycota
(a) because of their social nature
(b) because they form basidia
(c) because of the presence of an ascocarp
(d) because they are dikaryotic

_____ 7. Which statement is *not* true of the imperfect fungi?
(a) They reproduce sexually by means of conidia
(b) They form an ascocarp
(c) Sex organs are present in the form of oogonia and antheridia
(d) All of the above are false

_____ 8. A relationship between two organisms in which both members benefit is said to be
(a) parasitic
(b) saprotrophic
(c) mutualistic
(d) heterotrophic

_____ 9. An organism that is made up of a fungus and an associated green alga or cyanobacterium is known as a
(a) sac fungus
(b) lichen
(c) mycorrhizal root
(d) club fungus

_____ 10. A mutualistic association between a plant root and a fungus is known as a
(a) sac fungus
(b) lichen
(c) mycorrhizal root
(d) club fungus

EXERCISE 14

Fungi

POST-LAB QUESTIONS

14.1 Phylum Zygomycota: Zygosporangium-Forming Fungi

1. Observe this photo at the right of a portion of a fungus you examined in lab.

   a. What is structure a?

   b. Are the contents of structure a haploid or diploid?

2. Distinguish between a *hypha* and a *mycelium*.

3. Examine the below photomicrographs of a fungal structure you studied in lab. Is this structure labeled **b** in the photomicrograph below the product of sexual or asexual reproduction?

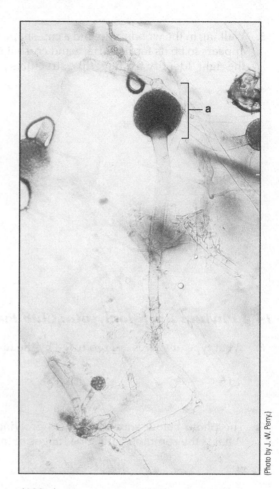

(130×).

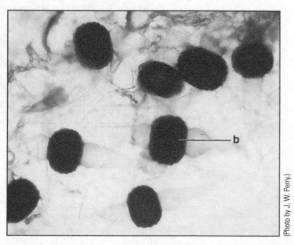

(120×).

14.3 Phylum Ascomycota: Sac Fungi

4. Distinguish among an *ascus*, an *ascospore*, and an *ascocarp*.

5. Walking in the woods, you find a cup-shaped fungus. Back in the lab, you remove a small portion from what appears to be its fertile surface and crush it on a microscope slide, preparing the wet mount that appears at the right. Identify the fingerlike structures present on the slide.

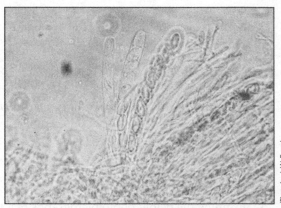

(287×).

14.4 Phylum Basidiomycota: Club Fungi

6. What type of spores are produced by the fungus pictured at the right?

7. The photo below shows a fungus growing on a dead hemlock tree. What is the common name of a fungus of this sort? Be specific.

14.5 "Imperfect Fungi"

8. Explain the name "imperfect fungi."

Food for Thought

9. Give the correct singular or plural form of the following words in the blanks provided.

	Singular	Plural
a.	hypha	_____
b.	_____	mycelia
c.	zygospore	_____
d.	_____	asci
e.	basidium	_____
f.	_____	conidia

10. Lichens are frequently the first colonizers of hostile growing sites, including sunbaked or frozen rock, recently hardened lava, and even gravestones. How can lichens survive in habitats so seemingly devoid of nutrients and under such harsh physical conditions?

Bryophytes and Seedless Vascular Plants

OBJECTIVES

After completing this exercise, you will be able to

1. define *alternation of generations, isomorphic, heteromorphic, dioecious, antheridium, archegonium, hygroscopic, protonema, tracheophyte, rhizome, mutualism, sporophyte, sporangium, gametophyte, gametangium, antheridium, archegonium, epiphyte, strobilus, node, internode, frond, sorus, annulus, hygroscopic,* and *chemotaxis;*

2. diagram the plant life cycle to show alternation of generations;

3. distinguish between alternation of isomorphic and heteromorphic generations;

4. list evidence supporting the evolution of land plants from green algae;

5. recognize charophytes, mosses, and liverworts;

6. identify the sporophytes, gametophytes, and associated structures of liverworts and mosses (those in **boldface** in the procedure sections);

7. describe the function of the sporophyte and the gametophyte;

8. postulate why bryophytes live only in environments where free water is often available;

9. recognize whisk ferns, club mosses, horsetails, and ferns when you see them, placing them in the proper taxonomic phylum;

10. identify the structures of the fern allies and ferns in **boldface;**

11. describe the life cycle of ferns;

12. explain the mechanism by which spore dispersal occurs from a fern sporangium;

13. describe the significant differences between the life cycles of the bryophytes, and the club mosses and ferns.

INTRODUCTION

This is the first in a series of exercises that addresses true plants—the kingdom Plantae.

Plants evolved from the green algae, specifically now extinct relatives of living green algae called **charophytes.** Evidence for this includes identical food reserves (starch), the same photosynthetic pigments (chlorophylls a and b, carotenes, and xanthophylls), similarities in structure of their flagella, and the way cell division takes place. Some biologists have gone so far as to suggest that the land plants are nothing more than highly evolved green algae.

One major mystery is the origin of a feature common to *all* true plants (but *not* charophytes), **alternation of generations.** In alternation of generations, two distinct phases exist: A diploid **sporophyte** alternates with a haploid **gametophyte**, as summarized in Figure 15-1.

As animals, we find the concept of alternation of generations difficult to envision. But think of it as the existence of two body forms of the same organism. The primary reproductive function of one

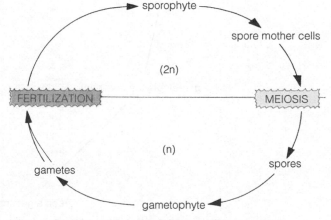

Figure 15-1 Summary of alternation of generations.

263

body form, the gametophyte, is to produce gametes (eggs and/or sperm) by *mitosis*. The primary reproductive function of the other, the sporophyte, is to produce spores by *meiosis*.

A fundamental distinction exists between the green algae and land plants with respect to alternation of generations. Although some green algae have alternation of generations, the sporophyte and gametophyte look *identical*. To the naked eye, they are indistinguishable. This is called alternation of **isomorphic** generations. (*Iso-* comes from a Greek word meaning "equal"; *morph-* is Greek for "form.")

By contrast, the alternation of generations in land plants is **heteromorphic** (*hetero-* is Greek for "different"). The sporophytes and gametophytes of land plants, including the bryophytes, are distinctly different from one another. The contrast between isomorphic and heteromorphic alternation of generations is modeled in Figure 15-2.

During the course of evolution, two major lines of divergence took place in the plant kingdom. The plants in one line had as their dominant phase the gametophytic generation, meaning that the sporophyte was never free living but was permanently attached to and dependent on the gametophyte for nutrition. Today these plants are represented by the bryophytes, mosses and their relatives. It seems that this line exemplifies *dead-end evolution*— no other group of plants present today arises from it.

In the other line of evolution, the sporophyte led an independent existence, the gametophyte being quite small and inconspicuous. Figure 15-3 summarizes these two evolutionary lines.

As you will see in this exercise, the land plants known as liverworts and mosses are often fairly dissimilar in appearance. They do share similarities, including the following:

1. Both exhibit alternation of heteromorphic generations in which the gametophyte is the dominant organism. The sporophyte remains attached to the gametophyte, deriving most of its nutrition from the gametophyte.
2. Both are dependent on water for fertilization, since their sperm must swim to a nonmotile egg.
3. Both lack true vascular tissues, xylem and phloem, and hence are relatively small organisms.

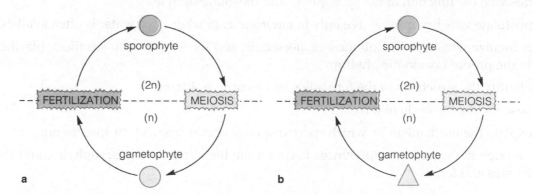

Figure 15-2 Types of alternation of generations. (**a**) Isomorphic. (**b**) Heteromorphic.

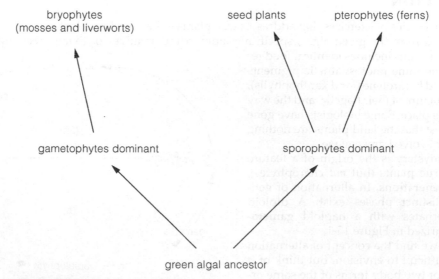

Figure 15-3 Evolution of land plants.

A major distinction exists between the bryophytes (liverworts and mosses) and the fern allies, ferns, gymnosperms, and flowering plants: Whereas the dominant and conspicuous portion of the life cycle in the bryophytes is the **gametophyte** (the gamete-producing part of the life cycle), in all other plants that you will examine in this and subsequent exercises, it is the **sporophyte** (that portion of the life cycle producing spores).

A second, and perhaps more important, distinction also exists between bryophytes and the fern allies, ferns, gymnosperms, and flowering plants: The latter group contains *vascular tissue*. Vascular tissues include *phloem*, the tissue that conducts the products of photosynthesis, and *xylem*, the tissue that conducts water and minerals. As a result of the presence of xylem and phloem, fern allies, ferns, gymnosperms, and flowering plants are sometimes called **tracheophytes.**

Push gently on the region of your throat at the base of your larynx (voicebox or Adam's apple). If you move your fingers up and down, you should be able to feel the cartilage rings in your *trachea* (windpipe). To visualize its structure, imagine that your trachea is a pipe with donuts inside of it (Figure 15-4). This same arrangement exists in some cell types within the xylem of plants. Thus, early botanical microscopists called plants having such an arrangement *tracheophytes*.

In this exercise, you will study the tracheophytes that lack seeds. Two phyla of plants make up the seedless vascular plants:

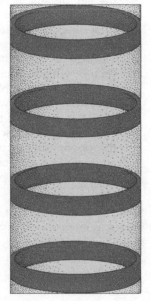

Figure 15-4 Three-dimensional representation of the cartilage in your trachea and some xylem cells of vascular plants.

Phylum	Common Names
Lycophyta	Club mosses
Moniliophyta	Whisk ferns, horsetails, and ferns

Whisk ferns and horsetails are very unlike what we often think of as ferns, but molecular evidence confirms their relationship to more typical ferns. We will study most closely the life cycle of the ferns, since they are common in our environment and have gametophytes and sporophytes that beautifully illustrate the concept of alternation of generations.

15.1   Phylum Charophyta—Ancestors of True Plants *(About 10 min.)*

Before beginning study of true plants (kingdom Plantae), let's look at their closest algal relatives. Charophytes (kingdom Protista) called stoneworts, are within a lineage that millions of years ago gave rise to the land plants.

Stoneworts are interesting organisms found "rooted" in brackish and fresh waters, particularly those high in calcium. (Two of your lab manual authors—the Perrys—have ponds that are delightfully full of stoneworts.) They get their common names by virtue of being able to precipitate calcium carbonate over their surfaces, encrusting them and rendering them somewhat stony and brittle. (The suffix *-wort* is from a Greek word meaning "herb.")

MATERIALS

Per student:

- small culture dish
- dissecting microscope

Per lab room:

- living culture (or preserved) *Chara* or *Nitella*
- demonstration slide of *Chara* with sex organs

PROCEDURE

1. From the classroom culture provided, obtain some of the specimen and place it in a small culture dish partially filled with water.

2. Observe the specimen with a dissecting microscope. Note that the stoneworts resemble what we would think of as a plant. They are divided into "stems" and "branches" (Figure 15-5).

3. Search for flask-shaped and spherical structures along the stem. These are the sex organs. If none is present on the specimen, observe the demonstration slide (Figure 15-6) that has been selected to show these structures.

4. The flask-shaped structures are **oogonia**, each of which contains a single, large egg (Figure 15-7). Notice that the oogonium is covered with cells that twist over the surface of the gametangium. Because of the presence of these cells, the oogonium is considered to be a *multicellular* sex organ. Multicellular sex organs are present in all land plants.

Based on your study of previously examined specimens, would you say the egg is motile or nonmotile?

Figure 15-5 The stonewort, *Chara*, with sex organs (2×).

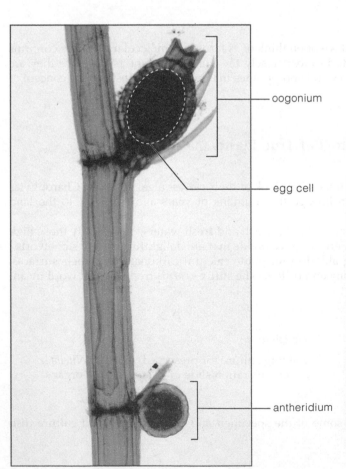

Figure 15-6 *Chara*, showing sex organs (30×). (Photo by J. W. Perry.)

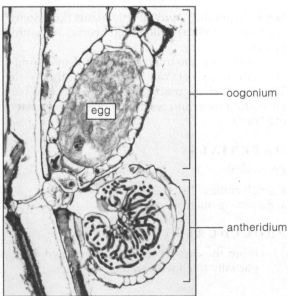

Figure 15-7 Antheridia and oogonia of *Chara* (80×).

5. Now find a spherical **antheridium** (Figure 15-7). Like the oogonium, the antheridium is covered by cells and is also considered to be multicellular. Cells within the interior of the antheridium produce numerous flagellated sperm cells.

Fertilization of eggs by sperm produces a *zygote* within the oogonium. The zygote-containing oogonium eventually falls off from the parent plant. The zygote can remain dormant for some time before the nucleus undergoes meiosis in preparation for germination. Apparently, three of the four nuclei produced during meiosis disintegrate.

15.2   Phylum Hepatophyta: Liverworts *(About 10 min.)*

Liverworts!? What in the world is a liverwort? Knowing why these plants are called liverworts is a start to knowing them.

The Greek suffix *-wort* means "herb," a nonwoody plant. The "liver" portion of the common name comes about because the plants look a little like a lobed human liver. These are small plants that you have likely overlooked (until now!) because they're inconspicuous and don't look like anything important. You'll find them growing on rocks that are periodically wetted along clean freshwater streams.

MATERIALS

Per student:

- living or preserved *Marchantia* thalli, with gemma cups, antheridiophores, archegoniophores, and mature sporophytes
- dissecting microscope

PROCEDURE

1. Examine the living or preserved plants of the liverwort, *Marchantia*, that are on demonstration. The body of this plant is called a **thallus** (the plural is *thalli*), because it is flattened and has little internal tissue differentiation (Figure 15-8).
2. The thallus is the gametophyte portion of the life cycle.

Is it haploid or diploid? _____

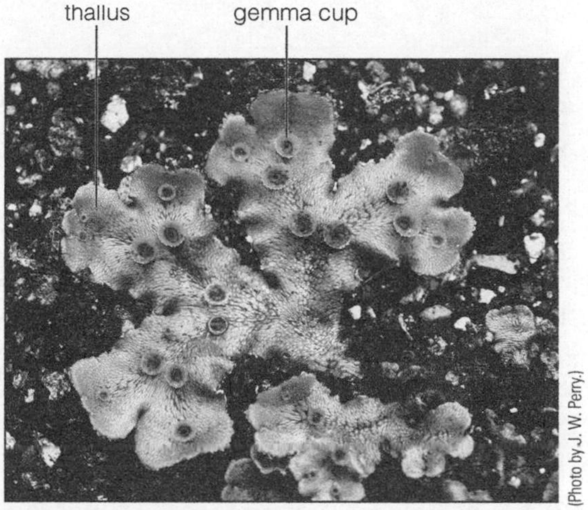

3. Obtain and examine one thallus of the liverwort. Notice that the thallus is lobed. Centuries ago, herbalists believed that plants that looked like portions of the human anatomy could be used to treat ailments of that part of the body. This plant reminded them of the liver, and hence the plant was called a liverwort.

Figure 15-8   Thallus of a liverwort with gemma cups (1×).

(Photo by J. W. Perry.)

4. Now place the thallus on the stage of a dissecting microscope. Looking first at the top surface, find the *pores* that lead to the interior of the plant and serve as avenues of exchange for atmospheric gases (CO_2 and O_2; Figure 15-8).
5. Identify **gemma cups** (Figure 15-8) on the upper surface. Look closely within a gemma cup to find the **gemmae** inside (the singular is *gemma*). Gemmae are produced by *mitotic* divisions of the thallus. They are dislodged by splashing water. If they land on a suitable substrate, they grow into new thalli; hence they are a means of asexual reproduction.
6. Are gemmae haploid or diploid? _____
7. Turn the plant over to find the hairlike **rhizoids** that anchor the organism to the substrate.

As noted above, the thalli are the gametophyte generation. Similar to higher animals (humans, as an example), gametes are produced in special organs on the plant. In this liverwort, the organs are contained within elevated branches. The general name for these elevated branches is *gametangiophores*. (The suffix *-phore* is derived from a Greek word meaning "branch.") There are male and female gametangiophores.

8. On a male thallus, find the male **antheridiophores.** They look like umbrellas (Figure 15-9a). The sperm-producing male sex organs called **antheridia** (the singular is *antheridium*) are within the flattened splash platform of the umbrella.
9. On female plants, find the female **archegoniophores** (Figure 15-9b); these look somewhat like an umbrella that has lost its fabric and has only ribs remaining. Underneath these ribs are borne the female gametangia, **archegonia** (singular, *archegonium*).

(Photos by J. W. Perry.)

Figure 15-9 Liverwort thalli with gametangiophores. (**a**) Male thalli with antheridiophores. (**b**) Female thalli with archegoniophores (0.5×).

<table>
<tr><td>**15.3**</td><td>**Experiment: Effect of Photoperiod**
(About 10 min. to set up; requires 6 weeks to complete)</td></tr>
</table>

Photoperiod, the length of light and dark periods, exerts a significant effect upon most living things. Flowering in many plants is dependent on photoperiod, some plants being classified as "long day (short night)," others as "short day (long night)," and still others as "day neutral." Flowering is a reproductive event. The reproductive behavior of many animals, such as deer, is also dependent on day length.

This experiment examines the effect of photoperiod on the production of gametangiophores and gemma cups in the liverwort, *Marchantia*. One possible experimental hypothesis is that *reproduction of* Marchantia *is stimulated by long days (short nights).*

MATERIALS

Per student group:

- vigorously growing colony of the liverwort *Marchantia polymorpha*
- pot label
- china or other marker

Per lab room:

- growth chambers set to long day (16 hr light, 8 hr darkness) and short day (8 hr light, 16 hr darkness) conditions

PROCEDURE

1. In Table 15-1, write a prediction about the effect of day length on *Marchantia* reproduction.
2. Obtain a colony of the liverwort, *Marchantia polymorpha*. On the pot label, write your name, date, and the experimental condition (long day or short day) to which you are going to subject the colony.
3. Place the colony in the growth chamber that is designated as "short day" or "long day," depending on which condition has been assigned to your experimental group. You will be responsible for watering the colonies on a regular basis. Failure to do so will result in unreliable experimental results and/or death of the plants.
4. Observe the colonies weekly over a period of six weeks, recording your observations in Table 15-1.
5. After six weeks of experimental treatment, compare observations with other experimental groups using the same and different independent variables.
6. List the results and make conclusions, writing them in Table 15-2.

Did the experimental results support your hypothesis? _____

Can you suggest any reasons why it might be beneficial to the liverwort to display the results that the different experimental groups found?

_____

_____

TABLE 15-1 The Effect of Photoperiod on the Reproduction of the Liverwort, *Marchantia polymorpha*

Prediction:

Week	Observations of the Effect of _____ (Long or Short) Day Illumination
1	
2	
3	
4	
5	
6	

TABLE 15-2 Summary of Results: The Effect of Photoperiod on the Reproduction of the Liverwort *Marchantia polymorpha*

Photoperiod	Results
Long day	
Short day	
Conclusions:	

15.4   Phylum Bryophyta: Mosses *(About 40 min.)*

Mosses are plants that you've probably noticed before. They grow on trees, rocks, and soil. Mosses have a wide variety of human uses, the most common ones being as a soil additive to potting soil and as a major component of peat, which is used for fuel in some parts of the world.

MATERIALS

Per student:

- *Polytrichum*, male and female gametophytes, the latter with attached sporophytes
- prepared slide of moss antheridial head, l.s.
- prepared slide of moss archegonial head, l.s.
- prepared slide of moss sporangium (capsule), l.s.
- glass microscope slide
- coverslip

- compound microscope
- dissecting microscope
- dissecting needle

Per student pair:

- dH₂O in dropping bottle

Per student group (table):

- moss protenemata growing on culture medium

PROCEDURE

Refer to Figure 15-16 as you study the life cycle of the mosses.

1. Obtain a living or preserved specimen of a moss that grows in your area. The hairy-cap moss, *Polytrichum*, is a good choice because one of the ten species in this genus is certain to be found near wherever you live in North America.

Lacking vascular tissues, the mosses do not have true roots, stems, or leaves, although they do have structures that are rootlike, stemlike, and leaflike and perform the same functions as the true organs.

2. Identify the rootlike **rhizoids** at the base of the plant. What function do you think rhizoids perform?

_____

_____

3. Notice that the leaflike organs are arranged more or less radially about the stemlike *axis*. You are examining the **gametophyte generation** of the moss. In terms of the life cycle of the organism, what function does the gametophyte serve?

_____

_____

Figure 15-10 Colony of male gametophytes (0.5×).

(Photo by J. W. Perry.)

Polytrichum is **dioecious,** meaning that there are separate male and female plants. (Some other mosses are monoecious.) The male gametophyte can usually be distinguished by the flattened rosette of leaflike structures at its tip.

4. Examine a male gametophyte (Figures 15-10, 15-16a), noting this feature. Embedded within the rosette are the male sex organs, **antheridia** (singular, *antheridium*).

5. With the low- and medium-power objectives of your compound microscope, examine a prepared slide with the antheridia of a moss (Figure 15-16c). Identify the *antheridia* (Figure 15-11).

6. Study a single antheridium (Figure 15-11), using the high-dry objective. Locate the jacket layer surrounding the sperm-forming tissue that, with maturity, gives rise to numerous biflagellate *sperm*. Scattered among the antheridia, find the numerous sterile *paraphyses*. These do not have a reproductive role (and hence are called sterile) but instead function to hold capillary water, preventing the sex organs from drying out.

7. Examine a female gametophyte of *Polytrichum* (Figure 15-16b). Before the development of the sporophyte, the female gametophyte can usually be distinguished by the absence of the rosette at its tip. Nonetheless, the apex of the female gametophyte contains the female sex organs, **archegonia** (singular, *archegonium*).

8. Obtain a prepared slide of the archegonial tip of a moss (Figures 15-12a, 15-16d). Start with the low-power objective of your compound microscope to gain an impression of the overall organization. Find the sterile *paraphyses* and the *archegonia*.

9. Switch to the medium-power objective to study a single archegonium, identifying the long **neck** and the slightly swollen base. Within the base, locate the **egg cell** (Figure 15-16d).

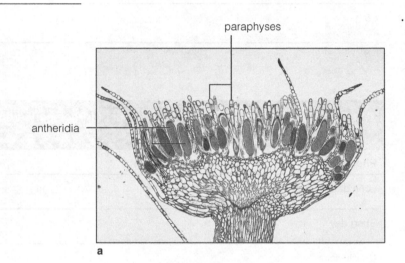

paraphyses

antheridia

a

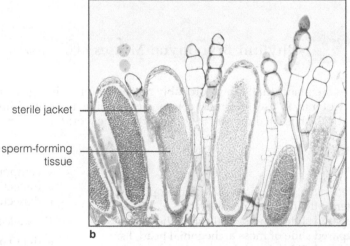

sterile jacket

sperm-forming tissue

b

(Photos by J. W. Perry.)

Figure 15-11 Longitudinal section of antheridial head of a moss (a) 8×. (b) 93×.

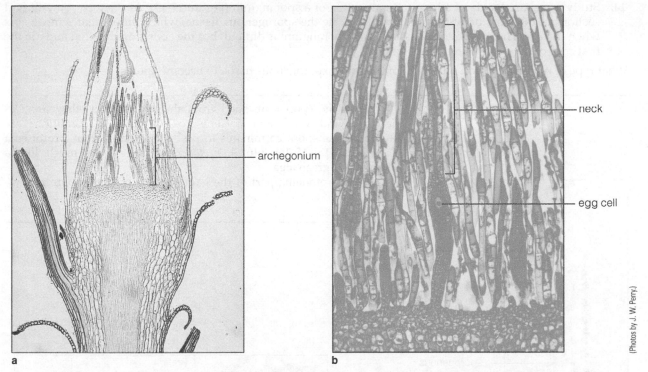

a   b

Figure 15-12 (a) Longitudinal section of archegonial head of a moss (22×). (b) Detail of archegonium (88×).

(Photos by J. W. Perry.)

Remember, you are looking at a *section* of a three-dimensional object. The archegonium is very much like a long-necked vase, except that it's solid. The egg cell is like a marble suspended in the middle of the base.

The central core of the archegonial neck contains cells that break down when the egg is mature, liberating a fluid that is rich in sucrose and that attracts sperm that are swimming in dew or rainwater. (Sperm are capable of swimming only short distances and so male plants must be close by.) Fertilization of the egg produces the diploid zygote (Figure 15-16e), the first cell of the **sporophyte generation.**

Numerous mitotic divisions produce an **embryo** (embryo sporophyte, Figure 15-16f), which differentiates into the **mature sporophyte** (Figure 15-16g) that protrudes from the tip of the gametophyte. What is the function of the sporophyte (that is, what does it do, or what reproductive structures does it produce)?

_____

_____

10. Grasp the stalk of the sporophyte and detach it from the gametophyte. The base of the stalk absorbs water and nutrients from the gametophyte.
11. At the tip of the sporophyte, locate the **sporangium** covered by a papery hood. The cover is a remnant of the tissue that surrounded the archegonium and is covered with tiny hairs—hence the common name, hairy-cap moss, for *Polytrichum*.
12. Remove the hood to expose the sporangium (Figure 15-16h). Notice the small cap at the top of the sporangium.
13. Remove the cap and observe the interior of the sporangium with a dissecting microscope. Find the *peristomal teeth* that point inward from the margin of the opening.

sporophytes   gametophytes

(Photo by J. W. Perry.)

Figure 15-13 Colony of female gametophytes with attached sporophytes (0.25×).

The peristomal teeth are **hygroscopic,** meaning that they change shape as they absorb water. As the teeth dry, they arch upward, loosening the cap over the spore mass. The teeth may subsequently shrink or swell, thus regulating how readily the spores inside may escape.

14. Study a prepared slide of a longitudinal section of a sporangium (Figure 15-14). At the top, you will find sections of the peristomal teeth. Internally, locate the **sporogenous tissue,** which differentiates into *spores* when mature. The sporogenous tissue of the sporangium is diploid, but the spores are haploid and are the first cells of the gametophyte generation.

What type of nuclear division must take place for the sporogenous tissue to become spores?

When the spores are shed from the sporangium, they are carried by wind and water currents to new sites. If conditions are favorable, the spore germinates to produce a filamentous **protonema** (Figures 15-15, 15-16j; plural, *protonemata*) that looks much like a filamentous green alga.

Is the protonema part of the gametophyte or sporophyte generation?

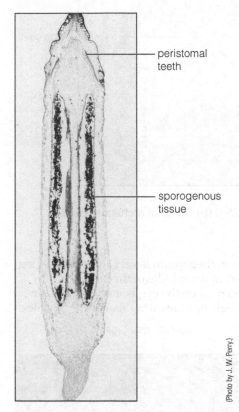

peristomal teeth

sporogenous tissue

(Photo by J. W. Perry.)

Figure 15-14 Longitudinal section of sporangium of a moss (12×).

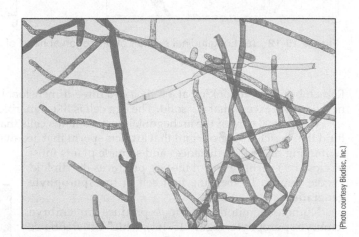

(Photo courtesy Biodisc, Inc.)

Figure 15-15 Moss protonema, whole mount (70×).

| 15.5 | **Experiment: Effect of Light Quality on Moss Spore Germination and Growth** (About 20 min. to set up; requires 8 weeks to complete) |

What causes plants to grow and reproduce? One significant environmental trigger is light. Both light *quantity* (the length of illumination) and light *quality* (the wavelengths or color of light) trigger growth and reproduction in organisms. In this experiment, you will grow moss protonemata from spores and examine the effect of different wavelengths of light on germination of spores and subsequent growth of the protonema. One hypothesis this simple experiment might address is that *moss spores germinate and protonemata grow best in short (blue) wavelength light.*

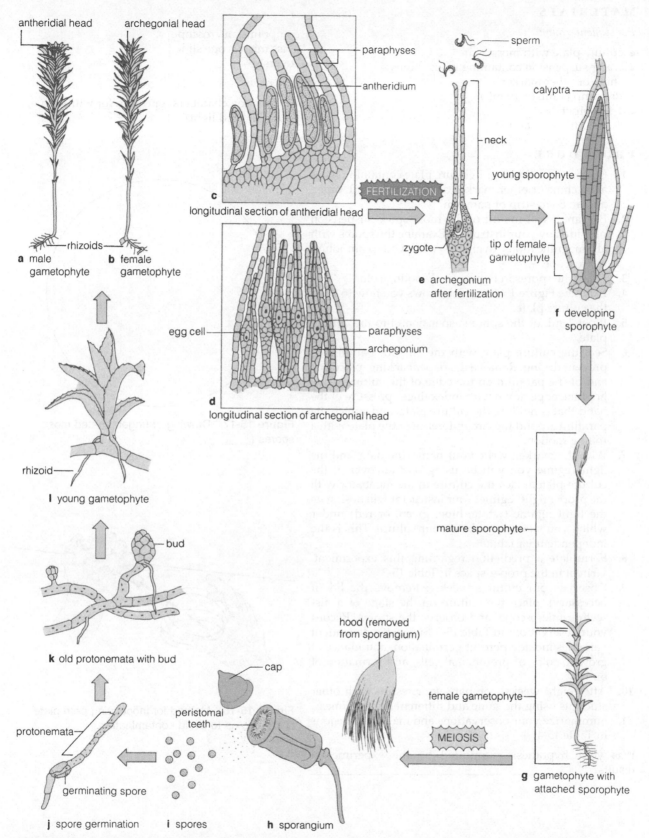

antheridial head

archegonial head

paraphyses

antheridium

sperm

calyptra

c longitudinal section of antheridial head

FERTILIZATION

neck

young sporophyte

a male gametophyte **b** female gametophyte

rhizoids

zygote

e archegonium after fertilization

tip of female gametophyte

f developing sporophyte

egg cell

paraphyses

archegonium

d longitudinal section of archegonial head

rhizoid

l young gametophyte

mature sporophyte

bud

k old protonemata with bud

hood (removed from sporangium)

cap

female gametophyte

peristomal teeth

MEIOSIS

protonemata

germinating spore

j spore germination   **i** spores   **h** sporangium

g gametophyte with attached sporophyte

Figure 15-16   Life cycle of representative moss. (Green structures are 2n.) (Modified from Carolina Biological Supply Company diagrams.)

MATERIALS

Per student group:

- culture plate with moss agar
- water suspension containing moss spores
- china or other marker
- Pi-pump (1 mL or 10 mL)
- 1-mL pipet
- 8-cm strip of parafilm

- compound microscope
- glass microscope slide
- coverslip

Per lab room:

- 4 incubation chambers (one each for white, blue, green, and red light)

PROCEDURE

1. Obtain the following: 1 culture plate containing moss agar, china or other marker, a Pi-pump, a 1-mL pipet, and an 8-cm strip of parafilm.
2. Prepare a wet mount of the moss spore suspension provided by your instructor. Examine the spores with all magnifications available with your compound microscope.
3. Draw the spores in Figure 15-17, noting color.
4. Examine Figure 15-18, which shows you how to open the culture plate.
5. Pipet 1 mL of the spore suspension onto the culture plate.
6. Seal the culture plate with the strip of parafilm to prevent drying: Remove the paper backing, place one end of the parafilm on the edge of the culture plate, hold the edge down with index finger pressure of the hand that is holding the culture plate, and stretch the parafilm around the circumference of the plate with a rolling motion.
7. With the marker, write your name, the date, and the light regime you will be using on the cover of the culture plate. Place the culture in the incubator with the proper light regime. Your instructor will designate the light regime (white, blue, green, or red) under which you should incubate your culture. This is the independent variable.
8. Formulate a prediction regarding this experiment. Write it in the proper space in Table 15-3.
9. Observe your cultures weekly: Remove the lid (if necessary), place the culture on the stage of a dissecting microscope, and observe the spores. Record your observations in Table 15-3. Suggested dependent variables include percent germination, abundance of growth, color of protonemal cells, and formation of buds.
10. After eight weeks, compare your results with other students using the same and different light regimes.
11. Summarize your observations and make conclusions in Table 15-4.

Was your hypothesis supported by the experimental results? _____

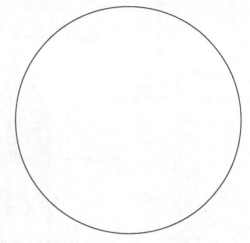

Figure 15-17 Drawing of ungerminated moss spores (_____×).

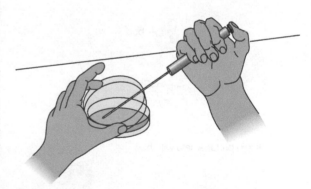

Figure 15-18 Method for inoculating petri plate culture so as to avoid contamination.

TABLE 15-3 Effect of Light Quality on Moss Spore Germination and the Growth of Protonemata

Prediction:

Week	Observations of Dependent Variables (list them here)
1	
2	
3	
4	
5	
6	
7	
8	

TABLE 15-4 Summary of Results on the Effect of Light Quality on the Germination of Moss Spores and Growth of Protonemata

Color of Light	Results
White	
Blue	
Green	
Red	
Conclusions	

Lycopodium and *Selaginella* are the most commonly found genera in the phylum Lycophyta, the club mosses. The common name of this phylum comes from the presence of the so-called **strobilus** (plural, *strobili*), a region of the stem specialized for the production of spores. The strobilus looks like a very small club. Strobili are sometimes called cones.

MATERIALS

Per lab room:

- *Lycopodium*, living, preserved, or herbarium specimens with strobili
- *Lycopodium* gametophyte, preserved (optional)
- *Selaginella*, living, preserved, or herbarium specimens

- *Selaginella lepidophylla*, resurrection plant, 2 dried specimens
- culture bowl containing water

A. Lycopodium

The club moss *Lycopodium* is a small, forest-dwelling plant. These plants are often called ground pines or trailing evergreens. The scientific name comes from the appearance of the rhizome—it looked like a wolf's paw to early botanists. *Lycos* is Greek for "wolf," and *podium* means "foot."

PROCEDURE

1. Observe the living, preserved, or herbarium specimens of *Lycopodium* (Figure 15-19).
2. On your specimen, identify the true **roots, stems,** and **leaves.**
3. The adjective *true* is used to indicate that the organs contain vascular tissue—xylem and phloem.
4. Identify the **rhizome** to which the upright stems are connected. In the case of *Lycopodium*, the rhizome may be either beneath or on the soil surface, depending on the species. Notice that leaves cover the rhizome, as are the upright stems.
5. At the tip of an upright stem, find the **strobilus** (Figure 15-19). Look closely at the strobilus. As you see, it too is made up of leaves, but these leaves are much more tightly aggregated than the sterile (nonreproductive) leaves on the rest of the stem.

strobilus

Figure 15-19   Sporophyte of *Lycopodium* (0.6×).

The leaves of the strobilus produce *spores* within a *sporangium*, the spore container. As was pointed out in the introduction, the plant you are looking at is a diploid sporophyte.

Is the sporophyte haploid (n) or diploid (2n)? _____

Since the spores are haploid, what process must have taken place within the sporangium?

As maturation occurs, the internodes between the sporangium-bearing leaves elongate slightly, the sporangium opens, and the spores are carried away from the parent plant by wind. If they land in a suitable habitat, the spores germinate to produce small and inconspicuous subterranean gametophytes, which bear sex organs—antheridia and archegonia. For fertilization and the development of a new sporophyte to take place, free water must be available, since the sperm are flagellated structures that must swim to the archegonium.

6. If available, examine a preserved *Lycopodium* gametophyte (Figure 15-20).

B. Selaginella

Selaginella is another genus in the phylum Lycophyta. These plants are usually called spike mosses. (This points out the problem with common names—the common name for the phylum is club mosses, but individual genera have their own common names.) Some species are grown ornamentally for use in terraria.

PROCEDURE

1. Examine living representatives of *Selaginella* (Figure 15-21).
2. Look closely at the tips of the branches and identify the rather inconspicuous **strobili.**
3. Examine the specimens of the resurrection plant (yet another common name!), a species of *Selaginella* sold as a novelty, often in grocery stores. A native of the southwestern United States, this plant grows in environments that are subjected to long periods without moisture. It becomes dormant during these periods.
4. Describe the color and appearance of the dried specimen.

   _____

   _____

5. Now place a dried specimen in a culture bowl containing water. Observe what happens in the next hour or so, and describe the change in appearance of the plant.

   _____

   _____

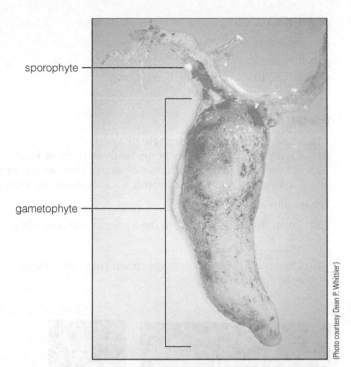

sporophyte

gametophyte

(Photo courtesy Dean P. Whittier)

Figure 15-20 Gametophyte with young sporophyte of *Lycopodium* (2×).

a

strobili

(Photos by J. W. Perry.)

b

Figure 15-21 **(a)** Sporophyte of *Selaginella* (0.4×). **(b)** Higher magnification showing strobili at branch tips (0.9×).

15.7   Phylum Moniliophyta, Subphylum Psilophyta: Whisk Ferns *(About 10 min.)*

The whisk ferns consist of only two genera of plants, *Psilotum* (the *P* is silent) and *Tmesipteris* (the *T* is silent). Neither has any economic importance, but they have intrigued botanists for a long time, especially because *Psilotum* resembles the first vascular plants that colonized the earth. Current evidence indicates that the psilophytes are not directly related to those very early land plants, but instead are ferns. We'll examine only *Psilotum*.

MATERIALS

Per lab room:

- *Psilotum*, living plant
- *Psilotum*, herbarium specimen showing sporangia and rhizome

- *Psilotum* gametophyte
- dissecting microscope

PROCEDURE

1. Observe first the **sporophyte** in a potted *Psilotum* (Figure 15-22). Within the natural landscapes of the United States, this plant grows abundantly in parts of Florida and Hawaii.
2. If the pot contains a number of stems, you can see how it got its common name, the whisk fern, because it looks a bit like a whisk broom. Closely examine a single aerial stem.

What color is it? _____

3. From this observation, make a conclusion regarding one function of this stem.

_____

4. Observe the herbarium specimen (mounted plant) of *Psilotum*. Identify the nongreen underground stem, called a **rhizome.**

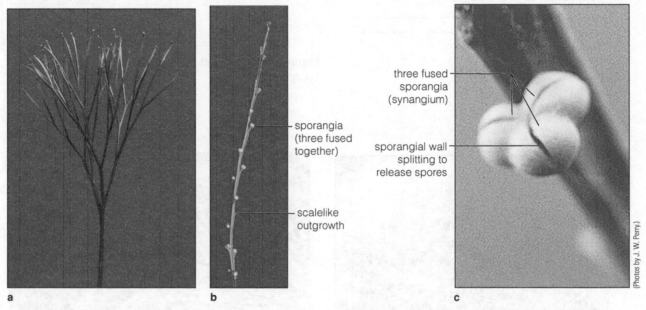

Figure 15-22 Sporophytes of *Psilotum*. (**a**) Single stem without sporangia. (**b**) Portion of stem with sporangia (0.3×). (**c**) Sporangia (10×).

5. *Psilotum* is unique among vascular plants in that it lacks roots. Absorption of water and minerals takes place through small rhizoids attached to the rhizome. Additionally, a fungus surrounds and penetrates into the outer cell layers of the rhizome. The fungus absorbs water and minerals from the soil and transfers them to the rhizome. This beneficial association is similar to the mycorrhizal association between plant roots and fungi described in the fungi exercise. A beneficial relationship like that between the fungus and *Psilotum* is called a **mutualistic symbiosis** or, more simply, **mutualism.**
6. On either the herbarium specimen or the living plant, identify the tiny scalelike outgrowths that are found on the aerial stems. Because these lack vascular tissue, they are not considered true leaves. Their size precludes any major role in photosynthesis.
7. The plant you are observing is a sporophyte. Thus, it must produce spores. Find the three-lobed structures on the stem (Figure 15-22b). Each lobe is a single **sporangium** containing spores.
8. If these spores germinate after being shed from the sporangium, they produce a small and infrequently found gametophyte that grows beneath the soil surface. The gametophyte survives beneath the soil thanks to a symbiotic relationship similar to that for the sporophyte's rhizome.
9. If a gametophyte is available, observe it with the aid of a dissecting microscope (Figure 15-23). Note the numerous **rhizoids.** If you look very carefully, you may be able to distinguish **gametangia** (sex organs; singular, *gametangium*).

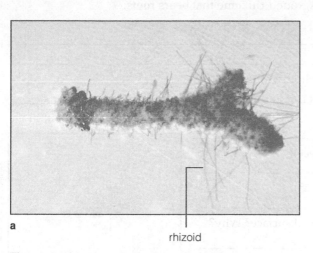

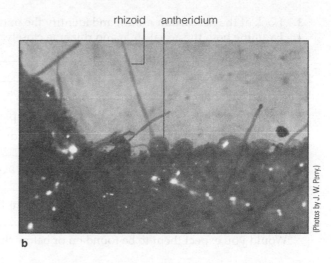

rhizoid   antheridium

(Photos by J. W. Perry.)

a

rhizoid

b

Figure 15-23 (a) *Psilotum* gametophyte (7.5×). (b) Higher magnification showing antheridia (23×).

10. The male sex organs are **antheridia** (Figure 15-23b; singular, *antheridium*); the female sex organs are **archegonia** (singular, *archegonium*). Both antheridia and archegonia are on the same gametophyte. Is the gametophyte dioecious or monoecious? _____

11. Fertilization of an egg within an archegonium results in the production of a new sporophyte.

15.8   Phylum Moniliophyta, Subphylum Sphenophyta: Horsetails *(About 10 min.)*

During the age of the dinosaurs, tree-sized representatives of this phylum flourished. But like the dinosaurs, they too became extinct. Only one genus remains. However, the remnants of these plants, along with others that were growing at that time, remain in the form of massive coal deposits. *Speno-*, a Greek prefix meaning "wedge," gives the phylum its name, presumably because the leaves of these plants are wedge shaped.

A single genus, *Equisetum*, is the only living representative of this subphylum. Different species of *Equisetum* are common throughout North America. Many are highly branched, giving the appearance of a horse's tail, and hence the common name. (*Equus* is Latin for "horse"; *saeta* means "bristle.")

MATERIALS

Per lab room:

- *Equisetum*, living, preserved, or herbarium specimens with strobili
- *Equisetum* gametophytes, living or preserved

PROCEDURE

1. Examine the available specimens of *Equisetum* (Figure 15-24). Depending on the species, it will be more or less branched.
2. Note that the plant is divided into **nodes** (places on the stem where the leaves arise) and **internodes** (regions on the stem between nodes). Identify the small leaves at the node.

If the specimen you are examining is a highly branched species, don't confuse the branches with leaves. The leaves are small, scalelike structures, often brown. (They *do* have vascular tissue, so they are true leaves.) Distinguish the leaves.

strobilius

leaves
at node

branches

(Photo by J. W. Perry.)

Figure 15-24 Sporophyte of the horsetail, *Equisetum* (0.2×).

3. Look at the herbarium mount and identify the underground **rhizome** that bears **roots.**
4. Examine both the aerial stem and rhizome closely.

Do both have nodes? _____

Do both have leaves? _____

Which portion of *Equisetum* is primarily concerned with photosynthesis? _____

5. Find the **strobilus** (Figure 15-24). Where on the plant is it located? _____

Based on the knowledge you've acquired in your study of *Lycopodium,* what would you expect to find within the strobilus? _____

When spores fall to the ground, what would you expect them to grow into after germination? _____

6. Now observe the horsetail gametophytes (Figure 15-25). What color are they? _____

Would you expect them to be found on or below the soil surface? Why?

_____

_____

Figure 15-25 Horsetail gametophyte (5×).

(Photo courtesy Dean Whittier.)

15.9	Phylum Moniliophyta, Subphylum Pterophyta: Ferns *(About 45 min.)*

The variation in form of ferns is enormous, as is their distribution and ecology. We typically think of ferns as inhabitants of moist environments, but some species are also found in very dry locations. Tropical species are often grown as houseplants.

This subphylum gets its name from the Greek word *pteri,* meaning "fern."

MATERIALS

Per student:

- fern sporophytes, fresh, preserved, or herbarium specimens
- prepared slide of fern rhizome, c.s.
- fern gametophytes, living, preserved, or whole mount prepared slides
- fern gametophyte with young sporophyte, living or preserved
- microscope slide

- compound microscope
- dissecting microscope

Per lab room:

- squares of fern sori, in moist chamber (*Polypodium aureum* recommended)
- demonstration slide of fern archegonium, median l.s.
- other fern sporophytes, as available

PROCEDURE

1. Obtain a fresh or herbarium specimen of a typical fern **sporophyte.** As you examine the structures described in this experiment, refer to Figure 15-33, a diagram of the life cycle of a fern.

The sporophyte of many ferns (Figures 15-26, 15-33a) consists of *true* roots, stems, and leaves; that is, they contain vascular tissue.

2. Identify the horizontal stem, the **rhizome** (which produces true **roots**), and upright leaves. The leaves of ferns are called **fronds** and are often highly divided.

Ferns, unlike bryophytes, are vascular plants, their sporophytes containing xylem and phloem. Let's examine this vascular tissue.

3. Obtain a prepared slide of a cross section of a fern stem (rhizome). Study it using the low-power objective of your compound microscope. Use Figure 15-27 as a reference.
   (a) Find the **epidermis, cortex,** and **vascular tissue.**
   (b) Within the vascular tissue, distinguish between the **phloem** (of which there are outer and inner layers) and the thick-walled **xylem** sandwiched between the phloem layers.
4. Now examine the undersurface of the frond. Locate the dotlike **sori** (singular, *sorus;* Figure 15-26).

Figure 15-26 (a) Morphology of a typical fern (0.25×). (b) Enlargement showing the sori (0.5×).

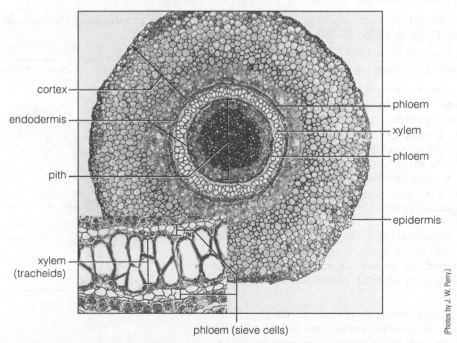

Figure 15-27 Cross section of a fern rhizome (29×). Inset shows higher magnification of vascular tissue (80×).

5. Each sorus is a cluster of **sporangia.** Using a dissecting microscope, study an individual sorus. Identify the sporangia (Figures 15-28, 15-33b).

Each sporangium contains *spores* (Figure 15-33c). Although the sporangium is part of the diploid (sporophytic) generation, spores are the first cells of the haploid (gametophytic) generation.

What process occurred within the sporangium to produce the haploid spores? _____

6. Obtain a single sorus-containing square of the hare's foot fern (*Polypodium aureum*).
   (a) Place it sorus-side up on a glass slide (DON'T ADD A COVER-SLIP), and examine it with the low-power objective of your compound microscope.
   (b) Note the row of brown, thick-walled cells running over the top of the sporangium, the **annulus** (Figures 15-28, 15-33b).

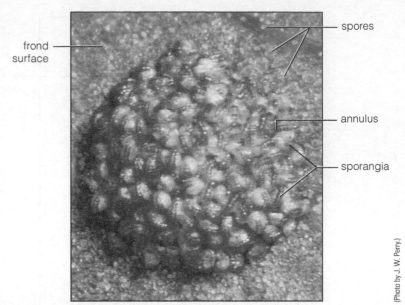

Figure 15-28 Sorus containing sporangia (30×).

The annulus is **hygroscopic.** Changes in moisture content within the cells of the annulus cause the sporangium to crack open. Watch what happens as the sporangium dries out.

As the water evaporates from the cells of the annulus, a tension develops that pulls the sporangium apart. Separation of the halves of the sporangium begins at the thin-walled lip cells. As the water continues to evaporate, the annulus pulls back the top half of the sporangium, exposing the spores. The sporangium continues to be pulled back until the tension on the water molecules within the annulus exceeds the strength of the hydrogen bonds holding the water molecules together. When this happens, the top half of the sporangium flies forward, throwing the spores out. The fern sporangium is a biological catapult!

If spores land in a suitable environment, one that is generally moist and shaded, they germinate (Figures 15-33d and e), eventually growing into the heart-shaped adult **gametophyte** (15-33f).

7. Using your dissecting microscope, examine a living, preserved, or prepared slide whole mount of the gametophyte (Figures 15-29, 15-33f).

What color is the gametophyte? _____

What does the color indicate relative to the ability of the gametophyte to produce its own carbohydrates?

_____

   (a) Examine the undersurface of the gametophyte. Find the **rhizoids,** which anchor the gametophyte and perhaps absorb water.
   (b) Locate the gametangia (sex organs) clustered among the rhizoids (Figure 15-33f). There are two types of gametangia: **antheridia** (Figure 15-33g), which produce the flagellated *sperm,* and **archegonia** (Figures 15-30, 15-33h), which produce *egg cells.*

8. Study the demonstration slide of an archegonium (Figure 15-31). Identify the **egg** within the swollen basal portion of the archegonium. Note that the *neck* of the archegonium protrudes from the surface of the gametophyte.

The archegonia secrete chemicals that attract the flagellated sperm, which swim in a water film down a canal within the neck of the archegonium. One sperm fuses with the egg to produce the first cell of the sporophyte generation, the *zygote* (Figure 15-33i). With subsequent cell divisions, the zygote

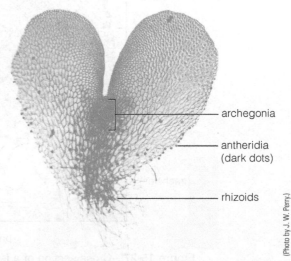

Figure 15-29 Whole mount of fern gametophyte, undersurface (10×).

develops into an embryo (embryo sporophyte). As the embryo grows, it pushes out of the gametophyte and develops into a young sporophyte (Figure 15-33j).

9. Obtain a specimen of a young sporophyte that is attached to the gametophyte (Figures 15-32, 15-33j). Identify the **gametophyte** and then the *primary leaf* and *primary root* of the young sporophyte. As the sporophyte continues to develop, the gametophyte withers away.

10. Examine any other specimens of ferns that may be on demonstration, noting the incredible diversity in form. Look for sori on each specimen.

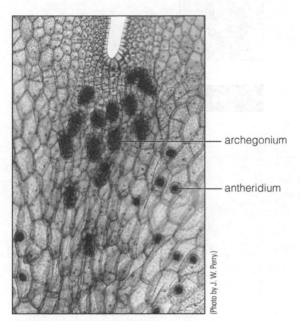

Figure 15-30   Gametangia (antheridia and archegonia) on undersurface of fern gametophyte (57×).

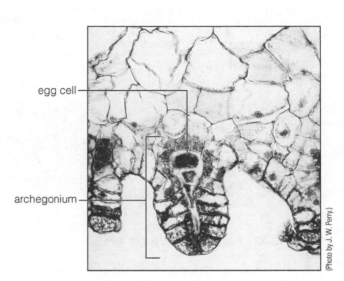

Figure 15-31   Longitudinal section of a fern archegonium (218×).

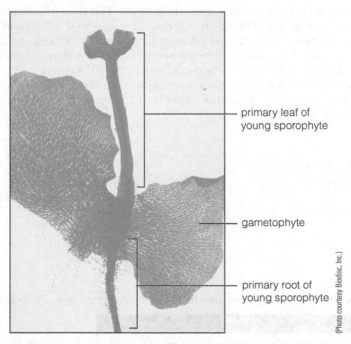

Figure 15-32   Fern gametophyte with attached sporophyte (12×).

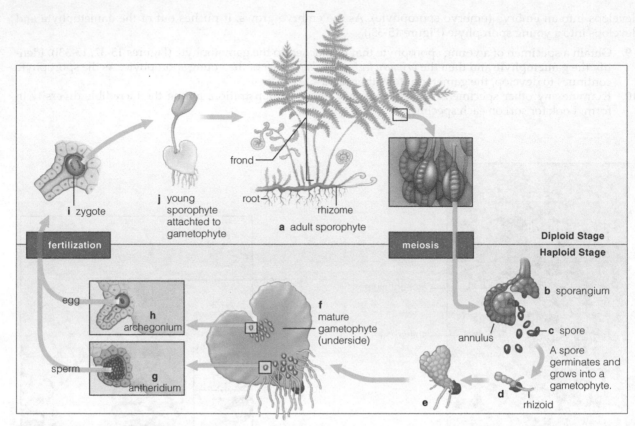

Figure 15-33 Life cycle of a typical fern.

Experiment: Fern Sperm Chemotaxis *(About 60 min.)*

Chemotaxis is a movement in response to a chemical gradient. This is an important biological process as it allows organisms to find their way or locate things. A dog following a scent is a common example of chemotaxis. How does a sperm cell shed from an antheridium find an archegonium-containing egg cell? The archegonium secretes a sperm-attracting chemical. The farther away the sperm from the archegonium, the more dilute the chemical. The sperm swim up the chemical concentration gradient, right to the egg cell.

   In this experiment, you will determine whether certain chemical substances evoke a chemotactic response by fern sperm. This experiment addresses the hypothesis that *some chemicals will attract sperm more strongly and for greater duration than others.*

MATERIALS

Per experimental group (pair of students):

- 6 sharpened toothpicks
- 5 test solutions
- dissecting needle
- depression slide

- dissecting microscope
- petri dish culture containing *C-Fern*™ gametophytes

PROCEDURE

Work in pairs.

   1. Obtain six wooden toothpicks whose tips have been sharpened to a fine point.

> **Caution**
>
> *Do not touch the sharpened end with your fingers!*

2. To identify the toothpicks, place one to five small dots along the side of the toothpicks near the unsharpened end so that each toothpick has a unique marking (one dot, two dots, and so on). One toothpick remains unmarked.
3. Dip the sharpened end of the toothpick with one dot into test solution 1. Insert it with the sharpened end up in the foam block toothpick holder.
4. Repeat step 3 with the four remaining toothpicks and four other test solutions. Leave the unmarked toothpick dry.

Which toothpick serves as the control? _____

5. Obtain a depression slide and pipet one drop of Sperm Release Buffer (SRB) into the depression.
6. Obtain a 12- to 18-day-old culture of *C-Fern*™ gametophytes. Remove the cover, place it on the stage of a dissecting microscope, and observe the gametophytes using transmitted light.
7. Identify the two types of gametophytes present—the smaller thumb-shaped male gametophytes that have many bumps on them and the larger, heart-shaped bisexual gametophytes (Figure 15-34).
8. Using a dissecting needle, pick up 7–10 male gametophytes and place them into the SRB in the depression slide well. They don't need to be completely submerged.

Caution

If you wound or damage a gametophyte, discard it.

9. Place the petri dish cover on the stage of the dissecting microscope, edges up, and then place the gametophyte-containing depression slide on the cover. (This keeps the gametophytes cooler, particularly important if your dissecting microscope has an incandescent light.) Adjust the magnification so it is at least 12×.
10. Within 5 minutes, you should be able to observe sperm being released from the antheridia.
11. Wait another 5 minutes to ensure that a sufficient number of sperm have been released, and then begin testing for chemotactic response by following steps 12–15.
12. While using 12–20× magnification, focus on the TOP surface of the sperm suspension droplet in an area devoid of male gametophytes.

Caution

As you make the tests, do not insert the toothpick fully into the drop—just touch the surface briefly (Figure 15-35).

13. Remove one test toothpick from the block and, while looking through the oculars of the dissecting microscope, gently and briefly touch the sharpened end of the toothpick to the surface of the drop (Figure 15-35).
14. Observe whether any chemotactic response takes place. Record your observation in Table 15-5.
15. Repeat the procedure with the remaining toothpicks, stirring the sperm suspension (if necessary) with a dissecting needle after each test to redistribute the sperm.
16. Place your data on the lab chalk or marker board. Your instructor will pool the data for all experimental groups.

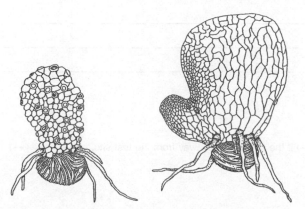

Figure 15-34 Mature male (left) and bisexual (right) *C-Fern*™ gametophytes.

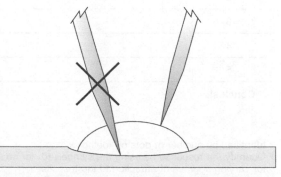

Figure 15-35 Method for applying test substances. Proper technique is important. Gently and briefly touch the end of the toothpick to the surface of the sperm suspension. *Note:* The size of the drop of suspension is exaggerated to show detail.

17. Make a conclusion about the chemotactic properties of the various chemical test substances, and note it in Table 15-5.

Examine the chemical structures of each test solution (Figure 15-36). Relate the biological (chemotactic) response you observed to any of the chemical structural differences of the test substances. ______________________

____

____

What is the advantage to using the pooled data to draw your conclusion? ______________

____

____

____

What do you hypothesize the naturally occurring chemical produced by mature archegonia to be?

____

____

____

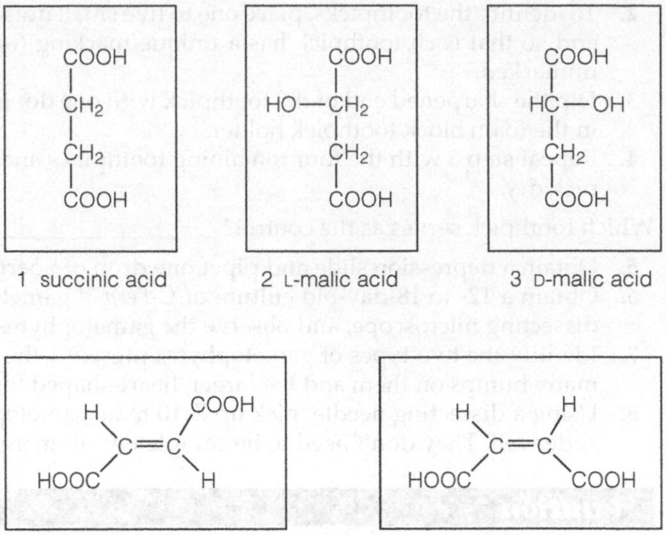

1 succinic acid 2 L-malic acid 3 D-malic acid

4 fumaric acid 5 maleic acid

Figure 15-36 Chemical structures of test substance.

TABLE 15-5 My Data: Chemotactic Response of *C-Fern*™ Sperm

Prediction:

	Swarming Response[b]			
	Intensity (low, medium, high)		**Duration (short, medium, long)**	
Test Substance[a]	**My Data**	**Pooled Data**	**My Data**	**Pooled Data**
0				
1				
2				
3				
4				
5				

Conclusion:

[a] Identified by number of dots on toothpick.
[b] Quantify the response by using the symbols (0) for no response, (–) if the sperm move away from the test site, and (+, ++, +++) for varying degrees of attraction to the test site.

_____ 1. Land plants are believed to have evolved from
   (a) mosses
   (b) ferns
   (c) charophytes
   (d) fungi

_____ 2. In the bryophytes, the sporophyte is
   (a) the dominant generation
   (b) dependent on the gametophyte generation
   (c) able to produce all of its own nutritional requirements
   (d) both a and c

_____ 3. Liverworts and mosses utilize which of the following pigments for photosynthesis?
   (a) chlorophylls a and b
   (b) carotenes
   (c) xanthophylls
   (d) all of the above

_____ 4. Which statement *best* describes the concept of alternation of generations?
   (a) One generation of plants is skipped every other year
   (b) There are two phases, a sporophyte and a gametophyte
   (c) The parental generation alternates with a juvenile generation
   (d) A green sporophyte phase produces food for a nongreen gametophyte

_____ 5. Alternation of heteromorphic generations
   (a) is found only in the bryophytes
   (b) is common to all land plants
   (c) is typical of most green algae
   (d) occurs in the liverworts but not mosses

_____ 6. An organ that is hygroscopic
   (a) is sensitive to changes in moisture
   (b) is exemplified by the peristomal teeth in the sporophyte of mosses
   (c) may aid in spore dispersal in mosses
   (d) is all of the above

_____ 7. Gemmae function for
   (a) sexual reproduction
   (b) water retention
   (c) anchorage of a liverwort thallus to the substrate
   (d) asexual reproduction

_____ 8. Sperm find their way to the archegonium
   (a) by swimming
   (b) due to a chemical gradient diffusing from the archegonium
   (c) as a result of sucrose being released during the breakdown of the neck canal cells of the archegonium
   (d) by all of the above

_____ 9. A protonema
   (a) is part of the sporophyte generation of a moss
   (b) is the product of spore germination of a moss
   (c) looks very much like a filamentous brown alga
   (d) produces the sporophyte when a bud grows from it

_____ 10. The suffix *-phore* is derived from a Greek word meaning
   (a) branch
   (b) moss
   (c) liverwort
   (d) male

_____ 11. The sporophyte is the dominant and conspicuous generation in
   (a) the fern allies and ferns
   (b) gymnosperms
   (c) flowering plants
   (d) all of the above

_____ 12. A tracheophyte is a plant that has
   (a) xylem and phloem
   (b) a windpipe
   (c) a trachea
   (d) the gametophyte as its dominant generation

_____ 13. *Psilotum* lacks
   (a) roots
   (b) a mechanism to take up water and minerals
   (c) vascular tissue
   (d) alternation of generations

_____ 14. Spore germination followed by cell divisions results in the production of
   (a) a sporophyte
   (b) an antheridium
   (c) a zygote
   (d) a gametophyte

_____ 15. Which phrase *best* describes a plant that is an epiphyte?
- (a) a plant that grows upon another plant
- (b) a parasite
- (c) a plant with the gametophyte generation dominant and conspicuous
- (d) a plant that has a mutually beneficial relationship with another plant

_____ 16. Club mosses
- (a) are placed in the subphylum Pterophyta
- (b) are so called because of the social nature of the plants
- (c) do not produce gametophytes
- (d) are so called because most produce strobili

_____ 17. The resurrection plant
- (a) is a species of *Selaginella*
- (b) grows in the desert Southwest of the United States
- (c) is a member of the phylum Lycophyta
- (d) all of the above

_____ 18. Which statement is *not* true?
- (a) Nodes are present on horsetails
- (b) The rhizome on horsetails bears roots
- (c) The internode of a horsetail is the region where the leaves are attached
- (d) Horsetails are members of the subphylum Sphenophyta

_____ 19. In ferns
- (a) xylem and phloem are present in the sporophyte
- (b) the sporophyte is the dominant generation
- (c) the leaf is called a frond
- (d) all of the above are true

_____ 20. The spores of a fern are
- (a) produced by mitosis within the sporangium
- (b) diploid cells
- (c) the first cells of the gametophyte generation
- (d) both a and b

EXERCISE 15

Bryophytes and Seedless Vascular Plants

POST-LAB QUESTIONS

Introduction

1. Explain why water must be present for the bryophytes to complete the sexual portion of their life cycle.

2. Green algae are believed to be the ancestors of the bryophytes. Cite four distinct lines of evidence to support this belief.

    a.

    b.

    c.

    d.

3. Complete this diagram of a "generic" alternation of generations.

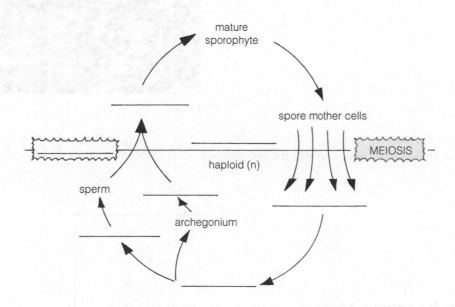

4. Describe in your own words the difference between a sporophyte and a gametophyte.

5. List two features that distinguish the seedless vascular plants from the bryophytes.

   a.

   b.

6. After consulting a biological or scientific dictionary, explain the derivation from the Greek of the word *symbiosis*.

15.2 Phylum Hepatophyta: Liverworts

7. While the plant shown here is one that you didn't study specifically in lab, you should be able to identify it. Is this the gametophyte or the sporophyte of the plant?

(1×).

8. In what structure does meiosis occur in the bryophytes?

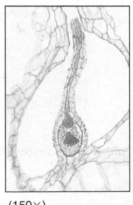

15.4 Phylum Bryophyta: Mosses

9. Identify the type of gametangium shown here.

(150×).

10. Walking along a stream in a damp forest, you see the plants shown at the right. Is this the sporophyte or the gametophyte?

(0.38×).

11. a. What are the golden stalks seen in this photo?

   b. Are they the products of meiosis or fertilization?

(0.25×).

12. Identify the plants in the photo at the right as male or female, gametophyte or sporophyte, moss or liverwort.

(0.25×).

15.6 *Phylum Lycophyta: Club Mosses*

13. Both *Lycopodium* and *Equisetum* have strobili, roots, and rhizomes. What did you learn in this exercise that enables you to distinguish these two plants?

Examine the photo at the right.

a. Give the genus of this plant.

_____

b. Label the structure indicated.

(0.3×).

14. Some species of *Lycopodium* produce gametophytes that grow beneath the surface of the soil, while others grow on the soil surface. Basing your answer on what you've learned from other plants in this exercise, predict how each respective type of *Lycopodium* gametophyte might obtain its nutritional needs.

15.7 *Phylum Moniliophyta, Subphylum Psilophyta: Whisk Ferns*

15. Using a biological or scientific dictionary, or a reference in your textbook, determine the meaning of the root word *psilo*, and relate it to the appearance of *Psilotum*.

15.8 *Phylum Moniliophyta, Subphylum Sphenophyta: Horsetails*

16. Examine the photo on the right. You studied this genus in lab, but this is a different species. This species has two separate stems produced at different times in the growing season. The stem on the left is a reproductive branch, while that on the right is strictly vegetative. Based on the characteristics obvious in this photo, identify the plant, give its scientific name, and then identify the labeled structures.

Scientific name

_____

A

_____

B

_____

C

_____

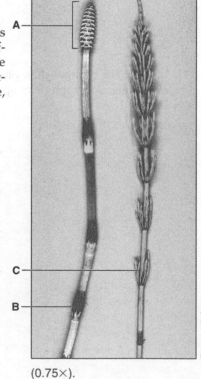

(0.75×).

17. Explain the distinction between a *node* and an *internode*.

15.9 *Phylum Moniliophyta, Subphylum Pterophyta: Ferns*

18. While rock climbing, you encounter the plant shown in the photo on the right growing out of a crevice.

    a. Is this the sporophyte or the gametophyte?

    b. What special name is given to the leaf of this plant?

(Photo by J. W. Perry.)

(0.5×).

19. The environments in which ferns grow range from standing water to very dry areas. Nonetheless, all ferns are dependent upon free water in order to complete their life cycles. Explain why this is the case.

Seed Plants I: Gymnosperms

OBJECTIVES

After completing this exercise, you will be able to

1. define *gymnosperm, pulp, heterosporous, homosporous, pollination, fertilization, dioecious, monoecious*;

2. describe the characteristics that distinguish seed plants from other vascular plants;

3. produce a cycle diagram of heterosporous alternation of generations;

4. list uses for conifers;

5. recognize the structures in **boldface** and describe the life cycle of a pine;

6. distinguish between a male and a female pine cone;

7. describe the method by which pollination occurs in pines;

8. describe the process of fertilization in pines;

9. recognize members of the four phyla: Coniferophyta, Cycadophyta, Ginkgophyta, and Gnetophyta.

INTRODUCTION

The development of the seed was a significant event in the evolution of vascular plants. Seeds have remarkable ability to survive under adverse conditions. This is one reason for the dominance of seed plants today.

Let's examine the characteristics of seeds and seed plants.

1. All seed plants produce **pollen grains.** Pollen grains serve as carriers for sperm. This characteristic is one factor accounting for the widespread distribution of seed plants. As a benefit of pollen production, the sperm of seed plants do *not* need free water to swim to the egg. Thus, seed plants can reproduce in harsh climates where nonseed plants are much less successful.

2. Almost all seeds have some type of **stored food** for the embryo to use as it emerges from the seed during germination. (The sole exception is orchid seeds, which rely on symbiotic association with a fungus to obtain nutrients.)

3. All seeds have a **seed coat,** a protective covering enclosing the embryo and its stored food.

A seed coat and stored food are particularly important for survival. An embryo within a seed is protected from an inhospitable environment. Consider a seed produced during a severe drought. Water is necessary for the embryo to grow. If none is available, the seed can remain dormant until growing conditions are favorable. When germination occurs, a ready food source is present to get things underway, providing nutrients until the developing plant can produce its own carbohydrates by photosynthesis.

4. Like the bryophytes, fern allies, and ferns, seed plants exhibit **alternation of generations.**

5. All seed plants are **heterosporous;** that is, they produce *two* types of spores. Bryophytes and most fern allies and ferns are **homosporous,** producing only *one* spore type.

Examine Figure 16-1, a diagram of the heterosporous alternation of generations.

Gymnosperms are one of two groups of seed plants. *Gymnosperm* translates literally as "naked seed," referring to the production of seeds on the *surface* of reproductive structures. This contrasts with the situation in the angiosperms (see Exercise 17), whose seeds are contained within a fruit.

The general assemblage of plants known as gymnosperms contains plants in four separate phyla:

Phylum	Common Name
Coniferophyta	Conifers
Cycadophyta	Cycads
Ginkgophyta	Ginkgos
Gnetophyta	Gnetophytes or vessel-containing gymnosperms

16.1 Phylum Coniferophyta: Conifers *(About 90 min.)*

By far the most commonly recognized gymnosperms are the conifers. Among the conifers, perhaps the most common is the pine (*Pinus*). Many people think of conifers and pines as one and the same. However, while all members of the genus *Pinus* are conifers, not all conifers are pines. Examples of other conifers include spruce (*Picea*), fir (*Abies*), and even some trees that lose their leaves in the fall such as larch.

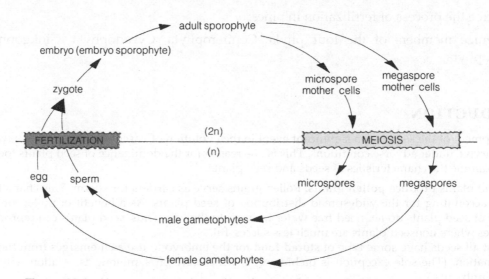

Figure 16-1   Heterosporous alternation of generations.

The conifers are among the most important plants economically, because their wood is used in building construction. Millions of hectares (1 hectare = 2.47 acres) are devoted to growing conifers for this purpose, not to mention the numerous plantations that grow Christmas trees. In many areas conifers are used for **pulp,** the moistened cell wall–derived cellulose used to manufacture paper.

The structures and events associated with reproduction in pine (*Pinus*) will be studied here as a representative conifer.

MATERIALS

Per student:

- cluster of male cones
- prepared slide of male strobilus, l.s., with microspores
- young female cone
- prepared slide of female strobilus, l.s., with megaspore mother cell
- prepared slide of pine seed, l.s.
- compound microscope
- dissecting microscope
- single-edged razor blade

Per student group (table):

- young sporophyte, living or herbarium specimen

Per lab room:

- demonstration slide of female strobilus with archegonium
- demonstration slide of fertilization
- pine seeds, soaking in water
- pine seedlings, 12 weeks old
- pine seedlings, 36 weeks old

Figure 16-2 Cluster of male cones shedding pollen (0.75×).

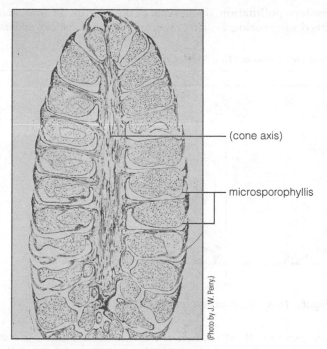

— (cone axis)

— microsporophyllis

Figure 16-3 Male pine cone, l.s. (12×).

PROCEDURE

As you do this activity, examine Figure 16-12, representing the life cycle of a pine tree. To refresh your memory, look at a specimen of a small pine tree. This is the adult **sporophyte** (Figure 16-12a). Identify the stem and leaves. You probably know the main stem of a woody plant as the trunk. The leaves of conifers are often called needles because most are shaped like needles.

A. Male Reproductive Structures and Events

1. Obtain a cluster of **male cones** (Figures 16-2, 16-12b). The function of male cones is to produce **pollen;** consequently, male cones are typically produced at the ends of branches, where the wind currents can catch the pollen as it is being shed.
2. Note all the tiny scalelike structures that make up the male cones. These are *microsporophylls*.

Translated literally, a sporophyll would be a "spore-bearing leaf." The prefix *micro-* refers to "small." But the literal interpretation "small, spore-bearing leaf" does not convey the full definition; biologists use *microsporophyll* to mean a leaf that produces *male* spores, called *microspores*. These develop into winged, immature male *gametophytes* called **pollen grains.** (Why they are immature is a logical question. The male gametophyte is not mature until it produces sperm.)

3. Remove a single microsporophyll and examine its lower surface with a dissecting microscope. Identify the two **microsporangia,** also called **pollen sacs** (Figure 16-12c).
4. Study a prepared slide of a longitudinal section of a male cone (also called a male *strobilus*), first with a dissecting microscope to gain an impression of the cone's overall organization and then with the low-power objective of your compound microscope. Identify the *cone axis* bearing numerous **microsporophylls** (Figures 16-3).
5. Switch to the medium-power objective to observe a single microsporophyll more closely. Note that it contains a cavity; this is the **microsporangium** (also called a *pollen sac*), which contains numerous *pollen grains*.

As the male cone grows, several events occur in the microsporophylls that lead to the production of pollen grains. Microspore mother cells within the microsporangia undergo meiosis to form *microspores*. Cell division within the microspore wall and subsequent differentiation result in the formation of the pollen grain.

6. Examine a single **pollen grain** with the high-dry objective (Figures 16-4, 16-12d). Identify the earlike *wings* on either side of the body.
7. Identify the four cells that make up the body: The two most obvious are the **tube cell** and smaller **generative cell.** (The nucleus of the tube cell is almost as large as the entire generative cell.) Note that this male gametophyte is made up of only four cells!

In conifers, **pollination**, the transfer of pollen from the male cone to the female cone, is accomplished by wind and occurs during spring. Pollen grains are caught in a sticky *pollination droplet* produced by the female cone.

B. Female Reproductive Structures and Events

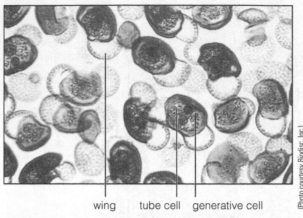

wing   tube cell   generative cell

(Photo courtesy Biodisc, Inc.)

Figure 16-4   Pine pollen grains (400×).

(Photo by J. W. Perry.)

Figure 16-5   Female cones of pine (1×).

Development and maturation of the female cone take 2–3 years; the exact time depends on the species. Female cones are typically produced on higher branches of the tree. Because the individual tree's pollen is generally shed downward, this arrangement favors crossing between *different* individuals.

1. Obtain a young **female cone** (Figures 16-5, 16-12e), and note the arrangement of the cone scales. Unlike the male cone, the female cone is a complex structure, each scale consisting of an **ovuliferous** scale fused atop a *sterile bract*.
2. Remove a single scale-bract complex (Figure 16-12f). On the top surface of the complex find the two **ovules**, the structures that eventually will develop into the **seeds.**
3. Examine a prepared slide of a longitudinal section of a female cone (Figures 16-6, 16-12g) first with the dissecting microscope. Note the spiral arrangement of the scales on the cone axis. Distinguish the smaller *sterile bract* from the *ovuliferous scale.*
4. Now examine the slide with the low-power objective of your compound microscope. Look for a section through an ovuliferous scale containing a very large cell; this is the **megaspore mother cell** (Figures 16-7, 16-12g). The tissue surrounding the megaspore mother cell is the **megasporangium.** Protruding inward toward the cone axis are "flaps" of tissue surrounding the megasporangium, the **integument.** Find the integuments and the opening between them, the **micropyle.**

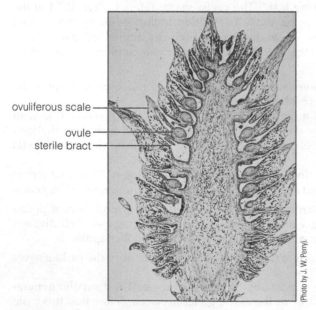

ovuliferous scale —
ovule —
sterile bract —

(Photo by J. W. Perry).

Figure 16-6   Female pine cone, l.s. (5×).

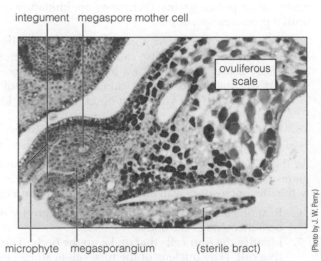

integument   megaspore mother cell

ovuliferous scale

microphyte   megasporangium   (sterile bract)

(Photo by J. W. Perry.)

Figure 16-7   Portion of female cone showing megaspore mother cell (75×).

Think for a moment about the three-dimensional nature of the ovule: It's much like a short vase lying on its side on the ovuliferous scale. The neck of the vase is the integument, the opening the micropyle. The integument extends around the base of the vase. If you poured liquid rubber inside the base of a vase, suspended a marble in the middle, and allowed the rubber to harden, you'd have a model of the megasporangium and the megaspore mother cell. Figure 16-8 gives you an idea of the three-dimensional structure.

The megaspore mother cell undergoes meiosis to produce four haploid *megaspores* (Figure 16-12h), but only one survives; the other three degenerate. The functional megaspore repeatedly divides mitotically to produce the multicellular **female gametophyte** (Figure 16-12i). At the same time, the female cone is continually increasing in size to accommodate the developing female gametophytes. (Remember, there are numerous ovuliferous scale/sterile bract complexes on each cone.)

The female gametophyte of pine is produced _____ (within *or* outside of) the megasporangium.

5. Examine the demonstration slide of **archegonia** (Figures 16-9, 16-12i) that have developed within the female gametophyte. Identify the single large **egg cell** that fills the entire archegonium. (The nucleus of the egg cell may be visible as well. The other generally spherical structures are protein bodies within the egg.)

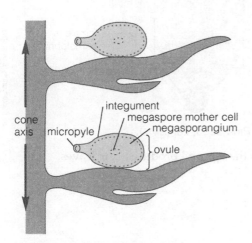

Figure 16-8 Ovule and ovuliferous scale/sterile bract complex.

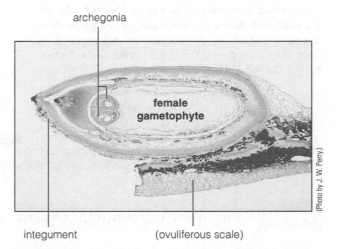

Figure 16-9 Pine ovule with female gametophyte and archegonia, l.s. (12×).

Recall that the pollen grains produced within male cones are caught in a sticky pollination droplet produced by the female cone. As the pollination droplet dries, the pollen grain is drawn through the micropyle and into a cavity called the *pollen chamber*.

Fertilization—the fusion of egg and sperm—occurs after the *pollen tube,* an outgrowth of the pollen grain's tube cell, penetrates the megasporangium and enters the archegonium. The generative cell of the pollen grain has divided to produce two sperm, one of which fuses with the egg. (The second sperm nucleus degenerates.)

6. Examine the demonstration slide of fertilization in *Pinus*. Identify the **zygote,** the product of fusion of egg and sperm (Figures 16-10, 16-12j).

After fertilization, numerous mitotic divisions of the zygote take place, eventually producing an **embryo** (*embryo sporophyte*). Fertilization also triggers changes in the integument, causing it to harden and become the seed coat.

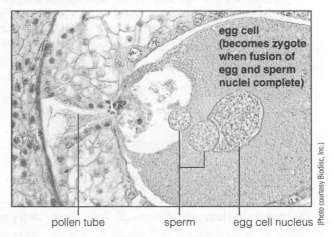

Figure 16-10 Fertilization in pine, l.s. (175×).

7. With the low-power objective of your compound microscope, study a prepared slide of a longitudinal section through a pine seed (Figures 16-11, 16-12k). Starting from the outside, identify the **seed coat, megasporangium** (a very thin, papery remnant), the female gametophyte, and **embryo** (embryo sporophyte).

8. Within the embryo, identify the **hypocotyl-root axis** and numerous **cotyledons,** in the center of which is the **epicotyl.** (*Hypo-* and *epi-* are derived from Greek, meaning "under" and "over," respectively. Thus, these terms refer to orientation with reference to the cotyledons.) The female gametophyte will serve as a food source for the embryo sporophyte when germination takes place.

9. Obtain a pine seed that has been soaked in water to soften the seed coat. Remove it and make a freehand longitudinal section with a sharp razor blade.

10. Identify the papery remnant of the **megasporangium,** the white **female gametophyte,** and **embryo** (embryo sporophyte).

How many cotyledons are present? _____

_____

11. Examine the culture of pine seeds that were planted in sand 12 weeks ago. Note the germinating seeds (Figure 16-12l).

12. Identify the **hypocotyl-root axis, cotyledons, female gametophyte,** and **seed coat.**

The cotyledons serve two functions. One is to absorb the nutrients stored in the female gametophyte during germination. As the cotyledons are exposed to light, they turn green. What then is the second function of the cotyledons?

_____

_____

_____

13. Finally, examine the 36-week-old sporophyte seedlings. Notice that the cotyledons eventually wither away as the epicotyl produces new leaves.

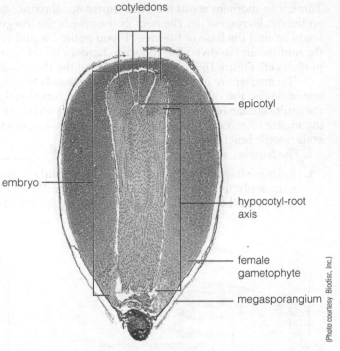

Figure 16-11 Pine seed, l.s. (24×).

(Photo courtesy Biodisc, Inc.)

16.2   Phylum Cycadophyta: Cycads *(About 10 min.)*

During the age of the dinosaurs (240 million years ago), cycads were extremely abundant in the flora. In fact, botanists often call this "the age of the cycads."

    Zamia and *Cycas* are two cycads most readily available in North America. In nature, these two species are limited to the subtropical regions. They're often planted as ornamentals in Florida, Gulf Coast states, and California.

    All cycads have separate male and female plants, unlike pine.

    The female cones of some other genera become extremely large, weighing as much as 30 kg!

MATERIALS

Per lab room:

■ demonstration specimens of *Zamia* and/or *Cycas*

PROCEDURE

1. Examine the demonstration specimen of *Zamia* and/or *Cycas* (Figure 16-13). Both have the common name *cycad.* Do these plants resemble any of the conifers you know? _____

_____

2. Notice the leaves of the cycads. They more closely resemble the leaves of the ferns than those of the conifers.

16.3   Phylum Ginkgophyta: Ginkgo *(About 10 min.)*

A single species of this phylum is all that remains of what was once a much more diverse assemblage of plants. *Ginkgo biloba* is sometimes called a living fossil because it has changed little in the last 80 million years. In fact, at one time it was believed to be extinct; the Western world knew it from the fossil record before living trees were discovered in the Buddhist temple courtyards of remote China. Today *Ginkgo* is a highly prized ornamental tree that is commonly planted in our urban areas. The tree has a reputation for being resistant to most insect pests and atmospheric pollution. The seeds are ground up and sold as a memory-enhancing dietary supplement, although the supplement's ability to do that is in dispute.

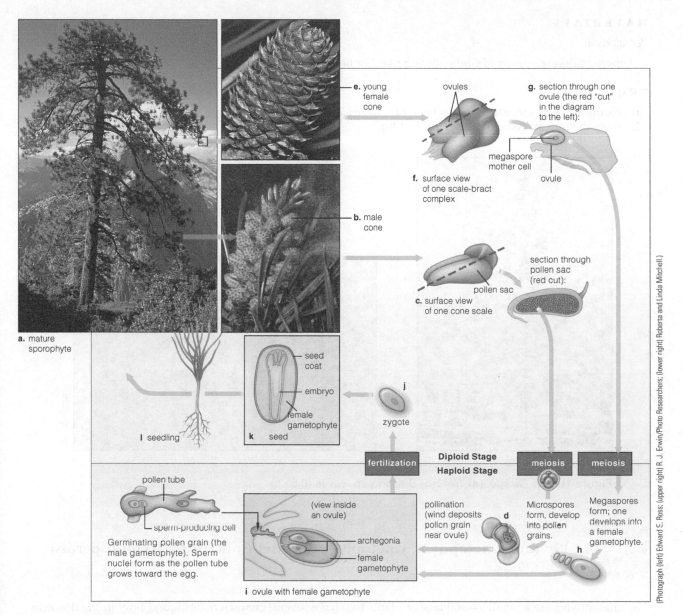

Figure 16-12 Pine life cycle.

Within the image:

- **e.** young female cone
- ovules
- **g.** section through one ovule (the red "cut" in the diagram to the left):
- megaspore mother cell
- ovule
- **f.** surface view of one scale-bract complex
- **b.** male cone
- section through pollen sac (red cut):
- pollen sac
- **c.** surface view of one cone scale
- seed coat
- embryo
- female gametophyte
- **j** zygote
- **k** seed
- **l** seedling
- **a.** mature sporophyte
- pollen tube
- sperm-producing cell
- Germinating pollen grain (the male gametophyte). Sperm nuclei form as the pollen tube grows toward the egg.
- (view inside an ovule)
- archegonia
- female gametophyte
- **i** ovule with female gametophyte
- fertilization
- **Diploid Stage**
- **Haploid Stage**
- meiosis
- meiosis
- pollination (wind deposits pollen grain near ovule)
- **d**
- Microspores form, develop into pollen grains.
- Megaspores form; one develops into a female gametophyte.
- **h**

(Photograph [left] Edward S. Ross; (upper right) R. J. Erwin/Photo Researchers; (lower right) Roberta and Linda Mitchell.)

a

b

Figure 16-13 Cycads. (**a**) *Cycas* (0.01×). (**b**) *Zamia* (0.1×).

(Photos by J. W. Perry.)

MATERIALS

Per lab room:

- demonstration specimen of *Ginkgo* (living plant or herbarium specimen)

PROCEDURE

1. Examine the demonstration specimen of *Ginkgo biloba*, the maidenhair tree (Figure 16-14a).
2. Note the fan-shaped leaves (Figure 16-14b).

(Photos by J. W. Perry.)

a   b

Figure 16-14   *Ginkgo.* (**a**) Tree. (**b**) Branch with leaves (0.5×).

16.4 Phylum Gnetophyta: Gnetophytes (Vessel-Containing Gymnosperms)
(About 10 min.)

The gnetophytes are a small assemblage of plants that have several characteristics found only in the flowering plants (angiosperms). Their reproductive structures look much more like flowers than cones. "Double fertilization", a feature thought to be unique to the flowering plants, has been reported in one gnetophyte species. And the water-conducting xylem tissue contains vessels in all species. Consequently, many scientists now believe these plants are very closely related to the flowering plants. Most species are found in desert or arid regions of the world. In the desert Southwest of the United States, *Ephedra* is a common shrub known as Mormon tea, because its stems were once harvested by Mormon settlers in Utah and used to make a tea.

MATERIALS

Per lab room:

- demonstration specimen of *Ephedra* and/or *Gnetum* (living plant or herbarium specimen)

PROCEDURE

1. Examine the demonstration specimen of *Ephedra* (Figure 16-15).
2. A second representative gnetophyte is the genus *Gnetum* (pronounced "neat-um"). A few North American college and university greenhouses—and some tropical gardens—keep these specimens. *Gnetum* is native to Brazil, tropical west Africa, India, and Southeast Asia. Different species vary in form from vines to trees (Figures 16-16a, b). Examine the living specimen if one is on display, noting particularly the broad, flat leaves (Figure 16-16c).

Figure 16-15 *Ephedra.* (**a**) Several plants (1×). (**b**) Close-up of stems (0.5×).

(Photos by J. W. Perry.)

a   b   c

Figure 16-16 *Gnetum.* (**a**) A species that is a vine (0.02×). (**b**) A species that is a tree (0.02×). (**c**) Leaves (0.38×).

(Photos by J. W. Perry.)

_____ 1. Which statement is *not* true about conifers?
(a) Conifers are gymnosperms
(b) All conifers belong to the genus *Pinus*
(c) All conifers have naked seeds
(d) Conifers are heterosporous

_____ 2. Seed plants
(a) have alternation of generations
(b) are heterosporous
(c) develop a seed coat
(d) are all of the above

_____ 3. A pine tree is
(a) a sporophyte
(b) a gametophyte
(c) diploid
(d) both a and c

_____ 4. The male pine cone
(a) produces pollen
(b) contains a female gametophyte
(c) bears a megasporangium containing a megaspore mother cell
(d) gives rise to a seed

_____ 5. The male gametophyte of a pine tree
(a) is produced within a pollen grain
(b) produces sperm
(c) is diploid
(d) is both a and b

_____ 6. Which of these are produced directly by meiosis in pine?
(a) sperm cells
(b) pollen grains
(c) microspores
(d) microspore mother cells

_____ 7. An ovule
(a) is the structure that develops into a seed
(b) contains the microsporophyll
(c) is produced on the surface of a male cone
(d) is all of the above

_____ 8. The process by which pollen is transferred to the ovule is called
(a) transmigration
(b) fertilization
(c) pollination
(d) all of the above

_____ 9. Which statement is true of the female gametophyte of pine?
(a) It's a product of repeated cell divisions of the functional megaspore
(b) It's haploid
(c) It serves as the stored food to be used by the embryo sporophyte upon germination
(d) All of the above are true

_____ 10. The seed coat of a pine seed
(a) is derived from the integuments
(b) is produced by the micropyle
(c) surrounds the male gametophyte
(d) is divided into the hypocotyl-root axis and epicotyl

EXERCISE 16

Seed Plants I: Gymnosperms

POST-LAB QUESTIONS

Introduction

1. What survival advantage does a seed have that has enabled the seed plants to be the most successful of all plants?

2. In the diagram below of a seed, give the ploidy level (n or 2n) of each part listed.

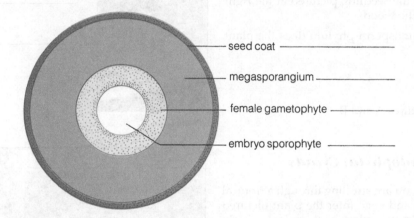

3. Distinguish between a homosporous and a heterosporous type of life cycle.

16.1 Phylum Coniferophyta: Conifers

4. List four uses for conifers.
   a.

   b.

   c.

   d.

5. While snowshoeing through the winter woods, you stop to look at a tree branch pictured at the right. Specifically, what are the brown structures hanging from the branch?

(0.5×).

6. Distinguish between *pollination* and fertilization.

7. Are antheridia present in conifers? _____

   Are archegonia present? _____

8. Suppose you saw the seedling pictured at the right while walking in the woods.

   a. To which gymnosperm phylum does the plant belong?

   b. Identify structures A and B.

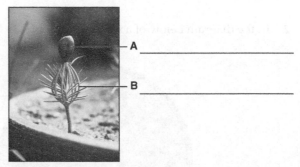

A _____

B _____

(0.5×).

16.2 *Phylum Cycadophyta: Cycads*

9. On spring break, you are strolling through a tropical garden in Florida and encounter the plant pictured at the right. Would the xylem of this plant contain tracheids, vessels, or both?

(0.01×).

16.3 *Phylum Ginkgophyta: Ginkgo*

10. A friend of yours picks up a branch like the one pictured at the right. Knowing that you have taken a biology course and studied plants, she asks you what the plant is. Identify this branch, giving your friend the full scientific name for the plant.

(0.75×).

Seed Plants II: Angiosperms

OBJECTIVES

After completing this exercise, you will be able to

1. define *angiosperm, fruit, pollination, double fertilization, endosperm, seed, germination, annual, biennial, perennial;*

2. describe the significance of the flower, fruit, and seed for the success of the angiosperms;

3. identify the structures of the flower;

4. recognize the structures and events (those in **boldface**) that take place in angiosperm reproduction;

5. describe the origin and function of fruit and seed;

6. identify the characteristics that distinguish angiosperms from gymnosperms.

INTRODUCTION

The **angiosperms,** Phylum Anthophyta, are seed plants that produce flowers. *"Antho"* means flower, and *"phyta"* plant. The word *angiosperm* literally means "vessel seed" and refers to the seeds borne within a fruit.

There are more flowering plants in the world today than any other group of plants. Assuming that numbers indicate success, flowering plants are the most successful plants to have evolved thus far.

The most important characteristic that distinguishes the Anthophyta from other seed plants is the presence of flower parts that mature into a **fruit,** a container that protects the seeds and allows them to be dispersed without coming into contact with the rigors of the external environment. In many instances, the fruit also contributes to the dispersal of the seed. For example, some fruits stick to fur (or clothing) of animals and are brushed off some distance from the plant that produced them. Animals eat others. The undigested seeds pass out of the digestive tract and fall into environments often far removed from the seeds' source.

Our lives and diets revolve around flowering plants. Fruits enrich our diet and include such things as apples, oranges, tomatoes, beans, peas, corn, wheat, walnuts, pecans . . . the list goes on and on. Moreover, even when we are not eating fruits, we're eating flowering plant parts. Cauliflower, broccoli, potatoes, celery, and carrots all are parts of flowering plants.

Biologists believe that flower parts originated as leaves modified during the course of evolution to increase the probability for fertilization. For instance, some flower parts are colorful and attract animals that transfer the sperm-producing pollen to the receptive female parts.

Figure 17-15 depicts the life cycle of a typical flowering plant. Refer to it as you study the specimens in this exercise.

Note: This exercise provides two alternative paths to accomplish the objectives described above, a traditional approach (17.1–17.2) and an investigative one (17.3). Your instructor will indicate which alternative you will follow.

17.1   External Structure of the Flower *(About 20 min.)*

The number of different kinds of flowers is so large that it's difficult to pick a single example as representative of the entire division. Nonetheless, there is enough similarity among flowers that, once you've learned the structure of one representative, you'll be able to recognize the parts of most.

MATERIALS

Per student:

- flower for dissection (gladiolus or hybrid lily, for example)
- single-edged razor blade
- dissecting microscope

PROCEDURE

1. Obtain a flower provided for dissection.
2. At the base of the flower, locate the swollen stem tip, the **receptacle,** upon which the whorls of floral parts are arranged.
3. Identify the **calyx,** comprising the outermost whorl. Individual components of the calyx are called **sepals.** The sepals are frequently green (although not always). The calyx surrounds the rest of the flower in the bud stage (see Figure 17-15a).
4. Moving inward, locate the next whorl of the flower, the usually colorful **corolla** made up of **petals.**

It is usually the petals that we appreciate for their color. Remember, however, that the evolution of colorful flower parts was associated with the presence of color-visioned *pollinators,* animals that carry pollen from one flower to another. The colorful flowers attract those animals, enhancing the plants' chance of being pollinated, producing seeds, and perpetuating their species.

Both the calyx and corolla are sterile, meaning they do not produce gametes.

5. The next whorl of flower parts consists of the male, pollen-producing parts, the **stamens** (also called *microsporophylls,* "microspore-bearing leaves"; Figure 17-15b). Examine a single stamen in greater detail. Each stamen consists of a stalklike **filament** and an **anther.** The anther consists of four **microsporangia** (also called *pollen sacs*).
6. Next locate the female portion of the flower, the **pistil** (Figures 17-15a and c). A pistil consists of one or more **carpels,** also called *megasporophylls,* "megaspore-bearing leaves."

If the pistil consists of more than one carpel, they are usually fused, making it difficult to distinguish the individual components. However, you can usually determine the number of carpels by counting the number of lobes of the stigma.

7. How many carpels does your flower contain? _____
8. Identify the different parts of the pistil (Figure 17-15c): at the top, the **stigma,** which serves as a receptive region on which pollen is deposited; a necklike **style;** and a swollen **ovary.** The only members of the plant kingdom to have ovaries are angiosperms!
9. With a sharp razor blade, make a section of the ovary. (Some students should cut the pistil longitudinally; others should cut the ovary crosswise. Then compare the different sections.)
10. Examine the sections with a dissecting microscope, finding the numerous small **ovules** within the ovary. The diagram in Figure 17-15c is oversimplified; many flowers have more than one ovule per ovary.
11. In Figure 17-1, make and label two sketches: one of the cross section and the other of a longitudinal section of the ovary. Notice that the ovules are completely enclosed within the ovary. After fertilization, the ovules will develop into *seeds,* and the ovary will enlarge and mature into the *fruit.*

There are two groups of flowering plants, monocotyledons and dicotyledons. The number of flower parts indicates which group a plant belongs to. Generally, monocots have the flower parts in threes or multiples of three. Dicots have their parts in fours or fives or multiples thereof.

12. Count the number of petals or sepals in the flower you have been examining. Are you studying a monocot or dicot?

_____

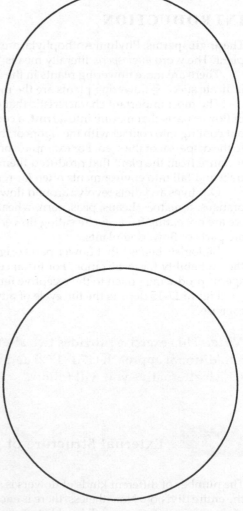

Figure 17-1 Drawings of cross section and longitudinal section of an ovary.
Label: ovules

A. Male Gametophyte (Pollen Grain) Formation in the Microsporangia *(About 20 min.)*

The male gametophyte in flowering plants is the pollen grain; a microscopic sperm-producing structure.

MATERIALS

Per student:

- prepared slide of young lily anther, c.s.
- prepared slide of mature lily anther (pollen grains), c.s.
- *Impatiens* flowers, with mature pollen
- glass microscope slide
- coverslip
- compound microscope

Per student pair:

- 0.5% sucrose, in dropping bottle

PROCEDURE

1. With the low-power objective of your compound microscope, examine a prepared slide of a cross section of an immature anther (Figures 17-2, 17-15d). Find sections of the four **microsporangia** (also called *pollen sacs*), which appear as four clusters of densely stained cells within the anther.

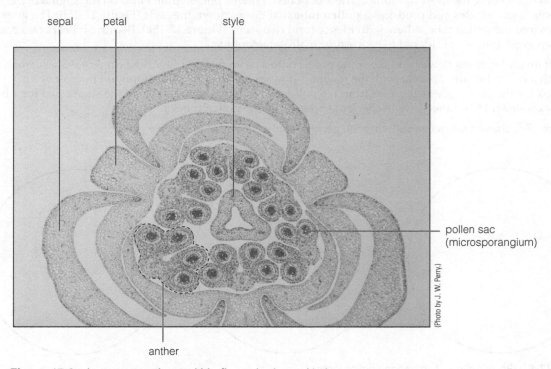

Figure 17-2 Immature anthers within flower bud, c.s. (4×).

2. Study the contents of a single microsporangium. Depending on the stage of development, you will find either diploid *microspore mother cells* (Figure 17-15e) or haploid *microspores* (Figures 17-15f and g).
3. Obtain a prepared slide of a cross section of a mature anther (Figure 17-3). Observe it first with the low-power objective, noting that the walls have split open to allow the pollen grains to be released, as shown in Figure 17-15i.
4. Pollen grains are immature *male gametophytes*, and very small ones at that. Switch to the high-dry objective to study an individual pollen grain more closely (Figure 17-4).
5. The pollen grain consists of only two cells. Identify the large **tube cell** and a smaller, crescent-shaped **generative cell** that floats freely in the cytoplasm of the tube cell (Figure 17-15h).
6. Note the ridged appearance of the outer wall layer of a pollen grain. Within the ridges and valleys of the wall, glycoproteins are present that appear to play a role in recognition between the pollen grain and the stigma.

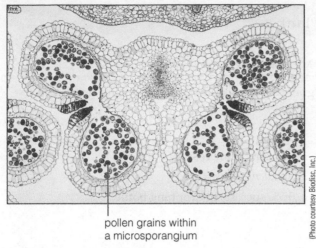

pollen grains within
a microsporangium

(Photo courtesy Biodisc, Inc.)

Figure 17-3 Mature anther, c.s. (21×).

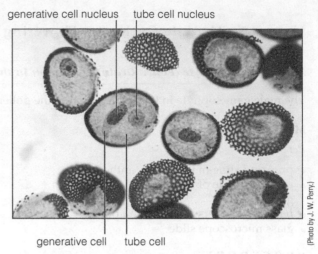

generative cell nucleus   tube cell nucleus

generative cell   tube cell

(Photo by J. W. Perry.)

Figure 17-4 Pollen grain, c.s. (145×).

Transfer of pollen from the microsporangia to the stigma, called **pollination,** occurs by various means—wind, insects, and birds being the most common carriers of pollen. When a pollen grain lands on the stigma of a compatible flower, it germinates and produces a **pollen tube** that grows down the style (Figure 17-15j). The generative cell flows into the pollen tube, where it divides to form two *sperm* (Figure 17-15k). Because it bears two gametes, the pollen grain is now considered to be a *mature* male gametophyte.

7. Obtain an *Impatiens* flower and tap some pollen onto a clean microslide. Add a drop of 0.5% sucrose, cover with a coverslip, and observe with the medium-power objective of your compound microscope.

8. Look for the *pollen tube* as it grows from the pollen grain. (You may wish to set the slide aside for a bit and re-examine it 15 minutes later to check the progress of the pollen tubes.)

In Figure 17-5, draw a sequence showing the germination of an *Impatiens* pollen grain.

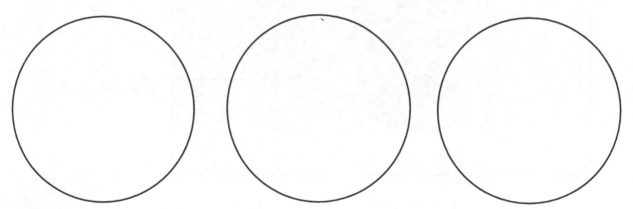

Figure 17-5 Germination of *Impatiens* pollen grain.

B. Female Gametophyte Formation in the Megasporangia *(About 20 min.)*

The female gametophyte is formed inside each ovule within the ovary of the flower. Like the male gametophyte, the female gametophyte in flowering plants consists of only a few cells.

MATERIALS

Per student:

- prepared slide of lily ovary, c.s., megaspore mother cell
- compound microscope

Per lab room:

- demonstration slide of lily ovary, c.s., 7-celled, 8-nucleate gametophyte
- demonstration slide of lily ovary, c.s., double fertilization

PROCEDURE

1. With the medium-power objective of your compound microscope, examine a prepared slide of a cross section of an ovary (Figures 17-6, 17-15l).

2. Find the several **ovules** that have been sectioned. One ovule will probably be sectioned in a plane so that the very large, diploid **megaspore mother cell** is obvious (Figures 17-7, 17-15m).

3. The megaspore mother cell is contained within the **megasporangium,** the outer cell layers of which form two flaps of tissue called **integuments.** Identify the structures in boldface print. After fertilization, the integuments develop into the *seed coat.*

4. Identify the **placenta,** the region of attachment of the ovule the ovary wall.

As the megasporangium develops, the integuments grow, enveloping the megasporangium. However, a tiny circular opening remains. This opening is the **micropyle.** (The micropyle is obvious in Figure 17-15n.) Remember, the micropyle is an opening in the ovule. After pollination, the pollen grain germinates on the surface of the stigma, and the pollen tube grows down the style, through the space surrounding the ovule, through the micropyle, and penetrates the megasporangium (Figure 17-15p).

5. Identify the micropyle present in the ovule you are examining.

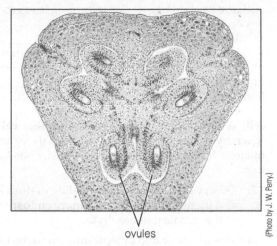

ovules

Figure 17-6 Lily ovary, c.s. (19×).

(Photo by J. W. Perry.)

Considerable variation exists in the next sequence of events; we describe the pattern found in the lily.

The diploid megaspore mother cell undergoes meiosis, producing four haploid *nuclei* (Figure 17-15n). (Note that cytokinesis does *not* follow meiosis, and thus only nuclei—not cells—are formed.) The cell containing the four nuclei (the old megaspore mother cell) is now called the **female gametophyte** (also called the *embryo sac).*

Three of these four nuclei fuse. Thus, the female gametophyte contains one triploid (3n) nucleus and one haploid (n) nucleus (Figure 17-15o). Subsequently, the nuclei undergo two *mitotic* divisions, forming eight nuclei in the female gametophyte. Cell walls form around six of the eight nuclei; the large cell remaining—the **central cell**—contains two nuclei, one of which is triploid, the other haploid. These two nuclei are called the **polar nuclei.** At the micropylar end of the ovule there are three cells, all haploid. One of these is the **egg cell.** Opposite the micropylar end are three 3n cells. This stage of development is often called the *seven-celled, eight-nucleate female gametophyte* (Figures 17-8, 17-15p). Fertilization takes place at this stage.

6. Study the demonstration slide of the seven-celled, eight-nucleate female gametophyte. Identify the **placenta, integuments, micropyle, egg cell, central cell,** and **polar nuclei** (Figures 17-8, 17-15p).

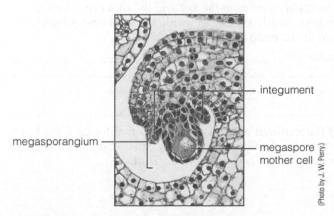

megasporangium

integument

megaspore mother cell

(Photo by J. W. Perry.)

Figure 17-7 Megaspore mother cell within megasporangium (94×).

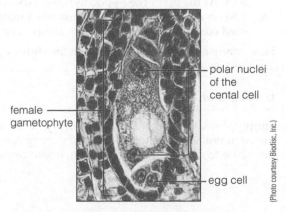

female gametophyte

polar nuclei of the cental cell

egg cell

(Photo courtesy Biodisc, Inc.)

Figure 17-8 Seven-celled, eight-nucleate female gametophyte (170×).

As the pollen tube penetrates the female gametophyte, it discharges the sperm; one of the sperm nuclei fuses with the haploid egg nucleus, forming the *zygote*. Figure 17-15q shows the female gametophyte after fertilization has occurred.

The zygote is a _____ (haploid, diploid) cell.

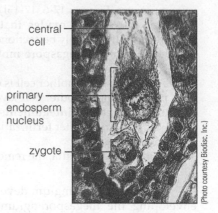
Figure 17-9 Double fertilization (170×).

The other sperm nucleus enters the central cell and fuses with the two polar nuclei, forming the primary endosperm nucleus (Figure 17-15q). Thus, the primary endosperm nucleus is _____ (haploid, diploid, triploid, tetraploid, pentaploid).

The cell containing the primary endosperm nucleus (the old central cell) is now called the **endosperm mother cell.** Traditionally, the process whereby one sperm nucleus fuses with the egg nucleus and the other sperm nucleus fuses with the two polar nuclei has been called **double fertilization.**

7. Observe the demonstration slide of double fertilization (Figure 17-9), and identify the **zygote, primary endosperm nucleus,** and **central cell** of the female gametophyte.

Numerous mitotic and cytoplasmic divisions of the endosperm mother cell form the **endosperm,** a tissue used to nourish the embryo sporophyte as it develops within the seed.

C. Embryogeny (About 10 min.)

The zygote undergoes mitosis and cytokinesis to produce a two-celled **embryo** (also called the *embryo sporophyte*). Numerous subsequent divisions produce an increasingly large and complex embryo.

MATERIALS

Per lab room:

■ demonstration slides of *Capsella* embryogeny: globular embryo, emerging cotyledons, torpedo-shaped embryo, mature embryo

PROCEDURE

1. Observe the series of four demonstration slides, which show the stages of embryo development in the female gametophyte of *Capsella* (Figure 17-10).
2. The first slide shows the so-called globular stage (Figure 17-10a), in which a chain of cells (the suspensor) attaches the embryo, a spherical mass of cells, to the wall of the female gametophyte (embryo sac). The very enlarged cell at the base of the suspensor is the *basal cell* and is active in uptake of nutrients that the developing embryo will use. Note the endosperm within the female gametophyte.
3. The second slide shows the heart-shaped stage (Figure 17-10b). Now you can distinguish the emerging **cotyledons** (seed leaves). In many plants, the cotyledons absorb nutrients from the endosperm and thus serve as a food reserve to be used during seed germination.
4. Further development of the embryo has occurred in the third slide, the torpedo stage (Figure 17-10c). Notice that the entire embryo has elongated. Find the *cotyledons*. Between the cotyledons, locate the **epicotyl,** which is the *apical meristem of the shoot*. Beneath the epicotyl and cotyledons find the **hypocotyl-root axis.** At the tip of the hypocotyl-root axis, locate the *apical meristem of the root* and the **root cap** covering it.
5. The final slide (Figure 17-10d) shows a mature embryo, neatly packaged inside the **seed coat.** Identify the seed coat and other regions previously identified in the torpedo stage.

How many cotyledons were there in the slides you examined? _____

Thus, is *Capsella* a monocot or dicot? _____

D. Fruit and Seed (About 20 min.)

Simply stated, a **fruit** is a matured ovary, while a **seed** is a matured ovule. Bear in mind that for each seed produced, a pollen grain had to fertilize the egg cells in the ovules. Fertilization not only causes the integuments of the ovule to develop into a seed coat, it also causes the ovary wall to expand into the fruit.

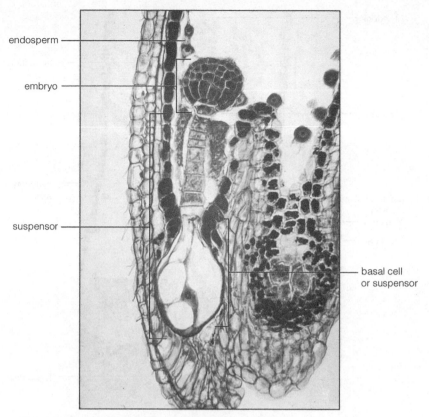

endosperm

embryo

suspensor

basal cell
or suspensor

a Globular embryo stage (232×).

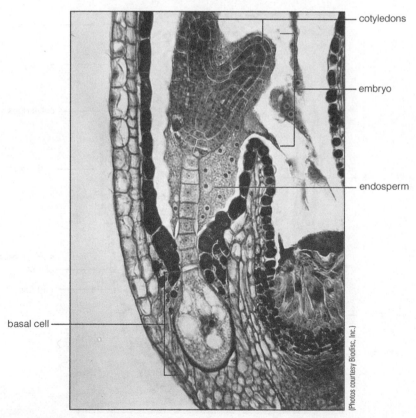

cotyledons

embryo

endosperm

basal cell

(Photos courtesy Biodisc, Inc.)

b Heart-shaped stage (240×).

Figure 17-10 Embryogeny in *Capsella* (shepherd's purse). *Continues on page 314.*

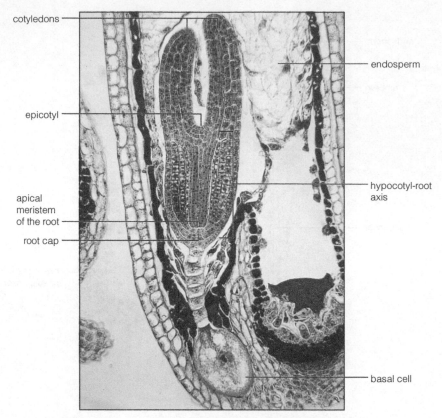

cotyledons

endosperm

epicotyl

apical
meristem
of the root

hypocotyl-root
axis

root cap

basal cell

c Torpedo stage (170×).

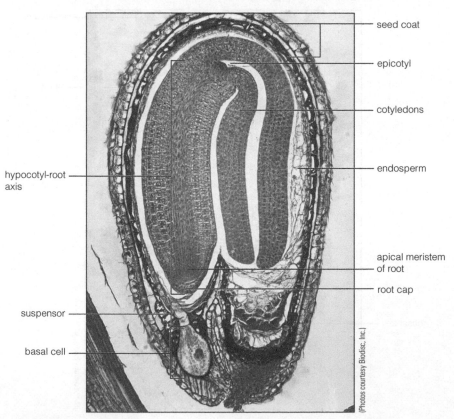

seed coat

epicotyl

cotyledons

hypocotyl-root
axis

endosperm

apical meristem
of root

root cap

suspensor

basal cell

(Photos courtesy Biodisc, Inc.)

d Mature embryo stage (113×).

Figure 17-10 Embryogeny in *Capsella* (shepherd's purse). *Continued.*

MATERIALS

Per student:

- bean fruits
- soaked bean seeds
- iodine solution (I₂KI), in dropping bottle

Per lab room:

- herbarium specimen of *Capsella,* with fruits
- demonstration slide of *Capsella* fruit, c.s.

PROCEDURE

1. Examine the herbarium specimen and demonstration slide of the fruits of *Capsella.* On the herbarium specimen, identify the **fruits,** which are shaped like the bag that shepherds carried at one time (Figure 17-11; the common name of this plant is "shepherd's purse").

2. Now study the demonstration slide of a cross section through a single fruit (Figure 17-12). Note the numerous **seeds** in various stages of embryo development.

3. Obtain a bean pod and carefully split it open along one seam.

    The pod is a matured ovary and thus is a _____.

    Find the *sepals* at one end of the pod and the shriveled *style* at the opposite end. The "beans" inside are

    _____.

4. Note the point of attachment of the bean to the pod. This is the placenta, a term shared by the plant and animal kingdom that describes the nutrient bridge between the "unborn" offspring and parent.

5. In Figure 17-13, draw the split-open bean pod, labeling it with the correct scientific terms. Figure 17-15r shows a section of a typical fruit.

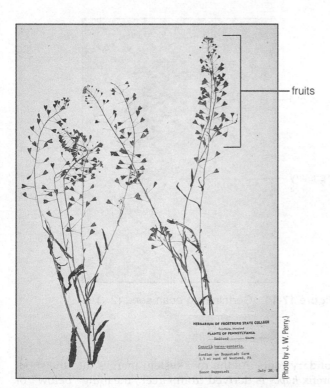

Figure 17-11 Herbarium specimen of *Capsella,* shepherd's purse (0.25×).

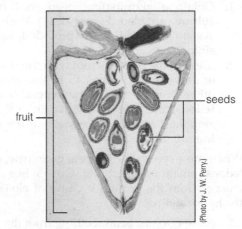

Figure 17-12 Cross section of *Capsella* fruit (8×).

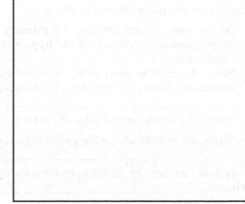

Figure 17-13 Drawing of an open bean pod.
Labels: fruit, seeds, placenta

6. Closely study one of the beans from within the pod or a bean that has been soaked overnight to soften it. Find the scar left where the seed was attached to the fruit wall.

7. Near the scar, look for the tiny opening left in the seed coat.

What is this tiny opening? _____ (*Hint:* The pollen tube grew through it.)

8. Remove the seed coat to expose the two large **cotyledons.** Split the cotyledons apart to find the **epicotyl** and **hypocotyl-root axis.** During maturation of the bean embryo, the cotyledons absorb the endosperm. Thus, bean cotyledons are very fleshy because they store carbohydrates that will be used during seed germination. Add a drop of iodine I₂KI to the cotyledon.

What substance is located in the cotyledon? _____
(*Hint:* Return to Exercise 4 if you've forgotten what is stained by I₂KI.)

E. Seedling *(About 20 min.)*

When environmental conditions are favorable for growth (adequate moisture, oxygen, and proper temperatures), the seed germinates; that is, the seedling (young sporophyte) begins to grow.

MATERIALS

Per student:

- germinating bean seeds
- bean seedlings

Per table:

- dishpan of water

PROCEDURE

1. Obtain a germinating bean seed from the culture provided (Figure 17-14). Wash the root system in the dishpan provided, *not* in the sink.

2. Identify the **primary root** with the smaller **secondary roots** attached to it. Emerging in the other direction will be the **hypocotyl,** the **cotyledons,** the **epicotyl** (above the cotyledons), and the first **true leaves** above the epicotyl. Identify these structures.

When some seeds (like the pea) germinate, the cotyledons remain *below* ground. Others, like the bean, emerge from the ground because of elongation of the hypocotyl-root axis.

3. Now, obtain a bean seedling from the growing medium. Be careful as you pull it up so as not to damage the root system. Wash the root system in the dishpan provided.

Your seedling should be in a stage of development similar to that shown in Figure 17-15s. Much of the growth that has taken place is the result of cellular elongation of the parts present in the seed.

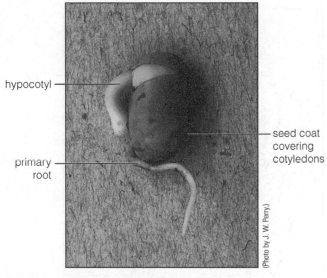

Figure 17-14 Germinating bean seed (2×).

4. On the root system, identify the **primary** and **secondary roots.** As the root system merges into the shoot (aboveground) system, find the **hypocotyl.** (The prefix *hypo-* is derived from Greek, meaning "below" or "underneath.")

5. Next, identify the cotyledons. Notice their shriveled appearance. Now that you know the function of the cotyledons from your previous study, why do you suppose the cotyledons are shriveled?

_____

6. Above the cotyledons, find that portion of the stem called the **epicotyl.**

Knowing what you do about the prefix *hypo-,* speculate on what the prefix *epi-* means.

_____

As noted earlier, the seedling has *true leaves.* Contrast the function of the cotyledons ("seed leaves") with the true leaves. _____

_____

Depending on the stage of development, your seedling may have even more leaves that have been produced by the shoot apex.

The amount of time between seed germination and flowering largely depends on the particular plant. Some plants produce flowers and seeds during their first growing season, completing their life cycle in that growing season. These plants are called **annuals.** Marigolds are an example of an annual. Others, known as **biennials,** grow vegetatively during the first growing season and do not produce flowers and seeds until the second growing season (carrots, for example). Both annuals and biennials die after seeds are produced.

Perennials are plants that live several to many years. The time between seed germination and flowering (seed production) varies, some requiring many years. Moreover, perennials do not usually die after producing seed, but flower and produce seeds many times during their lifetime.

17.3 Experiment: An Investigative Study of the Life Cycle of Flowering Plants

Recent developments with a fast-growing and fast-reproducing plant in the mustard family allow you to study the life cycle of a flowering plant over the course of only 35 days. The plants were bred by a plant scientist at the University of Wisconsin–Madison and are sometimes called Wisconsin Fast Plants™. You will grow these rapid-cycling *Brassica rapa* (abbreviated RCBr for *Rapid Cycling Brassica rapa*) plants from seeds, following the growth cycle and learning about plant structure, the life cycle of a flowering plant, and adaptive mechanisms for pollination.

Because this investigation occurs over several weeks, all Materials and Procedures are listed by the day the activity begins, starting from day 0. Record your measurements and observations in tables at the end of this exercise.

A. Germination

Day 0 (About 30 min.)

MATERIALS

Per student:

- one RCBr seed pod
- metric ruler
- 3 × 5 in. index card
- transparent adhesive tape
- petri dish
- paper toweling or filter paper disk to fit petri dish
- growing quad *or* two 35-mm film canisters
- 12 N-P-K fertilizer pellets (24 if using canister method)
- 4 paper wicks (growing quad) *or* 2 cotton string wicks (canister method)
- disposable pipet with bulb
- pot label (2 if using canister method)

- marker
- forceps
- model seed
- 500-mL beaker
- 2 wide-mouth bottles (film canister method)

Per student group (4):

- 2-L soda bottle bottom or tray
- moistened soil mix
- watertight tray
- water mat

Per lab room:

- light bank

PROCEDURE

The life cycle of a seed plant has no beginning or end; it's a true cycle. For convenience, we'll start our examination of this cycle with germination of nature's perfect time capsule, the seed. **Germination** is the process of switching from dormancy to growth.

1. Obtain an RCBr seed pod that has been sandwiched between two strips of transparent adhesive tape. This pod is the **fruit** of the plant. A fruit is a matured ovary. Measure and record the length of the pod in Table 17-1.
2. Harvest the seeds by scrunching up the pod until the seeds come out. Then carefully separate the halves of the pod to expose the seeds.
3. Roll a piece of adhesive tape into a circle with the sticky side out. Attach the tape to an index card and then place the seeds on the tape. (This keeps the small seeds from disappearing.) You may wish to handle the seeds with forceps. Count the number of seeds and record the number in Table 17-1.

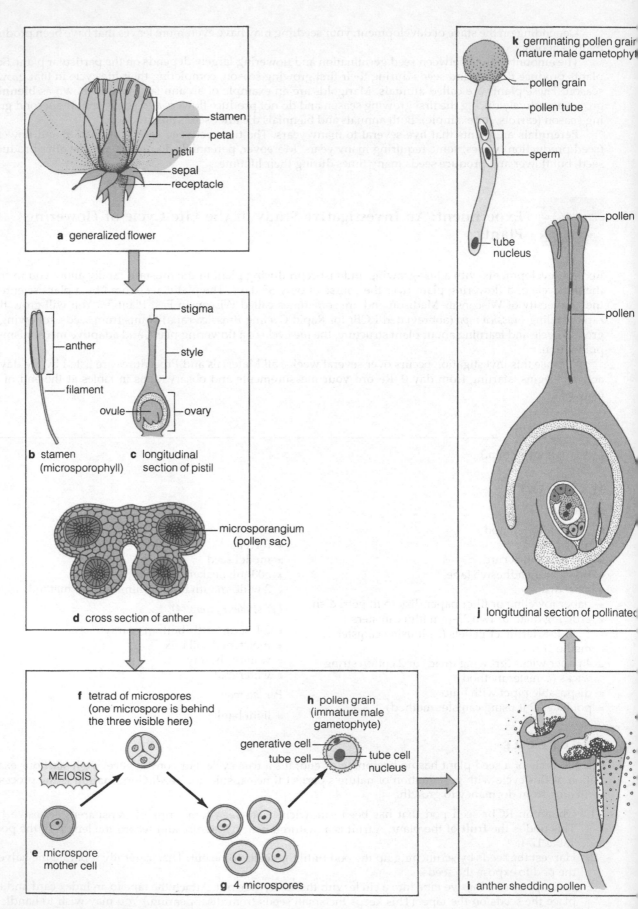

Figure 17-15 Life cycle of an angiosperm. [Color scheme: Haploid (n) structures are yellow, diploid (2n) are green, gold, or red; triploid (3n) are purple.]

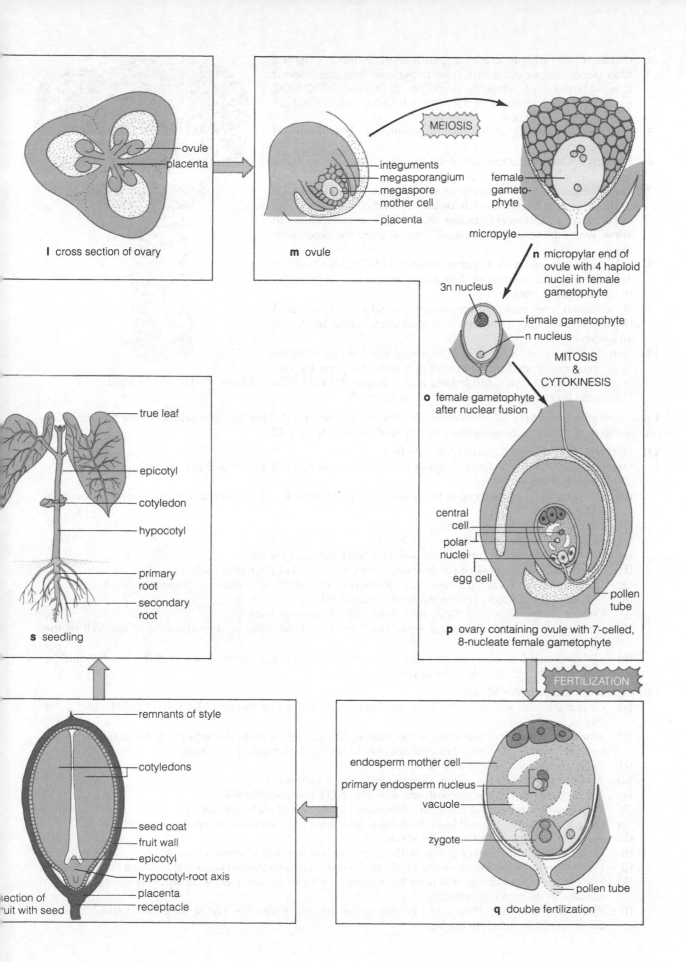

l cross section of ovary

- ovule
- placenta

m ovule

MEIOSIS

- integuments
- megasporangium
- megaspore mother cell
- placenta
- female gametophyte
- micropyle

n micropylar end of ovule with 4 haploid nuclei in female gametophyte

- 3n nucleus
- female gametophyte
- n nucleus

MITOSIS & CYTOKINESIS

o female gametophyte after nuclear fusion

- central cell
- polar nuclei
- egg cell
- pollen tube

p ovary containing ovule with 7-celled, 8-nucleate female gametophyte

FERTILIZATION

q double fertilization

- endosperm mother cell
- primary endosperm nucleus
- vacuole
- zygote
- pollen tube

s seedling

- true leaf
- epicotyl
- cotyledon
- hypocotyl
- primary root
- secondary root

section of fruit with seed

- remnants of style
- cotyledons
- seed coat
- fruit wall
- epicotyl
- hypocotyl-root axis
- placenta
- receptacle

4. Obtain a petri dish to use as a germination chamber. Obtain a filter paper disk or cut a disk from paper toweling and insert it into the larger (top) part of the petri dish. With a *pencil* (**DO NOT** use ink), label the bottom of the circle with your name, the date, and current time of day.

5. Moisten the toweling in the petri dish with tap water. Pour out any excess water.

6. Place five seeds on the top half of the toweling and cover with the bottom (smaller half) of the petri dish.

7. Place the germination chamber at a steep angle in shallow water in the base of a 2-L soda bottle or tray so that about 1 cm of the lower part of the towel is below the water's surface. This will allow water to be evenly "wicked" up to keep the seeds uniformly moist.

8. Set your experiment in a warm location (20–25°C is ideal). In Table 17-1, record the temperature and number of seeds placed in the germination plate.

9. To envision the germination process, obtain a model seed (Figure 17-16). A **seed** consists of a **seed coat, stored food,** and an **embryo.**

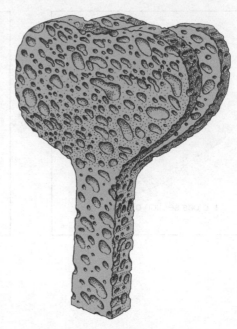

Figure 17-16 Model seed.

10. Germination in the RCBr seed starts with the imbibition of water. Toss your model seed into a beaker of water. Out pops the embryo, consisting of two **cotyledons** and a **hypocotyl-root axis.** (*Hypocotyl* literally means "below the cotyledons.")

You'll now plant seeds for additional studies. If your instructor indicates that you will be using growth quads, proceed to step 11. If you're using the canister method, skip to step 12.

11. *Option A:* Polystyrene Growth Quad Method
    (a) Obtain 4 diamond-shaped paper towel wicks, growth quad, moistened soil mix, 12 fertilizer pellets, and a pot label.
    (b) Drop 1 wick into each cell in the quad. The tip of the wick should protrude about 1 cm from the hole in the bottom of the quad.
    (c) Fill each cell *halfway* with soil.
    (d) Add 3 N-P-K fertilizer pellets to *each* cell.
    (e) Fill each cell to the top with *loose* soil. **DO NOT** compact the soil.
    (f) With your thumb, make a shallow depression in the soil at the top of each cell.
    (g) Drop 3 seeds into each depression. In Table 17-1, record the total number of seeds planted.
    (h) Sprinkle enough soil to cover the seeds in each cell.
    (i) Water very gently with a pipet until water drips from each wick.
    (j) Write your name and date on a pot label and insert the label against the side of one cell of the quad.
    (k) Place the quad on a wet water mat inside a tray under the light bank. The top of the quad should be approximately 5 cm below the lights.

12. *Option B:* Film Canister Method
    (a) Obtain 2 cotton string wicks, 2 film canisters (with holes in bottoms), 24 fertilizer pellets, and 2 pot labels.
    (b) Using tap water, wet the string wicks thoroughly and insert them through the holes in the canisters' bottoms. The wicks should extend about halfway up the length of the canister.
    (c) Fill each canister *halfway* with soil mix.
    (d) Add 12 N-P-K fertilize pellets atop the soil of each canister.
    (e) Fill each canister to the top with soil mix. **DO NOT** compact the soil.
    (f) With your thumb, make a shallow depression in the soil of each canister.
    (g) Drop 6 seeds atop the soil in each canister. Distribute the seeds so they're about equally spaced.
    (h) Sprinkle enough soil to cover the seeds.
    (i) Water the containers very gently with a pipet until water drips from each wick.
    (j) Write your name and date on pot labels and insert them into the soil against the side of each canister.
    (k) Fill 2 wide-mouth bottles that have the same diameter as the canisters with tap water and place each canister in its own water bottle.
    (l) Place the canisters in their water bottles under the light bank. The top of the canister should be approximately 5 cm below the lights.

Day 1 (About 10 min.)

MATERIALS

Per student:

- dissecting microscope
- disposable pipet with bulb
- tap water
- metric ruler

PROCEDURE

1. Water each quad cell or canister from the top using a pipet.
2. Observe your germination experiment in the petri dish chamber with a dissecting microscope. Note the **radicle** (primary root) and **hypocotyl** extending from a split in the brown **seed coat.** What colors are the radicle and the hypocotyl? _____
3. What does the color indicate about the source of carbohydrates for the germinating seeds?
   _____
4. Count the number of seeds that have germinated, and record it in Table 17-1. Measure the length of each hypocotyl and attached radicle. Record this in Table 17-1. Replace the cover and return your experiment to its place.

Day 2 (About 15 min.)

MATERIALS

Per student:

- germination experiment started on day 0
- disposable pipet with bulb
- tap water
- metric ruler
- forceps
- dissecting microscope
- 2 additional germination chambers (for gravitropism experiment)
- 2 RCBr seed pods

PROCEDURE

1. Water each quad cell or canister from the top using a pipet.
2. Observe your germination experiment. Count the number of germinated seeds and calculate the percentage of seeds that germinated. Record both numbers in Table 17-1. With a forceps, remove one of the germinating seeds from the toweling, place it on the stage of a dissecting microscope, and observe the numerous **root hairs** attached to the radicle. Root hairs greatly increase the surface area for absorption of water and minerals. Measure the lengths of the hypocotyls and attached radicles and record their average length in Table 17-1.

In what direction (up, down, or horizontal) are the hypocotyl and radicle growing?
   hypocotyl: _____
   radicle: _____
   This directional growth is called **gravitropism,** a plant growth response to gravity.

Day 3 (About 10 min.)

MATERIALS

Per student:

- germination experiment started on day 0
- metric ruler
- disposable pipet with bulb
- tap water

PROCEDURE

1. Water your plants in the quads or canisters from the top using a pipet. Is any part of the plant visible yet? _____ (yes or no) If yes, which part? _____
2. Observe your germination experiment. By this time, the seed coat should be shed, and the **cotyledons,** the so-called seed leaves that were packed inside the seed coat, should be visible. The cotyledons absorb food stored within the seed, making the food available for initial stages of growth before photosynthesis starts to provide carbohydrates necessary for growth.

What color are the cotyledons? _____
What function does this color indicate for the cotyledons? _____

3. Measure the length of the hypocotyls and attached radicles and record their average length in Table 17-1.

B. Growth

Day 4 or 5 (About 10 min.)

MATERIALS

Per student:

- forceps
- metric ruler

- dissecting microscope
- germination experiment started on day 0

PROCEDURE

1. Count the number of seedlings present, and record it in Table 17-1. Use a forceps to thin the plants to one per cell if you are using the quad growing method. If a cell has no plants, transplant one of the extra plants from another cell by inserting the tip of a pencil into the soil to form a hole and then inserting the transplant's root. Gently firm the soil around the transplant. Measure the height of each plant's shoot system—that is, from soil to the apex. Compute the average height and record it in Table 17-1.
2. Observe your germination experiment. Note that the radicle has produced additional roots. These are **lateral roots** (secondary roots). Remove one plant and observe it with the dissecting microscope.

Do lateral roots have root hairs? _____ (yes or no)
At this time, you may discard your germination experiment. Your instructor will indicate what you should do with the materials.

Day 7 (About 10 min.)

MATERIALS

Per student:

- metric ruler

PROCEDURE

1. By now your plant should have its first **true leaves.** Describe how the true leaves differ from the cotyledons.

_____
_____
_____

2. Measure the length of the true leaves, and record their average length in Table 17-1.
3. Measure the height of your plants from soil line to shoot apex, and record the average height in Table 17-1.

C. The Pollinator and Pollination

Pollen grains are actually sperm conveyors that deliver these male gametes to the egg-containing ovules within the ovary. Transfer of pollen from the anther to the stigma, called **pollination,** occurs by various means in different species of flowers, with wind, birds, and insects being the most common carriers of pollen. In the case of *Brassica*, the main pollinator is the common honeybee.

Pollinators and flowers are probably one of the best-known examples of **coevolution.** Coevolution is the process of two organisms evolving together, their structures and behaviors changing to benefit both organisms. Flowering plants and pollinators work together in nearly perfect harmony: The bee receives the food it needs for its respiratory activities from the flower, and the plant receives the male gametes directly to the female structure—it doesn't have to rely on the presence of water, which is necessary in lower plants, algae, and fungi, to get the gametes together.

Flowering plants have considerable advantage over gymnosperms. While gymnosperms also produce pollen grains, the pollinator in these plants is the wind. No organism carries pollen directly to the female structure. Rather, chance and the production of enormous quantities of pollen (which takes energy that might be used otherwise for growth) are integral in assuring reproduction.

In preparation for pollinating your plants' flowers, let's first study the pollinator, the common honeybee.

Day 10 (About 20 min.)

MATERIALS

Per student:

- metric ruler
- 3 toothpicks
- 3 dried honeybees
- dissecting microscope

Per student group (4):

- tube of glue
- scissors

PROCEDURE

1. Measure the lengths of the second set of true leaves and plant height, and record their averages in Table 17-1.
2. Obtain a dried honeybee and observe it with your dissecting microscope. Use Figure 17-17 to aid you in this study. Note the featherlike hairs that cover its body. As a bee probes for nectar, it brushes against the anthers and stigma. Pollen is entrapped on these hairs.
3. Closely study the foreleg of the bee, noticing the notch. This is the *antenna cleaner*. Quick movements over the antennae remove pollen from these sensory organs.
4. Examine the mid- and hind legs, identifying their *pollen brushes*, which are used to remove the pollen from the forelegs, thorax, and head.
5. On the hind leg, note the *pollen comb* and *pollen press*. The comb rakes the pollen from the midleg, and the press is a storage area for pollen collected.

Pollen is rich in the vitamins, minerals, fats, and proteins needed for insect growth. When the baskets are filled, the bee returns to the hive to feed the colony. Partially digested pollen is fed to larvae and to young, emerging bees.

As you can imagine, during pollen- and nectar-gathering at numerous flowers, cross-pollination occurs as pollen grains are rubbed off onto stigmas.

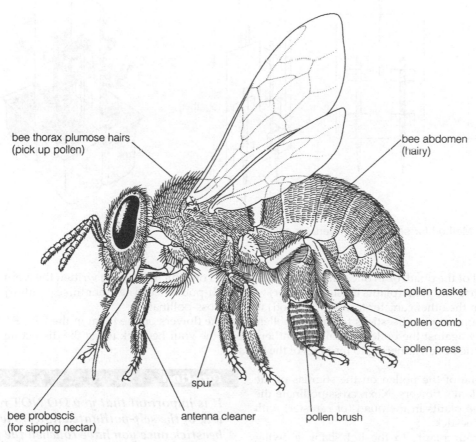

bee thorax plumose hairs (pick up pollen)

bee abdomen (hairy)

pollen basket

pollen comb

pollen press

pollen brush

spur

antenna cleaner

bee proboscis (for sipping nectar)

Figure 17-17 Structure of the honeybee.

On day 12, you will start pollinating your flowers. What better means to do it than to use a bee? (In this case, a dead bee that is attached to a stick.) Let's make some beesticks.

6. Add a drop of glue to one end of three round toothpicks. Insert the glue-bearing end of one stick into the top (dorsal) side of the thorax of a bee. Repeat with two more dried bees. Set your beesticks aside to dry.

D. Pollination

It's time to do a pollination study. Some species of flowers are able to **self-pollinate,** meaning that the pollen from one flower of a plant will grow on stigmas of other flowers of that same plant. Hence, the sperm produced by a single plant will fertilize its own flowers. Other species are **self-incompatible.** For fertilization to occur in the latter, pollen must travel from one plant to another plant of the same species. This is **cross-pollination.** In this portion of the exercise, you will determine whether RCBr is self-incompatible or is able to self-pollinate.

Day 12 (About 30 min.)

MATERIALS

Per student:

- beestick prepared on day 10
- metric ruler
- forceps

- index card divider
- pot label
- marker

PROCEDURE

1. Measure the height of each plant, and record the average height in Table 17-1.
2. Separate the cells of your quad or canister with a divider constructed with index cards as shown in Figure 17-18. This keeps the plants from brushing against each other.

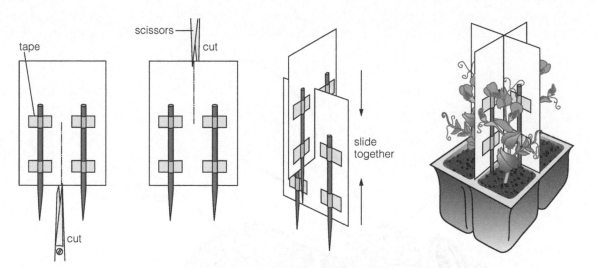

Figure 17-18 Method for separating flowers for pollination study.

3. In one cell of the quad, or in one canister, insert a pot label on which you have written the word ("SELF"), to indicate that this plant (plants if using canister) will be self-pollinated. The other three plants (quad method) or those in the other canister (canister method) will be cross-pollinated.
4. Now use one of your beesticks to pick up pollen from the flowers on the plant in the "SELF" cell. Rub the bee's body against the pollen-containing anthers. Examine your beestick using the dissecting microscope, and note the pollen adhering to the hairs of the bee's body.
5. Brush some of the pollen on the stigmas of the "SELF" plant's flowers. Next, cross-pollinate the other three plants in the quad (or canister) with the *same* beestick.
6. Return your plants to the light-bank growing area and discard the beestick.

> **Caution**
>
> *It is important that you DO NOT retouch any of the self-pollinated flowers with the beestick once you have touched the flowers of any of the other plants.*

Day 14 (About 15 min.)

MATERIALS

Per student:

- metric ruler
- beestick

PROCEDURE

1. Measure the height of each plant, and record their average height in Table 17-1.
2. Pollinate the newly opened flowers using the same techniques as described for day 12.
3. Return your plants to the light-bank growing area.

Day 17 (About 15 min.)

MATERIALS

Per student:

- metric ruler
- beestick

PROCEDURE

1. Measure the height of each plant, and record their average height in Table 17-1.
2. Count the number of flowers on each plant (opened and unopened), calculate the average number per plant, and record it in Table 17-1. Pinch off the shoot apex of each plant to remove all unopened buds and pinch off any side shoots the plant has produced.
3. Do a final pollination of your flowers, using a fresh beestick and taking the same precautions as on days 12 and 14.
4. Return your plants to the light-bank growing area.

E. Fertilization and Seed Development

Within 24 hours of pollination, a pollen grain that has landed on a compatible stigma will produce a pollen tube that grows down the style and through the micropyle of the ovule.

Within the pollen tube are two **sperm**, the male gametes. As the pollen tube enters the ovule, the two sperm are discharged. One fertilizes the haploid egg nucleus, forming a diploid **zygote.** The second sperm unites with two other nuclei within the ovule, forming a triploid nucleus known as the **primary endosperm nucleus.**

The fusion of one sperm nucleus to form the zygote and the other to form the primary endosperm nucleus is called **double fertilization.**

1. Re-examine Figure 17-9, which illustrates double fertilization.

What is the fate of cells formed by double fertilization? The zygote undergoes numerous cell divisions during a process called **embryogeny** (the suffix -*geny* is derived from Greek, meaning "production") to form the diploid embryo. The triploid primary endosperm nucleus divides numerous times without cytokinesis. The result is the formation of a multinucleate milky **endosperm** (each nucleus is 3n), which serves as the food source for the embryo as it grows. At the same time, changes are taking place in the cell layers of the ovule—they are becoming hardened and will eventually form the seed coat. Also, the ovary is elongating as it matures into the **fruit**—in this case, a pod.

2. Examine your plants, looking for small, elongated pods.
3. At this time, you should be able to tell whether RCBr can self-pollinate or is self-incompatible, because the fruit—a pod—will develop only if fertilization has taken place. Do the cross-pollinated plants have pods? (yes or no) _____
4. Does the self-pollinated plant have pods? (yes or no) _____
5. Make a conclusion regarding the need for cross-pollination in RCBr. _____

_____

The young embryo goes through a continuous growth process, starting as a ball of cells (the globular stage) and culminating in the formation of a mature embryo. The early stages are somewhat difficult to observe, but the later stages can be seen quite easily with a microscope. By day 24, many of the ovules will be at the globular stage, similar to that seen in Figure 17-10a.

Day 26 (About 30 min.)

MATERIALS

Per student:

- dissecting microscope
- 2 dissecting needles
- dH$_2$O in dropping bottle

- glass microscope slide
- iodine solution (I$_2$KI) in dropping bottle

PROCEDURE

1. Remove a pod from one of the plants, place it on the stage of your dissecting microscope, and tease it open using dissecting needles. RCBr flowers have pistils consisting of two carpels; each carpel makes up one longitudinal half of the pod, and the "seam" represents where the carpels fuse.
2. Carefully remove the ovules (immature seeds) inside the pod, placing them in a drop of water on a clean microscope slide.
3. Rupture the ovule by mashing it with dissecting needles and place the slide on the stage of a dissecting microscope.
4. Attempt to locate the embryo, which should be at the heart-shaped stage, similar to Figure 17-10b, and which may be turning green. A portion of the endosperm has become cellular and may be green, too, so be careful with your observation. The "lobes" of the heart-shaped embryo are the young cotyledons. Identify these. Find the trailing **suspensor,** a strand of eight cells that serves as a sort of "umbilical cord" to transport nutrients from the endosperm to the embryo.
5. Add a drop of I$_2$KI to stain the endosperm. What color does the endosperm stain? _____
6. What can you conclude about the composition of the endosperm? (*Hint:* If you've forgotten, consult Exercise 4.)

_____

Day 29 (About 20 min.)

MATERIALS

Per student:

- dissecting microscope
- 2 dissecting needles

- dH$_2$O in dropping bottle
- glass microscope slide

PROCEDURE

1. Remove another pod from a plant and examine it as you did on day 26.
2. Attempt to find the torpedo stage of embryo development, similar to that illustrated in Figure 17-10c.
3. Describe the changes that have taken place in the transition from the heart-shaped to the torpedo stage.

_____

Day 31 (About 20 min.)

MATERIALS

Per student:

- dissecting microscope
- 2 dissecting needles

- dH$_2$O in dropping bottle
- glass microscope slide

PROCEDURE

1. Again, remove and dissect a pod and ovules, and look for the embryo whose cotyledons have now curved, forming the so-called walking-stick stage.
2. Identify the **cotyledons** and the **hypocotyl-root axis** that has curved within the ovules' integuments.

Days 35–37 (About 20 min.)

MATERIALS

Per student:

- dissecting microscope
- 2 dissecting needles

PROCEDURE

The pods of your plants should be drying and have much the same appearance as the pod you started with about a month ago.

1. Remove seeds from mature pods and examine the seeds with your dissecting microscope. Note the bumpy **seed coat,** derived from the integument of the ovule. Look for a tiny hole in the seed coat. This is the **micropyle!**

2. Crack the seed coat and remove the embryo, identifying the cotyledons and hypocotyl-root axis. Separate the cotyledons and look between them, identifying the tiny **epicotyl,** which is the *apical meristem of the shoot.*

3. Harvest the remaining pods and count the number of seeds in each pod. Calculate the average number of seeds per pod and number of seeds per plant. Record this information in Table 17-1.

Congratulations! You have completed the life cycle study of a rapid cycling *Brassica rapa.* Virtually all of the 200,000 species of flowering plants have stages identical to that which you have examined over the course of the past 35–40 days. Most just take longer—from a few days to nearly a century longer!

Conclude your study by preparing the required graphs and answering the questions following Table 17-1.

4. In Figure 17-19, graph the increase in average length of the hypocotyl-root axis produced in the germination chamber.

Does length increase in a constant fashion (linearly), or does extension occur more quickly during a particular time period (logarithmically)? _____

Using the data from Table 17-1, graph in Figure 17-20 the rate of growth of your plants' shoot system as a function of time.

How does the number of seeds produced per pod compare with the number in the original pod from which you harvested your seed at the beginning of the study? _____

What is the *reproductive potential* (the number of possible plants coming from one parent plant) of your RCBr?_____

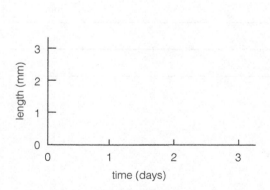

Figure 17-19 Growth of hypocotyl-root axis.

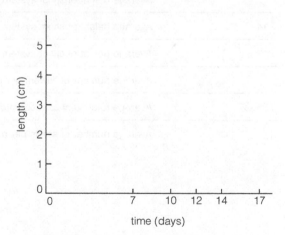

Figure 17-20 Growth of shoot system.

TABLE 17-1 Data for RCBr Life Cycle

Day	Parameter	Measurement
0	Pod length	mm
	Seed number (single pod)	
	Number of seeds sown in germination chamber	
	Temperature, °C	
	Number of seeds sown in quad (total)	
1	Germinated seeds (germination chamber)	
	Average length of hypocotyl-root axis	mm
2	Germinated seeds (germination chamber)	
	Percentage seeds germinated	
	Average length of hypocotyl-root axis	mm
3	Average length of hypocotyl-root axis	mm
4 or 5	Number of seedlings (quad)	
	Average height of shoot system	mm
7	Average length of first set of true leaves	mm
	Average height of shoot system	mm
10	Average length of second set of true leaves	mm
	Average height of shoot system	mm
12	Average number of leaves per plant	
	Average height of shoot system	mm
14	Average height of shoot system	mm
17	Average height of shoot system	mm
	Average number of flowers per plant	
35–37	Average total number seeds produced per plant	
	Average number of seeds per pod	

_____ 1. Plants that produce flowers are
(a) members of the Anthophyta
(b) angiosperms
(c) seed producers
(d) all of the above

_____ 2. All of the petals of a flower are collectively called the
(a) corolla
(b) stamens
(c) receptacles
(d) calyx

_____ 3. Which group of terms refers to the microsporophyll, the male portion of a flower?
(a) ovary, stamens, pistil
(b) stigma, style, ovary
(c) anther, stamen, filament
(d) megasporangium, microsporangium, ovule

_____ 4. A carpel is the
(a) same as a megasporophyll
(b) structure producing pollen grains
(c) component making up the anther
(d) synonym for microsporophyll

_____ 5. The portion of the flower containing pollen grains is
(a) the pollen sac
(b) the microsporangium
(c) the anther
(d) all of the above

_____ 6. Which group of terms is in the correct developmental sequence?
(a) microspore mother cell, meiosis, megaspore, female gametophyte
(b) microspore mother cell, meiosis, microspore, pollen grain
(c) megaspore, mitosis, female gametophyte, meiosis, endosperm mother cell
(d) all of the above

_____ 7. Where does germination of a pollen grain occur in a flowering plant?
(a) in the anther
(b) in the micropyle
(c) on the surface of the corolla
(d) on the stigma

_____ 8. Double fertilization refers to
(a) fusion of two sperm nuclei and two egg cells
(b) fusion of one sperm nucleus with two polar nuclei and fusion of another with the egg cell nucleus
(c) maturation of the ovary into a fruit
(d) none of the above

_____ 9. Ovules mature into _____, while ovaries mature into _____.
(a) seeds/fruits
(b) stamens/seeds
(c) seeds/carpels
(d) fruits/seeds

_____ 10. A bean pod is
(a) a seed container
(b) a fruit
(c) a part of the stamen
(d) both a and b

Name _____ Section Number _____

Seed Plants II: Angiosperms

POST-LAB QUESTIONS

Introduction

1. There are two major groups of seed plants, gymnosperms (Exercise 16) and angiosperms. Compare these two groups of seed plants with respect to the following features:

Feature	Gymnosperms	Angiosperms
a. type of reproductive structure	_____	_____
b. source of nutrition for developing embryo	_____	_____
c. enclosure of mature seed	_____	_____

17.1 External Structure of the Flower

2. What *event*, critical to the production of seeds, is shown at the right?

3. Distinguish between *pollination* and *fertilization*.

(0.84×).

4. Identify the parts of the trumpet creeper flower shown at the right.

   a.

   b.

   c.

   d.

   e.

(1×).

5. Examine the photo of the *Trillium* flower pictured at the right.

   a. Is *Trillium* a monocotyledon or dicotyledon?

   b. Justify your answer.

(0.75×).

6. Based on your observation of the stigma of the day-lily flower in the photo at the right, how many carpels would you expect to comprise the ovary?

(1×).

17.2 The Life Cycle of a Flowering Plant

7. The photo is a cross section of a (an) _____

   _____.

   The numerous circles within the four cavities are

   _____.

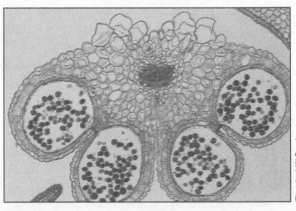

(74×).

8. The photo shows a flower of the pomegranate some time after fertilization. Identify the parts shown.

a. _____

b _____

c. _____

d. _____

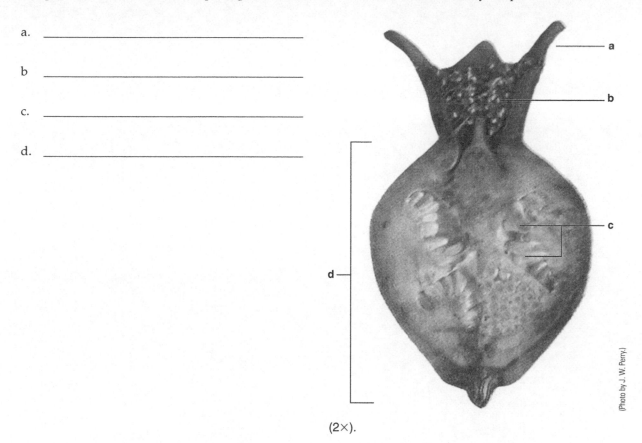

(2×).

(Photo by J. W. Perry.)

9. Some biologists contend that the term *double fertilization* is a misnomer and that the process should be called *fertilization* and *triple fusion*. Why do they argue that the fusion of the one sperm nucleus and the two polar nuclei is *not* fertilization?

Food for Thought

10. Your roommate says that you need vegetables and asks you to pick up tomatoes at the store. To your roommate's surprise, you say a tomato is not a vegetable, but a fruit. Explain.

Sponges, Cnidarians, Flatworms, and Rotifers

OBJECTIVES

After completing this exercise, you will be able to

1. define *larva, asymmetry, radial symmetry, bilateral symmetry, cephalization, coelom, pseudocoelom, monoecious, dioecious, invertebrates, spongocoel, osculum, spicules, choanocyte (collar cell), budding, sponging, polyp, tentacles, medusa, nematocyst, gastrovascular cavity epidermis, gastrodermis, mesoglea, acoelomate, regeneration, host, intermediate host, cuticle, scolex, proglottid, pseudocoelomate, cloaca,* and *bladder worm;*

2. describe the tissue and organ-system levels of organization;

3. list the characteristics of animals;

4. describe how the phyla Porifera and Cnidaria respectively show the cell-specialization and tissue levels of organization;

5. explain the basic body plan of members of the phyla Porifera and Cnidaria;

6. identify representatives of the classes Calcarea and Demospongiae of the phylum Porifera and classes Hydrozoa, Scyphozoa, and Anthozoa of the phylum Cnidaria;

7. identify structures (and indicate associate functions) of representatives of these classes;

8. explain the basic body plan of members of the phyla Platyhelminthes and Rotifera;

9. describe the natural history of members of the phyla Platyhelminthes and Rotifera;

10. identify representatives of the classes Turbellaria, Trematoda, and Cestoda of the phylum Platyhelminthes and representatives of phylum Rotifera;

11. identify structures (and indicate associated functions) or representatives of these groups;

12. outline the life cycles of human liver fluke;

13. outline the live cycle of *Ascaris;*

14. compare and contrast the anatomy and live cycles of free-living and parasitic flatworms.

INTRODUCTION

Animals are multicellular, oxygen-consuming, heterotrophic (heterotrophs feed on other organisms or on organic waste) *organisms that exhibit considerable motility.* Most animals are *diploid* and *reproduce sexually,* although asexual reproduction is also common. The animal life cycle includes a *period of embryonic development, often with a larva*—sexually immature, free-living form that grows and transforms into an adult—or equivalent stage. During the embryonic development of most animals, *three primary germ layers—ectoderm, mesoderm,* and *endoderm*—form and give rise to the adult tissues. Evidence accumulated from centuries of study supports the notion that animals evolved from a group of protistanlike ancestors distinct from those that gave rise to the plants and fungi.

Animal diversity is usually studied in the approximate order of a group's first appearance on the earth. Table 18-1 summarizes the trends evident in this sequence for the phyla considered in this and subsequent diversity exercises:

- In terms of habitat, marine animals precede freshwater and terrestrial forms.
- The organization of animal bodies is initially **asymmetrical**—cannot cut the body into two like halves—but rapidly becomes symmetrical, exhibiting first **radial symmetry**—any cut from the oral (mouth) to the other end through the center of the animal yields two like halves and then bilateral symmetry (Figure 18-1). In **bilateral symmetry,** a cut in only one plane, parallel to the main axis and dividing the body into right and

TABLE 18-1 Trends in the Evolution of Animals

Phylum	Common Name	Habitat	Body Organization	Level of Organization
Porifera	Sponges	Aquatic, mostly marine	None or crude radial symmetry	Cell specialization
Cnidaria	Cnidarians	Aquatic, mostly marine	Radial symmetry, incomplete gut	Tissue (ectoderm and endoderm only)
Platyhelminthes	Flatworms	Aquatic, moist terrestrial, or parasitic	Bilateral symmetry, head, incomplete gut	Organ system (all three primary germ layers)
Rotifera	Rotifers	Aquatic (mostly freshwater) or moist terrestrial	Bilateral symmetry, head, complete gut, pseudocoelom	Organ system
Annelida	Annelids some parasitic	Aquatic or moist terrestrial, bilateral symmetry, some parasitic	Bilateral symmetry, head, segmented, complete gut, coelom	Organ system
Mollusca	Molluscans	Aquatic or terrestrial, mostly marine	Unique body plan, complete gut, coelom	Organ system
Nematoda	Roundworms	Aquatic, moist terrestrial, or parasitic	Bilateral symmetry, head, complete gut, pseudocoelom	Organ system
Arthropod	Arthropods	Aquatic, terrestrial (land and air), some parasitic	Bilateral symmetry, head, complete gut, segmented, coelom	Organ system
Echinodermata	Echinoderms	Aquatic, marine	Radial symmetry (larvae are bilaterally symmetrical), no well-defined head, no segmentation, usually complete gut, coelom	Organ system
Chordata	Chordates	Aquatic, terrestrial (land and air)	Bilateral symmetry, head and postanal tail, segmentation, complete gut, coelom	Organ system

left pieces, yields like halves. Bilateral symmetry accompanies the movement of sensory structures and the mouth to the front end of the body to form a head, a process called **cephalization.** Spaces within the body are first "baskets" used to filter seawater, then incomplete guts (no anus) appear, and finally complete guts (with an anus) are suspended in an entirely enclosed ventral body space. This ventral cavity is called a **coelom** if it is lined by tissue derived from mesoderm, and a **pseudocoelom** if the mesodermal lining is incomplete (Figure 18-2).

- The level of organization of the animal body proceeds through aggregates of specialized cells, tissues formed by associations of similarly specialized cells, organs comprised of different tissues, and systems of functionally related organs.

- Body support moves from self-associating scaffolds in sponges to hydrostatic bodies to shells and exoskeletons or endoskeletons comprised of internal support structures.

- Sexual reproduction predominates in animals, but earlier forms are asexual as well. Earlier forms are often **monoecious** or *hermaphroditic*—individuals have both male and female reproductive structures. Latter

Body Support	Life Cycle	Special Characteristics
Spicules, spongin	Sexual reproduction (mostly monoecious), larva (amphiblastula), asexual reproduction (budding)	Spongocoel, osculum, gemmules, choanocytes (collar cells)
Hydrostatic gut	Sexual reproduction (monoecious or dioecious), larva (planula), asexual reproduction (budding and fragmentation)	Nerve net, nematocysts, two adult body forms—polyp (sessile) and medusa (free swimming)
Muscular	Sexual reproduction (mostly monoecious with internal fertilization); parasites have very complicated life cycles; some asexual reproduction (mostly transverse fission)	Brain, flame cells
Hydrostatic pseudocoelom	Sexual (dioecious with internal fertilization), some parthenogenesis, males much smaller than females or absent	Brain, flame cells, wheel organ, mastax, microscopic animals
Hydrostatic coelom	Sexual reproduction (monoecious or dioecious with mostly internal fertilization), larvae in some, asexual reproduction (mostly transverse fission)	Brain, nephridia, gills in some
Shells, floats (aquatic)	Sexual (monoecious or dioecious with external or internal fertilization), larvae in some	Brain, nephridium or nephridia, most have gills or lungs, external body divided into head, mantle, and foot
Hydrostatic pseudocoelom	Sexual reproduction (mostly dioecious with internal fertilization)	Brain, molt to grow
Exoskeleton of chitin	Sexual reproduction (mostly dioecious with internal fertilization), larvae, metamorphosis, some parthenogenesis	Brain, green glands or Malphigian tubules, gills, tracheae or book lungs, body regions (head, thorax, and abdomen), specialized segments form mouthparts and jointed appendages, molt to grow
Dermal endoskeleton of calcium carbonate (mesoderm)	Sexual reproduction (mostly dioecious with external fertilization), larvae, metamorphosis, high regenerative potential	Radial symmetry, water-vascular system
Most have an endoskeleton of cartilage or bone	Sexual reproduction (mostly dioecious with internal or external fertilization), some larvae, some metamorphosis	Brain, ventral heart, notochord, gill slits, and dorsal hollow nerve cord at some stage of life cycle

forms are **dioecious** with male and female reproductive structures in separate sexes. Fertilization of the egg by sperm is usually external in *sessile* (immobile and attached to a surface such as the sea bottom) animals and internal in most mobile forms, and *pathenogenesis* (the egg develops without fertilization) occurs sporadically.

Caution

Specimens are kept in preservative solutions. Use latex gloves whenever you handle a specimen. Wash any part of your body exposed to this solution with copious amounts of water. If preservative solution is splashed into your eyes, wash them with the safety eyewash bottle for 15 minutes. If you wear contact lenses during a dissection, wear safety goggles.

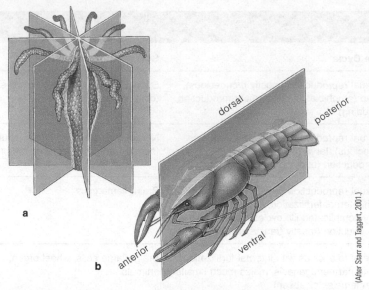

Figure 18-1 Examples of body symmetry. **(a)** Radial symmetry in a cnidarian polyp. **(b)** Bilateral symmetry in a crayfish, an arthropod.

(After Starr and Taggart, 2001.)

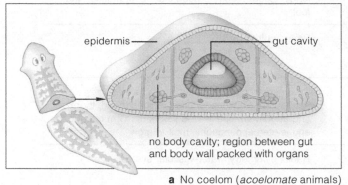

epidermis

gut cavity

no body cavity; region between gut and body wall packed with organs

a No coelom (*acoelomate* animals)

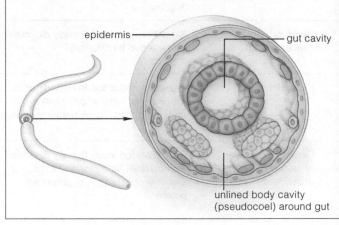

epidermis

gut cavity

unlined body cavity (pseudocoel) around gut

b Pseudocoel (*pseudocoelomate* animals)

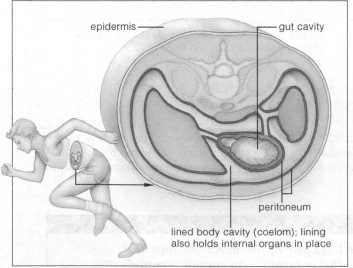

epidermis

gut cavity

peritoneum

lined body cavity (coelom); lining also holds internal organs in place

c Coelom (*coelomate* animals)

Figure 18-2 Development of ventral body cavity in animals. (After Starr and Taggart, 2001.)

Habitat: aquatic, mostly marine
Body Arrangement: asymmetrical or crude radial symmetry
Level of Organization: cell specialization
Body Support: spicules and/or spongin
Life Cycle: sexual reproduction (mostly monoecious),
                             larva (amphiblastula), asexual reproduction (budding and fragmentation)
Special Characteristics: spongocoel, osculum, gemmules, choanocytes (collar cells)

Aristotle in his *Scala Naturae* said of sponges, "In the sea there are things which it is hard to label as either animal or vegetable." This was a reference to their sensitivity to stimulation and their relative lack of motility.

Sponges evolved directly from protistanlike ancestors and diverged early from the main line of animal evolution (Figure 18-3). Most sponges are marine, and all forms are *aquatic.* As adults, sponges are *sessile,* but many disperse as free-swimming larvae. Although sponges possess a variety of different types of cells, they are atypical animals in their lack of definite tissues. They exhibit the *cell-specialization level of organization,* as there is a division of labor among the different cell types. Cells are organized into layers, but these associations of cells do not show all the characteristics of tissues. Sponges have a *crude radial symmetry or are asymmetrical.* Sponges along with the animals of most phyla are **invertebrates**—animals that do not posses a backbone or vertebral column.

Three body plans exist in sponges (Figure 18-4). They are referred to as *asconoid, syconoid,* and *leuconoid* and increase in the complexity of their structure in this order.

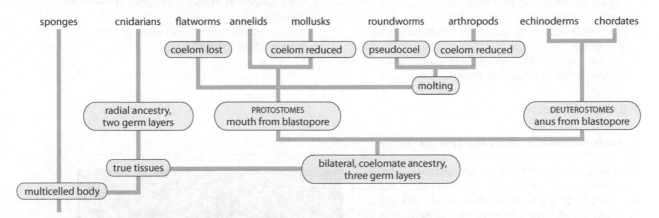

Figure 18-3 Presumed family tree for animals. (After Starr and Taggart, 2006.)

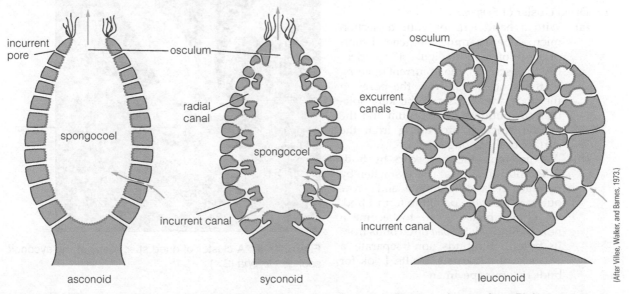

Figure 18-4 Three body plans of sponges. The blue arrows indicate the direction of water flow. Pink areas are lined by collar cells.

Sponges, Cnidarians, Flatworms, and Rotifers   **339**

Water enters a vase-shaped asconoid sponge through *incurrent pores*; flows directly to a large central internal cavity, the **spongocoel;** and exits through a larger opening, the **osculum.** A syconoid sponge is similar, except that the inner side of the body wall pockets in to form *radial canals* into which *incurrent canals* drain water through channels in their walls. In a leuconoid sponge, the radial canals are internalized as chambers within the sponge. These chambers empty into *excurrent canals* that in turn lead to the osculum. Concurrent with the development of increasingly complex patterns of water flow, the area lined with *collar cells,* which in asconoid sponges is the inner wall of the spongocoel, also becomes internalized, first lining the radial canals of syconoid sponges and then the chambers of leuconoid sponges.

MATERIALS

Per student:

- dissection microscope
- hand lens
- compound microscope, lens paper, a bottle of lens-cleaning solution (optional), a lint-free cloth (optional)
- clean microscope slide and coverslip (optional)

Per student pair:

- preserved specimen of *Scypha (Grantia)*
- prepared slides of
    Scypha spicules (optional)
    a longitudinal section of *Scypha*
    a whole mount of *Leucosolenia*
    a whole mount of teased commercial sponge fibers
    a whole mount of *Spongilla* spicules
    freshwater sponge gemmules

Per student group (4):

- 100-mL beaker of heatproof glass (optional)
- bottle of 5% sodium hydroxide (NaOH) or potassium hydroxide (KOH) (optional)
- waste container for 5% hydroxide solution (optional)
- hot plate (optional)
- squeeze bottle of dH_2O (optional)

Per lab room:

- demonstration collection of commercial sponges
- squeeze bottle of 50% vinegar and water
- demonstration of living or dried freshwater sponges
- boxes of different sizes of latex gloves
- box of safety glasses

PROCEDURE

A. Class Calcarea: Calcareous Sponges *(15 min.)*

Members of this class contain **spicules** (skeletal elements) composed of calcium carbonate ($CaCO_3$). All three body plans are represented in this group. You will study members of the genera *Scypha,* which all have the syconoid body plan, and *Leukosolenia,* which all have asconoid body plans.

1. Dried cluster of *Scypha.*
   (a) With a hand lens or your dissection microscope, examine the general morphology of *Scypha* (Figure 18-5). Identify the osculum, the excurrent opening of the spongocoel. Note the pores in the body wall. Observe the long spicules surrounding the osculum and the shorter spicules protruding from the surface of the sponge.
   (b) Asexual reproduction occurs by **budding.** Budding is a process whereby parental cells differentiate and grow outward from the parent to form a bud, sometimes breaking off to become a new individual. *Scypha* often buds from its base. If the buds don't separate, a cluster of individuals results. Look for buds on your specimen.

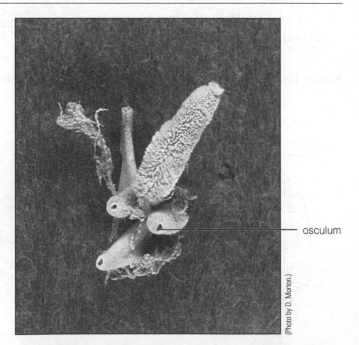

Figure 18-5 A cluster of dried specimens of the syconoid sponge *Scypha* (3×).

2. Examine a wet mount (w.m.) or a prepared slide of *Scypha* spicules with your compound microscope, using the medium-power objective. Draw what you observe in Figure 18-6. Follow these instructions to prepare the wet mount:

   (a) Put on a pair of safety goggles and place a small piece of a specimen of *Scypha* in a 100-mL heatproof beaker.

   (b) Carefully add enough 5% sodium hydroxide or potassium hydroxide solution to the beaker to cover the piece and bring this to a boil on the hot plate. Allow the beaker to cool.

   (c) Allow the remains of the sponge to settle to the bottom of the beaker and pour off the liquid into the waste container while retaining the residue.

   (d) Wash the residue by adding distilled water. After the remains of the sponge settle, pour off the water into the waste container.

   (e) Make a wet mount of the residue.

3. Longitudinal section of *Scypha*.

   (a) Examine a longitudinal section of *Scypha* (Figure 18-7) with a compound microscope. In your section, trace the path of water as it flowed into and through the sponge in nature (Figure 18-8). Identify the following structures and place them in the correct order for the water's circulation through the sponge by writing 1st, 2nd, and 3rd, and so on in the second column of Table 18-2.

   (b) Examine the **collar cells,** which appear as numerous small, dark bodies lining the radial canals. Each bears a flagellum that projects into the radial canal, but it's unlikely that you will be able to see the flagella, even with the high-dry objective. Examine the structure of a collar

Figure 18-6 Calcium carbonate spicules of *Scypha*, w.m. (_____ ×).

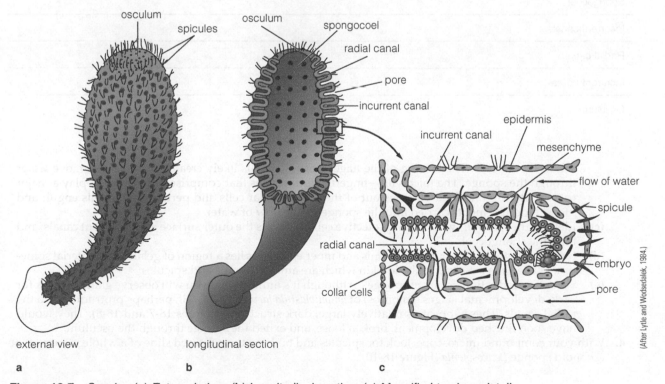

external view   longitudinal section

a   b   c

(After Lytle and Wodsedalek, 1984.)

Figure 18-7 *Scypha*. **(a)** External view. **(b)** Longitudinal section. **(c)** Magnified to show detail.

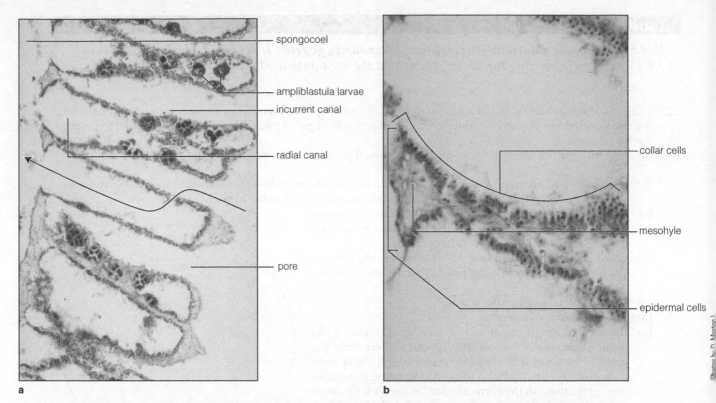

Figure 18-8 Longitudinal sections of *Scypha*. (a) The arrow indicates the path of water flow (96×). (b) 384×.

TABLE 18-2	Circulation of Water Through the Syconoid Sponge *Scypha*
Structure	**Order for Water Circulation**
Spongocoel	
Incurrent canals	
Radial canals	
Incurrent pores	
Osculum	

cell in Figure 18-9. The beating of the flagella of collar cells likely creates currents that move water through the sponge. The microvilli—fingerlike projections that comprise their collars—play a major role in filtering fine food particles out of the water. Collar cells and perhaps other cells engulf and digest food fragments carried into the sponge by the flow of water.

(c) Identify the *epidermis* with its flat protective cells. It covers the outer surface and incurrent canals, and lines the spongocoel.

(d) Note that between the outer epidermis and inner epidermis lies a region of gelatinous material sometimes called the *mesohyle*, embedded in which are amoeboid cells and spicules.

(e) Sexual reproduction occurs in sponges. Although it's unlikely that you will observe gametes, look for early developmental stages (embryos) of *amphiblastula larvae* in the wall, perhaps protruding into the radial canals. These appear as relatively large, dark structures (Figures 18-7 and 18-8). They would have soon finished development, broken loose, and exited the sponge through the osculum.

4. With your compound microscope, look for spicules and buds on the prepared slide of a whole mount of the asconoid sponge *Leucosolenia* (Figure 18-10).

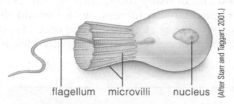

flagellum   microvilli   nucleus

Figure 18-9   Choanocyte (collar cell).

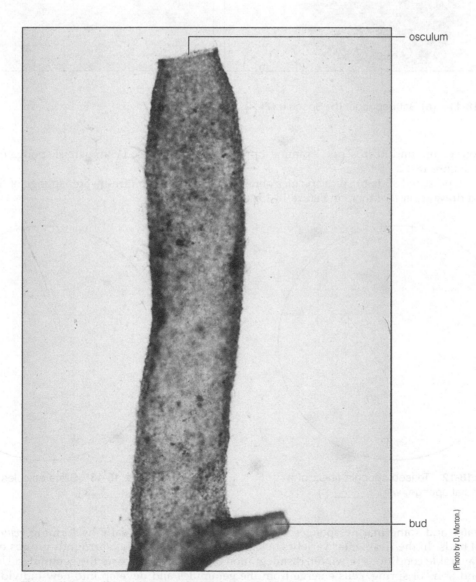

osculum

bud

Figure 18-10   Whole mount of *Leucosolenia* (34×).

B. Class Demospongiae: Commercial and Freshwater Sponges *(15 min.)*

This class contains the majority of sponge species. Most sponges in this class are supported by **spongin** (Figure 18-11b, a flexible substance chemically similar to human hair), by spicules of silica, or both. Silica is silicon dioxide, a major component of sand and of many rocks, such as flint and quartz. All sponges in this class are leuconoid and are able to hold and pass large volumes of water in and through their bodies because of their complex canal systems. Some sponges with skeletons composed only of spongin fibers were once commercially valuable as bath sponges, but synthetic sponges have largely replaced them in today's marketplace.

1. Look over the demonstration specimens of commercial sponges available in your lab room (Figure 18-11a).

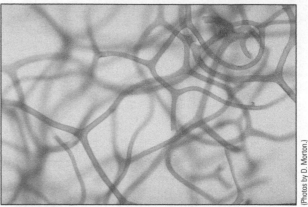

a   b

(Photos by D. Morton.)

Figure 18-11 **(a)** Bath sponge. **(b)** Spongin (74×).

2. With your compound microscope, examine a prepared slide of teased commercial sponge fibers and draw them in Figure 18-12.
3. Examine a prepared slide of silica spicules obtained from a common freshwater sponge of the genus *Spongilla* and draw several of them in Figure 18-13.

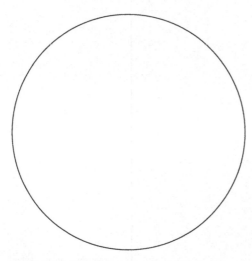

Figure 18-12 Teased spongin fibers of a commercial sponge, w.m. (_____ ×).

Figure 18-13 Silica spicules of *Spongilla*, w.m. (_____ ×).

4. Freshwater and some marine sponges in this class reproduce asexually by forming *gemmules,* resistant internal buds. In the freshwater varieties, a shell of silica and spicules surrounds groups of cells. During the unfavorable conditions of winter, disintegration of the sponge releases the gemmules. Under the favorable conditions of spring, cells emerge from the gemmules and develop into new individuals. Find and examine a gemmule on the prepared slide of freshwater sponge gemmules (Figure 18-14).

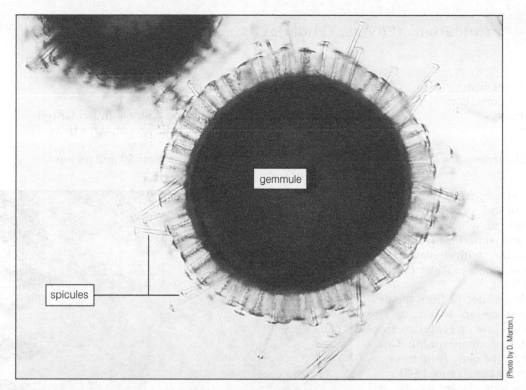

Figure 18-14　Freshwater sponge gemmule, w.m. (150×).

(Photo by D. Morton.)

5. Observe the demonstration of living freshwater sponges available in your lab room (Figure 18-15). They look like yellowish-brown, wrinkled scum with perforations and sometimes have a greenish tint due to the presence of symbiotic algae. They are fairly easy to find in nature; search for them in clear, well-oxygenated water, encrusting submerged plant stems and debris such as a waterlogged twig. Slow streams and the edges of ponds where there is some wave action are good places to find freshwater sponges.

Figure 18-15　Dried freshwater sponges, *Sporilla fragilis* with gemmules (0.5×).

(Photo by W. A. Yoder.)

Habitat: aquatic, mostly marine
Body Arrangement: radial symmetry, gut (incomplete)
Level of Organization: tissue (derived from ectoderm and endoderm)
Body Support: hydrostatic (gastrovascular cavity and epitheliomuscular cells)
Life Cycle: sexual reproduction (monoecious or dioecious), larva (planula),
                                  asexual reproduction (budding)

Special Characteristics: tissues, two adult body forms—polyp (sessile) and medusa
                                  (free swimming)—nematocysts, nerve net

This phylum contains some of the most beautiful organisms in the seas, many of which are brightly colored and superficially plantlike or flowerlike, with tentacles that move in response to water currents (Figure 18-16).

    Cnidarians derive from the protistanlike ancestors that gave rise to the main line of evolution from which other animals sprang. Cnidarians diverged early from this evolutionary lineage (Figure 18-3).

    These animals are *aquatic* and found mostly in shallow marine environments, with the notable exception of the freshwater hydras. Cnidarians have definite tissues and thus show a *tissue level of organization*. They are mostly *radially symmetrical*. Members of this phylum have a *gut* and a definite *nerve net*.

Figure 18-16   Pink-tipped sea anemone with mutalistic spotted cleaning shrimp (1×).

(Photo by W. A. Yoder.)

    Two main body forms exist in cnidarians (Figure 18-17). The usually sessile **polyp** is cylindrical in shape, with the oral end and **tentacles** directed upward and the other end attached to a surface. The free-swimming **medusa** (jellyfish) is bell- or umbrella-shaped, with the oral end and tentacles directed downward. Either or both body forms may be present in the Cnidarian life cycle. Also found throughout the group is a larval form called a *planula*.

    Cnidarians are efficient predators. All are carnivorous and possess tentacles that capture unwary invertebrate and vertebrate prey. Food is captured with the aid of stinging elements called **nematocysts** (Figure 18-18). The nematocysts are discharged by a combination of mechanical and chemical stimuli. The prey is either pierced by the nematocyst or entangled in its filament and pulled toward the mouth by the tentacles. In some species, nematocysts discharge a toxin that paralyzes the prey. The mouth opens to receive the food, which is deposited in the gut. Digestion is started in the gut cavity by enzymes secreted by its lining. It is completed inside the lining cells after they engulf the partially digested food.

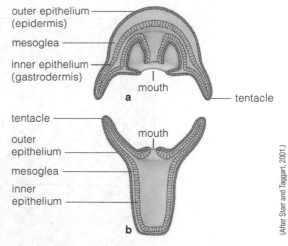

outer epithelium
(epidermis)

mesoglea

inner epithelium
(gastrodermis)

mouth

a

tentacle

tentacle

mouth

outer
epithelium

mesoglea

inner
epithelium

b

(After Starr and Taggart, 2001.)

Figure 18-17   Two body forms of Cnidarians.
(**a**) medusa. (**b**) polyp.

MATERIALS

Per student:

■ dissection microscope
■ compound microscope
■ clean microscope slide and coverslip

Per student pair:

■ prepared slides of
    whole mounts of *Hydra* (showing testes,
        ovaries, buds, and embryos)
    cross section of *Hydra*
■ whole mounts of *Obelia* (showing colonial polyps
    and medusa)

Per student group (4):

- small finger bowl
- scalpel or razor blade
- preserved specimen of *Gonionemus*
- prepared slides of
  a whole mount of the medusa (*ephyra*) of
  Aurelia
  a whole mount of planula larvae of *Aurelia*
  a whole mount of scyphistoma of *Aurelia*
  a whole mount of the strobila of *Aurelia*

Per lab room:

- demonstration specimens of *Physalia* and *Metridium*
- demonstration collection of corals
- boxes of different sizes of latex gloves

Per lab section:

- culture of live *Hydra*
- container of fresh stream water
- dropper bottle of glutathione
- culture of live copepods or cladocerans
- dropper bottle of vinegar

PROCEDURE

A. Class Hydrozoa: Hydrozoans

This is mostly a marine group, except for the hydras, which live in fresh water. A polyp, medusa, or both may be present in the life cycle of a hydrozoan. Marine polyps tend to be colonial.

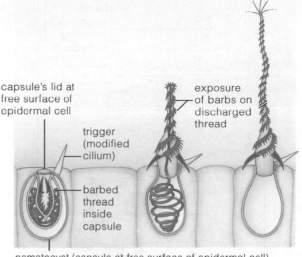

Figure 18-18   One type of nematocyst, a capsule with an inverted thread inside. This one has a bristlelike trigger (**a**). When prey (or predators) touch the trigger, the capsule becomes more "leaky" to water. As water diffuses inward, pressure inside the capsule increases and the thread is forced to turn inside out (**b**). The thread's tip extends to penetrate the prey, releasing a toxin as it does this (**c**).

(After Starr and Taggart, 2001.)

1. *Hydra* is a solitary polyp, and it does not have a medusae stage in its life cycle.
   (a) Place a live specimen of *Hydra* in a finger bowl of fresh stream water. Examine the live *Hydra* with a dissection microscope, along with prepared slides of whole mounts with the compound microscope, noting its polyp body form and other structural details (Figure 18-19).
   (b) Observe the tentacles at the oral end (Figure 18-19a). Find the elevation of the body at the base of the tentacles, in the center of which lies a mouth (Figure 18-19b). Note the swellings on the tentacles (Figure 18-19a); each is a stinging cell (*cnidocytes*) that contains a nematocyst. The clear area in the center of the body represents the gut or **gastrovascular cavity** (Figure 18-19d).
   (c) Look for evidence of sexual reproduction: one to several *testes* or an *ovary* on the body, both of which may occur on the same individual. Testes (Figure 18-19a) are conical and located more orally than the broad, flattened ovary (Figure 18-19b). You may be able to see a zygote or embryo marginally attached to or detached from the body of the *Hydra* (Figure 18-19c). Is *Hydra* monoecious or dioecious?

   _____

   (d) *Hydra* also reproduces asexually by budding (Figure 18-19d). Look for buds on your specimens.
   (e) Examine a cross section of *Hydra* (Figure 18-20) and locate the **epidermis.** The epidermis is derived from ectoderm, one of the primary germ layers. The epidermis contains *epitheliomuscular cells, interstitial cells,* and a few stinging cells. Locate the lining of the gastrovascular cavity, or **gastrodermis,** which is derived from the endoderm and composed of epitheliomuscular cells (sometimes called *nutritive muscular cells*), gland cells, and interstitial cells. Between these two layers is a thin gelatinous membrane called the **mesoglea.** In *Hydra,* the mesoglea is without cells.

The two layers of epitheliomuscular cells contract in opposite directions, although in the *Hydra* epidermis the longitudinally arranged contractile filaments are more developed than the circular filaments of the gastrodermis. How might two layers of epitheliomuscular cells function to maintain and change the body shape of a Cnidarian?

_____

_____

_____

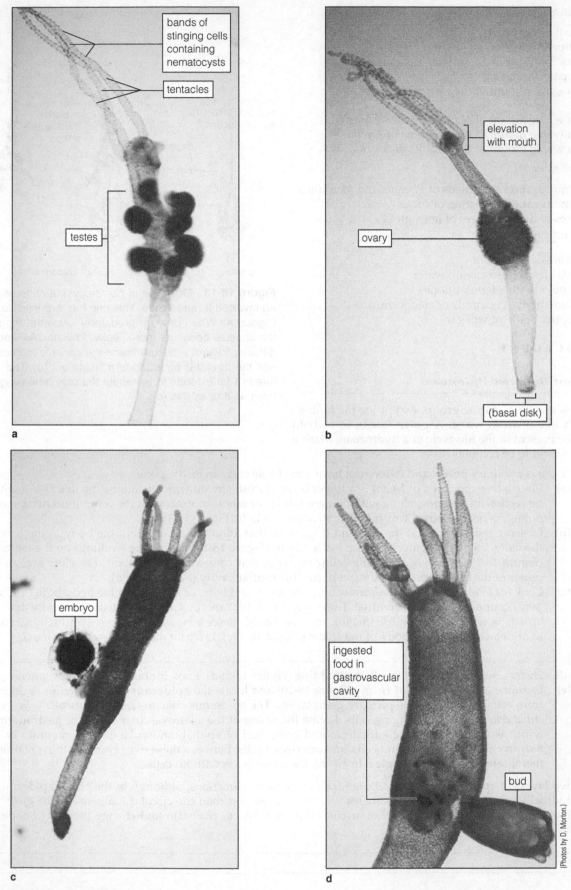

Figure 18-19 Whole mounts of *Hydra* with (**a**) testes, (**b**) an ovary, (**c**) an embryo, and (**d**) a bud (30×).

(Photos by D. Morton.)

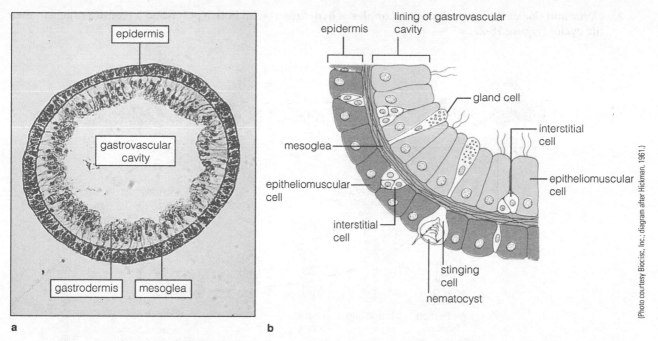

Figure 18-20 *Hydra.* (**a**) Cross section (100×). (**b**) Part of the wall.

What additional function can you suggest for the epithelio-muscular cells of the gastrodermis? (*Hint:* What happens to partially digested food?)

_____

_____

_____

_____

Interstitial cells can give rise to sperm, eggs, and any other type of cell in the body.

(f) If it is not already there, place the live *Hydra* in a small finger bowl to the dissection microscope. Add a drop of glutathione and a number of live copepodan or cladoceran crustaceans under the water. The glutathione stimulates the feeding response of *Hydra.* Look for feeding behavior by the *Hydra.* Note the action of the tentacles and the nematocysts. If you do not observe the discharge of nematocysts (Figure 18-21), place a drop of vinegar in the finger bowl and observe carefully. Describe what you observe.

_____

_____

_____

_____

_____

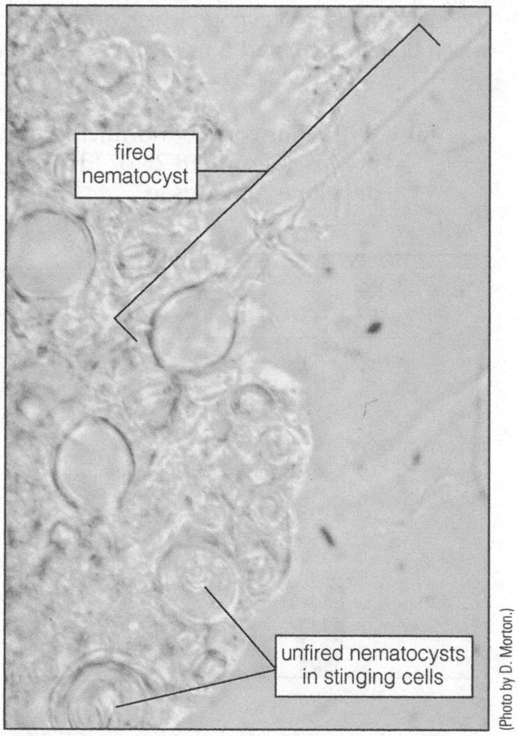

Figure 18-21 Portion of a tentacle of *Hydra* with fired and unfired nematocysts, w.m. (800×).

(g) Are any of your *Hydra* green? If so, use a scalpel or razor blade to remove a tentacle. Make a wet mount of the tentacle, being certain to crush it with the coverslip. What do you observe with your compound microscope? Can you explain the green color of your *Hydra?*

_____

_____

_____

2. *Obelia* and *Gonionemus* are common examples of hydrozoans with both a polyp and a medusae stage in their life cycles (Figure 18-22).

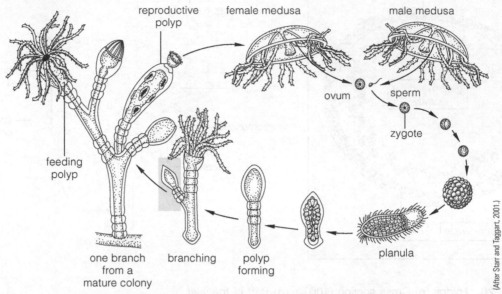

Figure 18-22 Life cycle of the marine hydrozoan *Obelia* that includes medusa and polyp stages. The medusa stage, produced asexually by a reproductive polyp, is free swimming. Fusion of gametes from male and female medusae leads to a zygote, which develops into a planula. The swimming or crawling planula develops into a polyp, which branches to start a new colony. As growth continues, new polyps are formed.

(a) With your compound microscope, examine a prepared slide of a whole mount of a colony of *Obelia* polyps (Figure 18-23a). Identify *feeding* and *reproductive polyps*. The reproductive polyps contain developing medusae.

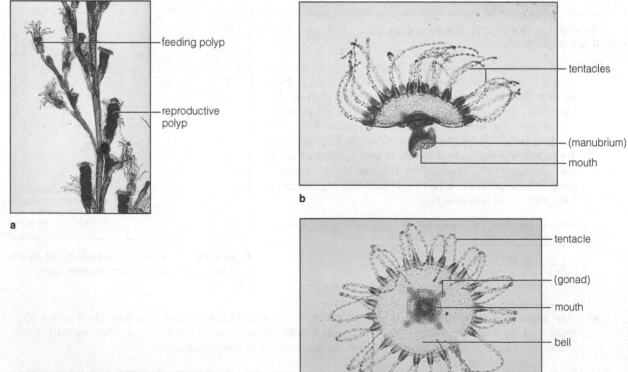

Figure 18-23 Whole mounts of *Obelia*.
(a) Colonial polyps (15×). (b) Side view of a medusa (92×). (c) Top view of a medusa (86×).

(b) Similarly observe a whole mount of a medusa and identify the *mouth*, the umbrellalike *bell*, and *tentacles* bearing bands of stinging cells (Figures 18-23b and c).

(c) Switch to a slide of a whole mount of a *Gonionemus* medusa. Compare your preserved specimen to the live medusa seen in Figure 18-24. In addition to the structures listed in Figure 18-23c, identify the muscular *velum* and the *adhesive pads* on the tentacles. Between the bases of the tentacles are balancing organs called *statocysts*. Medusa move by "jet propulsion." Contractions of the body bring about a pulsing motion that alternately fills and empties the cavity of the bell.

3. Look at the demonstration specimen of a preserved Portuguese man-of-war (*Physalia*) (Figure 18-25).

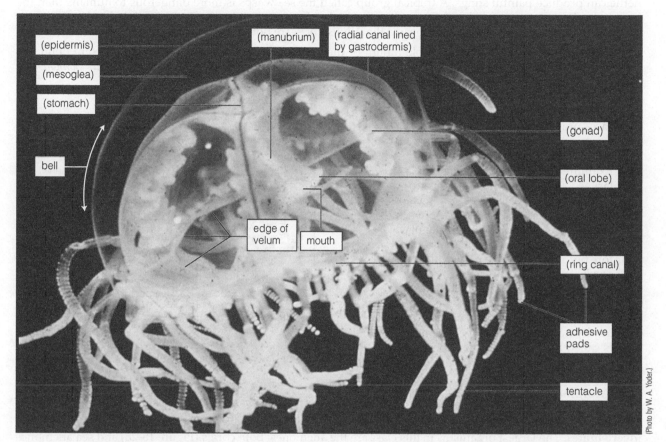

(epidermis)
(mesoglea)
(stomach)
bell
(manubrium)
(radial canal lined by gastrodermis)
(gonad)
(oral lobe)
edge of velum
mouth
(ring canal)
adhesive pads
tentacle
(Photo by W. A. Yoder.)

Figure 18-24 Live hydrozoan medusa, *Gonionemus* (4×).

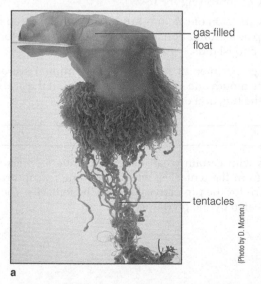

gas-filled float
tentacles
(Photo by D. Morton.)
a

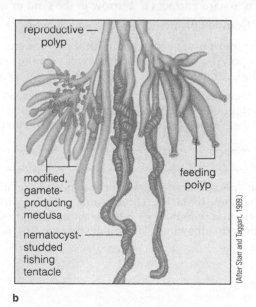

reproductive polyp
modified, gamete-producing medusa
nematocyst-studded fishing tentacle
feeding polyp
(After Starr and Taggart, 1989.)
b

Figure 18-25 The Portuguese man-of-war. (**a**) Preserved specimen. (**b**) The tentacles contain several types of modified polyps and medusae.

This organism is really a colonial hydrozoan composed of several types of modified polyps and medusae. Note the conspicuous gas-filled *float*. Its sting is very painful to humans, and a possibly fatal neurotoxin is associated with the nematocysts of the tentacles.

B. Class Scyphozoa: True Jellyfish

This marine class contains the true jellyfish. The medusa is the predominant stage of the life cycle and is generally larger than hydrozoan medusae. The nematocysts in the tentacles and oral arms of jellyfish such as the sea nettle can produce painful stings. A tropical group called the sea wasps is more dangerous to humans than the hydrozoan Portuguese man-of-war. Some forms are bioluminescent, producing flashes of light by chemical reactions. Bioluminescence may lure prey toward the jellyfish or scare potential predators.

1. Look at a prepared slide of a whole mount of the medusa (ephyra) of *Aurelia* with your compound microscope (Figure 18-26). The structure of this medusa is similar to that of the hydrozoan, *Gonionemus*. However, note that the *tentacles* are short and that four *oral arms* surround the square mouth. In this case, prey is captured with the oral arms instead of the tentacles. Also, a velum is absent.

2. Examine prepared slides of planula larvae, a scyphistoma, and a strobila (Figure 18-27). Sexes are separate, and fertilization is internal. The zygote develops into a planula larva, which swims for a while before becoming a sessile polyp called a *scyphistoma*. This is an active, feeding stage, equipped with tentacles and a mouth. At certain times of the year, young medusae are produced asexually by transverse budding. This stage is called a *strobila*, and the process is called strobilization.

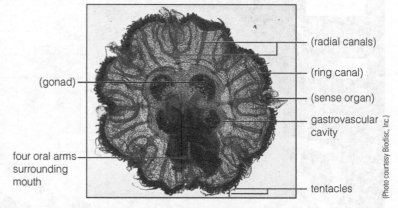

Figure 18-26 Scyphozoan jellyfish, *Aurelia*, w.m. (5×).

(gonad)

four oral arms surrounding mouth

(radial canals)

(ring canal)

(sense organ)

gastrovascular cavity

tentacles

(Photo courtesy Biodisc, Inc.)

C. Class Anthozoa: Corals and Sea Anemones

Class Anthozoa is a marine group in which all members are heavy-bodied polyps that feed on small fish and invertebrates. The largest organisms in this class are the anemones. Both corals (Figure 18-28) and sea anemones attach to hard surfaces or burrow in the sand or mud.

1. Examine the demonstration exoskeletons of a variety of corals (Figure 18-29).

The epidermis of each colonial coral polyp secretes an exoskeleton of calcium carbonate. After a polyp dies, another polyp uses the remaining skeleton of the dead polyp as a foundation and secretes its own skeleton. In this fashion, corals have built up various types of reefs, atolls, and islands.

2. Examine the demonstration of a preserved or living specimen of *Metridium,* the common sea anemone (Figure 18-30). Note its stout, stumplike body, with the mouth and surrounding tentacles at the oral end and the *pedal disk* at the other end. What do you think is the function of the pedal disk?

_____

_____

Some species of anemones form symbiotic relationships with certain small fishes and invertebrates that live among their tentacles. These animals receive protection from the tentacles, which for some reason do not discharge their nematocysts into the animals' bodies. In return for the protective cover of the tentacles, the anemones benefit—for example, by ingesting food particles not eaten by the predatory fish.

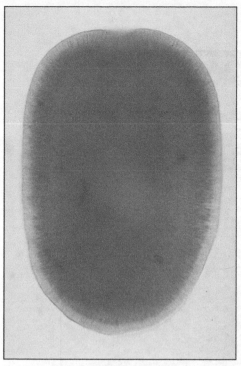

a

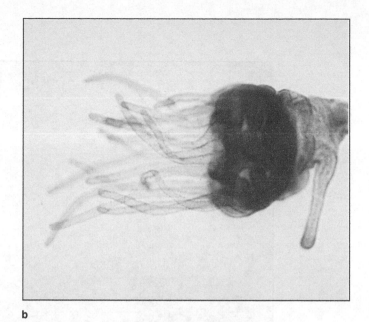

b

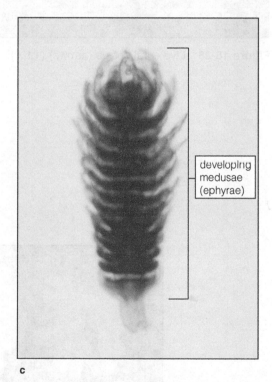

developing
medusae
(ephyrae)

c

Figure 18-27 Stages in the life cycle of *Aurelia*. (**a**) Planular larvae, w.m. (400×). (**b**) Scyphistoma, w.m. (50×). (**c**) Strobilia, w.m. (30×). (Photos by D. Morton.)

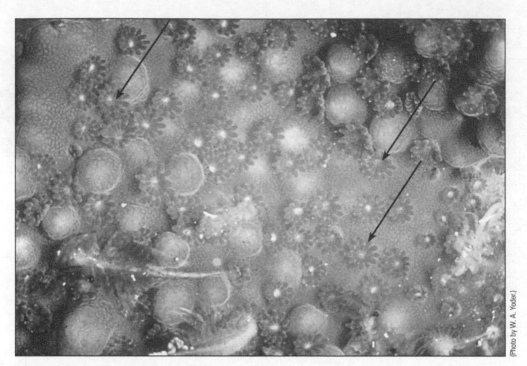

Figure 18-28 Live coral polyps (arrows) (1.5×).

(Photo by W. A. Yoder.)

Figure 18-29 Hard coral (0.30×).

(Photo by D. Morton.)

oral disk

mouth

column

tentacles

pedal disk

a

(Photo by D. Morton.)

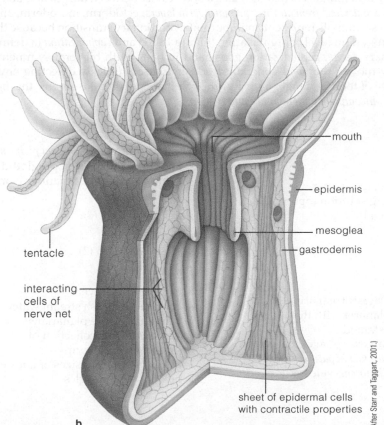

mouth

epidermis

mesoglea

gastrodermis

tentacle

interacting
cells of
nerve net

sheet of epidermal cells
with contractile properties

b

(After Starr and Taggart, 2001.)

Figure 18-30 (a) Top and side views of preserved specimens of the common sea anemone, *Metridium* (1.5×). (b) Organization of a sea anemone.

18.3   Flatworms (Phylum Platyhelminthes) *(About 75 min.)*

If you've ever turned over rocks and leaves in a clear, cool, shallow stream or roadside ditch, you may have observed representatives of phylum Platyhelminthes without realizing it. Do you recall any dark, smooth, soft-bodied, flattened or ovoid masses approximately 1 cm or less in length attached to the bottom of the rocks or leaves? This description fits that of a free-living flatworm or planarian. Parasitic flukes and tapeworms also belong to this phylum.

Habitat:	aquatic, moist terrestrial, or parasitic
Body Arrangement:	bilateral symmetry, acoelomate, head (cephalization), gut (incomplete)
Level of Organization:	organ system (tissues derived from all three primary germ layers); nervous system with brain; excretory system with flame cells; no circulatory respiratory, or skeletal systems
Body Support:	muscles
Life Cycle:	sexual reproduction (mostly monoecious with internal fertilization); parasites have very complicated life cycles; some asexual reproduction (mostly transverse fission)
Special Characteristics:	bilateral symmetry, head, organ systems

Flatworms have soft, flat, wormlike bodies with *bilateral symmetry* (Figure 18-1). This is the same type of symmetry we exhibit. Flatworms are **acoelomates** in that they lack a body cavity between the gut and body wall (Figure 18-2). The *tissues* of flatworms are *derived from all three primary germ layers:* ectoderm, mesoderm, and endoderm.

Organizationally, these animals represent an important evolutionary transition because they are the simplest forms to exhibit an organ-system level of organization (Figure 18-31) and *cephalization* (a definite head with sense organs). In an evolutionary sense, cephalization goes hand in hand with bilateral symmetry. When an animal with bilateral symmetry moves, the head generally leads, its sensors quickly detecting environmental stimuli that indicate the presence of food or danger ahead. Although there are definite organs, the *digestive system* is still of the *gastrovascular type, incomplete with a mouth and no anus.*

MATERIALS

Per student:

- dissection microscope
- compound microscope, lens paper, a bottle of lens-cleaning solution (optional), a lint-free cloth (optional)
- scalpel
- blunt probe

Per student pair:

- prepared slides of
  a whole mount of *Dugesia* (planaria)
  a whole mount of planaria with the digestive system stained
  a cross section of planaria (pharyngeal region)
  a whole mount of *Fasciola hepatica*
  a whole mount of *Clonorchis sinensis*

- whole mount(s) of *Taenia pisiformis*, with scolex and different types of proglottids
- a whole mount of a bladder worm (cysticercus) of *Taenia solium*

Per student group (4):

- penlight
- glass petri dish (2)

Per lab room:

- chunk of beef liver
- container of fresh stream water
- container of live planarians
- demonstration collection of preserved tapeworms
- boxes of different sizes of latex gloves
- box of safety goggles

PROCEDURE

A. Class Turbellaria: Planarians

Most species in this class are free-living flatworms, although a few species are parasitic or symbiotic with other organisms. Free-living flatworms are found in marine and fresh waters on the underside of submerged rocks,

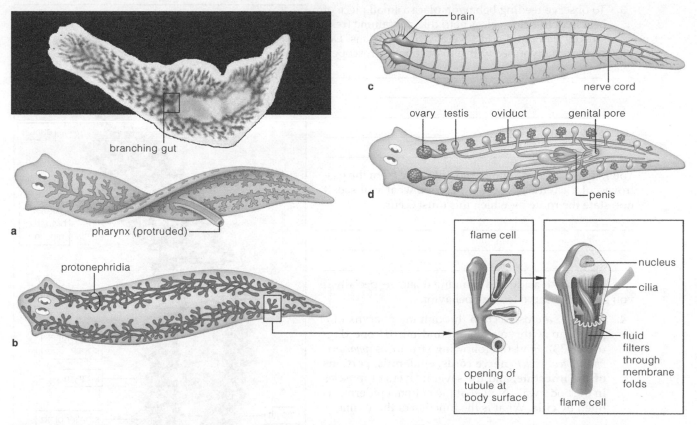

Figure 18-31 Organ systems of one type of flatworm. (**a**) Digestive system, which includes a pharynx that opens to the gut. The pharynx protrudes onto food sources, and then it retracts into its own chamber between feedings. (**b**) Water-regulating system, (**c**) nervous system, and (**d**) reproductive system. (After Starr, 2000.)

leaves, and sticks. They are especially abundant in cool freshwater streams and along the ocean shoreline. Some live in wet or moist terrestrial habitats.

Free-living freshwater planarians of the genus *Dugesia* are the representatives of this class typically studied in biology courses. They are relatively small flatworms, usually 2 cm or less in length. Planarians are monoecious—an individual has both male and female reproductive structures—but apparently do not self-fertilize. Individuals possess testes, ovaries, a penis, and a vagina; and copulation occurs. Planarians exhibit a high degree of **regeneration,** in which lost body parts are gradually replaced.

1. Microscopically examine a whole mount of a planarian. Refer to Figure 18-32 and locate the **head, eyespots,** and **auricles,** lateral projections of the head that are organs sensitive to touch and to molecules dissolved in the water. The **eyespots** are photoreceptors. They are sensitive to light but do not form images.

2. Obtain and similarly examine a whole mount of a planarian with an injected digestive system. Identify the muscular **pharynx** within the **pharyngeal pouch.** The *mouth* is located at the free end of the pharynx. The mouth opens into the pharyngeal cavity, which in turn is continuous with the cavity of the **intestine.** To feed, a free planarian protrudes the pharynx from its ventral (belly) side and sucks up organic morsels (insect larvae, small crustaceans, and other small living and dead animals). The intestine is elaborately branched and occupies much of the animal's body. What function does the extensive branching of the intestine facilitate?

_____

_____

_____

_____

_____

_____

3. To observe feeding behavior, place a small piece of beef liver in a clean glass petri dish containing fresh stream water and add several live planarians. Describe what you see with a dissection microscope.

_____

_____

_____

_____

You may observe the elimination of waste from the gastrovascular cavity. If you do, describe what you see; if not, state the route by which this must occur.

_____

_____

_____

Set the dish aside and examine it later, especially if you did not observe feeding behavior.

4. Examine a cross section through the pharynx of a planarian with your compound microscope (Figure 18-33). Find the following structures: *pharynx*, *pharyngeal pouch*, **nerve cords, epidermis, portions of the intestine,** and the several layers of muscles in the body wall. Note that the bottom epidermis is lined by **cilia.** What is the function of these cilia?

_____

_____

_____

Much of the rest of the body wall contains three layers of muscle: an outer circular layer, a middle diagonal layer, and an inner longitudinal layer. It is often difficult to distinguish all three layers.

5. Working in groups of four, use a dissection microscope to observe a living planarian in a clean glass petri dish filled with fresh stream water. Note its general size and shape. Touch the animal with a probe. What is its reaction? How does its shape change? Which muscle layers likely contract to produce this shape change?

_____

_____

_____

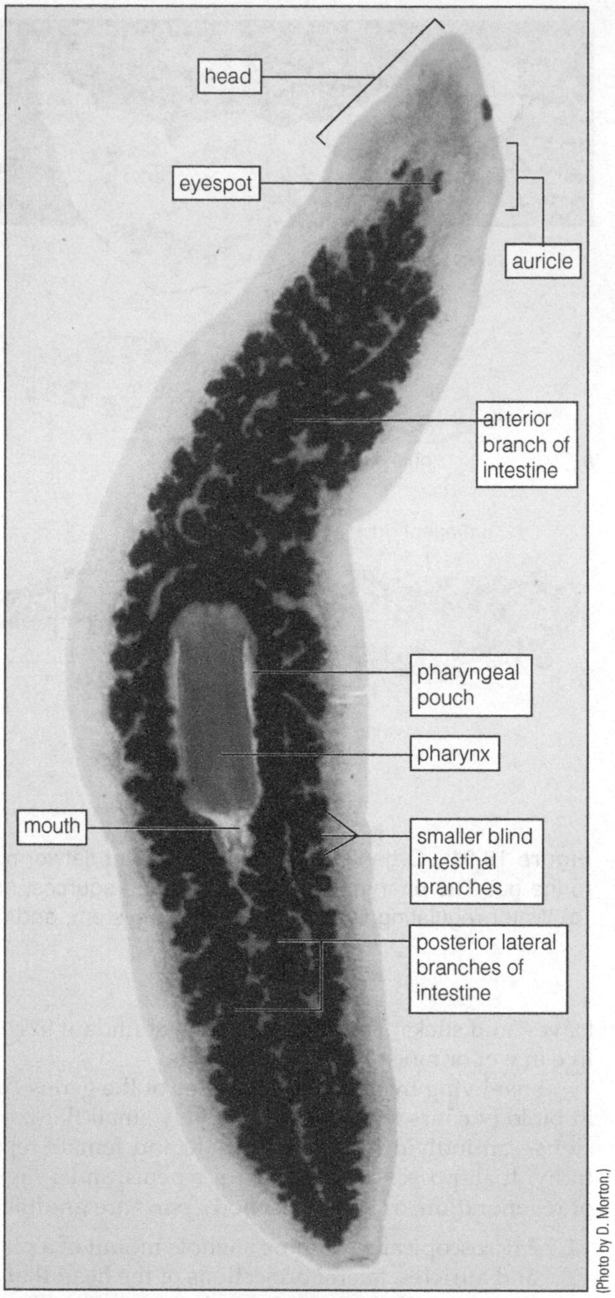

_____

Figure 18-32 Whole mount of planaria, with injected digestive system (45×).

B. Class Trematoda: Flukes

Flukes are small, leaf-shaped, parasitic flatworms with complex life cycles. They are generally internal parasites of vertebrates, including humans. A protoplasmic but noncellular cuticle covers the surface of their bodies. What is a likely explanation for the presence of this tough, resistant cuticle?

_____

_____

_____

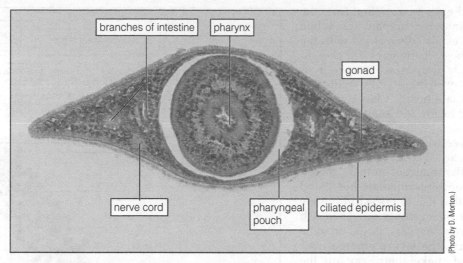

Figure 18-33 Cross section of planaria (150×).

1. Examine a whole mount of a sheep liver fluke, *Fasciola hepatica,* noting its general morphology. Identify the structures indicated in Figure 18-34.

2. Examine a whole mount of an adult *Clonorchis sinensis,* a human liver fluke. This animal is a common parasite in Asia, and humans are the final **host.** Identify the structures indicated in Figure 18-35. *Clonorchis* has two suckers for attachment to its host: an **oral sucker** surrounding the mouth and a **ventral sucker.** The digestive tract begins with the **mouth** and continues with a muscular **pharynx,** a short **esophagus,** and two blind pouches or **intestinal ceca.** Nitrogenous wastes are excreted at the tip of the back end through the **excretory pore.**

Clonorchis is monoecious. Find the following female reproductive structures: **ovary, yolk glands, seminal receptacle,** and **uterus.** Find the following male reproductive structures: **testes** and **seminal vesicle,** which may be visible in your specimen. During copulation, sperm is conveyed from the testes to the *genital pore* by paired vasa efferentia, which fuse to form the vas deferens. These ducts are difficult to see in most specimens. Copulating individuals exchange sperm at the genital pore, and the sperm is stored temporarily in the seminal receptacle. The yolk glands provide yolk for nourishment of the developing eggs. *Eggs* can be seen as black dots in the uterus, where they mature.

The life cycle of *Clonorchis sinensis* (Figure 18-36) involves two **intermediate hosts,** a snail and the golden carp. Intermediate hosts harbor sexually immature stages of a parasite.

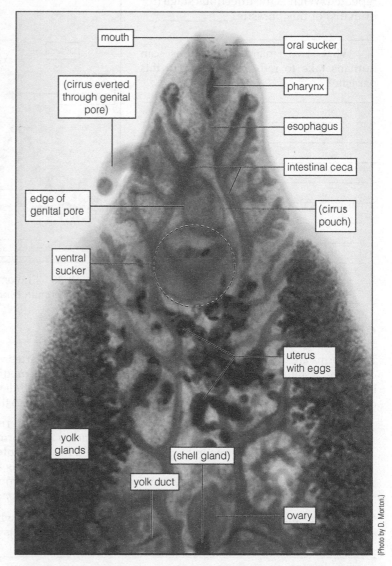

Figure 18-34 Anterior portion of *Fasciola hepatica,* the sheep liver fluke, w.m. (20×).

Fertilized eggs are expelled from the adult fluke into the human host's bile duct and are eventually voided with the feces of the host. If the eggs are shed in the appropriate moist or aquatic environment, they may be ingested by snails (the first intermediate host). A larva emerges from the egg inside the snail. The larva passes through several developmental stages and gives rise to a number of tadpolelike larvae. These free-swimming larvae leave the snail, encounter a golden carp (the second intermediate host), and burrow into this fish to encyst in muscle tissue. If raw or improperly cooked fish is eaten by humans, cats, dogs, or some other mammal, the larvae emerge from the cysts and make their way up the host's bile duct into the liver, where they mature into adult flukes. The cycle is then repeated. What is the infectious stage (for humans) of this parasite? _____

_____

What two main precautions can humans take to reduce exposure to this parasite?

_____
_____
_____
_____
_____
_____
_____
_____
_____
_____
_____

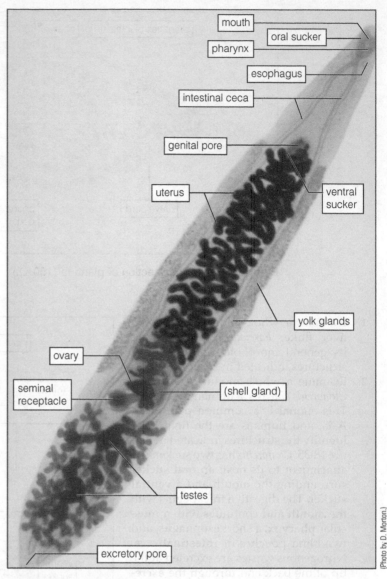

Figure 18-35 Photomicrograph of a whole mount of *Clonorchis sinensis*, a human liver fluke (20×).

C. Class Cestoda: Tapeworms

Tapeworms are internal parasites that inhabit the intestine of vertebrates. Their flattened, multiunit bodies have no digestive system; instead, they absorb across their body walls predigested nutrients provided by the host. Their bodies are essentially reproductive machines, with extensive, well-developed reproductive organs occupying much of each mature body unit. A thick cuticle, similar in construction to that found in trematodes, covers the body.

1. Look over the demonstration collection of preserved tapeworms taken from a variety of vertebrate hosts.
2. Examine a preserved specimen of *Taenia pisiformis*, a tapeworm of dogs and cats (Figure 18-37). Note that it is divided into three general regions: specialized head or **scolex**, *neck*, and **body**. The body is composed of successive units, the **proglottids**. New immature proglottids are added at the neck. As they move farther from the scolex, proglottids mature and, therefore, the hindmost gravid proglottids are the oldest.
3. Look at a prepared slide of a whole mount of a *Taenia pisiformis* scolex (Figure 18-38).

Observe that the scolex is covered with hooks and suckers. For what task does the tapeworm use these **hooks** and **suckers**? Note the **neck** and the new proglottids forming at its base.

_____
_____
_____

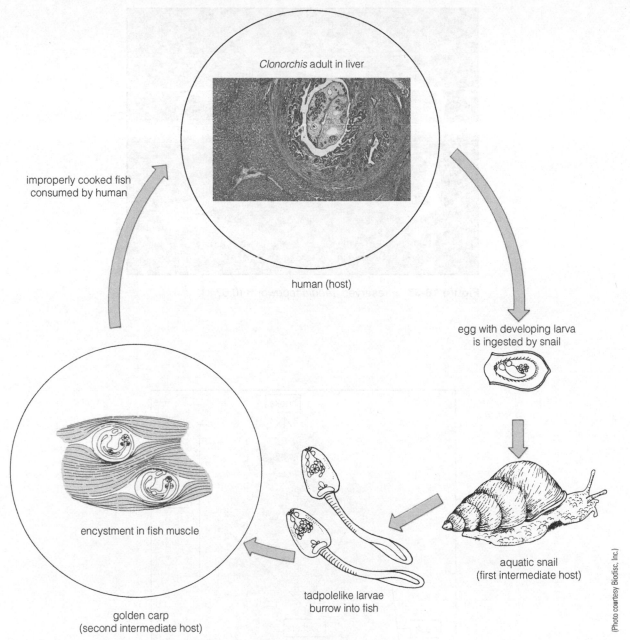

Figure 18-36 Life cycle of the human liver fluke, *Clonorchis sinensis*.

Labels in figure:

Clonorchis adult in liver

improperly cooked fish consumed by human

human (host)

egg with developing larva is ingested by snail

aquatic snail (first intermediate host)

encystment in fish muscle

tadpolelike larvae burrow into fish

golden carp (second intermediate host)

(Photo courtesy Biodisc, Inc.)

4. Find a whole mount of mature proglottids of *Taenia pisiformis* and examine them with a dissection microscope. Identify the structures indicated in Figure 18-39. The lateral **genital pore** opens into the **vagina** and receives sperm from the **vas deferens.** The *uterus* is usually visible as a line running top to bottom in the middle of the proglottid. The *ovaries* are represented by dark, diffuse masses of tissue lateral to the uterus at the vaginal end of the proglottid. The **testes** are paler, more diffuse masses of tissue lateral to the uterus at the other end. Tapeworms are monoecious. They are one of the few animals that can self-fertilize, perhaps a result of their isolation from each other. Flanking most of the reproductive organs are the **excretory canals.**
5. Locate a gravid proglottid of *Taenia pisiformis* (Figure 18-40). Gravid proglottids are almost completely filled with zygotes in an expanded uterus ready for release as eggs.
6. The life cycle of the pork tapeworm, *Taenia solium*, involves two host organisms, the pig and a human (Figure 18-41).

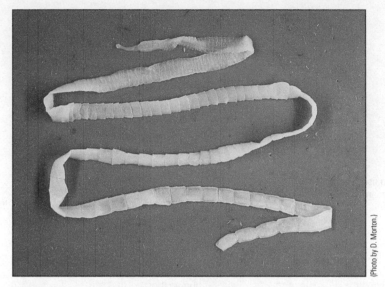

Figure 18-37 Preserved *Taenia* tapeworm (0.67×).

(Photo by D. Morton.)

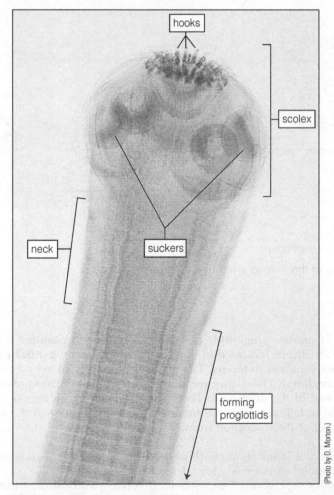

hooks

scolex

neck

suckers

forming
proglottids

(Photo by D. Morton.)

Figure 18-38 Tapeworm scolex (*Taenia pisiformis*), w.m.
(40×).

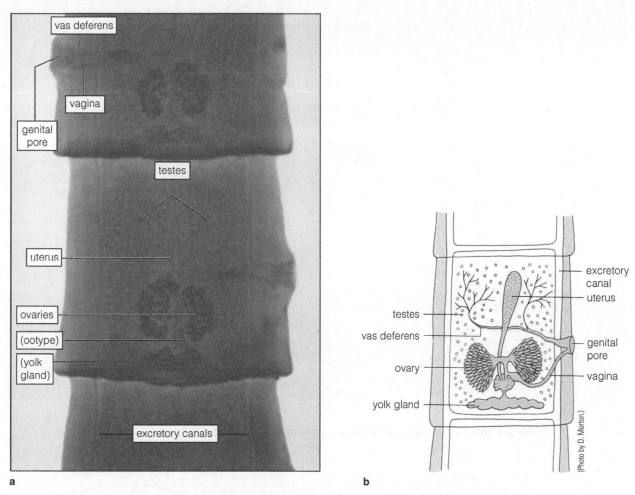

Figure 18-39   Tapeworm mature proglottids (*Taenia pisiformis*). (**a**) Whole mount (30×). (**b**) Diagram.

Figure 18-40   Tapeworm gravid proglottid (*Taenia pisiformis*), w.m. (30×).

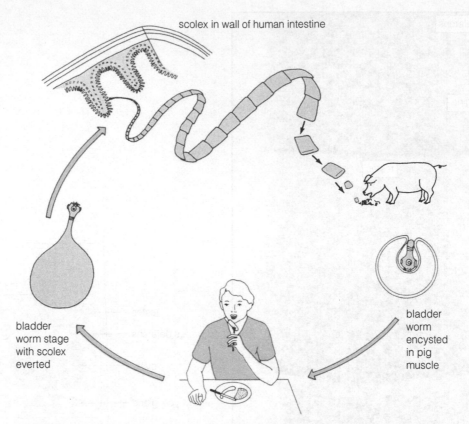

scolex in wall of human intestine

bladder
worm stage
with scolex
everted

bladder
worm
encysted
in pig
muscle

Figure 18-41   Life cycle of the pork tapeworm, *Taenia solium.*

Gravid proglottids containing embryos are voided with the feces of a human. If a pig ingests a proglottid, the embryos escape from the proglottid and bore into the intestinal wall and enter a blood or lymphatic vessel. From there they make their way to muscle, where they encyst as larvae known as **bladder worms.** If a human ingests improperly cooked pork, a bladder worm "pops" out of its encasement (bladder) and attaches its scolex to the intestinal wall. Here it matures and adds proglottids to complete the life cycle.

Which mammal is the host?

_____

Which mammal is the intermediate host?

_____

Examine the bladder worm stage of *Taenia solium* with the dissection microscope. Draw what you see in Figure 18-42.

Figure 18-42   Drawing of the bladder worm of *Taenia solium* (_____×).

18.4   Rotifers (Phylum Rotifera) *(About 10 min.)*

Rotifers are microscopic, the smallest animals. They are mostly bottom-dwelling, freshwater forms, and they are an important prey of larger animals in aquatic food chains.

Habitat: aquatic or moist terrestrial, mostly fresh water
Body Arrangement: bilateral symmetry,
                                           pseudocoelomate, head,
                                           gut (complete)

Level of Organization:	organ system (tissues derived from all three primary germ layers); nervous system with brain; excretory system with flame cells; no circulatory, respiratory, or skeletal systems
Body Support:	hydrostatic (pseudocoelom)
Life Cycle:	sexual (dioecious with internal fertilization), some parthenogenesis, males much smaller than females or absent
Special Characteristics:	microscopic animals, wheel organ, mastax

Rotifers are easily found in rain gutters and spouts, and in the slimy material around the bases of buildings. Their thin, usually transparent body wall is covered by a cuticle composed of protein. Also like the nematodes to be studied later, rotifers are pseudocoelomates. Rotifers are dioecious; however, many species have no males, and the eggs develop parthenogenetically.

MATERIALS

Per student:

■ clean microscope slide and coverslip
■ compound microscope

Per lab room:

■ live culture of rotifers
■ disposable transfer pipet

PROCEDURE

1. With a disposable transfer pipet, place drop from the rotifer culture on clean slide and make a wet mount.
2. Observe the microscopic rotifers with the compound microscope. The elongate, cylindrical body is divisible into the **trunk** and the **foot** (Figure 18-43).

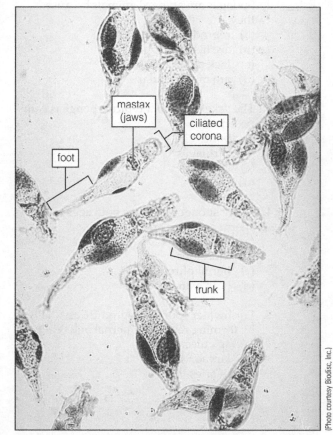

Figure 18-43 Prepared whole mount of rotifers (130×).

(Photo courtesy Biodisc, Inc.)

The foot bears two to several "sticky" *toes* for adhering to objects. The upper end of the trunk bears a **corona** of cilia. These cilia beat in such a fashion as to give the appearance of a wheel (or two) turning, hence the common references to this structure as the wheel organ and to these animals as wheel animals. The cilia create currents that sweep organic food particles into the mouth at the center of the wheel organ. The cilia also enable the organism to move. Inside the trunk, below the wheel organ, look for the movement of a grinding organ, the **mastax**. This structure is equipped with jaws made hard by a substance called chitin. Considering its location, what is the likely function of the mastax?

_____

_____

_____

_____ 1. Which of the following is characteristic of animals?
(a) heterotrophy
(b) autotrophy
(c) photosynthesis
(d) both b and c

_____ 2. Sponges are atypical animals because they
(a) reproduce sexually
(b) are heterotrophic
(c) lack definite tissues
(d) are multicellular

_____ 3. The "skeleton" of a typical sponge is composed of
(a) bones
(b) cartilages
(c) spicules
(d) scales

_____ 4. Many sponges produce both eggs and sperm and thus are
(a) monoecious
(b) dioecious
(c) hermaphroditic
(d) both a and c

_____ 5. Freshwater sponges reproduce asexually by forming resistant internal buds called
(a) spicules
(b) larvae
(c) ovaries
(d) gemmules

_____ 6. Cnidarians evolved from an ancestral group of
(a) sponges
(b) protists
(c) fungi
(d) plants

_____ 7. A free-swimming jellyfish has a body form known as a
(a) polyp
(b) strobila
(c) medusa
(d) hydra

_____ 8. Cnidarians capture their prey with the use of
(a) poison claws
(b) nematocysts
(c) oral teeth
(d) pedal disks

_____ 9. Which statement is true for scyphozoan medusae?
(a) They are the dominant body form
(b) They are generally smaller than hydrozoan medusae
(c) Most live in fresh water
(d) They are scavengers

_____ 10. Coral reefs, atolls, and islands are the result of the buildup of coral
(a) polyps
(b) medusae
(c) exoskeletons
(d) wastes

11. Cephalization is
    (a) the division of the trunk into a scolex, neck, and body
    (b) the presence of a head with sense organs
    (c) sexual reproduction involving self-fertilization
    (d) the ability to replace lost body parts

12. Flatworms
    (a) are radially symmetrical
    (b) possess a body cavity
    (c) reproduce by parthenogenesis
    (d) have tissues derived from all three primary germ layers

13. Free-living flatworms (planaria) move by using
    (a) cilia
    (b) flagella
    (c) pseudopodia
    (d) a muscular foot

14. The flukes and tapeworms are covered by a protective
    (a) cuticle
    (b) shell
    (c) scales
    (d) slime layer

15. The life cycle of the human liver fluke has _____ intermediate hosts.
    (a) one
    (b) two
    (c) three
    (d) four

16. Tapeworms do not have a
    (a) reproductive system
    (b) head
    (c) excretory system
    (d) digestive system

17. The oldest proglottid of a tapeworm is
    (a) the scolex
    (b) the one nearest its neck
    (c) the one in its middle
    (d) the one at its end

18. Bladder worms are found in
    (a) proglottids in human feces
    (b) proglottids in the human intestine
    (c) improperly cooked pork
    (d) none of the above

19. Nematodes and rotifers have a body cavity that is
    (a) called a coelom
    (b) called a pseudocoelom
    (c) completely lined by tissues derived from mesoderm
    (d) both a and c

20. Rotifers obtain their food by the action of
    (a) tentacles
    (b) cilia on the wheel organ
    (c) a muscular pharynx
    (d) pseudopodia

EXERCISE 18

Sponges, Cnidarians, Flatworms, and Rotifers

POST-LAB QUESTIONS

Introduction

1. List six characteristics of the animal kingdom.

2. Define *asymmetry* and *radial symmetry*.

3. Identify the symmetry (asymmetry, radial symmetry, or bilateral symmetry) of the following objects.
   a. fork
   b. jagged piece of rock
   c. rubber ball

18.1 Sponges (Phylum Porifera)

4. Define *cell specialization* and indicate how sponges exhibit this phenomenon.

5. Describe three ways that sponges reproduce.

6. Identify this photo; it is magnified 186×.

_____

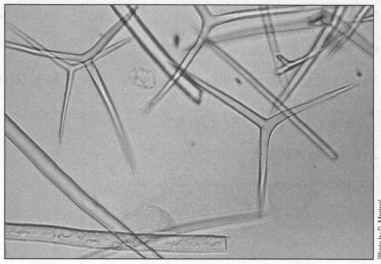

(Photo by D. Morton)

18.2 Cnidarians (Phylum Cnidaria)

7. Define the tissue level of organization and indicate how cnidarians exhibit this phenomenon.

8. Name and draw the two body forms of cnidarians.

9. Describe the means by which the cnidarians seize and eat faster organisms.

Food for Thought

10. What adaptations of sponges and cnidarians have helped them to endure for so long on earth?

18.3 Flatworms (Phylum Platyhelminthes)

11. How is bilateral symmetry different from radial symmetry?

12. List the characteristics of flatworms.

13. Distinguish between a host and an intermediate host.

14. Identify and label this illustration. It is magnified 9×.

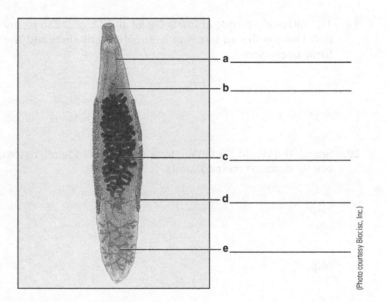

a _____

b _____

c _____

d _____

e _____

(Photo courtesy Biocisc, Inc.)

15. Describe several adaptations of parasitic flatworms to their external environment.

18.4 Rotifers (Phylum Rotifera)

16. *Pseudocoel* means "false body cavity." Why is this term applied to the body cavity of rotifers?

Food for Thought

17. Contrast the digestive systems of free-living and parasitic flatworms.

18. Explain the relationship between the tremendous numbers of eggs produced by flukes and tapeworms and the complexity of their life cycles.

19. The relatively simple animals *Hydra* and *Dugesia* can regenerate lost body parts, but humans generally cannot. Discuss this in terms of tissue differentiation and the comparative levels of structural complexity of these organisms.

20. Search the World Wide Web for sites that include references to flatworms or rotifers. List two addresses and briefly describe their contents.

    http://

    http://

Segmented Worms, Mollusks, Roundworms, and Joint-Legged Animals

After completing this exercise, you will be able to

1. define *coelom, coelomate, protostome, deuterostome, segmentation, peritoneum, setae, clitellum, cocoon, foot, mantle, pseudocoelomate, cloaca, bladder worm exoskeleton, hemolymph, chitin, carapace, sexual dimorphism, gills,* and *tracheae;*

2. differentiate between protostomes and deuterostomes;

3. describe the natural history of members of the phyla Annelida and Mollusca;

4. identify representatives of the classes Oligochaeta, Polychaeta, and Hirudinea of the phylum Annelida;

5. outline the life cycle of an earthworm;

6. identify representatives of the classes Bivalvia, Gastropoda, and Cephalopoda of the phylum Mollusca;

7. outline the life cycle of a freshwater clam;

8. identify structures (and indicate their associate functions) of the representatives of these phyla classes;

9. explain the term *ecdysis;*

10. identify representatives of the phylum Nematoda;

11. identify representatives of the subphyla Crustacea, Hexapoda, Myriapoda, and Chelicerata of the phylum Arthropoda;

12. identify structures (and indicate their associated functions) of the representatives of these phyla and classes.

INTRODUCTION

Some time after the origin of bilateral symmetry, the animals that subsequently evolved developed a true ventral body cavity or **coelom.** A true body cavity is lined by tissue derived entirely from mesoderm. The appearance of the coelom has great evolutionary significance, for it accommodates, cushions, and protects organs; allows an increase in overall body size and flexibility; and contributes to the development of systems primarily derived from mesoderm.

Coelomates, animals possessing a coelom, took two divergent evolutionary paths based on whether the *blastopore*—the first opening in a developing embryo—became the mouth (**protostomes**) or anus (**deuterostomes**) (Table 19-1). During subsequent evolutionary change, the coelom was reduced (mollusks and joint legged animals), lost (flatworms), or modified to a pseudocoel (rotifers and roundworms).

TABLE 19-1 Evolution of Coelomates		
Path	**Blastopore Becomes**	**Phyla**
Protostomes	Mouth	Mollusca, Annelida, and Arthropoda
Deuterostomes	Anus	Echinodermata, Hemichordata, and Chordata

19.1 Annelids (Phylum Annelida) *(45 min.)*

Habitat:	aquatic or moist terrestrial, some parasitic
Body Arrangement:	bilateral symmetry, coelomate, head, body segmented, gut (complete)
Level of Organization:	organ system (tissues derived from all three primary germ layers), nervous system with brain, circulatory system (closed), excretory system (nephridia), respiratory system (gills in some), no skeletal system
Body Support:	hydrostatic (coelom)
Life Cycle:	sexual reproduction (monoecious or dioecious with mostly internal fertilization), larvae in some, asexual reproduction (mostly transverse fission)
Special Characteristics:	segmented, closed circulatory system

The most striking annelid characteristic is the division of the cylindrical trunk into a series of similar segments. *Annelids* were the first animals to evolve the condition of **segmentation.** Segmentation is internal as well as external, with segmentally arranged components of various organ systems and the body cavity. The coelom is more or less divided by *septa* into compartments within each segment, each lined by **peritoneum.** Unlike the proglottids of the cestodes, segments cannot function independently of each other. The *circulatory system* is *closed*, with the blood entirely contained in vessels. There is a well-developed *nervous system* with a central nervous system composed of two fused dorsal ganglia (brain) and ventral nerve cords. The circulatory system is not segmented. Segmentation is significant in evolutionary history because each segment or group of segments can become specialized. Although this is an important characteristic of most of the phyla we have yet to study, it is barely evident in the earthworm.

MATERIALS

Per student:

- pins
- blunt probe or dissecting needle

Per student pair:

- preserved earthworm
- dissection microscope
- dissection pan
- fine dissecting scissors
- *Eisenia foetida* cocoons

Per lab room:

- labeled demonstration dissection of the internal anatomy of the earthworm (optional)
- demonstration collection of living and preserved polychaetes living earthworms
- several preserved leeches
- live freshwater leeches in an aquarium
- large plastic bag for disposal of dissected specimens
- boxes of different sizes of latex gloves

A. Oligochaetes (Class Oligochaeta)

PROCEDURE

1. Refer to Figure 19-1 as you examine the living earthworms on demonstration in the lab. You probably already know several characteristics of oligochaetes because of past observations with the earthworm (*Lumbricus terrestris*), especially if you fish. Recalling past experiences with this organism, list as many features of the earthworm as you can.

    _____

    _____

    _____

    _____

2. Let's see how you did! Obtain a preserved earthworm and have a dissection microscope handy. Is its body bilaterally or radially symmetrical? _____

3. Examine the external anatomy (Figure 19-2). Note the segments.
    Count them and record the number: _____

    (a) Very few segments are added after hatching. Note the **prostomium,** a fleshy lobe that hangs over the **mouth;** both structures are part of the first segment.

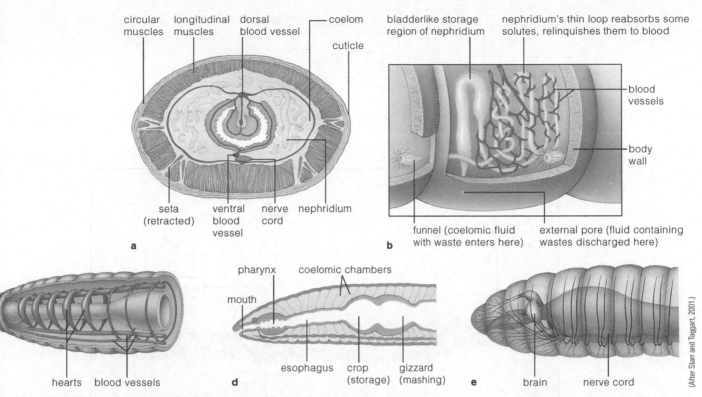

circular muscles · longitudinal muscles · dorsal blood vessel · coelom · cuticle

bladderlike storage region of nephridium · nephridium's thin loop reabsorbs some solutes, relinquishes them to blood

blood vessels

body wall

seta (retracted) · ventral blood vessel · nerve cord · nephridium

a

funnel (coelomic fluid with waste enters here) · external pore (fluid containing wastes discharged here)

b

pharynx · coelomic chambers

mouth

esophagus · crop (storage) · gizzard (mashing)

hearts · blood vessels

d

brain · nerve cord

e

(After Starr and Taggart, 2001.)

Figure 19-1 Earthworm body plan. (**a**) Midbody, transverse section. (**b**) A nephridium, one of many functional units that help maintain the volume and the composition of body fluids. (**c**) Portion of the closed circulatory system. The system is linked in its functioning with nephridia. (**d**) Part of the digestive system, near the worm's head end. (**e**) Part of the nervous system.

(b) Look for **setae,** tiny bristlelike structures on the ventral surface, with the dissection microscope. Take the specimen between your fingers and feel for them. What is their likely function?

_____

_____

_____

_____

(c) Use the dissection microscope to help you find the following. A pair of small **excretory pores** is found on the lateral or ventral surfaces of each segment, except the first few and the last. On the sides of segment 14, the openings (*female pores*) of the oviducts can be seen. The openings (*male pores*) of the sperm ducts, with their swollen lips, can be found on segment 15. The **clitellum** is the enlarged ring beginning at segment 31 or 32 and ending with 37. This glandular structure secretes a slimy band around two copulating individuals. The **anus** is the vertical slit in the terminal segment.

4. *Dissection of earthworm internal anatomy.*

(a) Place your worm in a dissection pan and stick a pin through each end of the specimen. With fine scissors, make an incision just to the right of the dorsal black line (blood vessel), 10 segments anterior to the anus, and cut superficially to the mouth. (You can use the prostomium as an indicator of the dorsal side of the animal if the black line cannot be seen.) Gently pull the incision apart and note that the coelom is divided into a series of compartments by septa. Carefully cut the septa free of the body wall with a scalpel and pin both sides to the pan at 5- to 10-segment intervals. Cover the specimen with about 0.5–1.0 cm of tap water and examine the internal anatomy (Figure 19-3).

(b) Use your blunt probe or dissection needle to find the male reproductive organs, the light-colored **seminal vesicles.** These can be found in segments 9–12. They are composed of three sacs, within which are the *testes.* A sperm duct leads from the *seminal* vesicles to the male pore.

(c) Find two pairs of female reproductive organs—the white, spherical **seminal receptacles** in segments 9 and 10. There are ovaries in segment 13. An oviduct runs from each ovary to the female pore.

(d) Locate the organs of the digestive tract. The *buccal cavity* occupies the first three segments. The muscular **pharynx** is the swelling in the digestive tract just behind the mouth. This organ is responsible for sucking soil particles containing food into the mouth. The *esophagus* extends from the pharynx through

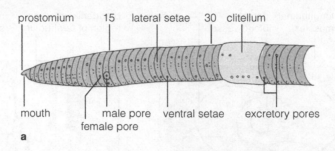

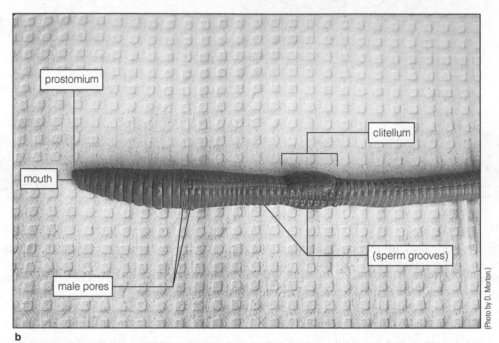

Figure 19-2 (a) External anatomy of the earthworm (numbers refer to the segment numbers). (b) External view of preserved earthworm.

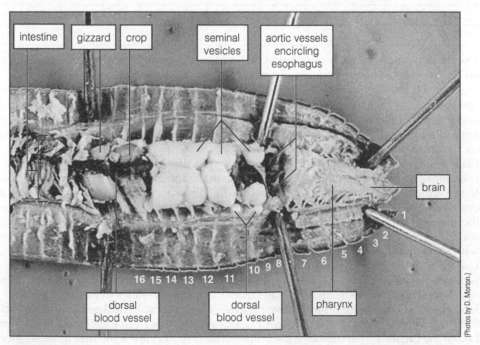

Figure 19-3 Dissection of the anterior end of an earthworm. The first 16 segments are numbered.

segment 14. The *crop* is a large, thin-walled sac in segments 15 and 16 that serves as a temporary storage organ for food. The **gizzard,** a grinding organ, is posterior to the crop. The **intestine,** the site of digestion and absorption, extends from the gizzard to the anus.

(e) Trace the parts of the circulatory system. The **dorsal blood vessel** runs along the dorsal side of the digestive tract and contains blood that flows toward the head. Five pairs of **aortic vessels** ("hearts") encircle the esophagus in segments 7–11, connecting the dorsal blood vessel with the **ventral blood vessel** (Figure 19-4). The ventral vessel runs along the ventral side of the digestive tract, carrying blood backward. The primary pumps of the circulatory system are the dorsal and ventral vessels. The dorsal vessel contains valves to prevent blood from flowing backward.

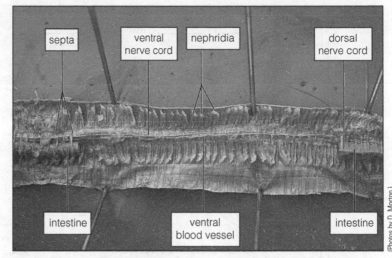

Figure 19-4 Dissection of the posterior end of an earthworm. The intestine has been partially removed to expose the ventral nerve cord.

(Photos by D. Morton.)

(f) Find the two-lobed **brain** (cerebral ganglia) on the anterior, dorsal surface of the pharynx in segment 3. The **ventral nerve cord** arises from the brain and extends along the floor of the body cavity to the last segment. Although they are difficult to locate, paired lateral nerves branch from the segmental ganglia along the nerve cord.

(g) Locate a pair of **nephridia**—coiled white tubes—at the base of every segment but the first three and last one. These structures constitute the "kidneys" of the earthworm (Figure 19-4).

(h) Dispose of your dissected specimen in the large plastic bag provided for this purpose.

5. Use a dissection microscope to examine cocoons from *Eisenia foetida,* another species of earthworm. Identify eggs and embryos.

The earthworm is monoecious—individuals have both male and female reproductive tracts. Copulation in earthworms usually takes 2–3 hours. Individuals face belly to belly in opposite directions and come together at their clitellums. These organs secrete a slimy substance that forms a band around the worms. Sperm are transferred between the participants, and the sperm are stored temporarily in the seminal receptacles. A few days after the individuals separate, each worm secretes a **cocoon.** The cocoon receives eggs and stored sperm from the seminal receptacles. The eggs are fertilized, and the cocoon moves forward as the worm backs out. The cocoon then slips off the front end of the worm and is deposited in the soil. In *Lumbricus terrestris,* a single young earthworm completes development using the other fertilized eggs as food, eventually breaks free of the cocoon, and becomes an adult in several weeks. As long as the stored sperm lasts, the earthworm continues to form cocoons.

6. Use your dissection microscope to examine a prepared slide of a cross section of the earthworm (Figure 19-5). First look at the body wall. A thin **cuticle** covers the body. The **epidermis,** which lies beneath it, secretes the cuticle, and internal to this are an **outer circular layer** and an **inner longitudinal layer** of muscle. Identify the other labeled structures in Figure 19-5.

B. Polychaetes (Class Polychaeta)

The polychaetes are abundant marine annelids with dorsoventrally flattened bodies. They are most common at shallow depths in the intertidal zone at the seashore and are prey for a variety of marine invertebrates.

Examine the polychaetes on demonstration in the lab. Although they resemble oligochaetes in many ways, notice that the polychaetes have fleshy, segmental outgrowths called **parapodia** (Figure 19-6). These structures are equipped with numerous setae, the characteristic for which they get their class name, *Polychaeta,* meaning "many bristles." Notice also the relatively complex **head.**

C. Leeches (Class Hirudinea)

The leeches are dorsoventrally flattened annelids that are usually found in fresh water.

1. Examine the preserved specimens of leeches (Figure 19-7). They lack parapodia and setae. They are predaceous or parasitic; parasitic species suck blood with the aid of a muscular pharynx. The segments

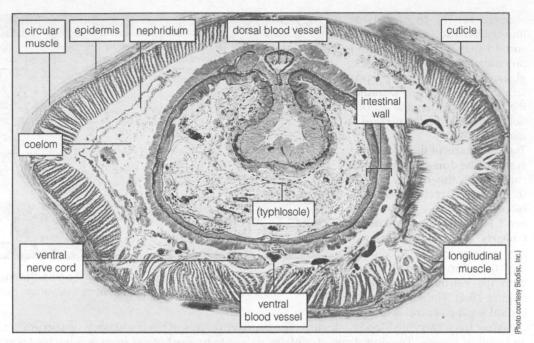

Figure 19-5 Cross section of an earthworm (20×).

(Photo courtesy Biodisc, Inc.)

Labels in Figure 19-5: circular muscle, epidermis, nephridium, dorsal blood vessel, cuticle, coelom, intestinal wall, (typhlosole), ventral nerve cord, longitudinal muscle, ventral blood vessel

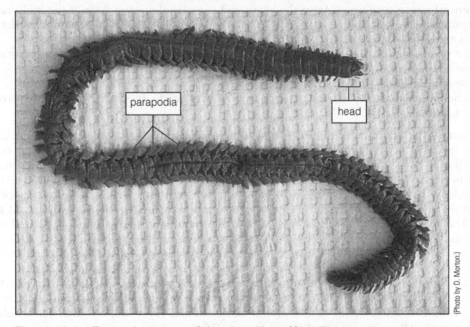

Figure 19-6 External anatomy of the clamworm, *Neanthes*.

(Photo by D. Morton.)

Labels in Figure 19-6: parapodia, head

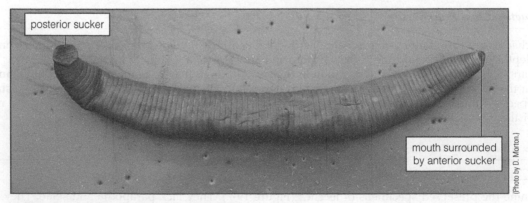

Figure 19-7 A leech (class Hirudinea).

(Photo by D. Morton.)

Labels in Figure 19-7: posterior sucker, mouth surrounded by anterior sucker

surrounding the mouth are modified into a small **anterior sucker.** The mouth is supplied with jaws, made hard with the substance chitin, used to wound the host and supply the leech with blood. Parasitic leeches secrete an anticoagulant into the wound to prevent the blood from clotting. A **posterior sucker** helps the leech attach to the host or prey.

2. Observe the movement of living freshwater leeches as they swim through the water in an aquarium. Explain their movements in terms of the contraction and relaxation of circular and longitudinal layers of muscle that comprise part of their body wall.

_____

_____

_____

_____

19.2   Mollusks (Phylum Mollusca) *(45 min.)*

Habitat: aquatic or terrestrial, mostly marine
Body Arrangement: head and unique body plan, gut (complete)
Level of Organization: organ system (tissues derived from all three primary germ layers), nervous system with brain, excretory system (nephridium or nephridia), circulatory system (open), respiratory system (most have gills or lungs), no skeletal system
Body Support: shells, floats (aquatic)
Life Cycle: sexual (monoecious or dioecious with external or internal fertilization), larvae in some
Special Characteristics: coelomate; open circulatory system; respiratory system; external body divided into head, mantle, and foot

This is the second largest animal phylum and includes snails and nudibranchs, slugs, bivalves—clams, mussels, and oysters, octopuses, cuttlefish, and squids. Bodies are typically soft (*molluscus* in Latin means "soft") and often covered by protective shells. Apart from the squids and some octopuses (and the land snails and slugs, which have adopted a terrestrial existence), *mollusks inhabit shallow marine and fresh waters*, where they crawl along or burrow into the soft substrate. Many species are valued foods for humans.

    Two characteristics distinguish the mollusks from other animals. They possess a ventral, muscular **foot** for movement and a dorsal integument, the **mantle,** which in different groups of mollusks secretes shells, functions in gaseous exchange, and facilitates movement. Even forms without shells usually have mantles.

MATERIALS

Per student:

- scalpel
- blunt probe or dissecting needle
- compound microscope, lens paper, a bottle of lens-cleaning solution (optional), a lint-free cloth (optional)

Per student pair:

- preserved freshwater clam
- dissection pan
- prepared slide of a cross section of clam gill
- dissection microscope
- glass petri dish
- prepared slide of glochidia

Per lab room:

- labeled demonstration dissection of a freshwater clam (optional) the internal anatomy of a squid
- demonstration collection of bivalve shells, and living and preserved specimens
- freshwater aquarium with snails (for example, *Physa*)
- demonstration collection of gastropod shells, and living and preserved specimens
- several preserved squids, octopuses, and chambered nautiluses
- large plastic bag for disposal of dissected specimens
- boxes of different sizes of latex gloves
- box of safety goggles

PROCEDURE

A. Bivalves (Class Bivalvia)

Bivalves are so named because they have shells composed of *two* halves or *valves*. Included in this class are the clams, mussels, oysters, and scallops. They have a hatchet-shaped (laterally flattened) body and a foot used for burrowing in soft substrates.

1. *External features of the freshwater clam.*
   (a) Obtain and examine a freshwater clam in a dissection pan (Figure 19-8). Each **valve** has a hump on the dorsal surface (back), the **umbo,** which points toward the front of the organism. The umbo is the oldest part of the valve.
   (b) Encircling the umbo are concentric rings of annual growth; the raised portions (*growth ridges*) represent periods of restricted winter growth. Determine the age of your clam by counting the number of ridges formed by the growth rings. Age of clam: _____

2. *Dissection of the freshwater clam.*
   (a) Remove the left valve by cutting the *anterior* and *posterior adductor muscles* that hold the valves together. To do this, first identify the anterior adductor muscle in Figure 19-9 and then

Figure 19-8 External view of a freshwater clam.

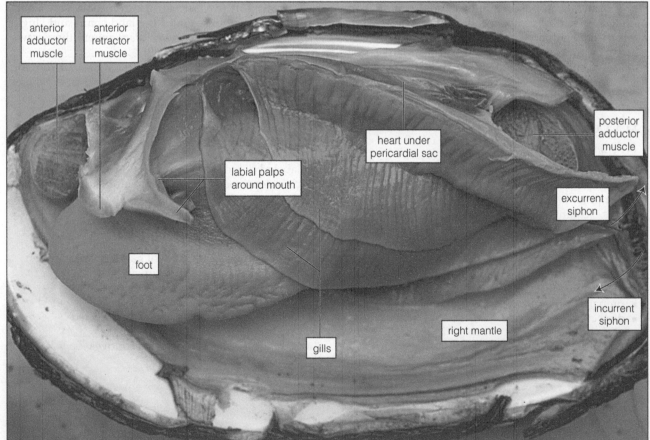

Figure 19-9 Dissection of a freshwater clam. The left valve and mantle have been removed. Arrows indicate the path of water flow into the incurrent siphon and out the excurrent siphon.

slip a scalpel between the valves at its approximate location feeling for a firm structure with the scalpel edge. Once you find it, cut back and forth. Repeat this process for the posterior adductor muscle. Once these muscles are severed, slide the scalpel between the valve to be removed and the tissues that line its inner surface. With the flat side of the scalpel, hold the tissues down while you remove the valve.

(b) The membrane that lines the inner surface of each valve is the **mantle,** which secretes the calcium carbonate shell. Pearls form as a result of irritations (usually grains of sand or small pebbles) in the mantle and sometimes are found embedded in the *pearly layer,* the inner of three layers that comprise the shell. Note that the mantle forms **incurrent** and **excurrent siphons** at the rear end.

(c) Remove the mantle on the left side by cutting it free with a scalpel. The relatively simple body plan includes a ventral foot and a dorsal visceral mass. Identify the **foot** and two sets of **gills,** one on each side of the *visceral mass.* The gills function in gaseous exchange. Water enters the incurrent siphon and flows over the gills, ultimately leaving through the excurrent siphon. Gills also trap food particles contained in incoming water; these are transported to the mouth by the cilia on the gills. Find the **labial palps,** which are situated around the mouth.

(d) The circulatory system is *open,* so the blood passes from arteries leading from the heart through body spaces called sinuses. Blood is returned to the heart by veins draining the mantle and gills. Find the **heart** in the pericardial sac between the visceral mass and the hinge between the valves.

(e) Dispose of your dissected specimen in the large plastic bag provided for this purpose.

(f) Examine the labeled dissection prepared by your instructor. Identify the **mouth, esophagus, stomach, intestine, digestive gland** (liver), **anus, nephridium** (kidney), and **gonad** (Figure 19-10).

3. Examine the demonstration slide of a transverse section of the gill with your compound microscope. Identify the structures labeled in Figure 19-11.

4. Fertilization is internal in bivalves. Sperm taken in through the female's incurrent siphon fertilize eggs in the gills. In many bivalves, *glochidia* larvae develop from the fertilized eggs and after several months leave the gill area via the excurrent siphons to parasitize fish by clamping onto the gills or fins with their jawlike valves. The fish grows tissue over the parasite, which remains attached for 2–3 months. The glochidium then breaks out of the fish to develop into an adult on the lake or river bottom. Microscopically look at glochidia in a prepared slide (Figure 19-12).

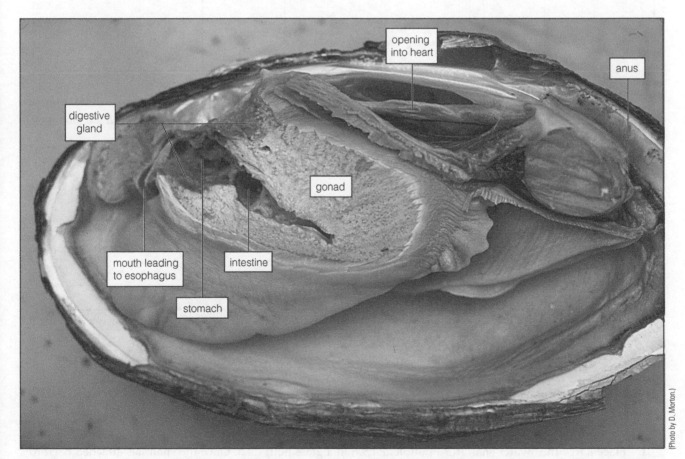

Figure 19-10 Further dissection of freshwater clam. The left gills and part of the visceral mass have been removed.

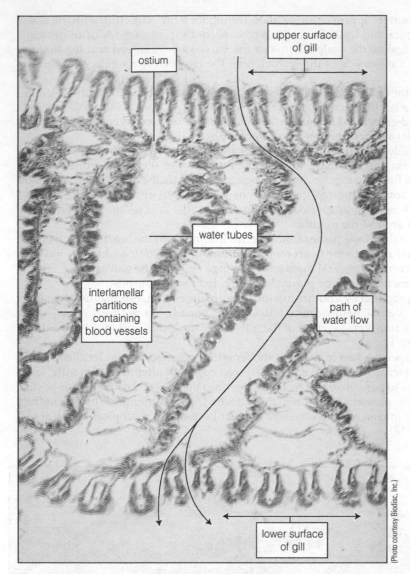

Figure 19-11 Cross section of clam gill (150×).

Labels: ostium, upper surface of gill, water tubes, interlamellar partitions containing blood vessels, path of water flow, lower surface of gill

(Photo courtesy Biodisc, Inc.)

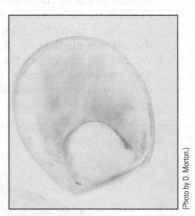

Figure 19-12 Whole mount of glochidia (250×).

(Photo by D. Morton.)

5. *Assorted bivalves*. Examine the assortment of bivalve shells and preserved specimens on demonstration. List the features they have in common.

_____

_____

_____

List some of the differences.

_____

_____

_____

How can you tell which live on sandy bottoms?

_____

B. Snails, Slugs, and Nudibranchs (Class Gastropoda)

Slugs and nudribranchs (Figure 19-13) have no shell; the former are terrestrial, the latter marine. Nudibranchs are very ornate gastropods, often brightly colored, and frequently adorned with numerous fleshy projections.

1. Place a live freshwater snail (for example, *Physa*) in a glass petri dish. Once the snail has attached itself to the glass, invert the petri dish and use the dissection microscope to observe it moving along a slimy path.

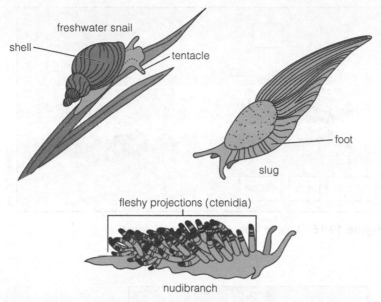

Figure 19-13 Gastropods.

A *slime gland* in the front of the foot secretes mucus. Briefly describe the muscular contractions of the foot that allow the animal to glide over this mucus. _____

_____

Looking at the foot through the glass, you can see the **mouth.** Describe its location. _____

_____

2. In snails with spirally coiled shells, the shell either spirals to the snail's right or left as it grows. Which way does your snail's shell spiral? _____

C. Squids, Octopuses, and Nautiluses (Class Cephalopoda)

Several significant modifications of the basic molluskan body plan occur in cephalopods (Figure 19-14), which are adapted for rapid swimming and catching prey. The foot of the squid has become divided into four pairs of arms and two tentacles, each with suckers, and a siphon. The tentacles, with their terminal suckers, shoot out to grasp prey. The muscular mantle draws water into its chamber when it relaxes.

Subsequent contraction of the mantle wall seals the cavity and forces water out of the siphon, resulting in jet propulsion propelling the animal through the water. Two lateral fins provide guidance. In addition, there are two large eyes that superficially resemble those of vertebrates. A relatively large brain coordinates all of these activities.

Nautiluses have shells, but shells are reduced in squids and absent in octopuses. The shell of the squid lies dorsally beneath the mantle and is called the pen.

1. Examine the external morphology of a squid. Identify the structures labeled in Figure 19-15.
2. Identify those structures indicated in Figure 19-16 on the demonstration dissection of the internal anatomy of the squid.
3. Look at the variety of cephalopod shells and preserved specimens on display in your lab room.

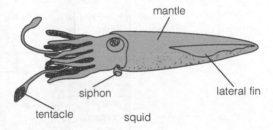

squid

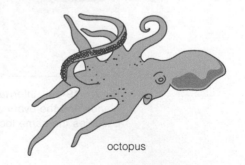

octopus

chambered nautilus

Figure 19-14 Cephalopods.

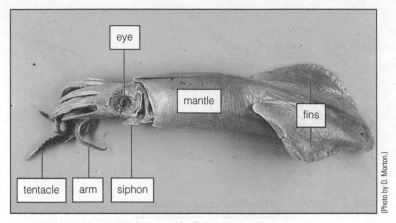

Figure 19-15 Lateral view of a preserved squid.

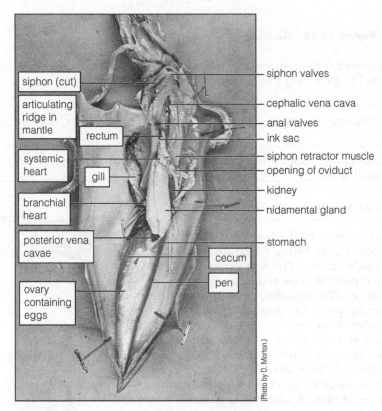

Figure 19-16 Internal anatomy of female squid. Nidamental glands and the ovary are absent in male specimens, and a penis occupies the same location as the oviduct.

Caution

Specimens are kept in preservative solutions. Use latex gloves whenever you handle a specimen. Wash any part of your body exposed to this solution with copious amounts of water. If preservative solution is splashed into your eyes, wash them with the safety eyewash bottle for 15 minutes. If you wear contact lenses during a dissection, wear safety goggles.

Habitat:	aquatic, moist terrestrial, or parasitic
Body Arrangement:	bilateral symmetry, pseudocoelomate, head, gut (complete)
Level of Organization:	organ system (tissues derived from all three primary germ layers); nervous system with brain; no circulatory, respiratory, or skeletal systems
Body Support:	hydrostatic (pseudocoelom)
Life Cycle:	sexual reproduction (mostly dioecious with internal fertilization)
Special Characteristics:	pseudocoelomate, complete gut

Nematodes or *roundworms*, like rotifers, are **pseudocoelomates,** which means they have a false body cavity or *pseudocoel*. This term is somewhat unfortunate, because there *is* a body cavity. It is not considered a true body cavity because tissues derived from mesoderm incompletely line it. Also, in an evolutionary sense it is likely derrived from the body cavity of ancestral coelomates, Unlike flatworms, roundworms have a complete gut with both a mouth and an anus. Together, the pseudocoel and the complete digestive system comprise a tube-within-a-tube arrangement, a body plan that is seen in most of the phyla yet to be discussed. Roundworms have slender, cylindrical bodies that taper at both ends. A tough, nonliving cuticle of protein covers them. This cuticle is shed all at once when roundworms molt, a characteristic they share with joint-legged animals. Molting, or ecdysis, is the periodic loss of the outer covering of the integument during growth. Unlike our epidermis, which in two to three weeks renews itself one cell at a time, the exoskeleton of joint-legged animals and the cuticle of roundworms are shed all at once. Actually, our hairs, which are grown in follicles derived from the epidermis, are likewise shed as a whole.

Parasitic roundworms include *Ascaris* (species), the hookworms, *Trichinella* worms of mammals (Figure 19-17), pinworms of humans, and the filarial worms of humans that cause elephantiasis by obstructing lymphatic vessels.

MATERIALS

Per student:

- dissection microscope
- dissecting needle

Per student pair:

- finger bowl

Per lab room:

- collection of free-living roundworms
- tray of moist soil
- container of fresh stream water
- demonstration dissections of
    a male *Ascaris*
    a female *Ascaris*
- demonstration microscope of a cross section of
    a male *Ascaris*
    a female *Ascaris*
- boxes of different sizes of latex gloves

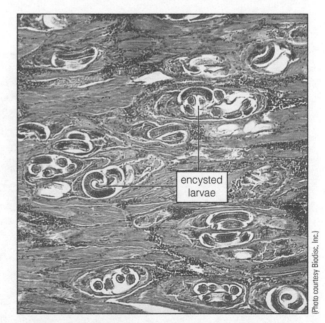

(Photo courtesy Biodisc, Inc.)

Figure 19-17 *Trichinella spiralis* larvae encysted in muscle (46×). This organism causes the disease trichinosis, which is usually contracted by eating undercooked pork or wild game.

PROCEDURE

1. Examine the demonstration collection of free-living roundworms.
2. Place some moist soil in a finger bowl with some fresh stream water. Using a dissection microscope, look for translucent roundworms in the soil sample (Figure 19-18).

    Describe the behavior of any roundworms you find and make a note of any visible anatomical features they possess.

_____

_____

_____

_____

_____

_____

_____

_____

B. *Ascaris*

Ascaris is a common parasite of humans, pigs, horses, and other mammals. Adults range from 15 to 40 cm in length, with males being smaller and having a hook shape to the posterior part of the body.

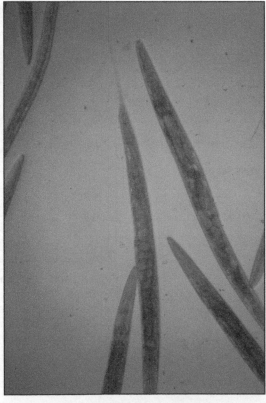

(Photo by W. A. Yoder.)

Figure 19-18 Free-living nematodes, *Rhabditella* females, w.m. (350×).

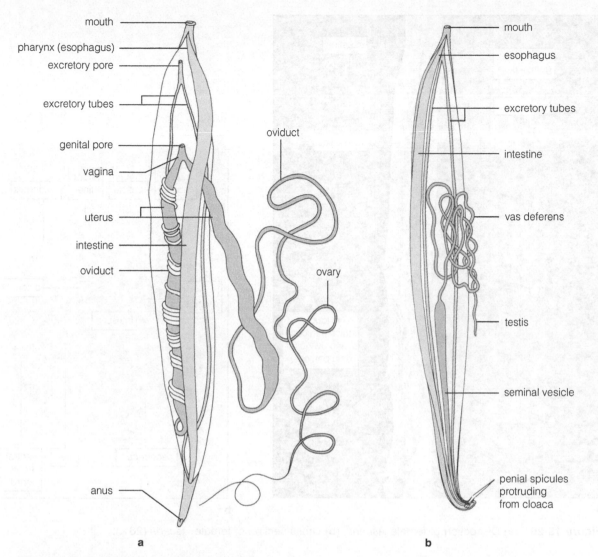

Figure 19-19 *Ascaris.* (**a**) Female. (**b**) Male.

1. Examine a female *Ascaris* (Figure 19-19a). At the front end, the triangular *mouth* is surrounded by three lips. The opening on the ventral surface near the other end of the body is the *anus*. A *genital pore* is located about one-third the distance from the mouth.
2. Look at the demonstration dissection and demonstration cross section of a female *Ascaris* and identify the structures labeled in Figures 19-20a and b. The paired **lateral lines** each contain an *excretory tube*. These tubes open to the outside through a single **excretory pore** located near the mouth on the ventral surface. The **ovaries** are the threadlike ends of a Y-shaped reproductive tract. The ovary leads to a larger **oviduct** and a still larger **uterus.** The two uteri converge to form a short, muscular **vagina** that opens to the outside through the **genital pore.**
3. Examine a male *Ascaris* (Figure 19-19b). The lips and mouth are the same as the female's. However, the male has a *cloaca*—an exit for both the reproductive and digestive systems—instead of an anus. A pair of *penial spicules* protrudes from this opening.

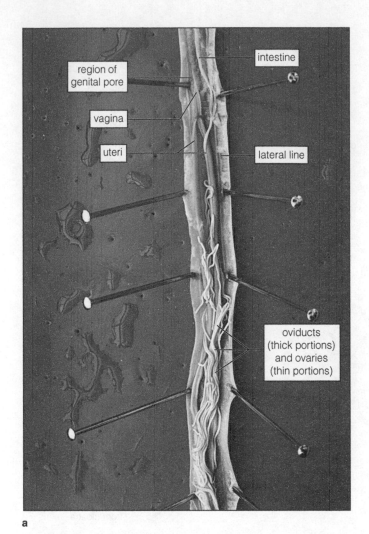

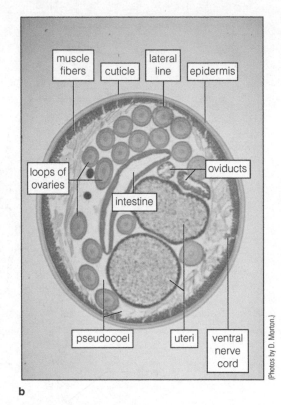

(Photos by D. Morton.)

a b

Figure 19-20 (a) Dissection of female *Ascaris*. (b) Cross section of female *Ascaris* (26×).

4. Look at the demonstration dissection and demonstration cross section of a male *Ascaris* and identify the structures labeled in Figures 19-21a and b. The reproductive system superficially resembles that of the female. It consists of a single tubular structure that includes a threadlike **testis** joined to a larger **vas deferens,** an even larger **seminal vesicle,** and terminally a short, muscular ejaculatory duct that empties into the cloaca.
5. After copulation, the eggs are fertilized in the female oviduct. The eggs, surrounded by a thick shell, are expelled through the genital pore to the outside; as many as 200,000 may be shed into the host's intestinal tract per day. Examine a prepared slide with whole *Ascaris* eggs and draw what you see in Figure 19-22.

The eggs are eliminated in the host's feces. They are very resistant and can live for years under adverse conditions. Infection with *Ascaris* results from ingesting eggs in food or water contaminated by feces or soil. The eggs pass through the stomach and hatch in the small intestine. The small larvae migrate through the bloodstream, exit the circulatory system into the respiratory system, and crawl up the respiratory tubes and out the trachea into the throat. There they are swallowed to reinfect the intestinal tract and mature. Sometimes larvae exit the nasal passages rather than being swallowed. The entire journey of the larva from intestine back to intestine takes about 10 days.

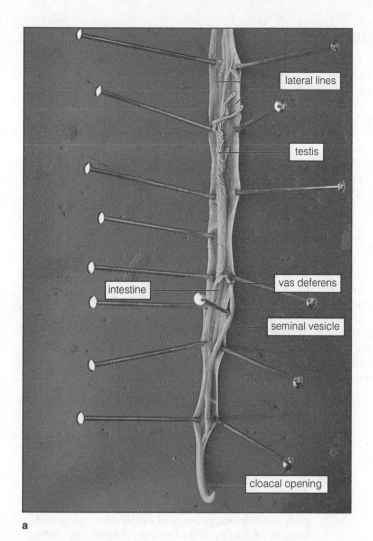

lateral lines

testis

intestine

vas deferens

seminal vesicle

cloacal opening

a

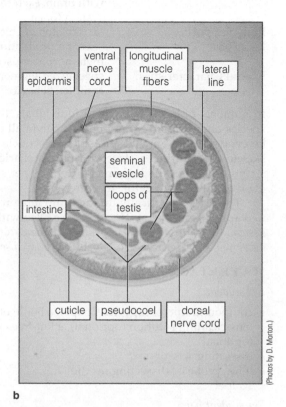

ventral
nerve
cord

longitudinal
muscle
fibers

lateral
line

epidermis

seminal
vesicle

loops of
testis

intestine

cuticle

pseudocoel

dorsal
nerve cord

(Photos by D. Morton.)

b

Figure 19-21 (a) Dissection of male *Ascaris*. (b) Cross section of male *Ascaris* (35×).

Figure 19-22 Drawing of the *Ascaris* eggs
(_____×).

Habitat:	aquatic, terrestrial (land and air), some parasitic
Body Arrangement:	bilateral symmetry, coelomate, body regions (head, thorax, and abdomen), segmented (specialized segments form mouthparts and jointed appendages), gut (complete)
Level of Organization:	organ system (tissues derived from all three primary germ layers), nervous system with brain, excretory system (green glands or Malphigian tubules), circulatory system (open), respiratory system (gills, trachae, or book lungs)
Body Support:	exoskeleton of chitin
Life Cycle:	sexual reproduction (mostly dioecious with internal fertilization), larvae, metamorphosis, some parthenogenesis
Special Characteristics:	specialized segments, exoskeleton of chitin, metamorphosis, flight in some, complex behavior

In numbers of both species and individuals, this is the largest animal phylum. There are more than 1 million species of arthropods, more in fact than in all of the other animal phyla combined. Arthropods are believed to be evolved from annelid ancestors, the cuticle becoming thicker and harder to form an **exoskeleton** (external skeleton) of **chitin**—a structural polysaccharide. Like the annelids, the bodies of arthropods are segmented. Arthropod ancestors had one pair of legs per segment. In modern arthropods, some of these appendages have evolved into other useful structures.

The body is typically divided into three regions: the *head, thorax,* and *abdomen.* The mouthparts are modified appendages (legs). The coelom is filled with the blood, or **hemolymph,** and connected to an *open circulatory system.* In addition to the exoskeleton, the evolutionary success of arthropods is due in large part to their *highly developed central nervous system, well-developed sense organs,* and *complex behavior.*

MATERIALS

Per student:

- compound microscope, lens paper, a bottle of lens-cleaning solution (optional), a lint-free cloth (optional)
- dissecting scissors
- blunt probe or dissecting needle
- scalpel

Per student pair:

- dissection microscope
- preserved crayfish
- dissection pan
- grasshopper (*Romalea*)

- prepared slide of a whole mount of grasshopper mouthparts

Per lab room:

- demonstration collections of living and preserved and mounted crustaceans living and preserved and mounted insects living and preserved centipedes living and preserved millipedes living and preserved and mounted spiders, horseshoe crabs, scorpions, ticks, and mites
- large plastic bag for disposal of dissected specimens
- boxes of different sizes of latex gloves

PROCEDURE

A. Crustaceans (Subphylum Crustacea)

Some of our most prized seafoods are crustaceans. This group contains shrimps, lobsters, crabs, and crayfish—and animals of considerably less culinary interest, including water fleas, sand fleas, isopods (sow or pill bugs), ostracods, copepods, and barnacles. Most species are marine, but some live in fresh water. Most isopods occupy moist areas on land. For the most part, crustaceans are carnivorous, scavenging, or parasitic.

Crustaceans have two pairs of antennae on the head, and their exoskeleton is hardened on top and sometimes on the sides to form a protective **carapace.**

1. Examine the demonstration specimens of crustaceans on display in the lab.
2. *External features of a crayfish.*
   (a) Obtain a crayfish, a fairly typical crustacean, and rinse it in water at the sink. Place the rinsed crayfish in a dissection pan.
   (b) Recognize the two major body regions (Figure 19-23): an anterior **cephalothorax,** composed of a fused head and thorax, and a posterior *abdomen.* Note the **carapace.** The portion of the carapace that extends between the eyes is the **rostrum.**
   (c) Identify the pair of large **compound eyes** (comprised of many closely packed photosensory units) each one atop a movable stalk, a pair of **antenna,** and two pairs of smaller and shorter **antennules.**
   (d) Find the **mouth,** which is comprised of three pairs of mouthparts—one pair of **mandibles** and two pairs of **maxillae.** Three pairs of **maxillapeds** (Figure 19-24) surround the mouthparts.

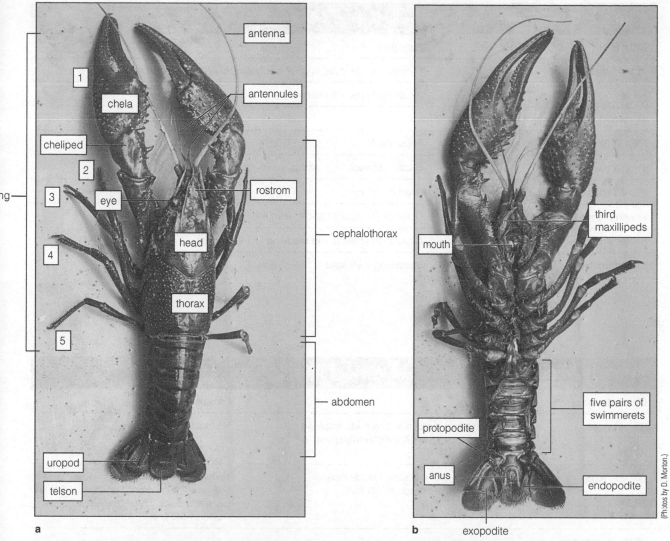

a

b

(Photos by D. Morton.)

Figure 19-23 (a) Dorsal and (b) ventral views of the external anatomy of the crayfish. The five walking legs are numbered in (a).

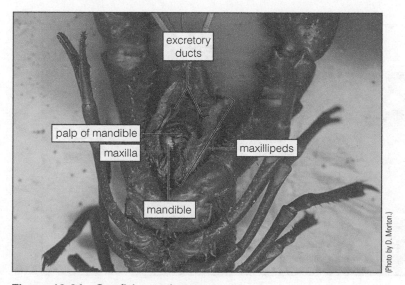

(Photo by D. Morton.)

Figure 19-24 Crayfish mouthparts.

TABLE 19-2 Function of Crayfish Legs and Modified Appendages

Structure	Function
Antennule	Contains a statocyst, an organ of balance, in the basal segment
Antenna	Vision and chemoreception
Mandible	Jaw
Maxilla (first)	Food handling
Maxilla (second)	Creates current for gaseous exchange in the gills
Maxillipeds (first, second, and third)	Food handling
Walking legs (1–5)	Defense (chelipeds) and movement
Swimmerets (1–5)	Copulation (males) and brooding young (female)
Uropod (combines with telson and other uropod to form tail fan)	Swimming backward

TABLE 19-3 Sexual Dimorphisms in Crayfish (Figure 19-25)

Characteristic	Male	Female
Swimmerets	Front two pairs of swimmerets enlarged for copulation and transferring sperm to the female	No modifications
Walking legs	Openings of the *sperm ducts* (sperm ducts) at the bases of 5th pair	*Seminal receptacle* in the middle between the bases of the 4th and 5th pairs and discs covering the oviducts at bases of 3rd pair
Width of abdomen (broad or narrow)		

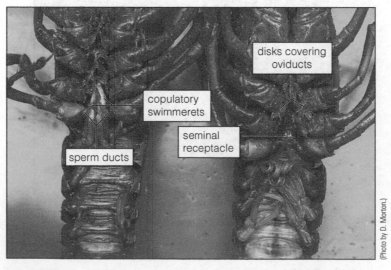

Figure 19-25 Ventral views of (left) a male, (right) a female

(e) Locate the five pairs of large **walking legs;** the first, called the **chelipeds,** bear large pincers called **chelae.** Continuing toward the tail, count the five pairs of abdominal **swimmerets.** Near the end of the abdomen, locate the **uropods,** a pair of large, flattened lateral appendages located on both sides of the *telson,* which is an extension of the last abdominal segment. The **anus** opens onto the ventral surface of the telson.

(f) Table 19-2 summarizes in order from front to back down one side of the body those structures that are legs or modified appendages. Review them in order and familiarize yourself with their functions.

(g) Crayfish exhibit **sexual dimorphism** (morphological differences between male and female). Look at a number of specimens of both sexes, observe any differences in the characteristics listed in Table 19-3 (and shown in Figure 19-25), and fill in any blanks.

3. *Crayfish dissection.*

(a) Locate the **gills** within the branchial chambers by carefully cutting away the lateral flaps of the carapace with your scissors (Figure 19-26). The gills are feathery structures containing blood channels that function in gaseous exchange.

(b) With scissors, superficially cut forward from the rear of the carapace to just behind the eyes. With your scalpel, carefully separate the hard carapace from the thin, soft, underlying tissue. Next, remove the gills to reveal the internal organs (Figure 19-27).

Figure 19-26 Internal anatomy of the crayfish with side of carapace removed.

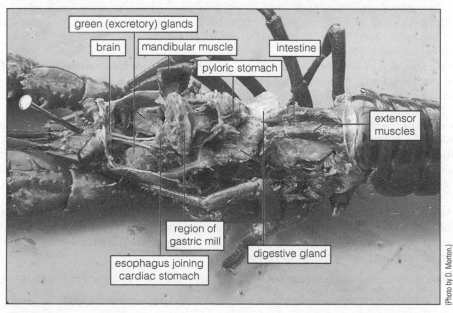

Figure 19-27 Internal anatomy of the crayfish with carapace, gills, and heart removed.

(c) Note the two longitudinal bands of *extensor muscles* that run dorsally through the thorax and abdomen. In the abdomen, find the large *flexor muscles* lying ventrally below the extensor muscles and intestine. What is the function of these two sets of muscles?

_____

_____

_____

(d) Locate the small, membranous heart or its location just posterior to the stomach. Remember, the circulatory system is open; from the arteries, blood flows into open spaces, or *sinuses*, before returning to the heart through openings in the wall of this organ.

(e) Identify the organs of the upper digestive tract. The mouth leads to the tubular **esophagus,** which empties into the **cardiac stomach.** At the border between the cardiac stomach and the **pyloric stomach** is the **gastric mill,** a grinding apparatus composed of three chitinous teeth. What role in digestion do you think this structure plays?

_____

_____

_____

(f) Note the large **digestive gland.** This organ secretes digestive enzymes into the cardiac stomach and takes up nutrients from the pyloric stomach. Absorption of nutrients from the tract continues in the intestine, which runs from the abdomen to the *anus.*

(g) Locate the **green glands,** the excretory organs situated ventrally in the head region near the base of the antennae. A duct leads to the outside from each green gland.

(h) The gonads lie beneath the heart. They will be obvious if your specimen was obtained during the reproductive season. If not, they will be small and difficult to locate. In a female, try to find the *ovaries* just beneath the heart. In the male, two white *testes* will occupy a similar location.

(i) Remove the organs in the cephalothorax to expose the *ventral nerve cord.* Observe the segmental ganglia and their paired lateral nerves. Trace the nerve cord forward to locate the **brain.** Note the nerves leading from the brain to the eyes, antennules, antennae, and mouthparts. Dispose of your dissected specimen in the large plastic bag provided for this purpose.

B. Insects (Subphylum Hexapoda)

Of the more than 1 million species of arthropods, more than 850,000 of them are insects. These animals occupy virtually every kind of terrestrial and freshwater habitat. This extraordinary proliferation of species is almost certainly due to the evolution of wings. Insects were the first organisms to fly and thus were able to exploit a variety of opportunities not available to other animals.

1. Examine insects on demonstration in the lab. Groups of insects you are probably familiar with—dragon flies and damsel flies; grasshoppers, crickets, and katydids; cockroaches; "praying" mantids; termites; cicadas, aphids, leafhoppers, and spittlebugs; "true" bugs; beetles; "true" flies; fleas; butterflies and moths; and wasps, bees, and ants—each belong to one of the orders of insects.

2. *External features of a typical insect.*
    (a) Obtain a preserved grasshopper of the genus *Romalea* and place it in a dissection pan (Figure 19-28).
    (b) Observe that the body consists of a **head, thorax,** and **abdomen.** The thorax is composed of three segments, each of which bears a pair of legs. The middle and back segments also each bear a pair of wings. The abdomen, unlike the situation in crustaceans, has no appendages.
    (c) On the sides of the thorax and abdomen find the **spiracles,** tiny openings into the **tracheae,** or breathing tubes, which course through the body and function in gaseous exchange.
    (d) Note that the **head** of the grasshopper contains a pair of **compound eyes** between which are three simple eyes, the **ocelli,** light-sensitive organs that do not form images. A single pair of antennae distinguishes the insects from the crustaceans (which have two) and the arachnids (which have none).
    (e) Locate the feeding appendages (Figure 19-29)—a pair of **mandibles** and two pairs of **maxillae,** the second pair fused together to form the lower lip, the **labium.** The upper lip, the **labrum,** covers the mandibles. Near the base of the labium is a tonguelike process called the **hypopharynx** (Figure 19-30).

3. Look at the demonstration slide of a whole mount of grasshopper mouthparts. Locate and label a mandible, a maxilla, the labium, and the labrum in Figure 19-30.

Mouthparts are extremely variable in insects, with modifications for chewing, piercing, siphoning, and sponging (Figure 19-31).

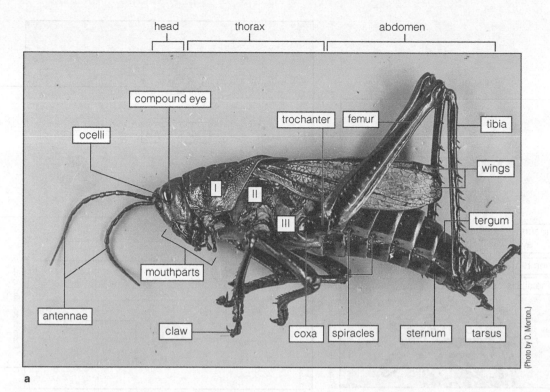

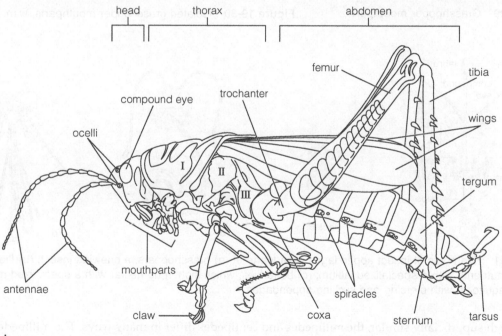

Figure 19-28 Photograph (**a**) and diagram (**b**) of the external anatomy and body form of a grasshopper, a typical insect.

C. Centipedes and Millipedes (Subphylum Myriapoda)

Centipedes are swift, predaceous arthropods adapted for running. The body is flattened with one pair of legs per segment, except for the head and the rear two segments (Figure 19-32a). The first segment of the body bears a pair of legs modified as *poison claws* for seizing and killing prey. In some tropical species (which can get as long as 20 cm), the poison can be dangerous to humans. Centipedes live under stones or the bark of logs. Their prey consists of other arthropods, worms, and mollusks. The common house centipede eats roaches, bedbugs, and other insects.

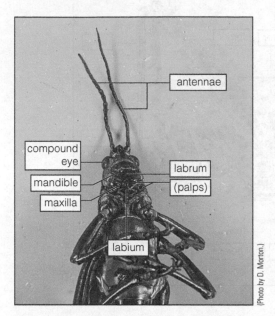

Figure 19-29 Grasshopper mouthparts.

(Photo by D. Morton.)

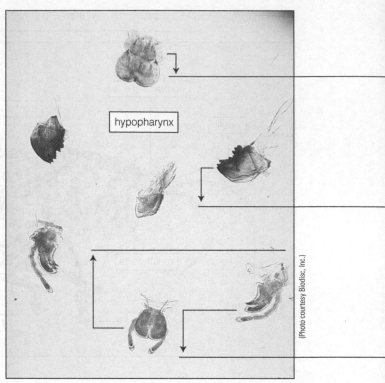

hypopharynx

(Photo courtesy Biodisc, Inc.)

Figure 19-30 Isolated grasshopper mouthparts, w.m. (7×).

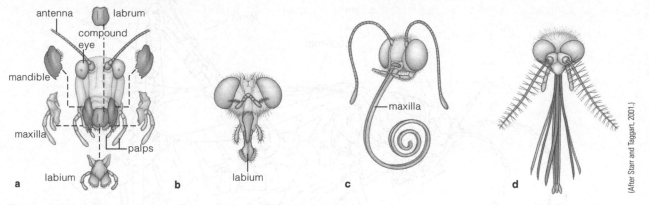

a
antenna   labrum
compound
eye
mandible
maxilla
palps
labium

b
labium

c
maxilla

d

(After Starr and Taggart, 2001.)

Figure 19-31 Examples of insect appendages. Headparts of (**a**) grasshoppers, a chewing insect; (**b**) flies, which sponge up nutrients with a specialized labium; (**c**) butterflies, which siphon up nectar with a specialized maxilla; and (**d**) mosquitoes, with piercing and sucking appendages.

Although superficially similar, the millipedes and centipedes differ in many ways. The millipedes typically have cylindrical bodies. All segments, except those of the short thorax, bear two pairs of legs. There are no poison claws (Figure 19-32b). Millipedes are very slow arthropods that live in dark, moist places. They scavenge on decaying organic matter. When handled or disturbed, they often curl up into a ball and secrete a noxious fluid from their scent glands as a means of defense.

1. Examine the centipedes on demonstration in the lab.
2. Similarly observe the millipedes on display in the lab.

D. Subphylum Chelicerata: Spiders, Horseshoe Crabs, Scorpions, Ticks, and Mites

Today most species are terrestrial, but the first chelicerates were marine. Horseshoe crabs represent these ancient marine forms. Their bodies and five pairs of walking legs are concealed under a tough carapace (Figure 19-33a).

The body of spiders and scorpions consists of a *cephalothorax* and *abdomen*. In ticks and mites, these parts are fused. The cephalothorax bears six pairs of appendages, the rear four of which are walking legs (Figure 19-33b).

Figure 19-32 (a) Live centipede. (b) Live millipede.

Figure 19-33 (a) Ventral view of a horseshoe crab. (b) Live spider. *Continues*

How many pairs of walking legs do insects have? _____

Chelicerates have no true mandibles or antennae, and their eyes are simple. In spiders, the first pair of appendages bear fangs. Each fang has a duct that connects to a poison gland. The second pair of appendages is used to chew and squeeze their prey. This second pair is also sensory, and males use them to store and transfer sperm to females. Most, but not all, spiders spin webs to catch their prey. They secrete digestive enzymes into the prey's (usually an insect's) body, and the liquefied remains are sucked up. Other chelicerates have front appendages specialized to their particular method of feeding. Ticks suck the blood of vertebrates and most mites are scavengers—for example, dust mites eat the shed skin cells that are a major component of house dust.

Examine the assortment of chelicerates on display in the lab.

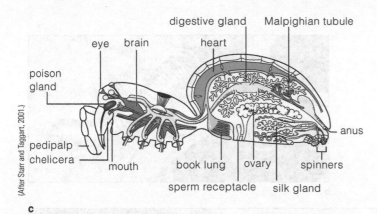

(After Starr and Taggart, 2001.)

c

Figure 19-33 (c) Internal anatomy of a spider. *Continued*

PRE-LAB QUESTIONS

_____ 1. The protostomes are animals whose
  (a) stomach is in front of the crop
  (b) mouth is covered by a fleshy lip
  (c) blastopore becomes a mouth
  (d) digestive tract is lined by mesoderm

_____ 2. The deuterostomes are animals whose
  (a) stomach is in front of the crop
  (b) mouth is covered by a fleshy lip
  (c) blastopore becomes the anus
  (d) digestive tract is lined by mesoderm

_____ 3. Further evolution of the protostomes
changed the coelom by
  (a) losing it
  (b) reducing it
  (c) it becoming a pseudocoel
  (d) all of the above

_____ 4. Earthworms exhibit segmentation, defined as the
  (a) division of the body into a series of
      similar segments
  (b) presence of a true coelom
  (c) difference in size of the male and
      female
  (d) presence of a "head" equipped with
      sensory organs

_____ 5. The copulatory organ of the earthworm is
the
  (a) penis
  (b) clitellum
  (c) gonopodium
  (d) vestibule

_____ 6. Leeches belong to the phylum
  (a) Arthropoda
  (b) Mollusca
  (c) Annelida
  (d) Cnidaria

_____ 7. One of the two major distinguishing
characteristics of mollusks is
  (a) the presence of three body regions
  (b) the mantle
  (c) segmentation of the body
  (d) jointed appendages

_____ 8. Functions of the mantle include
  (a) secreting the shell
  (b) forming peals around irritating
      grains of sand
  (c) forming the incurrent and excurrent
      siphons
  (d) all of the above

_____ 9. Which of the following is a gastropod?
  (a) clam
  (b) snail
  (c) squid
  (d) octopus

_____ 10. Which of the following is a cephalopod?
  (a) clam
  (b) snail
  (c) slug
  (d) octopus

_____ 11. Structures that are shed in one piece include
(a) the cuticle of roundworms
(b) the exoskeleton of arthropods
(c) human hair
(d) all of the above

_____ 12. Nematodes and rotifers have a body cavity that is
(a) called a coelom
(b) called a pseudocoelom
(c) completely lined by tissues derived from mesoderm
(d) both a and c

_____ 13. Filarial roundworms cause elephantiasis by
(a) encysting in muscle tissue
(b) promoting the growth of fatty tumors
(c) obstructing lymphatic vessels
(d) laying numerous eggs in the joints of their hosts

_____ 14. Digestive wastes of the male _Ascaris_ roundworm exit the body through the
(a) anus
(b) cloaca
(c) mouth
(d) excretory pore

_____ 15. The animal phylum with the most species is
(a) Mollusca
(b) Annelida
(c) Arthropoda
(d) Platyhelminthes

_____ 16. The body of arthropods includes
(a) a head
(b) a thorax
(c) an abdomen
(d) all of the above

_____ 17. Sexual dimorphism, as seen in the crayfish, is
(a) the presence of male and female individuals
(b) the production of eggs and sperm by the same individual
(c) another term for copulation
(d) the presence of observable differences between males and females

_____ 18. The grinding apparatus of the digestive system of the crayfish is/are the
(a) oral teeth
(b) gizzard
(c) pharyngeal jaw
(d) gastric mill

_____ 19. The insects were the first organisms to
(a) show bilateral symmetry
(b) exhibit segmentation
(c) fly
(d) develop lungs

_____ 20. Like the insects, the arachnids (spiders and so on) have
(a) three pairs of walking legs
(b) one pair of antennae
(c) true mandibles
(d) an abdomen

EXERCISE 19

Segmented Worms, Mollusks, Roundworms, and Joint-Legged Animals

POST-LAB QUESTIONS

Introduction

1. Indicate the differences between protostomes and deuterostomes.

2. List the animal phyla that are protostomes and those that are deuterostomes.

3. What happens to the coelom as each phyla of protostomes evolve?

4. Indicate the differences between protosomes and deuterosomes, and list the phyla of animals in each group.

5. Describe ecdysis in roundworms and joint-legged animals.

19.1 Annelids (Phylum Annelida)

6. Briefly describe the life cycle of an earthworm.

7. Label this diagram of an annelid and identify its class (Oligochaeta, Polychaeta, or Hirudinea).

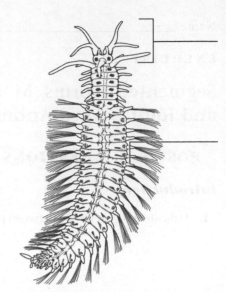

19.2 Mollusks (Phylum Mollusca)

8. Explain what causes the growth ridges found on the shells of bivalves.

9. List some structures that distinguish bivalves from the other molluscan classes.

10. What characteristics distinguish slugs from nudibranchs?

19.3 Roundworms (Phylum Nematoda)

11. *Pseudocoel* means "false body cavity." Why is this term applied to the body cavity of roundworms?

19.4 Joint-Legged Animals (Phylum Arthropoda)

12. Define *sexual dimorphism*, describe it in the crayfish, and list two other animals that exhibit this phenomenon.

13. Identify and label the structures noted in this diagram.

a. _____   e. _____   h. _____

b. _____   f. _____   i. _____

c. _____   g. _____   j. _____

d. _____

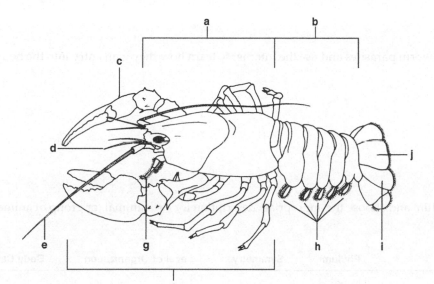

14. How has flight been at least partly responsible for the success of the insects?

15. List several differences between insects and spiders.

Food for Thought

16. Is it possible that some of the monsters reported by ancient mariners were giant squid or octopi? Search the Internet before giving and explaining your answer.

17. Define *segmentation* and indicate how the annelids and your body exhibit this phenomenon.

18. List the roundworm parasites and use the Internet to learn how they gain entry into the body. Describe that process here.

19. Write the phylum and choose the appropriate description for each animal (or group of animals) listed in the following table.

Animal	Phylum	Symmetry	Level of Organization	Body Cavity	Gut
Sponge	None				
Hydra and jellyfish	None				
Flatworm					
Rotifer					
Segmented worm					
Clam, snail, and octopus					
Roundworm					
Crayfish, insect, and spider					

20. Define *segmentation* and indicate how the annelids, arthropods, and your body exhibit this phenomenon.

Echinoderms, Invertebrate Chordates, and Vertebrates

After completing this exercise, you will be able to

1. define *dermal endoskeleton, water vascular system, notochord, pharyngeal gill slits, dorsal hollow nerve cord, postanal tail, invertebrate, vertebrate, vertebral column, cranium, vertebrae, cloaca, lateral line, placoid scale, operculum, atrium, ventricle, artery, vein, ectothermic, endothermic,* and *viviparous;*

2. describe the natural history of members of phylum Echinodermata and the invertebrate members of phylum Chordata;

3. identify some representatives of the echinoderm classes (sea stars, brittle stars, sea urchins, and sea cucumbers);

4. compare and contrast the invertebrate chordates (subphyla Urochordata and Cephalochordata);

5. identify structures (and indicate associated functions) of the representatives of these phyla and subphyla;

6. describe the basic characteristics of members of the subphylum Vertebrata;

7. identify representatives of the vertebrate classes, Cephalaspidomorphi, Chondrichthyes, Osteichthyes, Amphibia, Reptilia, Aves, and Mammalia;

8. identify structures (and indicate associated functions) of representatives of the vertebrate classes.

INTRODUCTION

The deuterostome (animals whose blastopore—first opening of developing embryo—becomes an anus) branch of coelomates leads to echinoderms and chordates, the phylum to which we belong. Although the adult forms of these two phyla look quite different, chordates are thought to have evolved from the bilaterally symmetrical larvae of ancestral echinoderms some 600 million years ago.

There is an additional small phylum, the Hemichordata, whose members (acorn worms) have both echinoderm and chordate characteristics. Acorn worms are *long, slender, wormlike* animals (Figure 20-1). Most *live in shallow seawater.* They *obtain food by burrowing in mud and sand or filter feeding.* Evidence of their former activity can be seen on exposed tidal flats where mud or sand that has passed through their bodies is left in numerous coiled, ropelike piles.

In terms of the numbers of individuals and species, chordates do not constitute a large phylum; however, chordates known as vertebrates have had a disproportionately large ecological impact. The **vertebrates** have the *four basic characteristics of chordates* listed in the previous exercise, plus a cranium and a **vertebral column.** The dorsal hollow nerve cord has differentiated into a *brain* and a *spinal cord.* Bones protect both of these structures, the brain by the bones of the **cranium** (braincase) and the spinal cord by the **vertebrae,** the bones that make up the vertebral column.

Like echinoderm larvae and sea squirt larvae and lancelets, vertebrates are *cephalized, bilaterally symmetrical,* and *segmented.* In adults, segmentation is most easily seen in the musculature, vertebrae, and ribs. The body is typically divided into a *head, neck, trunk,* and *tail.* If appendages are present, they are paired, lateral *thoracic appendages* (pectoral fins, forelimbs, wings, and arms) and *pelvic appendages* (pelvic fins, hindlimbs, and legs), which are linked to the vertebral column and function to support and to help move the body. Seven classes of vertebrates have representatives alive today (lampreys, cartilaginous fishes, bony fishes, amphibians, reptiles, birds, and mammals), and one class is completely extinct (armored fishes).

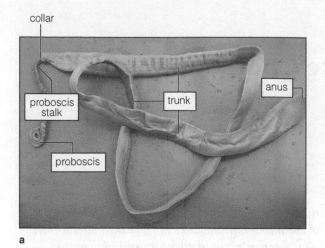

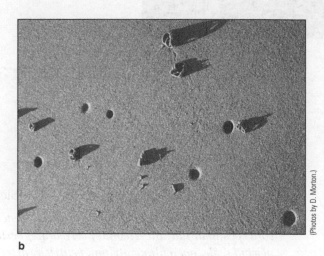

(Photos by D. Morton.)

a   b

Figure 20-1 (a) Preserved acorn worm. (b) Ropelike piles of sand formed by burrowing acorn worms.

20.1	Echinoderms (Phylum Echinodermata) *(45 min.)*

Habitat: aquatic, marine
Body Arrangement: radial symmetry (larvae are bilaterally symmetrical), coelomate,
                                  no well-defined head, no segmentation, gut (usually complete)
Level of Organization: organ system (tissues derived from all three primary germ layers),
                                  circular nervous system with radial nerves, reduced circulatory system,
                                  no respiratory and excretory system, dermal endoskeleton (mesoderm)
Body Support: endoskeleton of calcium carbonate
Life Cycle: sexual reproduction (mostly dioecious with external fertilization),
                                  larvae, metamorphosis, high regenerative potential
Special Characteristics: radial symmetry, no segmentation, water-vascular system, dermal
                                  endoskeleton

Caution

Specimens are kept in preservative solutions. Use latex gloves whenever you handle a specimen. Wash any part of your body exposed to this solution with copious amounts of water. If preservative solution is splashed into your eyes, wash them with the safety eyewash bottle for 15 minutes. If you wear contact lenses during a dissection, wear safety goggles.

Echinoderm means "spiny skin." These are marine animals; they live on the bottom of both shallow and deep seas. Their feeding methods range from trapping organic particles and plankton (for example, sea lilies and feather stars) to scavenging (sea urchins) and predatory behavior (sea stars).

The echinoderms exhibit five-part *radial symmetry* and a calcareous (contains calcium carbonate) **dermal endoskeleton** (internal skeleton) composed of many small plates. Much of the *coelom* is taken up by a unique *water-vascular system*, which functions in movement, attachment, respiration, excretion, food handling, and sensory perception.

MATERIALS

Per student:

- dissecting scissors
- blunt probe or dissecting needle

Per student pair:

- dissection microscope
- preserved sea star
- dissection pan

- dissection pins
- prepared slide of
  a cross section of a sea star arm (ray)
  a whole mount of bipinnaria larvae
- compound microscope, lens paper,
  a bottle of lens-cleaning solution (optional),
  and a lint-free cloth (optional)

Per lab room:

- collections of
  preserved or mounted brittle stars
  sea urchins, sea biscuits, and sand dollars
  (preserved specimens and skeletons)
  preserved sea cucumbers
- live specimens (optional)

- collection of preserved or mounted feather stars and sea lilies
- large plastic bag for disposal of dissected specimens
- boxes of different sizes of latex or vinyl gloves
- box of safety goggles

A. Sea Stars (Class Asteroidea)

The sea stars are familiar occupants of the sea along shores and coral reefs. They are slow-moving animals that sometimes gather in large numbers on bare areas of rock. Many are brightly colored. Some sea stars are more than 1 m in diameter.

PROCEDURE

1. Obtain a preserved specimen of a sea star and keep your specimen moist with water in a dissection pan. Note the **central disk** on the *aboral* side—the side without the mouth; the **five arms;** and the **sieve plate** (madreporite), a light-colored calcareous sieve near the edge of the disk between two arms (Figure 20-2a). The pores in the sieve plate open into the water-vascular system.

2. With a dissection microscope, note the many protective **spines** scattered over the surface of the body. Near the base of the spines are many small pincerlike structures, the **pedicellariae** (Figure 20-2b). The *valves* of these structures grasp objects that land on the surface of the body, such as potential parasites. Also among the spines are many soft, hollow **skin gills** (dermal branchiae) that communicate directly with the coelom and function in the exchange of gases and excretion of ammonia (a nitrogen-containing metabolic waste).

3. Locate the **mouth** on the oral side (Figure 20-3a). Note that an **ambulacral groove** extends from the mouth down the middle of the oral side of each arm. Numerous **tube feet** extend from the water-vascular system and occupy this groove. Each tube foot consists of a bulblike structure attached to a **sucker** (Figure 20-3b). The animal moves by alternating the suction and release mechanisms of the tube feet. Tube feet are also used to adhere to the shells of bivalves and to pry them open to get at their soft insides. The tube feet, along with the skin gills, also function in the exchange of gases and excretion of ammonia.

4. With your dissecting scissors, cut across the top of an arm about 1 cm from the tip. Next, cut out a rectangle of spiny skin by carefully cutting along each side of the arm to the central disk and then across the top of the arm at the edge of the disk. Observe the hard, calcareous plates of the dermal endoskeleton as you cut. Remove the rectangle of skin to uncover the large *coelom*, which contains the internal organs (Figure 20-4).

5. Cut around the *sieve plate* to remove the upper portion of the body wall of the *central disk*. The *mouth* connects via an extremely short esophagus to the clearly visible pouchlike **cardiac stomach.** The cardiac stomach opens into the upper *pyloric stomach*. A slender, short *intestine* leads from the upper side of the stomach to the *anus*. Find the two green, fingerlike **digestive glands** in each arm, which produce digestive enzymes and deliver them to the pyloric stomach.

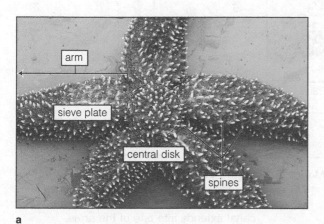

a

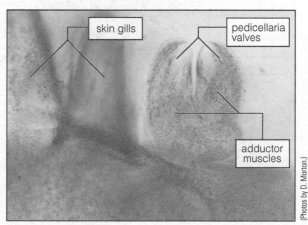

b

(Photos by D. Morton.)

Figure 20-2 (a) Aboral view of sea star. (b) Pedicellaria and skin gills, w.m. (205×).

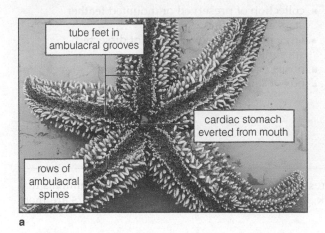

tube feet in ambulacral grooves

cardiac stomach everted from mouth

rows of ambulacral spines

a

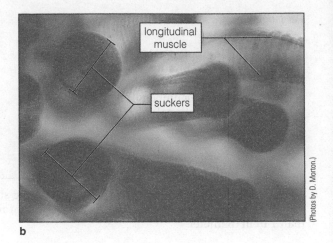

longitudinal muscle

suckers

(Photos by D. Morton.)

b

Figure 20-3 (**a**) Oral view of sea star. (**b**) Close-up of tube feet, w.m. (78×).

6. Identify the **dark gonads** that are located near the base of each arm. The sexes are separate but difficult to distinguish, except by microscopic examination. In nature, sperm and eggs are released into the seawater, and the fertilized eggs develop into bipinnaria larvae.

7. The water-vascular system is unique to echinoderms (Figure 20-5). Note that the *sieve plate* leads to a short **stone canal**, which in turn leads to the *circular canal* surrounding the mouth. Five *radial canals* lead from the circular canal into the ambulacral grooves. Each radial canal connects by short side branches with many pairs of tube feet.

8. The nervous system (not shown in Figure 20-3) is simple. A circular *nerve ring* surrounds the mouth, and a *radial nerve* extends from this into each arm, ending at a light-sensitive eyespot. There are no specific excretory organs. Dispose of your dissected specimen in the large plastic bag provided for this purpose.

9. Examine a prepared slide of a cross section of an arm with the compound microscope. Identify the structures labeled in Figure 20-6.

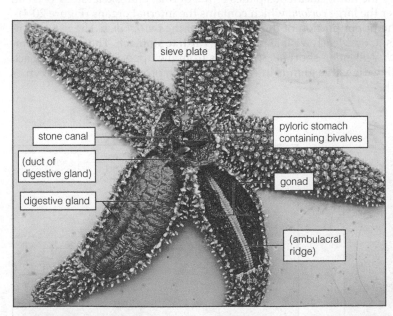

sieve plate

stone canal

(duct of digestive gland)

digestive gland

pyloric stomach containing bivalves

gonad

(ambulacral ridge)

Figure 20-4 Internal anatomy of a sea star. (Photo by D. Morton.)

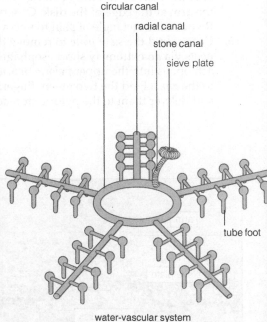

circular canal

radial canal

stone canal

sieve plate

tube foot

water-vascular system

Figure 20-5 The water-vascular system of a five-armed sea star. The ring canal is located in the central disk, and each radial canal extends into one of the arms.

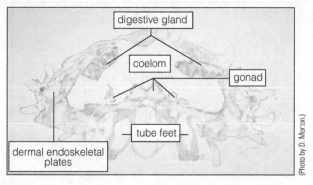

Figure 20-6 Cross section of a sea star arm (23×).

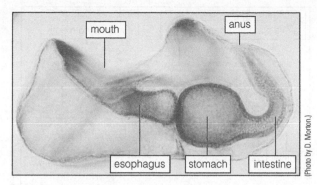

Figure 20-7 Bipinnaria larva (196×).

10. Examine a prepared slide of a whole mount of bipinnaria larvae. The digestive tract, including the *mouth, esophagus, stomach, intestine,* and *anus,* is clearly visible (Figure 20-7).
11. Observe preserved, plastic-mounted, or living specimens of sea stars.

B. Other Echinoderms

PROCEDURE

1. *Class Ophiuroidea* includes the brittle stars (Figure 20-8). These echinoderms have slender arms and move relatively rapidly, for echinoderms! In life, the rays are much more flexible than those of the sea stars. Examine preserved, plastic-mounted, or living specimens of brittle stars.
2. *Class Echinoidea* contains the sea urchins, animals that lack arms. They have long, movable spines, which surround the top and sides of a compact skeleton called the **test** (Figure 20-9). Sea urchins ingest food by means of a complex structure, *Aristotle's lantern,* which contains teeth. The sea biscuits and sand dollars are also members of this class. They have flat, disk-shaped bodies. Sea urchins, sea biscuits, and sand dollars, despite their lack of arms, all exhibit five-part radial symmetry and move by means of tube feet. Study preserved or living specimens and the tests (Figure 20-10) of members of this class.
3. *Class Holothuroidea* are the sea cucumbers (Figure 20-11). No spines are present. A collection of tentacles at one end surrounds the mouth. In some forms, the tube feet occur over the entire surface of the body. Sea cucumbers move in wormlike fashion as a result of the contractions of two layers of circular and longitudinal muscles. Examine preserved or living sea cucumbers.

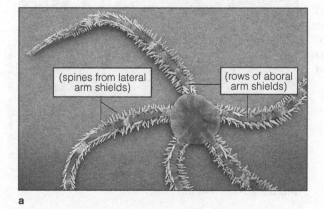

a

b

Figure 20-8 (a) Dried and (b) live brittle star.

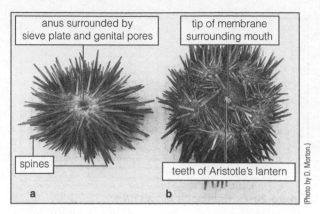

anus surrounded by sieve plate and genital pores

tip of membrane surrounding mouth

spines

teeth of Aristotle's lantern

a   b

(Photo by D. Morton.)

Figure 20-9 (a) Aboral and (b) oral views of a sea urchin.

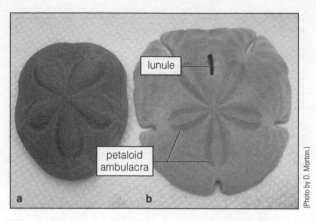

lunule

petaloid ambulacra

a   b

(Photo by D. Morton.)

Figure 20-10 (a) Sea biscuit and (b) sand dollar tests.

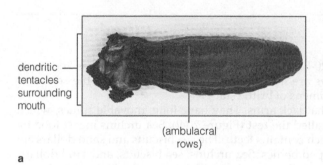

dendritic tentacles surrounding mouth

(ambulacral rows)

a

b

(Photos by D. Morton and W. A. Yoder.)

Figure 20-11 (a) Preserved and (b) live sea cucumber.

4. *Class Crinoidea* contains the graceful, flowerlike feather stars and sea lilies (Figure 20-12). These stationary animals attach to the sea floor by a long stalk. The arms are featherlike; each contains a ciliated groove and tentacles (derived from tube feet) to direct food to the mouth. There can be from 5 to over 200 arms, some of which reach 35 cm. *Class Concentricycloidea* contains the sea daisies. Examine any available specimens and illustrations of these classes.

(Photo by Chris Huss/The Wildlife Collection.)

Figure 20-12 Feather star.

Habitat:	aquatic, terrestrial (land and air)
Body Arrangement:	bilateral symmetry, coelomate, head and postanal tail, segmentation, gut (complete)
Level of Organization:	organ system (tissues derived from all three primary germ layers), nervous system with brain, circulatory system (closed with a ventral heart), organ systems highly developed, many have an endoskeleton (mesoderm)
Body Support:	most have an endoskeleton of cartilage or bone
Life Cycle:	sexual reproduction (mostly dioecious with internal or external fertilization), some larvae, some metamorphosis
Special Characteristics:	notochord, pharyngeal gill slits, dorsal hollow nerve cord, and a postanal tail at some stage of life cycle

The word *chordate* refers to one of the four major diagnostic characteristics shared by all members of this phylum at least some time in their life span, a **notochord, pharyngeal gill slits,** a **dorsal hollow nerve cord,** and a **postanal tail** (Table 20-1). Most chordates are also vertebrates—animals that possess a backbone or vertebral column—but a few are not. The latter animals belong to the subphlya Urochordata and Cephalochordata. They are **invertebrates** because they lack a vertebral column. All of the animal phyla you have studied so far are also invertebrates.

TABLE 20-1   Chordate Characteristics

Characteristic	Definition	Developmental Fate in Humans
Notochord	A dorsal, flexible rod that provides support for most of the length of the body	Center of intervertebral disks
Pharyngeal gill slits	Paired, lateral outpockets of pharynx that perforate through its wall	Auditory tubes and endocrine glands of neck
Dorsal hollow nerve cord	A tube of nervous tissue dorsal to notochord	Brain and spinal cord
Postanal tail	Posterior tail that extends past the anus	Present in embryo

MATERIALS

Per student:

- blunt probe

Per student pair:

- dissection microscope
- compound microscope, lens paper, a bottle of lens-cleaning solution (optional), and a lint-free cloth (optional)
- dissection pan

- preserved or mounted sea tunicates (sea squirts)
- preserved and plastic mounted lancelet specimens
- prepared slide of a whole mount of a lancelet
- prepared slide of a cross section of a lancelet in region of the pharynx
- boxes of different sizes of latex or vinyl gloves (optional)
- box of safety goggles (optional)

A. Tunicates (Subphylum Urochordata)

Tunicates are also called sea squirts. Both common names are descriptive of obvious features of these animals. A leatherlike "tunic" covers the adult, and water is squirted out of an excurrent siphon (Figure 20-13).

    Sea squirt larvae resemble tadpoles in general body form and have a notochord confined to the postanal tail. The notochord and tail degenerate when the larva becomes an adult tunicate. The larva also has a rudimentary

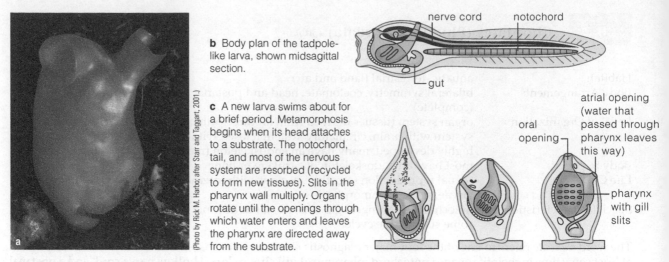

b Body plan of the tadpole-like larva, shown midsagittal section.

c A new larva swims about for a brief period. Metamorphosis begins when its head attaches to a substrate. The notochord, tail, and most of the nervous system are resorbed (recycled to form new tissues). Slits in the pharynx wall multiply. Organs rotate until the openings through which water enters and leaves the pharynx are directed away from the substrate.

(Photo by Rick M. Harbo; after Starr and Taggart, 2001.)

Figure 20-13 (a) Adult form of a sea squirt. (b) A sea squirt larva. (c) Metamorphosis of a larva into the adult.

brain and sense organs. These structures undergo reorganization into a nerve ganglion and nerve net in the adult. The number of pharyngeal gill slits increases in the adult.

1. Examine a preserved or plastic mounted specimen of a sea squirt with a dissection microscope. These animals are small, inactive, and very common marine organisms that inhabit coastal areas of all oceans. The adult is a filter feeder, capturing organic particles into an incurrent siphon by ciliary action. Individuals either float freely in the water (singly or in groups) or are sessile, attached to the bottom as branching individuals or colonies.

2. Water enters a sea squirt through the *incurrent siphon*. It travels into the *pharynx*, where it seeps through *gill slits* to reach a chamber, the *atrium*. The water eventually exits through the *excurrent siphon*. Back in the pharynx, food particles are trapped in sticky mucus and then passed to the digestive tract. Undigested materials are discharged from the anus into the atrium, to be expelled with water out the excurrent siphon. Identify as many of these structures as possible.

B. Lancelets (Subphylum Cephalochordata)

The lancelets are distributed worldwide and are especially abundant in coastal areas with *warm, shallow waters.* *Amphioxus* (Figure 20-14) is the commonly studied representative of this subphylum. The word *Amphioxus* means "sharp at both ends," and this certainly describes its *elongate, fishlike body form.* Lancelets reach 50–75 mm in length as adults. Despite their streamlined appearance, these organisms are not very active. They spend most of their time buried in the sand with their heads projecting while they filter organic particles from the water.

1. Examine a preserved lancelet with the dissection microscope (Figure 20-15). Note that the lancelets are translucent, with a low, continuous dorsal and caudal (tail) **fin.** Trace the flow of water from the **mouth**

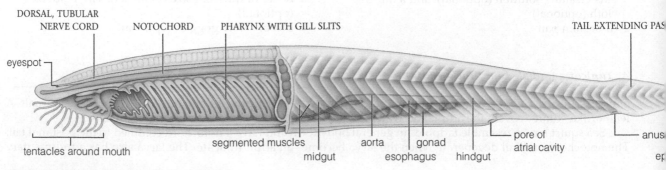

Figure 20-14 Cutaway view of a lancelet.

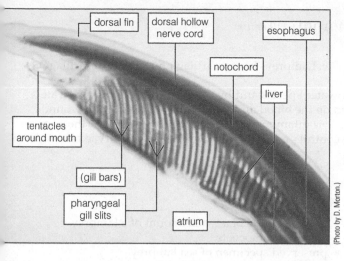
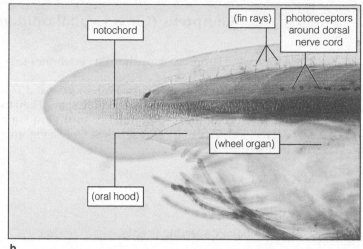

Whole mounts of (**a**) anterior end (30×) and (**b**) head (100×) of a lancelet.

(surrounded by **tentacles**), through the roughly 150 **pharyngeal gill slits** into the *atrium,* and out of the pore of the atrial cavity.

2. Now describe the capture of food and the path it takes through the digestive system, as was done previously for the tunicates. Unlike the tunicates, the *anus* in a lancelet opens externally.

_____

_____

_____

_____

_____

3. Find the flexible **notochord** extending nearly the full length of the individual; this structure persists into the adult stage. There is a small brain, with a **dorsal hollow nerve cord** that bears light-sensitive cells (the *eyespot*) at its anterior end.

4. Examine a cross section of a lancelet with a compound microscope and identify the structures indicated in Figure 20-16.

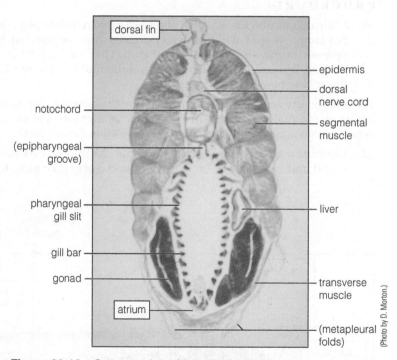

Figure 20-16 Cross section of lancelet in region of pharynx (20×).

Caution

Preserved specimens are kept in preservative solutions. Use latex gloves whenever you handle a specimen. Wash any part of your body exposed to this solution with copious amounts of water. If preservative solution is splashed into your eyes, wash them with the safety eyewash bottle for 15 minutes. If you wear contact lenses during a dissection, wear safety goggles.

The ancestors of lampreys were the first vertebrates to evolve. Lampreys have a cartilaginous (made of cartilage), primitive skeleton.

Lampreys are represented by both marine and freshwater species and by some species that use seawater and fresh water at some point in their life span. They feed on the blood and tissue of fishes by rasping wounds in their sides. A landlocked population of the sea lamprey is infamous for having nearly decimated the commercial fish industry in the Great Lakes. Only a vigorous control program that targeted lamprey larvae restored this industry.

MATERIALS

Per student:

■ scalpel
■ blunt probe or dissecting needle

Per student pair:

■ dissection microscope
■ dissection pan

Per student group:

■ prepared slide of a whole mount of a lamprey larva or ammocoete
■ preserved specimen of sea lamprey

Per lab room:

■ boxes of different sizes of vinyl or latex gloves
■ box of safety goggles

PROCEDURE

1. With a dissection microscope, observe a prepared slide of a whole mount of a lamprey larva or ammocoete. Sea lampreys spawn in freshwater streams. Ammocoetes hatch from their eggs and, after a period of development, burrow into the sand and mud. They are long-lived, metamorphosing into adults after as long as 7 years. Because of their longevity and dissimilar appearance, adult lampreys and ammocoetes were long thought to be separate species.

The ammocoete is considered by many to be the closest living form to ancestral chordates. Unlike a lancelet, it has an eye (actually two median eyes), internal ears (otic vesicles), and a heart, as well as several other typical vertebrate organs. Identify the structures labeled in Figure 20-17.

2. Examine a preserved sea lamprey (Figure 20-18). Note its slender, rounded body. The skin of the lamprey is soft and lacks scales. Look at the round, suckerlike **mouth**, inside of which are circular rows of horny,

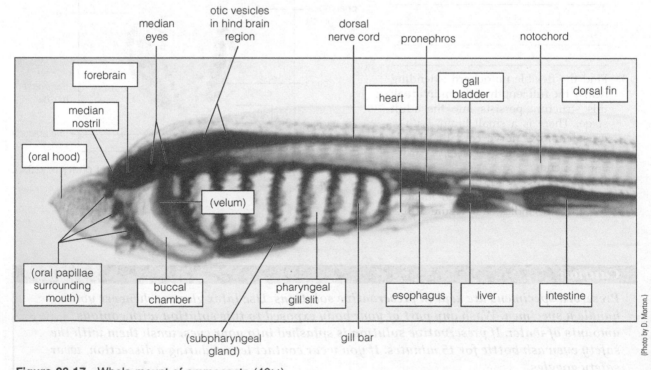

Figure 20-17 Whole mount of ammocoete (40×).

(Photo by D. Morton.)

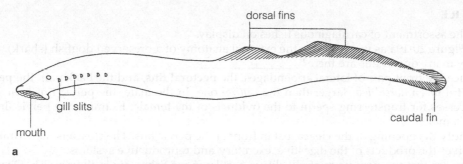

a

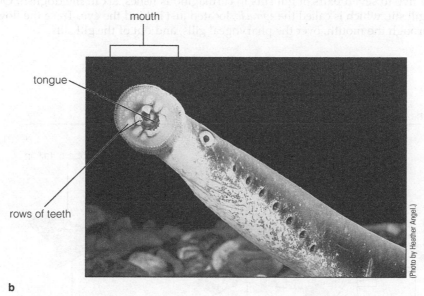

(Photo by Heather Angel.)

b

Figure 20-18 (a) Side view and (b) oral disk of a sea lamprey.

rasping **teeth** and a deep, rasping **tongue**. Although there are no lateral, paired appendages, there are two **dorsal fins** and a **caudal fin** (tail fin). Count and record below the number of pairs of external **gill slits**.

20.4   Cartilaginous Fishes (Class Chondrichthyes) *(10 min.)*

The first vertebrates to evolve jaws and paired appendages, according to the fossil record, were the heavily armored *placoderms* (class Placodermi). These fishes arose about 425 million years ago. Current hypotheses suggest that the ancestors of placoderms relatively quickly gave rise to the cartilaginous fishes.

Members of the class Chondrichthyes are *jawed fishes* with *cartilaginous skeletons* and *paired appendages*. The evolution of jaws and paired appendages and the evolution of more efficient respiration and a better developed nervous system, including sensory structures, were critical events in the history of vertebrate evolution. Together these adaptations allowed cartilaginous fishes to chase, catch, and eat larger and more active prey. This mostly marine group includes the carnivorous skates, rays, and sharks.

MATERIALS

Per student:

- scalpel
- forceps
- dissecting scissors

Per student pair:

- dissection microscope
- dissection pan

Per student group:

- preserved specimen of dogfish (shark)

Per lab room:

- collection of
  preserved cartilaginous fishes
  shark jaw with teeth
- boxes of different sizes of vinyl or latex gloves
- box of safety goggles

PROCEDURE

1. Look at the assortment of cartilaginous fishes on display.
2. Refer to Figure 20-19a as you examine the external anatomy of a preserved dogfish (shark).
   (a) How many dorsal fins are there? _____
   (b) Notice the front pair of lateral appendages, the **pectoral fins,** and the back pair, the **pelvic fins.** The **tail fin** has a *dorsal lobe* larger than the *ventral* one. In the male, the pelvic fins bear **claspers,** thin processes for transferring sperm to the oviducts of the female. Examine the pelvic fins of a female and a male.
   (c) Identify the opening of the **cloaca** just in front of the pelvic fins. The cloaca is the terminal organ that receives the products of the digestive, excretory, and reproductive systems.
   (d) There are five to seven pairs of gill slits in cartilaginous fishes, six in the dogfish. Observe the most anterior gill slit, which is called the *spiracle*, located just behind the eye. Trace the flow of seawater in nature through the mouth, over the pharyngeal gills, and out of the gill slits.

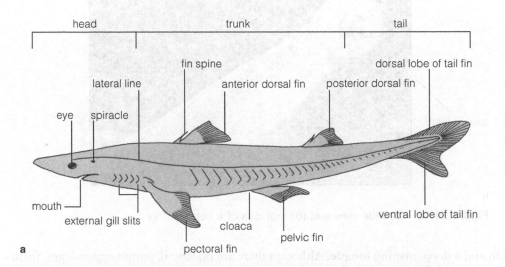

Figure 20-19 (**a**) External appearance of the dogfish (shark), *Squalus acanthias.* (After S. Wischnitzer, *Atlas and Dissection Guide for Comparative Anatomy*, 3rd ed., © 1967, 1972, 1979, W. H. Freeman and Company. Used by permission.) (**b**) Shark jaw.

(e) Locate the **nostrils,** which open into blind olfactory sacs; they don't connect with the pharynx, as your nostrils do. They function solely in olfaction (smell) in the cartilaginous fishes. The **eyes** are effective visual organs at short range and in dim light. There are no eyelids.

(f) Find the dashed line that runs along each side of the body. This is called the **lateral line** and functions to detect vibrations in the water. It consists of a series of minute canals perpendicular to the surface that contain sensory hair cells. When the hairs are disturbed, nerve impulses are initiated; their frequency enables the fish to locate the disturbance.

(g) Placoid scales, toothlike outgrowths of the skin, cover the body. Run your hand from head to tail along the length of the animal. How does it feel?

_____

_____

Now run your hand in the opposite direction along the animal. How does it feel this time?

_____

_____

Figure 20-20 Drawing of shark skin (_____×).

Cut out a small piece of skin with your scissors, pick it up with your forceps, and examine its surface with the dissection microscope. Draw what you see in Figure 20-20.

3. Examine the demonstration of a shark jaw with its multiple rows of teeth (Figure 20-19b).

20.5 Bony Fishes (Class Osteichthyes) _(10 min.)_

The bony fishes, the fishes with which you are most familiar, inhabit virtually all the waters of the world and are the largest vertebrate group. They are economically important, commercially and as game species. The ancestors of cartilaginous fish gave rise to the first bony fishes about 415 million years ago. As their name suggests, the _skeleton_ is at least _partly ossified (bony)_, and the flat _scales_ that cover at least some of the surface of most bony fishes are _bony_ as well. Gill slits are housed in a common chamber covered by a bony movable flap, the **operculum.**

MATERIALS

Per student:

- scalpel
- compound microscope, lens paper, a bottle of lens-cleaning solution (optional), and a lint-free cloth (optional)
- clean microscope slide and coverslip
- dH₂O in dropping bottle
- forceps
- blunt probe or dissecting needle

Per student pair:

- dissection pan

Per student group:

- preserved specimens of yellow perch

Per lab room:

- collection of preserved bony fishes
- boxes of different sizes of vinyl or latex gloves
- box of safety goggles

PROCEDURE

1. Examine the assortment of bony fishes on display.
2. Examine a yellow perch as a representative advanced bony fish (Figure 20-21).
   (a) Identify the _spiny_ **dorsal fin** located in front of the _soft_ **dorsal fin.** Paired **pectoral** and **pelvic fins** are also present, and behind the **anus** is an unpaired **anal fin.** The **tail fin** consists of _dorsal_ and _ventral lobes_ of approximately equal size. The fins, as in the cartilaginous fishes, are used to brake, steer, and maintain an upright position in the water.

Figure 20-21 Preserved yellow perch.

(b) Note the double **nostrils,** leading into and out of *olfactory sacs.* The eyes resemble those in sharks, with no eyelids. With forceps, pry open the operculum to see the organs of respiration, the **gills.**

(c) Find the **lateral line.** This functions similarly to that of the shark.

(d) Bony scales cover the body. Using forceps, remove a scale, make a wet mount, and examine it with the compound microscope. Locate the *annual rings,* which indicate the age of the fish. How are the annual rings of the fish scale analogous to the annual ridges of the mussel or clam shell?

_____

_____

_____

Draw the scale in Figure 20-22.

Figure 20-22 Drawing of yellow perch scale (_____×).

20.6   Amphibians (Class Amphibia) *(30 min.)*

The *amphibians* were the first vertebrates to assume a terrestrial existence, having evolved about 370 million years ago from a group of lobe-finned fishes. The paired appendages are modified as legs, which support the individual during movement on land. Respiration by one species or another or by the different developmental stages of one species is by lungs, gills, and the highly vascularized skin and lining of the mouth. There is a *three-chambered heart,* compared to the two-chambered heart of fishes, with two *circuits for the blood circulation,* compared to one circuit in fishes. Reproduction requires water, or at least moist conditions on land. Fertilized eggs hatch in water, and larvae generally live in water. The skeleton is bonier than that of the bony fishes, but a considerable proportion of it remains cartilaginous. The *skin* is usually *smooth and moist,* with mucous glands; scales are usually absent. This group of vertebrates includes the frogs, toads, salamanders, and tropical, limbless, burrowing, mostly blind caecilians.

MATERIALS

Per student:

- forceps
- dissecting scissors
- blunt probe or dissecting needle
- dissection pins

Per student pair:

- preserved leopard frog

Per lab room:

- collection of preserved bony fishes
- collection of preserved amphibians
- skeleton of frog
- skeleton of human
- boxes of different sizes of vinyl or latex gloves
- box of safety glasses

PROCEDURE

1. Examine preserved specimens of a variety of amphibians.
2. The leopard frog illustrates well the general features of the vertebrates and the specific characteristics of the amphibians. Obtain a preserved specimen and after rinsing it in tap water, examine its external anatomy (Figure 20-23).

    (a) Find the two *nostrils* at the tip of the head. These are used for inspiration and expiration of air. Just behind the eye locate a disklike structure, the *tympanum,* the outer wall of the middle ear. There is no external ear. The tympanum is larger in the male than in the female. Examine the frogs of other students in your lab.

    Is your frog a male or female?

    _____

    (b) At the back end of the body, locate the *cloacal opening.*

    (c) The forelimbs are divided into three main parts: the **upper arm, forearm,** and **hand.** The hand is divided into a **wrist, palm,** and **fingers** (digits). The three divisions of the hindlimbs are the **thigh, shank** (lower leg), and **foot.** The foot is further divided into three parts: the **ankle, sole,** and **toes** (digits).

3. Examine the internal anatomy of the frog.

    (a) With the scissors, cut through the joint at both angles of the jaw. Identify the structures labeled in Figure 20-24. The region containing the opening into the **esophagus** and the **glottis** is the **pharynx** (throat). The esophagus leads to the stomach and the rest of the digestive tract; the glottis leads into the blind respiratory tract.

    (b) Fasten the frog, ventral side up, with pins to the wax of a dissection pan. Lift the skin with forceps. Then make a superficial

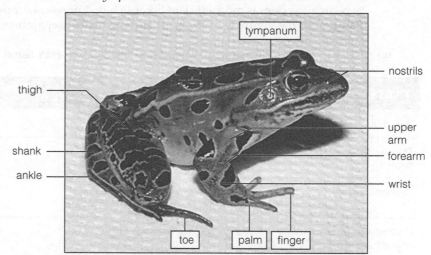

Figure 20-23   External anatomy of the frog. (Photo by D. Morton.)

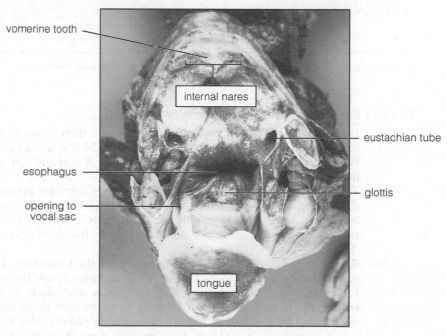

Figure 20-24   Oral cavity of the frog. (Photo by D. Morton.)

cut with your scissors from the lower abdomen forward, and just left or right of center, to the tip of the lower jaw. Pin back the skin on both sides to expose the large abdominal muscles. Lift these muscles with your forceps, and cut through the body wall with your scissors again from the lower abdomen to the tip of the lower jaw, cutting through the sternum (breastbone) but not damaging the internal organs. Pin back the body wall as you did the skin to expose the internal organs (Figure 20-25).

(c) Use Figure 20-25 to locate the *spleen* (Figure 20-25c) and the following organs of the digestive system: **stomach, small intestine, large intestine** (colon), **liver, gall bladder,** and **pancreas.** Trace the esophagus from its opening in the pharynx to the stomach. Parallel to its upper portion find the **bronchi** (the singular is *bronchus*), which lead toward the **lungs.** Are the bronchi dorsal or ventral to the esophagus?

_____

(d) Identify the two thin-walled **atria** and the thick-walled **ventricle** and the **conus arteriosus** of the heart along with and the right and left branches of the **truncus arteriosus** (Figure 20-9a). The latter structure carries blood to the paired **pulmocutaneous arches** and **systemic arches.** In frogs and other amphibians a *pulmocutaneous circuit* circulates blood to and from the lungs and the skin, and a *systemic circuit* services the rest of the body. Even though there is only one ventricle, because of the structure of the heart and its large arteries, most of the deoxygenated blood returning from the body is pumped into the pulmocutaneous circuit before the oxygenated blood is pumped into the systemic circuit. In comparison, a fish has a two-chambered heart with only one atrium and a one-circuit circulation. A two-circuit system is more efficient because blood pressure drops less after the blood is oxygenated. Partition of the ventricle of the heart and the further separation of the two circuits will continue in subsequent vertebrate classes.

(e) Refer to Figures 20-25 and 20-26 and find the major *arteries* listed in Table 20-2.

TABLE 20-2 Major Arteries of the Frog		
Arteries	**Location**	**Function Is to Convey Blood to**
Pulmocutaneous arches	First pair of vessels branching off truncus arteriosus	Lungs and skin
Systemic arches	Second pair of vessels branching off truncus arteriosus	The rest of the body
Common carotids (carotid arches)	Branch from the systemic arches	Head
Subclavians	Branch from the systemic arches	Arms
Dorsal aorta	Formed by the fusion of the systemic arches along dorsal body wall in mid abdomen	Lower half of the body
Common iliacs	Division of dorsal aorta in lower abdomen	Legs

Which of these arteries carries deoxygenated blood? _____

(f) Locate the **anterior** and **posterior vena cavae,** which return deoxygenated blood from the systemic circuit to the right atrium of the heart (Figures 20-25c and 20-26). Other veins return blood from the head, arms, and legs to one of the vena cavae. Right and left **pulmonary veins** return oxygenated blood from the lungs to the left atrium.

(g) The excretory and reproductive organs together comprise the *urogenital system* (Figure 20-27). Locate the urine-producing excretory organs, the pair of **kidneys** located dorsally in the body cavity. A duct, the **ureter,** leads from the kidney to the **urinary bladder** (also Figure 20-25a), an organ that stores urine for resorption of water from the urine into the circulatory system. The bladder empties into the *cloaca.*

(h) Refer to Figure 20-27a and, in the male, locate the **testes** (also Figure 20-25c); these organs produce sperm that are carried to the kidneys through tiny tubules, the *vasa efferentia.* The ureters serve a dual function in male frogs, transporting both urine and sperm to the cloaca. Find the vestigial female oviducts (also Figure 20-25b), which are located lateral to the urogenital system.

(i) Refer to Figure 20-27b and, in the female, find the *ovaries* (also Figure 20-25d); these organs expel eggs into the *oviducts.* The oviducts lead into the *uterus.* As in the male, the reproductive tract ends in the cloaca.

(j) Locate the *fat bodies* (Figures 20-25c and d), many yellowish, branched structures just above the kidneys. They store food reserves for hibernation and reproduction.

(k) The nervous system of vertebrates is composed of (1) the *central nervous system*, the brain and spinal cord, and (2) the *peripheral nervous system*, nerves extending from the central nervous system. Turn your frog over and remove the skin from the dorsal surface of the head between the eyes and along the vertebral column. With your scalpel, shave thin sections of bone from the skull, noting the shape and size of the *cranium*, until you expose the *brain*. Pick away small pieces of bone with your forceps to expose the entire brain. Use the same procedure to expose the vertebrae and *spinal cord*. Note the *cranial nerves* coming from the brain and the *spinal nerves* coming from the spinal cord. Spinal nerves are also easy to see running along the dorsal wall of the abdomen.

(l) Dispose of your dissected specimen in the large plastic bag provided for this purpose.

4. Comparison of frog and human skeleton.

(a) The skeleton of vertebrates consists of the following: (1) the **axial skeleton,** consisting of the bones of the **skull, sternum, hyoid bone, vertebral column,** and rib cage (when present); and (2) the **appendicular skeleton,** with the bones of its **girdles** and their appendages. The **pectoral girdle** consists of the paired **clavicle** and **scapula** bones (**suprascapular** in the bullfrog). The latter articulates—forms a joint—with the **humerus.** The **pubis, ischium,** and **ilium** bones form the pelvic girdle, which articulates with the femur. Compare the skeleton of the bullfrog and human (Figure 20-28), and identify the axial and appendicular portions of the skeleton.

(b) There are differences in the two skeletons, but their basic plan is remarkably similar. To a large extent, the size and shape of the different parts of the skeleton of a vertebrate correlate with body specializations and behavior. List some reasons for the differences in the girdles, appendages, and cranium of the frog and human skeletons.

_____

_____

_____

20.7   Reptiles (Class Reptilia) *(10 min.)*

Reptiles are the oldest group of vertebrates *adapted to living primarily on land,* although many do live in fresh water or seawater. They arose from the ancestors of amphibians about 300 million years ago. Their *skeleton* is bonier than that of amphibians. The *skin* is *dry* and covered by *epidermal scales*. There are virtually no skin glands. Reptiles lay *amniotic eggs* (Figure 20-29). Inside these eggs, embryos are suspended by fetal membranes in an internal aquatic environment surrounded by a shell to prevent their drying out.

There is no larval stage. The amniotic egg (characteristic of birds and some mammals as well as reptiles) and lack of larvae are adaptations to living an entire life cycle on land. The heart consists of two atria and a partially divided ventricle or two ventricles (in crocodiles). The nervous system, especially the brain, is more highly developed than that of amphibians.

Reptiles, like the fishes and amphibians, are largely **ectothermic,** without the capability to maintain a relatively high core-body temperature physiologically and with a greater dependence on gaining or losing heat from or to their external environment. Reptiles and other ectotherms have an adaptive advantage over endothermic mammals in warm and humid areas of the earth because they expend minimal energy on maintaining body temperature, which leaves more energy for reproduction. This advantage is reflected in the greater numbers of individuals and species of reptiles compared to mammals in the tropics. Mammals have the advantage in moderate to cold areas.

MATERIALS

Per student:

■ scalpel

Per student pair:

■ dissection pan

Per lab room:

■ collection of preserved reptiles
■ skeleton of snake
■ turtle skeleton or shell
■ boxes of different sizes of vinyl or latex gloves
■ box of safety goggles

PROCEDURE

1. Examine an assortment of preserved reptiles and note their diversity as a group. This diverse group of vertebrates includes the turtles, lizards, snakes, crocodiles, and alligators.

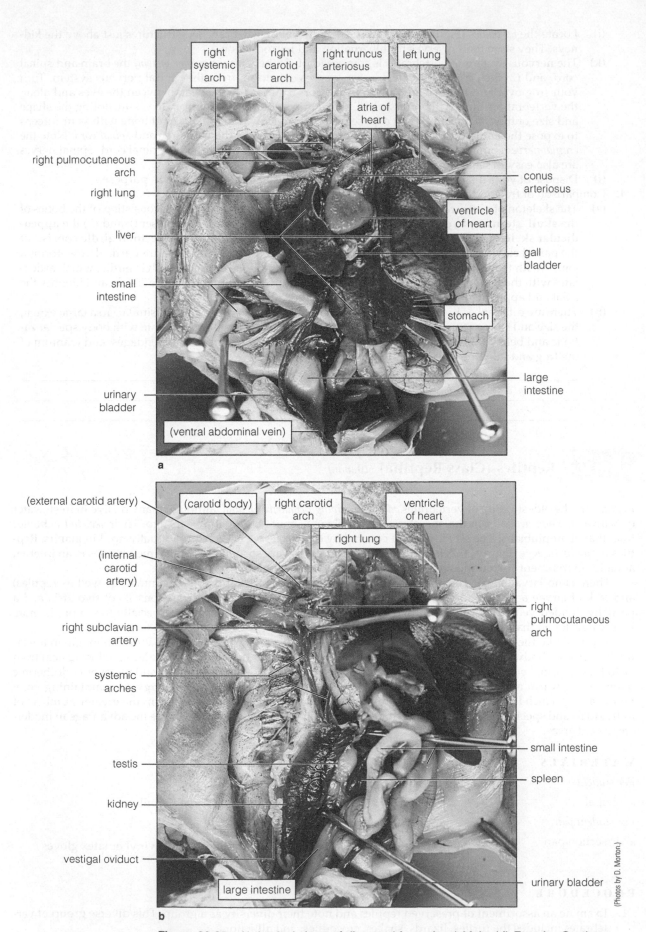

right systemic arch

right carotid arch

right truncus arteriosus

left lung

atria of heart

right pulmocutaneous arch

conus arteriosus

right lung

ventricle of heart

liver

gall bladder

small intestine

stomach

large intestine

urinary bladder

(ventral abdominal vein)

a

(external carotid artery)

(carotid body)

right carotid arch

ventricle of heart

right lung

(internal carotid artery)

right pulmocutaneous arch

right subclavian artery

systemic arches

testis

small intestine

spleen

kidney

vestigal oviduct

large intestine

urinary bladder

b

(Photos by D. Morton.)

Figure 20-25 Abdominal views of dissected frogs. **(a–c)** Male. **(d)** Female. *Continues.*

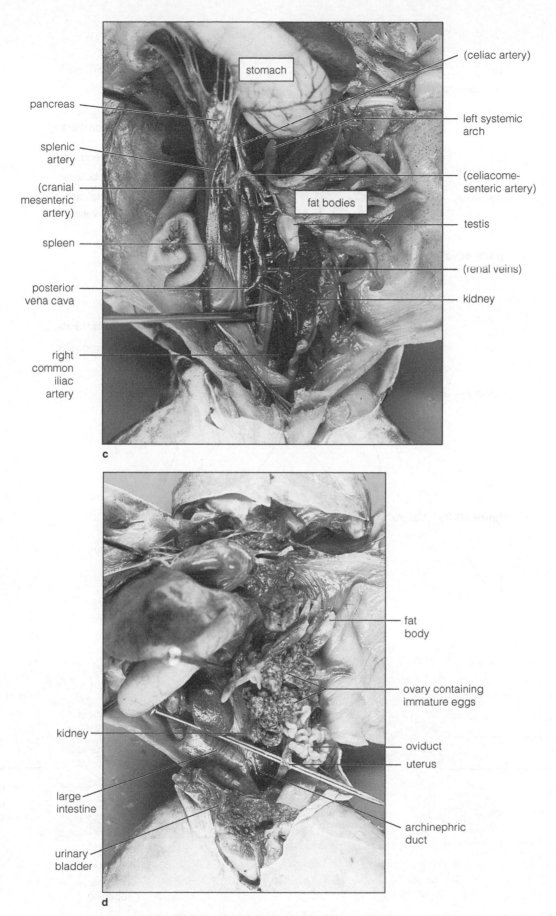

Figure 20-25 Abdominal views of dissected frogs. *Continued.*

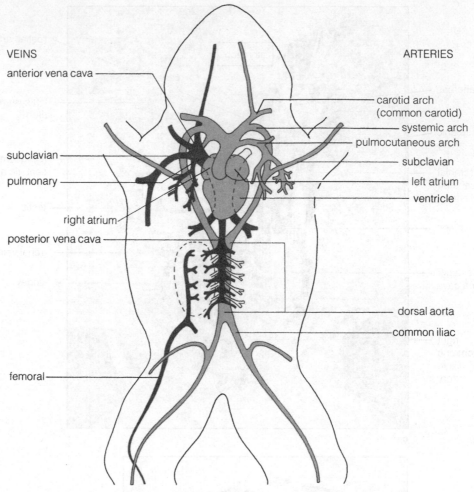

VEINS   ARTERIES

anterior vena cava

                                                                       carotid arch
                                                                       (common carotid)
                                                                       systemic arch
                                                                       pulmocutaneous arch
subclavian   subclavian
pulmonary   left atrium
                                                                       ventricle
right atrium

posterior vena cava

                                                                       dorsal aorta
                                                                       common iliac

femoral

Figure 20-26   Ventral view of the major blood vessels of the frog.

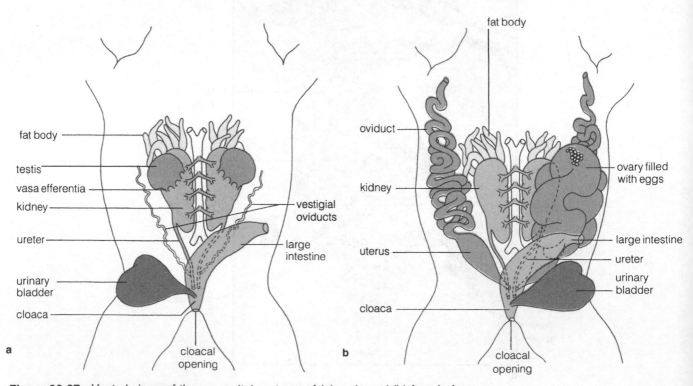

fat body   fat body

fat body   oviduct

testis   ovary filled
vasa efferentia   with eggs
kidney
                                                      kidney
                       vestigial
ureter   oviducts
                                                                       large intestine
                       large   uterus   ureter
urinary   intestine   urinary
bladder   bladder

cloaca   cloaca

a   b

         cloacal   cloacal
         opening   opening

Figure 20-27   Ventral views of the urogenital systems of (a) male and (b) female frogs.

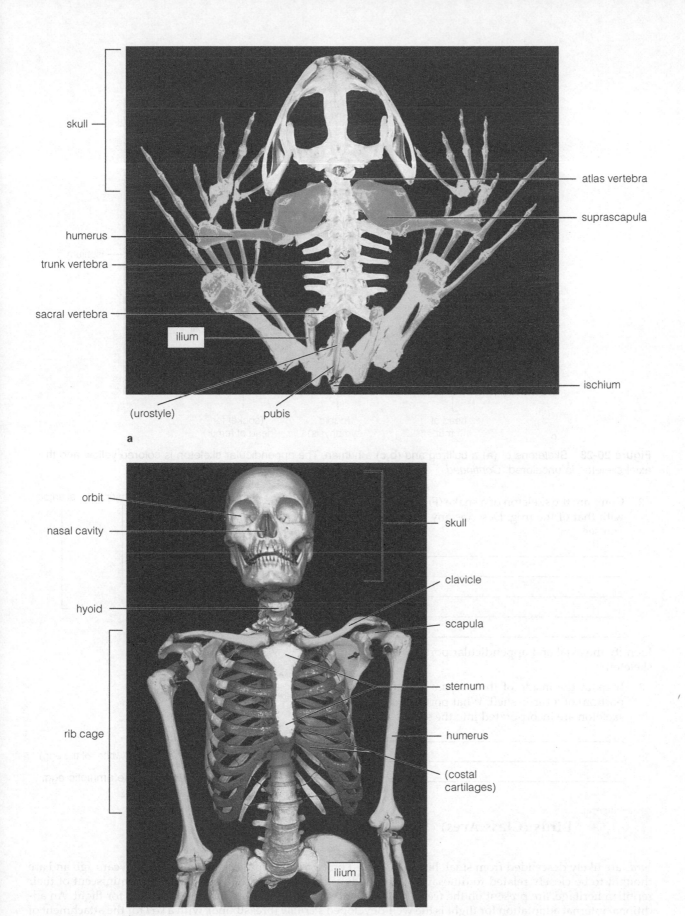

Figure 20-28 Skeletons of (**a**) a bullfrog and (**b,c**) a human. The appendicular skeleton is colored yellow and the axial skeleton is uncolored. (Photos by D. Morton.) *Continues.*

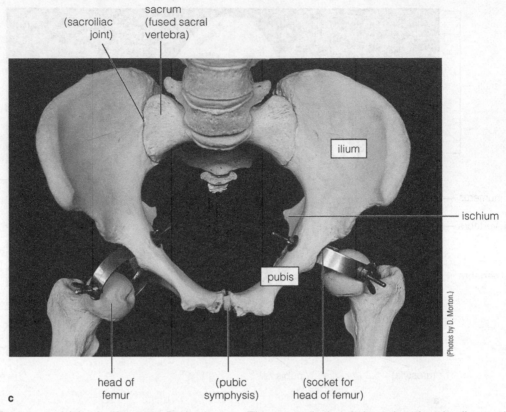

sacrum
(fused sacral
vertebra)

(sacroiliac
joint)

ilium

ischium

pubis

head of
femur

(pubic
symphysis)

(socket for
head of femur)

(Photos by D. Morton.)

c

Figure 20-28 Skeletons of (**a**) a bullfrog and (**b,c**) a human. The appendicular skeleton is colored yellow and the axial skeleton is uncolored. *Continued.*

2. Compare the skeleton of a snake (Figure 20-30) with that of the frog. Describe any differences you see.

_____

_____

_____

_____

_____

Identify the axial and appendicular portions of the skeleton.

3. Look at the inside of the *carapace* (the upper portion) of a turtle shell. What portions of the skeleton are incorporated into the shell?

_____

_____

_____

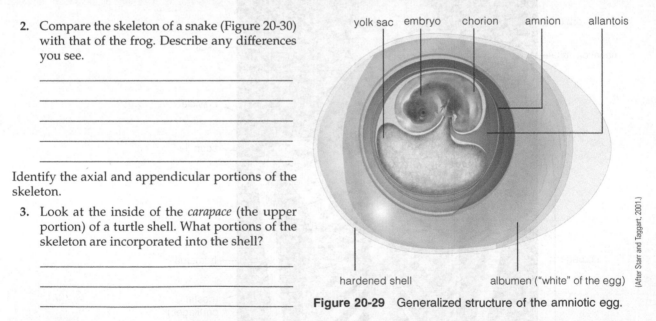

yolk sac   embryo   chorion   amnion   allantois

hardened shell   albumen ("white" of the egg)

(After Starr and Taggart, 2001.)

Figure 20-29 Generalized structure of the amniotic egg.

| **20.8** | **Birds (Class Aves)** *(10 min.)* |

Birds are likely descended from small bipedal (walked on two legs) reptiles about 160 million years ago and are thought to be closely related to dinosaurs. Their body is covered with *feathers*; and *scales*, reminiscent of their reptilian heritage, are present on the feet. The front limbs in most birds are modified as *wings* for flight. An additional internal adaptation for flight is the well-developed *sternum* (breastbone), with a *keel* for the attachment of powerful muscles for flight (Figure 20-31). Birds are **endothermic** vertebrates, capable of maintaining relatively high core-body temperatures physiologically.

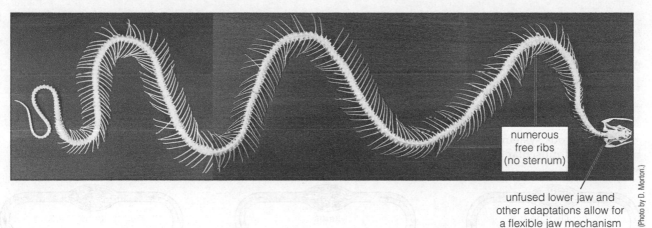

numerous
free ribs
(no sternum)

(Photo by D. Morton.)

unfused lower jaw and
other adaptations allow for
a flexible jaw mechanism

Figure 20-30   Snake skeleton.

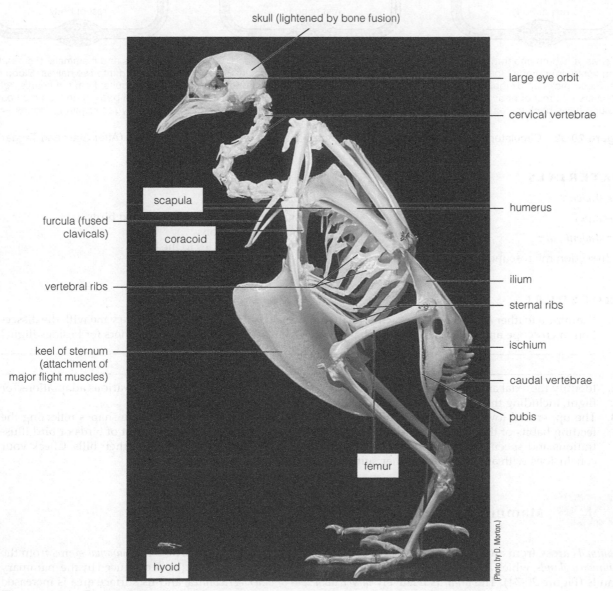

skull (lightened by bone fusion)

large eye orbit

cervical vertebrae

scapula

humerus

furcula (fused
clavicals)

coracoid

ilium

vertebral ribs

sternal ribs

ischium

keel of sternum
(attachment of
major flight muscles)

caudal vertebrae

pubis

femur

hyoid

(Photo by D. Morton.)

Figure 20-31   Bird skeleton.

The major *bones* of birds are *hollow* and contain *air sacs connected to the lungs.* Birds have a *four-chambered heart,* with two atria and two ventricles (Figure 20-32). This permits the complete separation of oxygenated and deoxygenated blood. Why is this circulatory arrangement an advantage for birds, as opposed to that found in amphibians and reptiles? (*Hint:* Recall that birds are endothermic and most can fly.)

_____

_____

The nervous system resembles that of reptiles. However, the brain is larger, permitting more sophisticated behavior and muscular coordination; the optic lobes are especially well developed in association with a keen sense of sight.

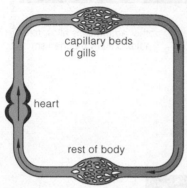 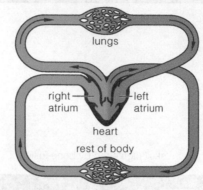 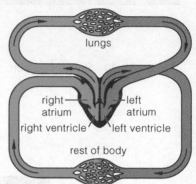

a In fishes, a two-chambered heart (atrium, ventricle) pumps blood in one circuit. Blood picks up oxygen in gills and delivers it to rest of body. Oxygen-poor blood flows back to heart.

b In amphibians, a heart pumps blood through two partially separate circuits. Blood flows to lungs, picks up oxygen, and returns to heart. It mixes with oxygen-poor blood still in heart, flows to rest of body, and returns to heart.

c In birds and mammals, the heart is fully partitioned into two halves. Blood circulates in two circuits: from the heart's right half to lungs and back, then from the heart's left half to oxygen-requiring tissues and back.

Figure 20-32 Circulatory systems of (**a**) fishes, (**b**) amphibians, and (**c**) birds and mammals. (After Starr and Taggart, 200

MATERIALS

Per student:

■ scalpel

Per student pair:

■ dissection microscope

Per lab room:

■ an assortment of feathers, stuffed birds, or bird illustrations
■ several field guides to the birds

PROCEDURE

1. Examine a feather and identify the **rachis** (shaft) and **vane** (Figure 20-33). Examine the vane with the dissection microscope and note the **barbs** and **barbules.** What else do birds use their feathers for besides flight?

_____

_____

2. Identify the axial and appendicular portions of the skeleton (Figure 20-31). Note the various adaptations for flight, including the elongated wrists and ankles.
3. The upper and lower jaws are modified as variously shaped *beaks* or *bills,* with the shapes reflecting the feeding habits of the species. No teeth are present in adults. Examine the assortment of birds or bird illustrations and speculate on their food preferences by studying the configurations of their bills. Check your conclusions with a field guide or other suitable source that describes the food habits.

| 20.9 | Mammals (Class Mammalia) *(10 min.)* |

Mammals arose from an ancient branch of reptiles over 200 million years ago. The term *mammal* stems from the *mammary glands,* which all mammals possess. Female mammals feed their young milk produced by the mammary glands (Figure 20-34). The *brain* is relatively *larger than that of other vertebrates,* and its surface area is increased by grooves and folds (Figure 20-35). Behavior is more complex and flexible. *Hair* is present during some portion of the life span. Mammals are equipped with a variety of modifications and outgrowths of the skin in addition to hair, all made of the protein keratin. They include horns, spines, nails, claws, and hoofs. Mammalian dentition is intricate, with teeth specialized for cutting, grooming, wounding, tearing, slicing, and grinding. The teeth in the upper and lower jaws match up.

Like birds, mammals are *endothermic* and the *heart* is *four-chambered*. The ear frequently has a cartilaginous outer portion called the *pinna*, which functions to collect sound waves. In addition to mammary glands, there are three other types of skin glands: *sebaceous, sweat, and scent glands.*

MATERIALS

Per student:

- scalpel

Per lab room:

- stuffed mammals or mammal illustrations
- mammalian placentas and embryos (in utero)

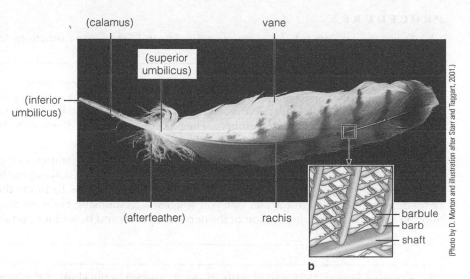

Figure 20-33 (a) Feather and (b) expanded view of the vane.

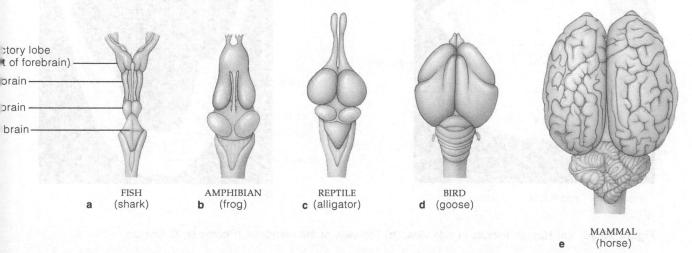

Figure 20-34 Female dog with nursing puppies.

Figure 20-35 Evolutionary trend toward an expanded, more complex brain, as suggested by comparing the brain of existing vertebrates: (**a**) shark; (**b**) frog; (**c**) alligator; (**d**) goose; and (**e**) horse. These dorsal views are not to the same scale. (After Starr and Taggart, 2001.)

PROCEDURE

1. Study the assortment of mammals on display and list as many functions for hair as you can.

_____

_____

2. Except for two species of monotremes that lay eggs (platypus and spiny anteater), most mammals are **viviparous:** Females bear live young after supporting them in the uterus with a _placenta,_ a connection between the mother and the embryo. The placenta exchanges gases with, delivers nutrients to, and removes wastes from the embryo, among other functions.

Some viviparous mammals (marsupials—kangaroos, opossums, and so on) give birth early, and along with spiny anteater hatchlings, continue development suckling milk in a skin pouch. Eutherian mammals give birth later, but the young of all mammals receive extended parental care. Examine the placentas and embryos in uteri of mammals on display in the lab. Can you suggest a relationship between the early development of mammals and the comparative sophistication of the nervous system and behavior of adult mammals?

_____

_____

3. Examine Figure 20-36 and identify the teeth present in the skull of the human skeleton. Fill in the numbers of each type in the upper and lower jaws and their functions in Table 20-3.

TABLE 20-3 Human Dentition

Type of Tooth	Number in Upper Jaw	Number in Lower Jaw	Function
Incisors			
Canines			
Premolars			
Molars			

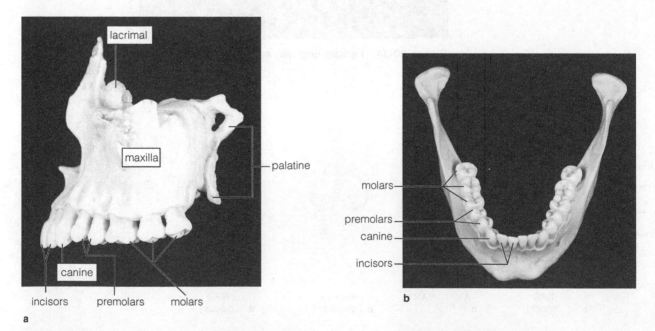

Figure 20-36 (a) Human maxilla in side view. (b) Top view of the mandible. (Photos by D. Morton.)

_____ 1. Echinoderms and chordates are
   (a) deuterostomes
   (b) protostomes
   (c) animals whose blastopore
       becomes a mouth
   (d) both b and c

_____ 2. The skin gills of sea stars are structures
   (a) for the exchange of gases and
       excretion of ammonia
   (b) of defense
   (c) for movement
   (d) that produce the skeletal elements

_____ 3. The tube feet of a sea star
   (a) bear tiny toes
   (b) are located only at the tips of
       the arms
   (c) function in movement
   (d) protect the organism from
       predatory fish

_____ 4. Aristotle's lantern is a tooth-bearing
   structure of the
   (a) sea stars
   (b) tunicates
   (c) acorn worms
   (d) sea urchins

_____ 5. The sieve plate, stone canal, circular
   canal, and radial canals of a sea star are
   structures of the
   (a) nervous system
   (b) water-vascular system
   (c) digestive system
   (d) excretory system

_____ 6. The dermal endoskeleton of echinoderms
   is made of
   (a) cartilage
   (b) bone
   (c) calcium carbonate
   (d) chitin

_____ 7. Which structure is _not_ a major diagnostic
   feature of the chordates?
   (a) dorsal hollow nerve cord
   (b) notochord
   (c) pharyngeal gill slits
   (d) vertebral column

_____ 8. The sea squirts and lancelets have an
   inner chamber that expels water. This
   chamber is called the
   (a) intestine
   (b) bladder
   (c) atrium
   (d) nephridium

_____ 9. Invertebrates
   (a) do not have a vertebral column
   (b) have a vertebral column
   (c) are all members of the phylum
       Chordata
   (d) do not include animals that are
       members of the phylum Chordata

_____ 10. Which animal is a chordate?
   (a) sea star
   (b) sea cucumber
   (c) acorn worm
   (d) lancelet

_____ 11. Animals that have the four basic
   characteristics of chordates plus a
   vertebral column are
   (a) invertebrates
   (b) hemichordates
   (c) cephalochordates
   (d) vertebrates

_____ 12. The ammocoete is the larva of
   (a) lampreys
   (b) sharks
   (c) bony fishes
   (d) amphibians

_____ 13. Placoid scales are characteristic of class
(a) Cephalaspidomorphi
(b) Chondrichthyes
(c) Osteichthyes
(d) Reptilia

_____ 14. The bony movable flap that covers the gills and gill slits is called the
(a) operculum
(b) cranium
(c) tympanum
(d) colon

_____ 15. Which group of animals is endothermic?
(a) fishes
(b) mammals
(c) reptiles
(d) amphibians

_____ 16. The bones that surround and protect the brain are collectively called the
(a) spinal cord
(b) vertebral column
(c) pelvic girdle
(d) cranium

_____ 17. The lampreys are unusual vertebrates in that they have no
(a) eyes
(b) jaws
(c) gill slits
(d) mouth

_____ 18. The structure of the cartilaginous and bony fishes that detects vibrations in the water is the
(a) anal fin
(b) operculum
(c) nostrils
(d) lateral line

_____ 19. Amphibians have
(a) a two-chambered heart
(b) a three-chambered heart
(c) a four-chambered heart
(d) none of the above

_____ 20. The amniotic egg of reptiles, birds, and mammals is an adaptation to
(a) carnivorous predators
(b) a life on land
(c) compensate for the short period of development of the young
(d) protect the young from the nitrogenous wastes of the mother during the formation of the embryo

EXERCISE 20

Echinoderms, Invertebrate Chordates, and Vertebrates

POST-LAB QUESTIONS

Introduction

1. What characteristic of vertebrates is missing in all invertebrates?

2. List the basic characteristics of vertebrates.

20.1 Echinoderms (Phylum Echinodermata)

3. Describe the various structures on the aboral "spiny skin" of a sea star and discuss their functions.

4. Describe the various functions of tube feet.

5. Identify the following structures on this aboral view of the tip of a sea star arm.

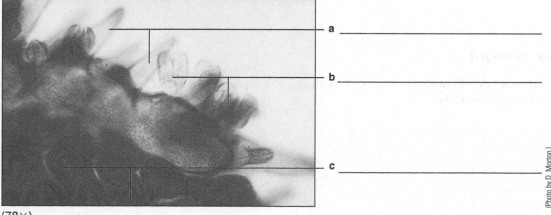

a _____

b _____

c _____

(Photo by D. Morton.)

(78×).

6. Describe reproduction in the sea star.

7. List the structures in the water-vascular system of a sea star.

20.2 Chordates (Phylum Chordata)

8. List the four major characteristics of chordates.

9. Why are sea squirts and lancelets referred to as invertebrate chordates?

10. Compare the presence or absence of the four chordate characteristics in adult sea squirts and lancelets.

11. Compare arthropods, echinoderms, and chordates in terms of their similarities and differences.

Food for Thought

12. Hypothesize an evolutionary relationship among the bipinnaria larva of echinoderms, larval and adult tunicates, and lancelets.

20.3 Jawless Fishes (Class Cephalaspidomorphi)

13. What is an ammocoete? What is its possible significance to the evolution of vertebrates?

14. How do lampreys differ from other vertebrates?

20.4 Cartilaginous Fishes (Class Chondrichthyes)

15. Explain how the adaptations of cartilaginous fishes make them better predators.

20.5 Bony Fishes (Class Osteichthyes)

16. Identify this vertebrate structure. What do the rings represent?

(Photo courtesy of Eiodisc, Inc.)

(24×).

20.7 Reptiles (Class Reptilia)

17. Describe an amniotic egg. What is its evolutionary significance?

20.9 Mammals (Class Mammalia)

18. List the unique characteristics of mammals.

19. Describe the evolution of the heart in the vertebrates.

Food for Thought

20. About 65 million years ago, an asteroid impact may have caused a mass extinction, which led to the demise of dinosaurs and created new opportunities for surviving plants and animals, including mammals. Create a short evolutionary scenario for the next human-made or natural mass extinction event.

Plant and Animal Organization

OBJECTIVES

After completing this exercise, you will be able to

1. define *vegetative, morphology, dicotyledon (dicot), monocotyledon (monocot), cotyledon, taproot system, node, internode, adventitious root, herb (herbaceous plant), woody (woody plant), heartwood, sapwood, growth increment (annual ring); tissue, organ, system, organism, histology, basement membrane, goblet cells, cilia, brush border of microvilli, keratinization, keratin, collagen fibers, elastic fibers, fibroblast, fat cells, lumen chondrocytes, lacunae, Haversian systems, osteocytes, actin filaments, myosin filaments, intercalated disk, neuron, cell body, axon, dendrites, neuroglia, body cavities—thoracic, abdominal, pleural, pericardial, pelvic, cranial* and *spinal;*

2. identify and give the function of the external structures of flowering plants (those in **boldface**);

3. identify and give the functions of the tissues and cell types of roots, stems, and leaves (those in **boldface**);

4. determine the age of woody branches;

5. discuss the high degree of organization present in animal structure and explain its significance;

6. recognize the four basic tissues and their common mammalian subtypes;

7. explain how the four basic tissues combine to make organs;

8. list each system of a mammal and its vital functions;

9. describe the basic plan of the mammalian body;

10. explain the layout of the body's major cavities;

11. locate the major organs in a mammal's body.

INTRODUCTION

Unicellular eukaryotic organisms contain organelles that carry on all functions vital to life. In multicellular organisms, each cell is specialized to emphasize certain activities, although most continue to carry on basic functions such as cellular respiration. In most animals, clusters of similarly specialized cells associate together along with any extracellular material they secrete and maintain to form **tissues.** Different subtypes from each of the four basic tissue types are combined much like quadruple-decker sandwiches to make the body's **organs.** Groups of related organs are strung together functionally, and usually structurally, to form **systems.** The levels of organization of an animal with systems can be summarized as follows:

Animal Organism

↑

Systems (e.g., digestive, respiratory)

↑

Organs (e.g., esophagus, stomach, trachea, lungs)

↑

Epithelial, Connective, Muscular, and Nervous Tissues

↑

Specialized Cells

↑

Organelles

The first part of this exercise introduces you to the external and internal structure of the **vegetative organs**—those not associated with sexual reproduction—of flowering plants. The organs are the roots, stems, and leaves. Flowers are the sexual reproductive organs, and were considered in detail in Exercise 17. In the second part of this exercise, we examine animal tissues, organs, and organ systems.

Each organ is usually distinguished by its shape and form, its **morphology.** But the cells and tissues comprising these three organs are remarkably similar. Each organ is covered by the protective dermal tissue; each possesses vascular tissue that transports water, minerals, and the products of photosynthesis; and each contains ground tissue, that which is covered by the dermal tissue and in which the vascular tissue is embedded. Examine Figure 21-1, which illustrates these relationships.

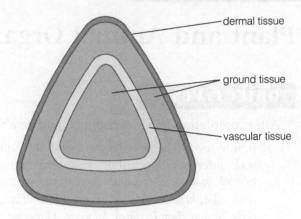

Figure 21-1   Model plant organ in cross section.

The organs of the plant body are more similar than they are dissimilar. In fact, for this reason the differences between a root and a stem or leaf are said to be *quantitative* rather than *qualitative.* That is, these differences are in the number and arrangement of cells and tissues, not the type. Consequently, the plant body is a continuous unit from one organ to the next.

We'll begin our study of plant and animal organization by examining the organs of two rather large assemblages of flowering plants, **dicotyledons** (dicots) and **monocotyledons** (monocots). **Cotyledons** are the leaves that are formed within the seed. Dicots have two seed leaves, monocots one. In addition to this fundamental difference, dicots and monocots have other dissimilarities, as shown in Figure 21-2. Refer to Figure 21-2 as you proceed through each section.

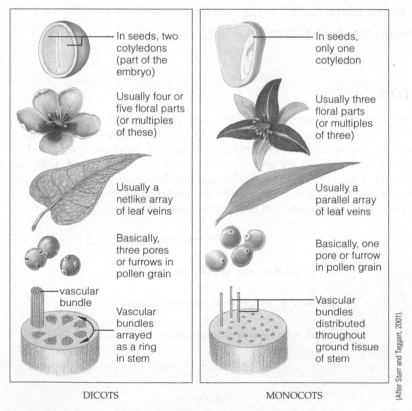

DICOTS   MONOCOTS

Figure 21-2   Comparison of dicot and monocot structure.

MATERIALS

Per student group (table):

- mature corn plant

Per lab room:

- living bean and corn plants in flats
- potted geranium and dumbcane plants
- dishpan half-filled with water

PROCEDURE

A. Dicotyledons

The common garden bean, *Phaseolus vulgaris*, is a representative dicot. Other familiar examples of dicotyledons (dicots) are sunflowers, roses, cucumbers, peas, maples, and oaks. Let's look at its external morphology. Label Figure 21-3 as you study the bean plant.

1. Obtain a bean plant by gently removing it from the medium in which it is growing. Wash the root system in the dishpan provided, not in the sink.

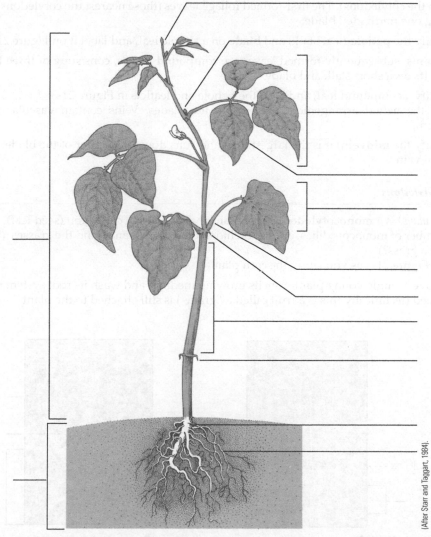

(After Starr and Taggart, 1984).

Figure 21-3　External structure of the bean plant.
Labels: root system, shoot system, primary root (taproot), lateral root, node, internode, axillary bud, terminal bud, cotyledon remnant, petiole, blade, simple leaf, compound leaf, leaflet

2. The plant consists of a **root system,** that portion usually below ground, and a typically aerial **shoot system.** Examine the root system first. The root system of the bean is an example of a **taproot system,** that is, one consisting of one large **primary root** (the **taproot**) from which **lateral roots** arise. Identify the taproot and lateral roots.

3. Now turn your attention to the shoot system, consisting of the stem and the leaves.

4. Identify the points of attachment of the leaves to the stem, called **nodes;** the regions between nodes are **internodes.**

5. Look in the upper angle created by the junction of the stem and leaf stalk to find the **axillary bud.** These buds give rise to branches and/or flowers.

6. Find the **terminal bud** at the very tip of the shoot system. The terminal bud contains an apical meristem that accounts for increases in length of the shoot system.

If lateral branches are produced from axillary buds, each lateral branch is terminated by a terminal bud and possesses nodes, internodes, and leaves, complete with axillary buds. As you see, the shoot system can be a highly branched structure.

7. Look several centimeters above the soil line for the lowermost node on the stem. If the plant is relatively young, you should find the **cotyledons** attached to this node. The cotyledons shrivel as food stored in them is used for the early growth of the seedling. Eventually the cotyledons fall off.

The cotyledons are sometimes called *seed leaves* because they are fully formed (although unexpanded) in the seed. By contrast, most of the leaves you're observing on the bean plant were immature or not present at all in the seed.

Now let's examine the other component of the shoot system, the foliage leaves (also called *true leaves* in contrast to the cotyledons). The first-formed foliage leaves (those nearest the cotyledons) are **simple leaves,** each leaf having one undivided blade.

8. Identify the **petiole** (leaf stalk) and **blade** on a simple leaf, and label it on Figure 21-3.

In bean plants, subsequently formed leaves are **compound leaves,** consisting of three **leaflets** per petiole. Each leaflet has its own short stalk and blade.

9. Identify a compound leaf, and label the petiole and leaflets in Figure 21-3.

10. Note the netted arrangement of veins in the blades. Veins contain vascular tissues—the xylem and phloem.

11. Identify the **midvein;** it is the largest vein and runs down the center of the blade, giving rise to numerous lateral veins.

B. Monocotyledons

Corn *(Zea mays)* is a **monocotyledon.** These plants have only one cotyledon (seed leaf). You're probably familiar with a number of monocots: lilies, onions, orchids, coconuts, bananas, and the grasses. (Did you realize that corn is actually a grass?)

Label Figure 21-4 as you study the corn plant.

1. Remove a single young plant from its growing medium and wash its root system in the dishpan. Note that the seed (technically this is a fruit called a "grain") is still attached to the plant.

soil line

soil line

fibrous adventitious root system

(Photos by J. W. Perry.)

a

b

Figure 21-4 External structure of a corn plant. (**a**) Seedling (0.5×). (**b**) Lower portion of mature plant (0.1×). **Labels:** root system, prop root, leaf sheath

2. Identify the **root system.** Note that there is no one particularly prominent root. In most monocots, the primary root is short lived and is replaced by numerous **adventitious roots.** Adventitious roots are roots that arise from places other than existing roots.
3. Trace the adventitious roots back to the corn grain.

Where do they originate?

_____

As the roots branch, they develop into a **fibrous root system,** one particularly well suited to prevent soil erosion.

4. Examine the mature corn plant. Identify the large **prop roots** at the base of the plant.

Where do prop roots arise from?

_____

Would you classify these as adventitious roots? (yes or no) _____

5. The shoot system of a young corn plant appears somewhat less complex than that of the bean. There seem to be no nodes or internodes. Look at the mature corn plant again. You should see that nodes and inter-nodes do indeed exist. In your young plant, elongation of the stem has not yet taken place to any appreciable extent.
6. Strip off the leaves of the young corn plant. Keep doing so until you find the shoot apex. (It's deeply embedded.)
7. Examine the leaves of the corn plant in more detail. Note the absence of a petiole and the presence of a **sheath** that extends down the stem. The leaf sheath adds strength to the stem. Look at the veins, which have a parallel arrangement. (Contrast this to the petioled, netted-vein arrangement of the bean leaves.)

21.2   The Root System _(About 15 min.)_

MATERIALS

Per student:

- single-edged razor blade
- clean microscope slide
- coverslip
- prepared slide of buttercup (_Ranunculus_) root, c.s.
- prepared slide of corn (_Zea_) root, c.s.
- compound microscope

Per student pair:

- dH₂O in dropping bottles

Per lab room:

- germinating radish seeds in large petri dishes
- demonstration slide of Casparian strip in endo-dermal cell walls

PROCEDURE

A. Living Root Tip

1. Obtain a germinating radish seed. Identify the **primary root.** Its fuzzy appearance is due to the numerous tiny **root hairs** (Figure 21-5). Root hairs increase the absorptive surface of the root tremendously.
2. Using a razor blade, cut off the seed and discard it, then make a wet mount of the primary root. (Add enough water so that no air surrounds the root. _Do not_ squash the root.)
3. Examine your preparation with the low-power objective of your compound microscope. (If you're having difficulty seeing the root clearly, increase its contrast by closing the microscope's diaphragm somewhat.) Locate the conical root tip.

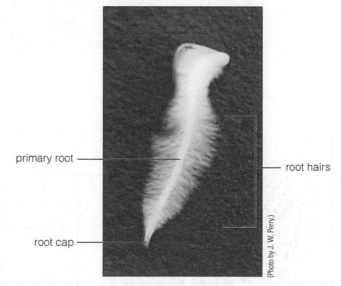

Figure 21-5   Living primary root (4×).

4. At the very end of the root tip, identify the protective **root cap** that covers the tip. As a root grows through the soil, the tip is thrust between soil particles. Were it not for the root cap, the apical meristem containing the dividing cells would be damaged.

5. Find the root hairs.

Do they originate all the way down to the root cap? (yes or no) _____

6. Examine the root hairs carefully.

What happens to their length as you observe them at increasing distance from the root tip?

_____

_____

     The youngest root hairs are the shortest. What does this imply regarding their point of origin and pattern of maturation?

_____

_____

B. Dicot Root Anatomy

Now let's see what the internal architecture of the root looks like.

1. Obtain a prepared slide containing a mature buttercup (*Ranunculus*) root in cross section. Refer to Figure 21-6 as you study this slide. Examine the slide first with the low-power objective of your compound microscope to gain an overall impression of the organization of the tissues present.
2. Starting at the edge of the root, identify the **epidermis.**
3. Moving inward, locate the **cortex** and the central **vascular column.** These regions represent the dermal, ground, and vascular tissue systems, respectively.
4. Switch to the medium-power objective for further study. Look at the outermost layer of cells, the **epidermis.**
5. Beneath the epidermis find the relatively wide **cortex,** consisting of parenchyma cells that contain numerous **starch grains.**

Based on the presence of starch grains, what would you suspect one function of this root might be?

_____

_____

6. Switch to the high-dry objective. Between the cells of the cortex, find numerous **intercellular spaces.**

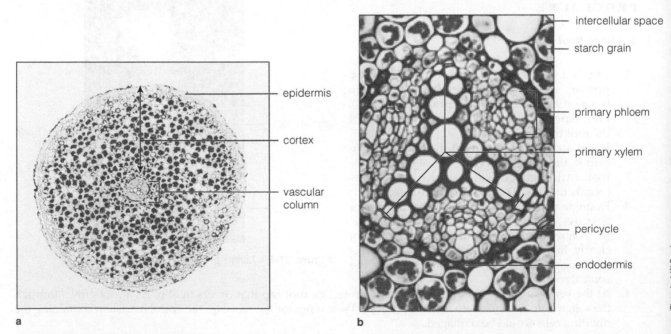

Figure 21-6 Cross section of buttercup (*Ranunculus*) root. (**a**) 45×. (**b**) Portion of cortex and vascular column (75×).

The innermost layer of the cortex is given a special name. This cylinder, a single cell thick, is called the **endodermis.**

7. Locate the endodermis on your slide and in Figure 21-6b.
8. Switch back to the medium-power objective, and focus your attention on the central vascular column. The cell layer immediately beneath the endodermis is the **pericycle.** Cells of the pericycle have the capacity to divide and produce lateral roots, like those you saw in the bean plant (Figure 21-3).
9. Finally, find the **primary xylem,** consisting of three or four ridges of thick-walled cells. (The stain used by most slide manufacturers stains the xylem cell walls red.) The xylem is the principal water-conducting tissue of the plant.
10. Between the "arms" of the xylem find the **primary phloem,** the tissue responsible for long-distance transport of carbohydrates produced by photosynthesis (known as photosynthates).

Primary xylem and phloem make up the *primary* vascular tissues, which are those produced by the apical meristems at the tips of the root and shoot. Later, we'll examine a *secondary* vascular tissue.

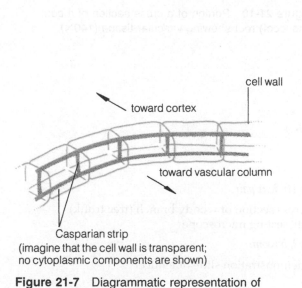

cell wall

toward cortex

toward vascular column

Casparian strip
(imagine that the cell wall is transparent; no cytoplasmic components are shown)

Figure 21-7 Diagrammatic representation of endodermal cells with Casparian strip.

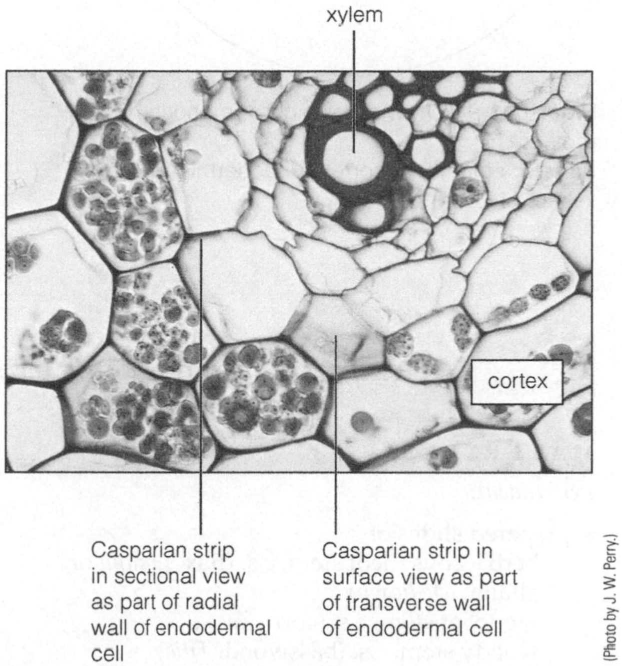

xylem

cortex

(Photo by J. W. Perry.)

Casparian strip in sectional view as part of radial wall of endodermal cell

Casparian strip in surface view as part of transverse wall of endodermal cell

Figure 21-8 Endodermal cells showing Casparian strip (296×).

C. Monocot Root Anatomy

Now that you have an understanding of the internal structure of a dicot root, let's compare and contrast it with that of a monocot. Draw and label the monocot root in Figure 21-9 as you proceed with your study.

1. Obtain a prepared slide of a cross section of a corn (*Zea mays*) root. Examine it with the low-power objective of your compound microscope. Draw the general shape of the corn root in Figure 21-9.
2. Identify and draw the **epidermis.**
3. Next locate the **cortex.** Count the number of cell layers and draw the cortex.

Do you find any starch grains in the cortex of the corn stem? _____

4. Now locate the **endodermis.** If the root you have is a mature one, you will see that the endodermis has very thick walls on three sides.
5. The vascular column of a monocot is quite different from a dicot in several ways. First note that the center is occupied by thin-walled cells. These cells make up a region called the **pith.**
6. Draw the pith.
7. Next note that the **xylem** of this monocot root consists of scattered cells. Look at Figure 21-10, a high magnification photomicrograph of the xylem and phloem of this root.
8. Draw the xylem.

Figure 21-9 Drawing of a corn (monocot) root, c.s. (____×).
Labels: epidermis, cortex, endodermis, pith, xylem, phloem

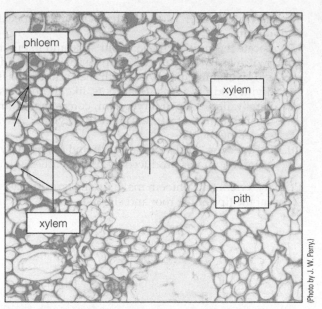

(Photo by J. W. Perry.)

Figure 21-10 Portion of a cross section of a corn (monocot) root showing vascular tissue (140×).

21.3   The Shoot System: Stems *(About 25 min.)*

MATERIALS

Per student:

- prepared slides of
    herbaceous dicot stem, c.s. (flax, *Linum;* or
    alfalfa, *Medicago*)
    monocot stem, c.s. (corn, *Zea*)
    woody stem, c.s. (basswood, *Tilia*)
- woody twig (hickory, *Carya;* or horse chestnut, *Aesculus*)
- metric ruler or meter stick
- compound microscope

Per student pair:

- cross section of woody branch (tree trunk)
- dissecting microscope

Per lab room:

- demonstration slide of lenticel

PROCEDURE

A. Dicot Stem: Primary Structure

Remember that the root and shoot are basically similar in structure; only the arrangement of tissues differs. Dicot stems have their vascular tissues arranged in a more or less complete ring of individual bundles of vascular tissue (called vascular bundles). Moreover, the ground tissue of dicots can be differentiated into two regions: **pith** and **cortex.**

You've probably heard the term *herb*. An *herb* is an **herbaceous plant,** one that develops no or very little wood. Herbaceous plants have only primary tissues. Woody plants develop secondary tissues—wood and bark.

Beans, flax, and alfalfa are examples of herbaceous dicots. Maples and oaks are woody dicots. Let's look at an herbaceous stem first.

1. Obtain a slide of an herbaceous dicot stem (flax or alfalfa). This slide may be labeled "herbaceous dicot stem." As you study this slide, refer to Figure 21-11, of a partial section of an herbaceous dicot stem.
2. Using the low- and medium-power objectives, identify the single-layered **epidermis** covering the stem, a multilayered **cortex** between the epidermis and **vascular bundles,** and the **pith** in the center of the stem.

3. Observe a vascular bundle with the high-dry objective (Figure 21-12). Adjacent to the cortex, find the **primary phloem.**
4. Just to the inside of the primary phloem, locate the thin-walled cells of the **vascular cambium.** (The vascular cambium is the lateral meristem that produces wood and secondary phloem, both secondary tissues. Despite your specimen being an herbaceous plant, a vascular cambium may be present. Generally, this meristem does not produce enough secondary tissue to result in the plant being considered woody.)
5. Locate the thick-walled cells of the **primary xylem.** The primary xylem is that vascular tissue closest to the pith. (The wall of the xylem cells is probably stained red.)

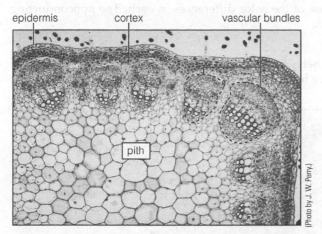

Figure 21-11 Portion of a cross section of an herbaceous dicot stem (75×).

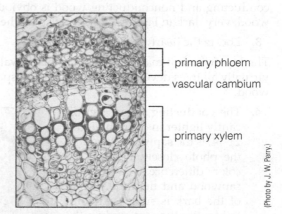

Figure 21-12 Vascular bundle of an herbaceous dicot stem (192×).

B. External Features of Woody Dicot Stems

Woody plants are those that undergo secondary growth, producing the secondary tissues. Both roots and shoots may have secondary growth. This growth occurs because of activity of the two meristems near the edge of the plant—the vascular cambium and the cork cambium. The vascular cambium produces secondary xylem (wood) and secondary phloem, while the cork cambium produces a portion of the bark called the **periderm.**

1. Examine a twig of hickory (*Carya*) or buckeye (*Aesculus*) that has lost its leaves. Label Figure 21-13 as you study the twig.
2. Find the large **terminal bud** at the tip of the twig. If your twig is branched, each branch has its own terminal bud.
3. Identify the shield-shaped **leaf scars** at each **node.** (Remember, a node is the region where a leaf attaches to the stem.) Leaf scars represent the point at which the leaf petiole was attached on the stem.
4. Within each leaf scar, note the numerous dots. These are the **vascular bundle scars.** Immediately above and adjacent to most leaf scars should be an **axillary bud.**
5. In the **internode** regions of the twig, locate the small raised bumps on the surface; these are **lenticels,** the regions of the periderm that allow for exchange of gases.

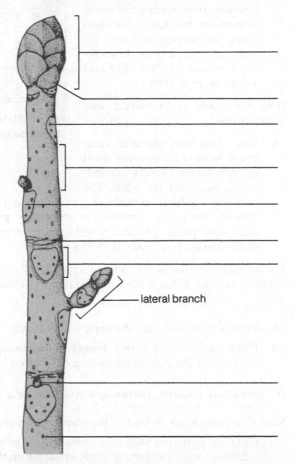

Figure 21-13 External structure of a woody stem. **Labels:** terminal bud, leaf scar, node, vascular bundle scar, axillary bud, internode, lenticel, bud scale, terminal bud scale scars

1. Examine a cross section of a tree trunk (Figure 21-14; *trunk* is the nonscientific term for any large, woody stem). A tiny region in the center of the stem is the **pith**. Most of the trunk is made up of **wood** (also called *secondary xylem*). Locate the pith and the wood.
2. Now identify the **bark**.

As a woody stem grows larger, it requires additional structural support and more tissue for transport of water. These functions are accomplished by the wood. However, it is only the outer few years' growth of wood that is involved in water conduction. In some tree species, like that pictured in Figure 21-14, the distinction between conducting and nonconducting wood is obvious because of the color differences in each. The nonconducting wood, very dark in Figure 21-14, is called the **heartwood**.

3. Locate the heartwood in your stem.

The heartwood of many species, like black walnut or cherry, is highly prized for furniture making. Although virtually all trees develop heartwood, not all species have heartwood that is *visibly* distinct from the conducting wood.

4. The conducting wood, significantly lighter in color in Figure 21-14, is the **sapwood**. Examine the photo closely, because the color difference between the sapwood and the inner layers of the bark is subtle. Attempt to identify the sapwood on the stem you are examining.
5. Count the **growth increments (annual rings)** within the wood to estimate the age of the stem when the section was cut. (In most woods, a growth increment includes *both* a light and a dark layer of cells.)

How old would you estimate your section to be? _____ years

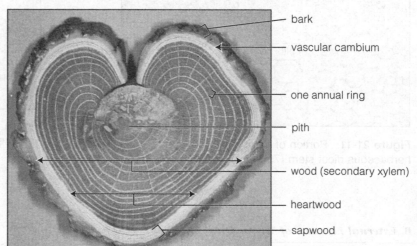

Figure 21-14 Cross section of a woody locust stem. This stem was injured several years into its growth, causing the unusual shape (0.5×).

(Photo by J. W. Perry.)

6. Now find the **vascular cambium** located *between* the most recently formed wood (secondary xylem) and the **bark.** The bark is everything *external* to the vascular cambium.
7. Identify the bark, consisting of **secondary phloem** and **periderm**. (As the vascular cambium produces secondary phloem to the outside, the primary phloem, cortex, and epidermis are sloughed off, much as dead skin on your body is shed.)

The periderm performs the same function as the epidermis before the epidermis ruptured as a result of the stem's increase in girth. What is the function of the periderm?

_____

8. Within the wood, find the **rays,** which appear as lines running from the center toward the edge of the stem.
9. Examine the wood with a dissecting microscope and locate the numerous holes in the wood. These are the cut ends of the water-conducting cells, often called **pores.**

D. Secondary Growth: Microscopic Anatomy of a Woody Dicot Stem

Now that you've got an idea of the composition of a woody stem, let's examine one with the microscope.

1. Obtain a prepared slide of a cross section of a woody stem (basswood, *Tilia*, or another stem). Examine it with the various magnifications available on the dissecting microscope. Use Figure 21-15 as a reference.
2. Starting at the edge, identify the **periderm** (darkly stained cells). Depending on the age of the stem, **cortex** (thin-walled cells with few contents) may be present just beneath the periderm. Now find the broad band of **secondary phloem** (consisting of cells in pie-shaped wedges). Identify the **vascular cambium** (a narrow band of cells separating the secondary phloem from secondary xylem), **secondary xylem (wood),** and pith

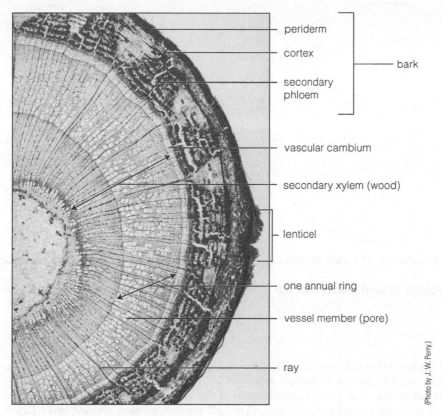

periderm
cortex
secondary phloem
bark
vascular cambium
secondary xylem (wood)
lenticel
one annual ring
vessel member (pore)
ray

(Photo by J. W. Perry.)

Figure 21-15 Cross section of a woody dicot stem (15×).

(large, thin-walled cells in the center). Count the number of growth increments (annual rings). Note the largest, thick-walled cells in the wood. These are the pores.

How old is this section? _____ years

3. Find the **rays** running through the wood. Rays are parenchyma cells that carry water and photosynthates *laterally* in the stem. (For the most part, the xylem and phloem carry substances *vertically* in the plant.)

Recall that the periderm replaces the epidermis as secondary growth takes place. The epidermis had stomata, which allowed the exchanging of gases between the plant and the environment. When the epidermis was shed, so were the stomata. But the need for exchange of gases still exists because the living cells require oxygen for respiration (Exercise 10). The plant has solved this problem by having special regions, **lenticels,** in the periderm. Lenticels are groups of cells with lots of intercellular space, in contrast to the tightly packed cells in the rest of the periderm.

4. Identify a lenticel on your slide. If none is found, examine the *demonstration* slide, specifically chosen to demonstrate this feature.

E. Structure of a Monocot Stem

Monocot and dicot stems differ in several ways. Monocot stems lack differentiation of the ground tissue into cortex and pith. Unlike the dicot stem whose vascular tissues are arranged in a ring, the vascular bundles of monocots are scattered throughout the ground tissue.

As you study the monocot stem, draw what you see in Figure 21-16.

1. Obtain a prepared slide of a monocot (corn, *Zea mays*) stem cross section. Observe the slide first using the low-power objective of your compound microscope, or perhaps even a dissecting microscope.
2. Draw its general appearance in Figure 21-16, labeling the **epidermis** on the outside, the thin-walled cells comprising the **ground tissue,** and the numerous **vascular bundles** scattered within the vascular tissue.
3. Switch to the medium-power objective of your compound microscope to study a single vascular bundle more closely. Draw the detail of one vascular bundle in Figure 21-17. (The vascular bundles of monocots look a lot like the face of a human or other primate.)

Figure 21-16 Drawing of a corn (monocot) stem, c.s. (____×).
Labels: epidermis, ground tissue, vascular bundles

Figure 21-17 Drawing of the vascular bundle of a corn (monocot) stem, c.s. (____×).
Labels: bundle sheath, xylem, phloem

4. Observe the thick-walled cells surrounding the vascular bundle. These cells make up the bundle sheath. (The walls of these cells are often stained red.)

5. Note the very large cells that give the impression of the vascular bundle's "eyes, nose, and mouth." These cells make up the water-conducting xylem. (The cell walls are often stained red.) Draw and label these xylem.

6. Finally, find the phloem, located on the vascular bundle's "forehead." (Phloem cell walls are typically stained green or blue.)

21.4   The Shoot System: Leaves *(About 20 min.)*

Leaves make up the second part of the shoot system. You examined the external morphology of the leaves in Section 21.1. Now let's look inside leaves, starting with a typical dicot leaf.

MATERIALS

Per student:

- prepared slide of dicot leaf, c.s. (lilac, *Syringa*)
- prepared slide of monocot leaf, c.s. (corn, *Zea*)
- compound microscope

PROCEDURE

A. Dicot Leaf Anatomy

1. Obtain a prepared slide of a cross section of a dicot leaf (lilac or other). Refer to Figure 21-18 as you examine the leaf.

2. Examine the leaf first with the low-power objective to gain an overall impression of its morphology. Note the size and orientation of the **veins** within the leaf. The veins contain the xylem and phloem.

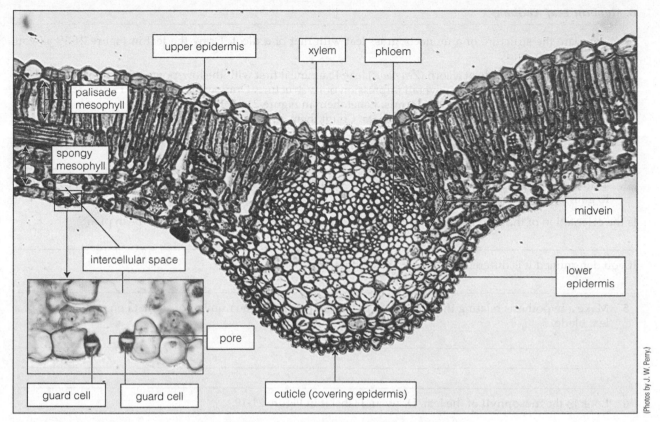

Figure 21-18 Cross section of a dicot leaf with midrib (105×); inset of a stoma (480×).

(Photos by J. W. Perry.)

3. Find the centrally located **midvein** within the **midrib,** the midvein-supporting tissue. You might think of the midvein as the major pipeline of the leaf, carrying water and minerals to the leaf and materials produced during photosynthesis to sites where they will be used during respiration.

4. Use the high-dry objective to examine a portion of the blade to one side of the midvein. Starting at the top surface of the leaf, find the **cuticle,** a waxy, water-impervious substance covering the **upper epidermis.** The epidermis is a single layer of tightly appressed cells.

5. The ground tissue of the leaf is represented by the **mesophyll** (literally "middle leaf"). In dicot leaves, the mesophyll is usually divided into two distinct regions; immediately below the upper epidermis, find the two layers of **palisade mesophyll.** These columnar-shaped cells are rich in chloroplasts.

6. Below the palisade mesophyll, find the loosely arranged **spongy mesophyll.** Note the large volume of **intercellular air space** within the spongy mesophyll.

Does the spongy mesophyll contain any chloroplasts? (yes or no) _____

What is one function that occurs within the spongy mesophyll? _____

7. In the **lower epidermis,** find a **stoma** (plural: *stomata*) with its **guard cells** and the **pore** (inset, Figure 21-18). Large **epidermal hairs** shaped somewhat like mushrooms are usually found on the lower epidermis.

Is the lower epidermal layer covered by a cuticle? _____

8. Compare the abundance of stomata within the lower epidermis with that in the upper epidermis.

Which epidermal surface has more stomata? _____

9. Examine the midvein in greater detail. Find the thick-walled xylem cells (often stained red).

10. Below the xylem, locate the **phloem** (usually stained green).

11. Now identify and examine the smaller veins within the lamina (blade). Note that these, too, contain both xylem and phloem.

Let's compare the structure of a monocot grass leaf with that of a dicot. Draw the leaf in Figure 21-19 as you proceed with your study.

1. Obtain a prepared slide of a corn (*Zea mays*) leaf. Examine it first with the low-power objective of your compound microscope to gain an overall impression of its structure. Draw what you see in Figure 21-19.
2. Identify the **upper** and **lower epidermis.** Label them in Figure 21-19.
3. Search on both epidermal layers for **stomata.** Count them on each layer.

Are there more, fewer, or about the same number on the upper epidermis versus the lower?

_____

4. Think about the orientation of the leaf blade on the plant. (If necessary, examine a living or dried corn plant to gain an impression of the orientation.)

Is the orientation of the corn leaves the same or different from that of a dicot, such as the bean plant?

_____

If you determined it is different, in what way is it different?

_____

5. Make a hypothesis relating the orientation of the leaf blade and the number of stomata on each surface of a leaf blade.

_____

_____

_____

6. Look at the **mesophyll** of the leaf. Draw and label it in Figure 21-19.

Figure 21-19 Drawing of a corn (monocot) leaf, c.s. (____×).
Labels: cuticle, epidermis, stoma, guard cells, mesophyll, bundle sheath, xylem, phloem

Is the mesophyll divided into palisade and spongy layers? (yes or no) _____

7. Note the **intercellular air spaces** within the mesophyll. Draw and label the intercellular air spaces.
8. Note that the bundle sheath cells are tightly packed, with no intercellular spaces separating them. The bundle sheath serves a function like that of the endodermis in the root.
9. Finally, identify the larger **xylem** cells and **phloem** cells within the vascular bundle. Draw and label these tissues in Figure 21-19.

21.5 Tissues *(50 min.)*

The study of tissues is called **histology.** The four basic tissues types are **epithelial tissue, connective tissue, muscle tissue,** and **nervous tissue.** Each type has subtypes. In your study of these subtypes, you will examine a variety of tissues and organs, all permanently mounted on glass microscope slides. Most slides have sections of tissues and organs, usually 6–10 μm thick. Other slides have whole pieces of organs that are either transparent or are teased (gently pulled apart) and spread on the surface of the slide until they are thin enough to see through. Some organs are simply smeared onto the surface of slides.

These prepared slides have been chosen to show not only the characteristics of each tissue but also how variations in the basic structure of each allows for the related but distinct functions of its subtypes. For the sake of simplicity, this exercise includes only the common subtypes of adult mammalian tissues.

MATERIALS

Per student:

- compound microscope, lens paper, a bottle of lens-cleaning solution (optional), a lint-free cloth (optional), a dropper bottle of immersion oil (optional)
- prepared slides of
   whole mount of mesentery (simple squamous epithelium)
   section of the cortex of the mammalian kidney
   section of trachea
   small intestine c.s. (preferably of the ileum)
   section of esophagus
   section of mammalian skin
   sections of contracted and distended urinary bladders

teased spread of loose (areolar) connective tissue
tendon, l.s.
compact bone ground c.s.
section of white adipose tissue
sections of the three muscle types
smear of the spinal cord of an ox (neurons)

Per lab room:

- demonstration of intercalated disks in cardiac muscle tissue
- 50-mL beaker three-fourths full of water
- 50-mL beaker three-fourths full of immersion oil
- 2 small glass rods

PROCEDURE

A. Epithelial Tissues

Epithelial tissues are widespread throughout the body, covering both the body's outer (epidermis of skin) and inner surfaces (ventral body cavities), and lining the inner surfaces of tubular organs (the small intestine, for example). Their main functions are *protection* and *transport* (such as secretion and absorption). Specialized functions include sensory reception and the maintenance of the body's gametes (egg and developing sperm).

Epithelial cells carry on rapid cell division in the adult, and various stages of mitosis are often seen in this tissue type. As a consequence, epithelia have the highest rates of cell turnover among tissues. For example, the lining of the small intestine replaces itself every 3–4 days. Epithelial tissues don't have blood vessels and are attached to the underlying connective tissue by an extracellular **basement membrane** that is difficult to see if it is not specially stained. Use the following questions to identify most subtypes of epithelial tissue:

QUESTION 1. What is the shape of the outermost cells?

There are three answers: *squamous* (scalelike or flat), *cuboidal*, or *columnar*. These choices describe the shape of the cells when viewed in a section perpendicular to the surface. In surface view, all three types are polygonal.

QUESTION 2. How many layers of cells are there?

Epithelial tissues are *simple* if there is one layer of cells; they are *stratified* if there are two or more. Figure 21-20 applies these definitions to squamous epithelia. *Pseudostratified* epithelial subtypes appear stratified but

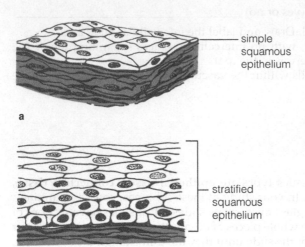

simple
squamous
epithelium

a

stratified
squamous
epithelium

b

Figure 21-20 Squamous epithelia. (a) Simple.
(b) Stratified.

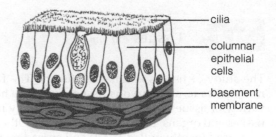

cilia

columnar
epithelial
cells

basement
membrane

Figure 21-21 Pseudostratified columnar
epithelium. It is columnar because of the
shape of the cells that reach the free surface.

really are not because their cells all rest on the basement membrane (Figure 21-21). As only the tallest cells reach the free surface, cell nuclei lie at different levels, giving the false appearance of several layers of cells.

In addition, the epithelium that lines the inside of the urinary bladder (and parts of other nearby urinary system organs) changes its subtype as it fills or empties with urine. This epithelium is called *transitional* to avoid confusion.

There are six common subtypes of epithelia: *simple squamous, simple cuboidal, simple columnar, pseudostratified columnar, stratified squamous,* and *transitional.* Refer to Figures 21-40, 21-41, and 21-42 as you find and examine these subtypes in the following slides.

1. *Whole mount of mesentery* (Figure 21-41). The mesentery is a fold of the abdominal wall and holds the intestines in place. This slide is a surface view of the **simple squamous epithelium,** which lines the inside of the ventral body cavities and the organs within.

What is the shape of the surface cells? _____

Draw what you see in Figure 21-22 and note the total magnification of the compound microscope you're using. Repeat this procedure for each subsequent drawing.

2. *Kidney* (Figure 21-42). With the help of Figure 21-23, find a *renal corpuscle.* A renal corpuscle is composed of a tuft of capillaries surrounded by a hollow capsule formed from the cup-shaped end of one of the kidney's functional units, the nephron. Identify a side view of the simple squamous epithelium that comprises the outer wall of the capsule. Around the renal corpuscle, you can see a number of transverse and oblique sections of the tubular portion of the nephron. Their walls are composed of **simple cuboidal epithelium.** The main function of the nephrons and the entire kidney is the production of urine.

3. *Trachea* (Figure 21-40). The trachea is a tubular organ that conveys air to and from the lungs. Note that its inner surface is lined by **pseudostratified columnar epithelium** (Figure 21-24). Locate the unicellular glands, which are called **goblet cells** because of their shape. They secrete mucus that is difficult to see unless it is specifically stained. Using high power, do you see the numerous hairlike structures that project from the surface of the columnar epithelial cells? These are **cilia,** which in life move synchronously, sweeping mucus, trapped bacteria, and debris up the trachea to the throat. When you clear your throat, you collect this mucus and swallow it.

Figure 21-22 Drawing of a surface view
of simple squamous epithelium (____×).

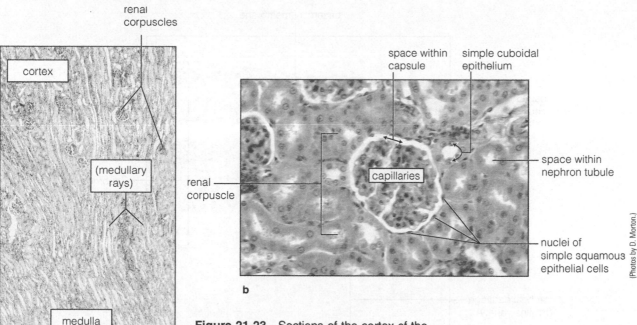

renal
corpuscles

cortex

(medullary
rays)

medulla

a

space within   simple cuboidal
capsule   epithelium

renal
corpuscle

capillaries

space within
nephron tubule

nuclei of
simple squamous
epithelial cells

(Photos by D. Morton.)

b

Figure 21-23 Sections of the cortex of the kidney, **(a)** low-power (25×) and **(b)** medium-power (183×) views.

4. *Small intestine* (Figure 21-41). The small intestine is a tubular organ that connects the stomach to the large intestine. Its inner surface is lined with **simple columnar epithelium** (Figure 21-25). Note that *goblet cells* are present. Don't confuse the relatively thinner **brush border of microvilli** with the cilia from the previous slide. Microvilli are primarily responsible for the large surface area of the small intestine. Individual microvilli can best be seen at the electron-microscopic level (Figure 21-26).

5. *Esophagus* (Figure 21-40). The tubular esophagus connects the throat to the stomach. Examine the epithelium lining the inner surface of the esophagus (Figure 21-27).

   Is the shape of the outermost cells squamous, cuboidal, or columnar? _____

   Are there one or many layers of cells in this tissue? _____

   Name this subtype of epithelial tissue. _____

6. *Skin.* The skin is divided into three layers (Figure 21-28). The *epidermis* is composed of stratified squamous epithelium, while the *dermis* and *hypodermis* (or subcutaneous layer) are connective tissue and will be studied later.

The epidermis is the most extreme example of a protective epithelium. The process of keratinization, whereby the cells transform themselves into bags of the protein keratin, causes strata in the epidermis. It is this substance that gives skin its tough, flexible, and water-resistant surface.

At one of the free edges of your section of mammalian skin, locate the **keratinized stratified squamous epithelium.** It will be stained bluer than the predominately pink connective tissue layers. Hair follicles and multicellular sweat glands may be present in the connective tissue layers. These structures grow into the connective tissue layers from the epidermis during the development of the skin.

7. *Urinary bladder* (Figure 21-42). Figure 21-29 shows two sections, one from a contracted bladder, the other from a distended bladder. Locate the **transitional epithelium** at the surface of each section. The transitional epithelium from the contracted bladder appears like stratified cuboidal epithelium. In the distended bladder, the transitional epithelium is thinner and looks like stratified squamous epithelium. Draw a high-power view of these two extremes in Figure 21-30.

B. Connective Tissue

Connective tissues occur in all parts of the body. They contain a large amount of material external to the cells, called the *extracellular matrix*. This matrix consists of *fibers* embedded in *ground substance*. Two basic kinds of fibers exist: collagen and elastic. Collagen fibers are tough, flexible, and inelastic, whereas elastic fibers stretch when pulled, returning to their original length when the pull is removed. Except for cartilage, connective tissues contain blood vessels. Similar to epithelia, connective tissue cells are capable of cell division in the adult, but at a reduced rate. Table 21-1 lists most connective tissue subtypes along with their matrix characteristics.

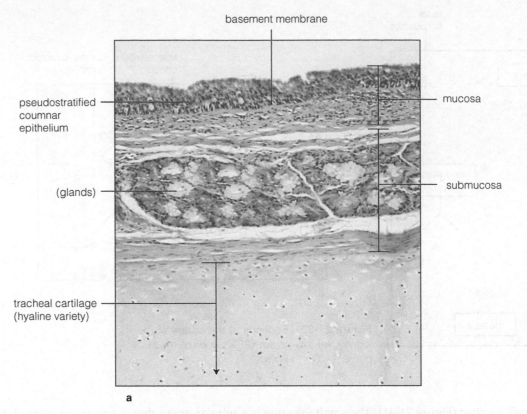

basement membrane

pseudostratified
coumnar
epithelium

mucosa

(glands)

submucosa

tracheal cartilage
(hyaline variety)

a

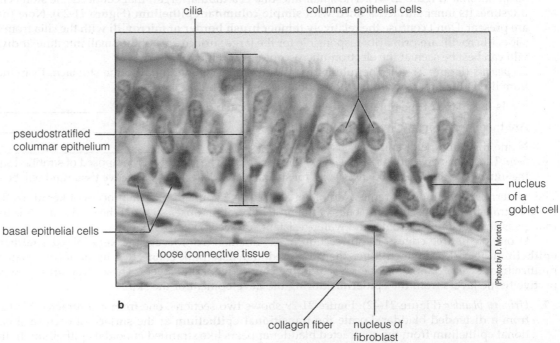

cilia

columnar epithelial cells

pseudostratified
columnar epithelium

nucleus
of a
goblet cell

basal epithelial cells

loose connective tissue

(Photos by D. Morton.)

b

collagen fiber

nucleus of
fibroblast

Figure 21-24 Sections of the inner surface of the trachea, **(a)** low-power (100×)
and **(b)** oil immersion (1000×) views.

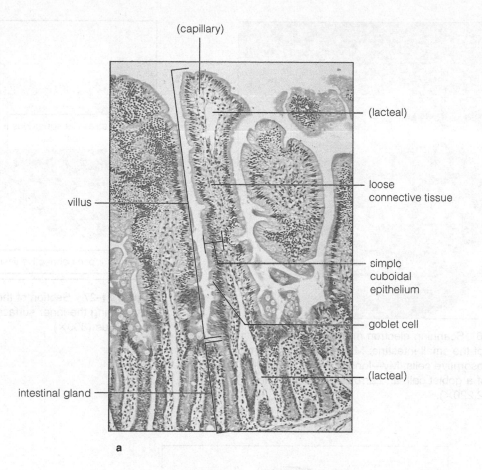

(capillary)

(lacteal)

villus

loose
connective tissue

simple
cuboidal
epithelium

goblet cell

(lacteal)

intestinal gland

a

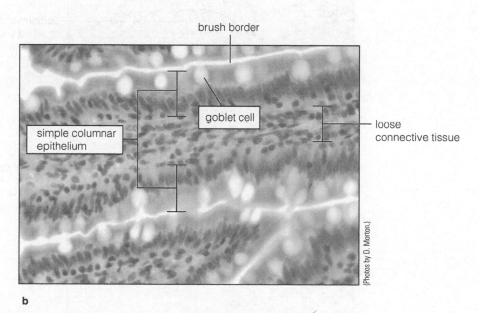

brush border

goblet cell

loose
connective tissue

simple columnar
epithelium

(Photos by D. Morton.)

b

Figure 21-25   Sections of the inner surface of the small intestine, (**a**) medium-power (250×) and (**b**) high-power (400×) views.

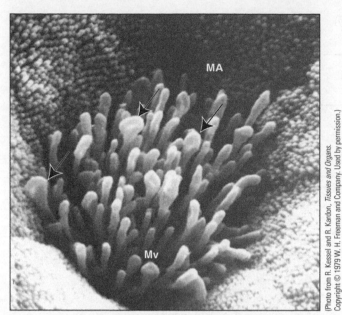

Figure 21-26 Scanning electron microscopic view of the surface of the small intestine: MA—small microvilli on surface of absorptive cells, Mv—longer, larger microvilli on surface of a goblet cell, and arrows indicate droplets of mucus (12,220×).

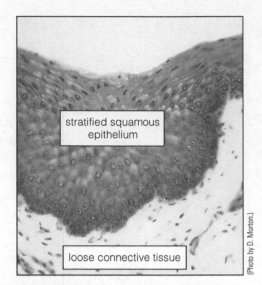

stratified squamous epithelium

loose connective tissue

Figure 21-27 Section of the epithelial tissue lining the inner surface of the esophagus (300×).

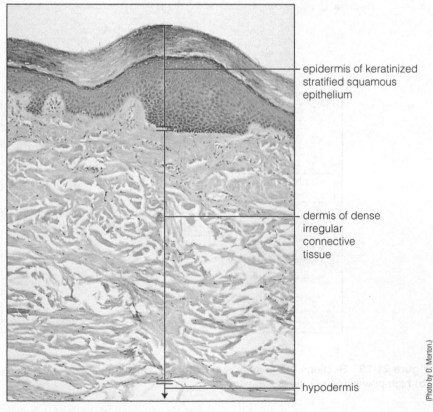

epidermis of keratinized stratified squamous epithelium

dermis of dense irregular connective tissue

hypodermis

Figure 21-28 Section of skin (30×).

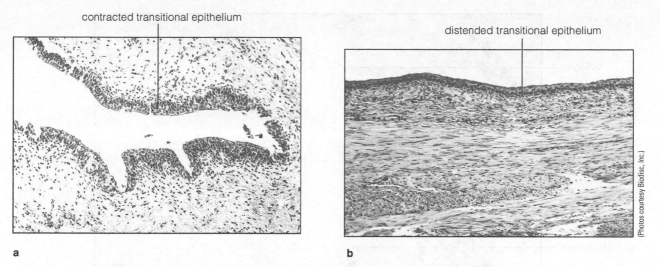

contracted transitional epithelium

distended transitional epithelium

a

b

Figure 21-29 Sections of (**a**) contracted and (**b**) distended urinary bladders (89×).

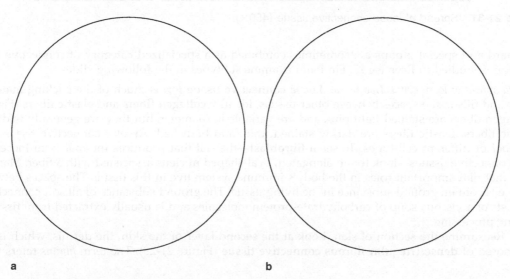

a

b

Figure 21-30 Drawings of high-power views of transitional epithelia from (**a**) contracted and (**b**) distended urinary bladders (_____×).

	TABLE 21-1 Connective Tissue Groups and Common Subtypes	
Group	**Subtypes**	**Matrix Contains**
Soft	Loose (also called areolar)	Few collagen and elastic fibers arranged apparently randomly
	Dense irregular	Many collagen or elastic fibers arranged apparently randomly
	Dense regular	Many collagen or elastic fibers arranged in a parallel fashion
Hard	Cartilage	Many collagen or elastic fibers and polymerized (or jellylike) ground substance
	Bone	Many collagen fibers and mineralized ground substance
Special	Adipose	Delicate collagen fibers
	Blood	Plasma

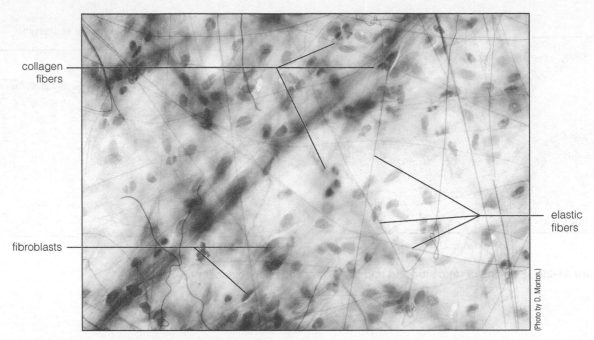

Figure 21-31 Spread of loose connective tissue (450×).

collagen fibers

elastic fibers

fibroblasts

(Photo by D. Morton.)

The hard and special groups are sometimes combined as a specialized category of connective tissue subtypes. Blood is studied in Exercise 25. Find and examine subtypes in the following slides.

1. *Teased spread of loose connective tissue.* **Loose connective tissue** forms much of the packing material of the body and fills in the spaces between other tissues. Identify **collagen fibers** and **elastic fibers** (Figure 21-31). Collagen fibers are stained light pink and are variable in diameter, but they are generally wider than the elastic fibers. Elastic fibers are darkly stained, thin, and branched. Areolar connective tissue contains a number of different cell types. To see a **fibroblast**—the cell that produces the matrix of loose and other soft connective tissues—look for an elongated, oval-shaped nucleus associated with a fiber. Many kinds of cells that play important roles in the body's immune system live in this tissue. The spaces between fibers and cells contain ground substance in the living tissue. The ground substance of all soft connective tissues consists of a viscous soup of carbohydrate-protein molecules and is usually extracted from tissue sections during processing.

2. *Skin.* Reexamine the section of skin. Look at the second layer of the skin, the dermis, which is primarily composed of **dense irregular fibrous connective tissue** (Figure 21-28). The term *fibrous* refers to collagen fibers.

The dermis cushions the body from everyday stresses and strains. Note that the looser irregular connective tissue of the hypodermis has a lower concentration of collagen fibers and islands of fat cells. It functions as a shock absorber, as an insulating layer, and as a site for storing water and energy. In animals that move their skin independently of the rest of the body (like cats), skeletal muscle tissue also is found in the hypodermis.

3. *Tendon.* A tendon connects a skeletal muscle organ to a bone organ. Tendons are composed predominantly of **dense regular fibrous connective tissue.** Examine a longitudinal section of a tendon (Figure 21-32).

Describe the density and arrangement of the fibers.

_____

_____

_____

This design makes tendons very strong, much like a rope composed of braided strings, which in turn are made of even smaller fibers. How are the fibroblasts oriented relative to the arrangement of the fibers?

_____

_____

4. *Trachea* (Figure 21-40). Re-examine the section of the trachea. Find a portion of one of the supporting cartilages, which prevent the wall from collapsing and closing the **lumen** (the space within any hollow organ). These cartilaginous structures are composed of **hyaline cartilage** tissue (Figure 21-33). Look for the **chondrocytes** (cartilage cells) that are located in small holes in the matrix called **lacunae** (singular, *lacuna*).

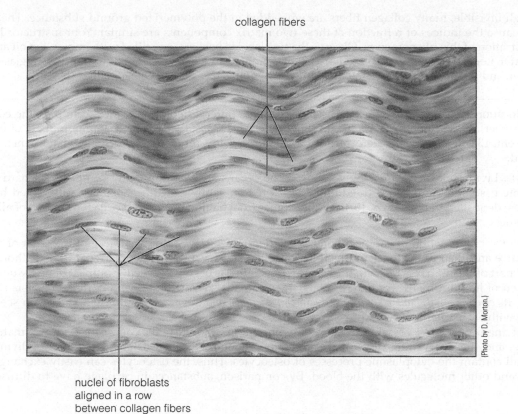

collagen fibers

nuclei of fibroblasts
aligned in a row
between collagen fibers

(Photo by D. Morton.)

Figure 21-32 Dense regular fibrous connective tissue from a longitudinal section of tendon (500×).

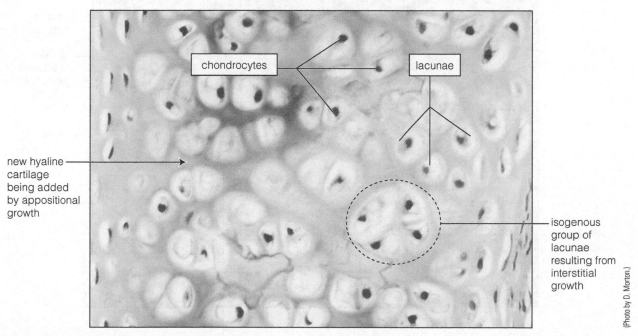

chondrocytes

lacunae

new hyaline
cartilage
being added
by appositional
growth

isogenous
group of
lacunae
resulting from
interstitial
growth

(Photo by D. Morton.)

Figure 21-33 Section of hyaline cartilage in the wall of the trachea (250×).

Although invisible, many collagen fibers are embedded in the polymerized ground substance. They cannot be seen because the indices of refraction of these two matrix components are similar. Your instructor has set up a demonstration of this phenomenon. Observe the two labeled beakers, one filled with immersion oil and the other with water. Look at the glass rod placed in each of them. The index of refraction is about 1.58 for glass, about 1.33 for water, and about 1.52 for immersion oil. In which fluid is it easier to see the glass rod?

In locations where cartilage has to be more durable (intervertebral disks, for example), the collagen content is higher and the fibers are visible. In other sites (such as outer ear flaps), large numbers of elastic fibers are present. Elastic cartilage is deformed by a small force and returns to its original shape when the force is removed.

5. *Bone.* Living bones are amazingly strong. Bone organs are predominantly composed of hard yet flexible **bone** tissue. Its hardness is due to minerals (predominantly calcium-phosphate salts called hydroxyapatites) deposited in the matrix. Its flexibility comes from having the highest collagen content of all connective tissues.

Examine the cross section of compact bone tissue (Figure 21-34). Grinding a piece of the shaft of a long bone with coarse and then finer stones until a thin wafer remains has produced this preparation. Although only the mineral part of the matrix is present, the basic architecture has been preserved. Compact bone tissue is primarily composed of longitudinally arranged **Haversian systems** (also called *osteons*). Locate a Haversian system and identify its *central canal, lamellae* (singular, *lamella*), *lacunae, and canaliculi*—little canals that you see connecting lacunae with each other and with the central canal.

In living bone, blood vessels and nerves are present in central canals, concentric layers of matrix form the lamellae, and the intervening rings of *lacunae* contain bone cells called **osteocytes.** In young living bone, the canaliculi contain the cytoplasmic processes of osteocytes. Thus, the osteocytes can easily exchange nutrients, wastes, and other molecules with the blood. By comparison, substances in cartilage have to diffuse across the

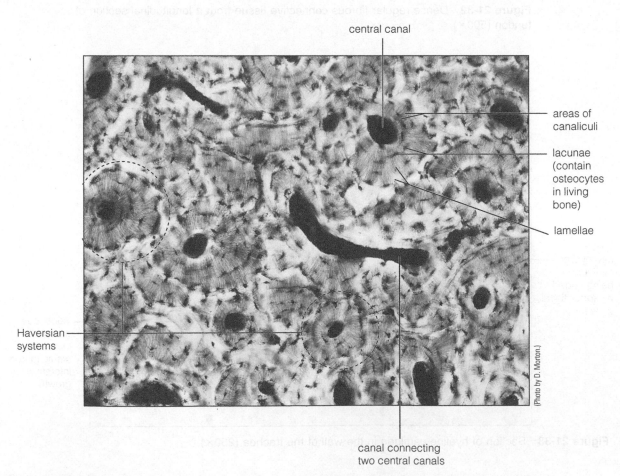

central canal

areas of canaliculi

lacunae (contain osteocytes in living bone)

lamellae

Haversian systems

canal connecting two central canals

(Photo by D. Morton.)

Figure 21-34 Ground cross section of the shaft of a long bone showing compact bone tissue (100×).

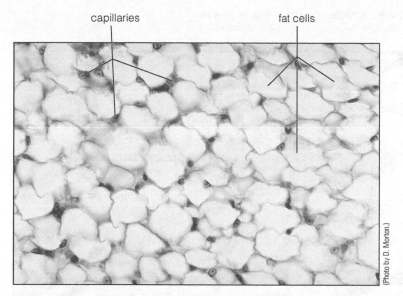

capillaries   fat cells

(Photo by D. Morton.)

Figure 21-35   Section of white adipose tissue (450×).

matrix between chondrocytes and blood vessels in the surrounding soft connective tissue. Which tissue, bone or cartilage, will heal quicker? Explain why you made this choice.

_____

_____

_____

6. *White adipose tissue.* As you see in Figure 21-35, **white adipose tissue** consists mainly of **fat cells.** However, careful examination of the section using high power shows them to be surrounded by delicate collagen fibers, fibroblasts, and capillaries. Because the fat has been lost during the slide preparation, the fat cells look empty. The primary function of this tissue is energy storage.

C. Muscle Tissue

Muscle tissue is contractile. Its cells (fibers) can shorten and produce changes in the position of body parts, or they try to shorten and produce changes in tension. Contraction results from interactions between two types of protein filaments: **actin** and **myosin.** Like epithelial tissue, muscle tissue is primarily cellular; but unlike both epithelial and connective tissues, its cells (fibers) do not normally divide in the adult. Therefore, dead fibers are typically not replaced. There are three subtypes of muscle tissue: *skeletal, cardiac,* and *smooth.*

Skeletal muscle tissue is under *voluntary control.* This means that your conscious mind can order skeletal muscles to contract, although most of their contractions are actually involuntary. Cardiac muscle and smooth muscle tissues are always under *involuntary control.* Involuntary control means contractions are directed at the unconscious level.

Examine your slide with sections of all three subtypes (Figure 21-36). Identify their characteristics (see Table 21-2). In skeletal and cardiac muscle fibers, actin and myosin filaments overlap to produce the alternating pattern of light and dark bands (cross or transverse striations) seen in these tissues. Only cardiac muscle cells are

TABLE 21-2	Characteristics of Muscle Tissue Subtypes			
Subtypes	**Organ Location**	**Position and Number of Nuclei in Fibers**	**Cross-Striated Fibers**	**Special Fiber Features**
Skeletal	Skeletal muscles	Peripheral and many	Yes	Long cylindrical shape
Cardiac	Heart	Central and usually one	Yes	Branches and intercalated disks
Smooth	Walls of internal organs	Central and one	No	Shape tapers at both ends

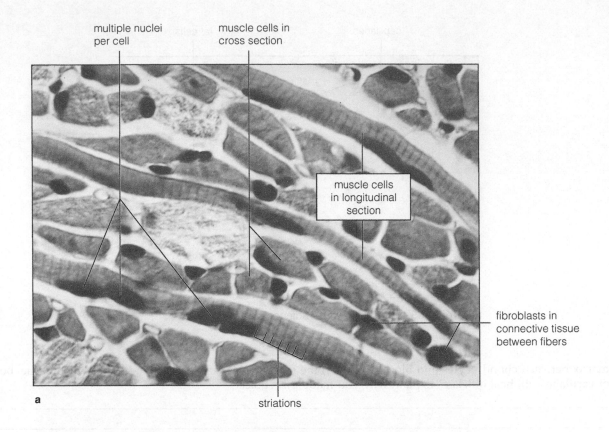

multiple nuclei per cell

muscle cells in cross section

muscle cells in longitudinal section

fibroblasts in connective tissue between fibers

a

striations

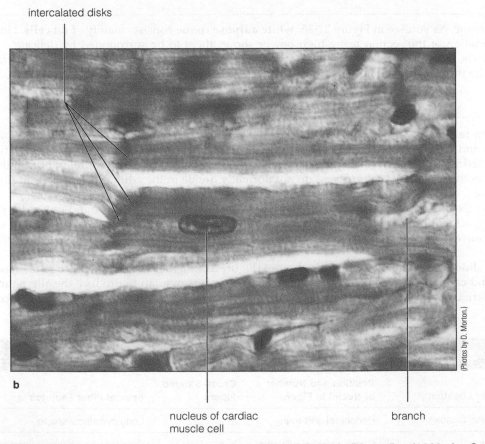

intercalated disks

(Photos by D. Morton.)

b

nucleus of cardiac muscle cell

branch

Figure 21-36 Sections of muscle tissue: (**a**) Skeletal (400×), (**b**) cardiac (1000×). *Continues.*

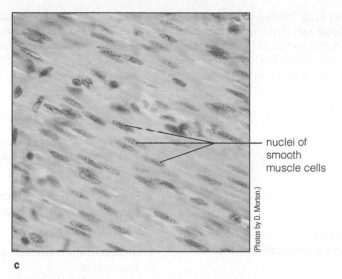

nuclei of
smooth
muscle cells

(Photos by D. Morton.)

c

Figure 21-36 *Continued.* Sections of muscle tissue: (**c**) smooth (400×).

branched. Where the branch of one fiber joins another, the cells are stuck together and in direct communication through a complex of cell-to-cell junctions. The complex is called an **intercalated disk.** Find an intercalated disk in your section, but if you have trouble seeing it, look at the demonstration set up by your instructor.

D. Nervous Tissue

Nervous tissue is found in the brain, spinal cord, nerves, and all of the body's organs. Its function is the point-to-point transmission of information. Messages are carried by impulses that travel along the functional unit of the nervous system, the **neuron** (or nerve cell). Like muscle cells, neurons do not divide in the adult, and therefore natural replacement is impossible.

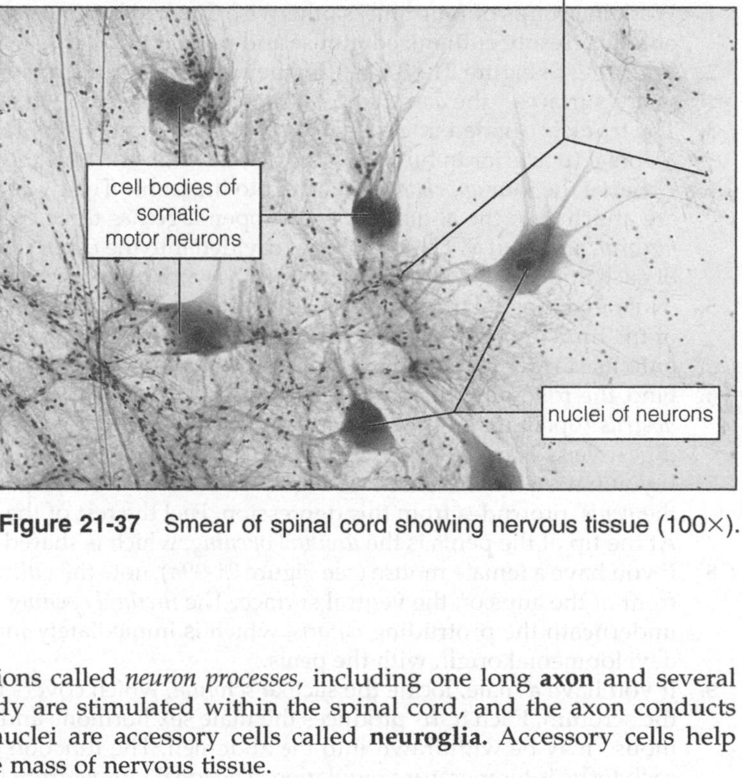

The largest cells in Figure 21-37 are *motor neurons*, which connect the spinal cord to muscle fibers or glands. Find a similar cell in your smear of an ox spinal cord. The motor neuron has a **cell body**

smaller nuclei of numerous or neuroglial accessory cells

cell bodies of somatic motor neurons

nuclei of neurons

(Photo by D. Morton.)

Figure 21-37 Smear of spinal cord showing nervous tissue (100×).

and a number of slender cytoplasmic extensions called *neuron processes*, including one long **axon** and several shorter **dendrites.** The dendrites and cell body are stimulated within the spinal cord, and the axon conducts impulses out of it. The cells with smaller nuclei are accessory cells called **neuroglia.** Accessory cells help neurons function and make up about half the mass of nervous tissue.

| 21.6 | Observation: Mouse Dissection *(50 min.)* |

There are 11 systems in mammals, and each one contains a number of organs. In this portion of the exercise, you will identify the vital functions of these systems and many of their constituent organs. Many of these organs are located in or near the major **body cavities.** The *ventral* (located toward the belly surface) and *dorsal* (located toward the back surface) *body cavities* of humans are typical of mammals. A muscular *diaphragm* separates the **thoracic cavity** and **abdominal** cavities. The thoracic cavity is further subdivided into two lateral (away from the midline) **pleural cavities** and a medial (at or near the midline). Preservation of animals for dissection changes the appearance and consistency of the body and its organs. There are obvious advantages to observing the insides of a mammalian body much as a surgeon does during an operation.

MATERIALS

Per group (4):

- scalpel
- dissecting scissors
- 4 dissecting pins
- dissecting needle
- blunt probe
- freshly euthanized mouse
- squeeze bottle containing 0.9% saline solution
- cotton balls
- dissection pan

- small plastic bag for disposal of organs and soiled cotton balls during the dissection

Per lab section:

- large plastic bag for disposal of the small plastic bags and the carcasses at the end of the exercise

Per lab room:

- demonstration dissection of the nervous system of the mouse
- boxes of different sizes of latex gloves

PROCEDURE

A. External Features

1. Work in groups of four unless otherwise directed by your instructor. After putting on a pair of latex gloves, obtain a freshly euthanized mouse and place it in the dissection pan.
2. Referring to Figure 21-38, identify the four major body regions—the *head, neck, trunk,* and *tail*—and the two major surfaces—the *dorsal surface* (back) and the *ventral surface* (belly).
3. The trunk is divided into a ventral (anterior in humans) portion, the *thorax* (supported by the rib cage), and a dorsal (posterior in humans) portion, the *abdomen.* Feel the rib cage and the soft abdomen.
4. Observe the four *appendages* attached to the trunk. The two *arms* are attached to the thorax, and the two *legs* are attached to the abdomen. Each appendage has three segments. The arm is divided into the *upper arm, forearm,* and *front foot,* and the leg is divided into the *thigh, shank,* and *hind foot.* There are five *digits* at the end of each appendage, although the thumb is reduced in size and has a *nail* rather than a *claw.*
5. Note that most of the mouse's body is covered by hair, which traps air and forms an insulating layer. Most of the human body surface is sparsely covered by hair and has sweat glands in the skin that secrete sweat onto its surface for cooling via evaporation. Mice have sweat glands only on the pads of their feet.
6. Find the following openings on the body surface: *mouth* (entrance to the digestive system), two *external nostrils* (openings of the respiratory system), two *external auditory canals* (in ears), and *anus* (exit from the digestive system)
7. If you have a male mouse (see Figure 21-39b), note the *preputial opening.* A portion of the copulatory organ, the *penis,* protrudes from this depression. Feel the rest of the penis under the skin of the preputial opening. At the tip of the penis is the *urethral opening,* which is shared by the reproductive and urinary systems.
8. If you have a female mouse (see Figure 21-39a), note the *vulva.* This opening of the reproductive system is in front of the anus on the ventral surface. The *urethral opening* is the exit of the urinary system and is located underneath the protruding *clitoris,* which is immediately in front of the vulva. The clitoris has a common developmental origin with the penis.
9. If you have a male, locate the saclike *scrotum,* which covers the anus ventrally. Feel for the paired testes in the scrotum. Each *testis* produces the male sex hormone and sperm. Unlike human males, the testes of the mouse may be withdrawn into the abdomen. The function of the scrotum and the muscles that attach to each testis is temperature regulation. The ideal temperature for sperm production is slightly less than body temperature.

If you have a mature female, find the well-developed teats of the *mammary glands.* How many do you count?

10. Exchange your mouse for one of the opposite sex from another group and examine the differences in external features.

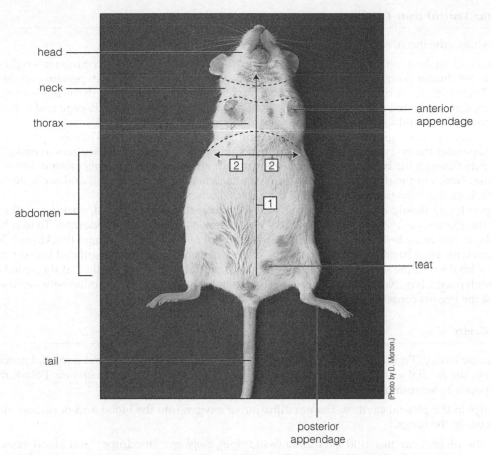

head

neck

thorax

abdomen

tail

anterior
appendage

teat

posterior
appendage

(Photo by D. Morton.)

Figure 21-38 How to open the ventral body cavities. The numbered arrows indicate incisions.

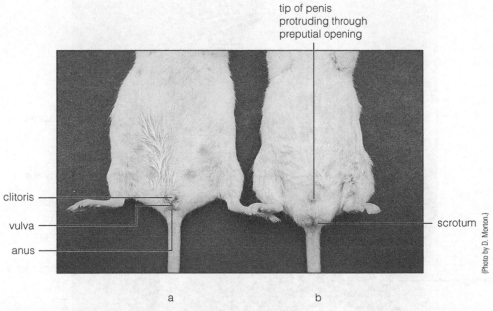

tip of penis
protruding through
preputial opening

clitoris

vulva

anus

scrotum

(Photo by D. Morton.)

a

b

Figure 21-39 Posterior halves of the ventral surfaces of (**a**) female and (**b**) male mice.

B. Opening the Ventral Body Cavities

Figure 21-38 shows the incisions (cuts) necessary to open the ventral body cavities.

1. With a scalpel, make a longitudinal incision through the skin just to the right (the mouse's right) of the midline. This cut should extend from the neck to the preputial opening or clitoris, depending on the sex of your mouse. Then with a pair of scissors pierce the body wall just below the ribs. With the blades angled upward to prevent damage to the internal organs, cut along the incision through the rib cage and collar bone. Turn the scissors around and likewise cut through the abdominal muscles.

2. Pull the sides of the longitudinal incision apart and look for the *diaphragm*. This thin partition of skeletal muscle separates the body cavities of the thorax and abdomen. Again use the scissors to make two perpendicular cuts through the body wall just below the diaphragm. Extend these cuts around and to the back of the mouse. Now cut the diaphragm away from its attachment to the body wall and separate the organs of the thoracic cavity from their attachment to the ventral body wall.

3. At this point, you should be able to fold back four triangular flaps of body wall, opening up the *ventral body cavities*, the *thoracic cavity* in the thorax, and the *abdominopelvic cavity* in the abdomen. To fold back the two upper flaps, you must break the ribs near their attachment to the *vertebral column* (backbone). You can easily do this with your hands. Pin the flaps to the dissecting tray. Rinse any coagulated blood from the body cavities with 0.9% saline solution and remove any excess fluid with cotton balls. Put the soiled cotton balls in the small plastic bag. Note that a thin, wet serosal membrane that reduces friction between the inner body wall and the organs contained therein lines each body cavity.

C. Thoracic Cavity

1. The thoracic cavity (Figure 21-40) is comprised of two lateral *pleural cavities* and a medial *pericardial cavity*. At the top, the medial walls of the pleural sacs join to form a septum, the *mediastinum*. Below, the mediastinum expands to surround the pericardial sac.

Locate the *lungs* in the pleural cavities. The net diffusion of oxygen into the blood and of carbon dioxide out of the blood occurs in the lungs.

2. Between the pleural cavities, find the *trachea* (windpipe), *esophagus* (foodpipe), and blood vessels traveling to and from the head in the mediastinum. Especially in young mice, you must remove the *thymus*, which

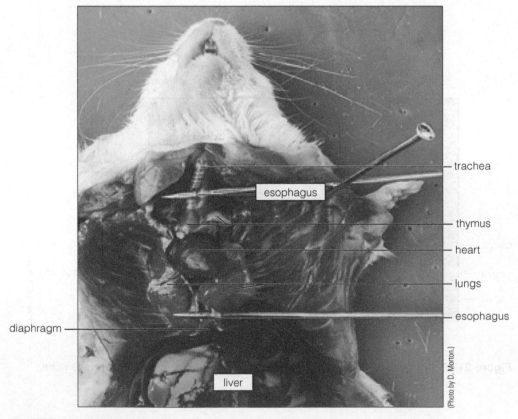

Figure 21-40   Ventral view of organs in thoracic cavity (2×).

covers the mediastinum. The thymus is important in the development of the immune defenses. The size and function of the thymus decrease with age. Dispose of the thymus in the small plastic bag.

3. Open the pericardial sac and examine the pump of the circulatory system, the *heart*. Just above and behind the heart, find where the trachea divides into a right and left *bronchus*. During the process of inspiration, the bronchi carry air from the trachea toward the lungs. Under the trachea and pericardial sac is the esophagus. Do you see where it pierces the diaphragm, passing into the abdominopelvic cavity?

D. Abdominopelvic Cavity

1. Referring to Figure 21-41, observe just posterior to the diaphragm the body's largest gland, the *liver*. It has several lobes. Lift them and look for a reddish brown sac, the *gall bladder*. The gall bladder stores *bile* secreted by the liver. When the gall bladder contracts, bile is squirted through a duct into the small intestine to aid in the digestion of fats.

2. Locate the *stomach* in the left side of the abdominopelvic cavity, under the liver. The stomach helps process swallowed food. Can you discover where the esophagus joins the stomach at its anterior end? Examine the *spleen*, a dark red organ located just to the left of the stomach. The spleen filters the blood, removes old blood cells, and plays a role in immunity.

3. Note where the *small intestine* arises from the lower end of the stomach. The small intestine completes the digestion of food and is the main site for absorption of the end products of digestion. It is the longest organ in the body, and its convolutions fill most of the abdominopelvic cavity. A double membrane, the mesentery, attaches the small intestine to the dorsal body wall.

4. To observe the *pancreas*, look for a diffuse organ in the first loop of the small intestine. This gland produces enzymes for the digestion of food and, as part of the endocrine system, secretes *insulin* and other hormones into the blood.

5. Find the lower end of the small intestine, the place where it empties into the *large intestine*. The large intestine is shaped like an upside-down U. To see it, lift the small intestine. Follow the large intestine until it joins the short *rectum*, which terminates at the anus. Return to the junction between the small and large intestines. Note the blind sac extending in the direction opposite from the large intestine. This is the *cecum*, which in humans is a relatively small structure. In humans a blind twisted tube, the *vermiform appendix*, is attached to the cecum.

6. Carefully remove the abdominal portion of the digestive tract by cutting (a) the esophagus between the diaphragm and stomach, (b) the dorsal mesentery, and (c) the descending portion of the large intestine (the third arm of the upside-down U). Do not disturb the urinary bladder, which lies in front of the cecum. Also, if you have a female, do not disturb the reproductive system, which is situated under the small intestine and in front of the rectum.

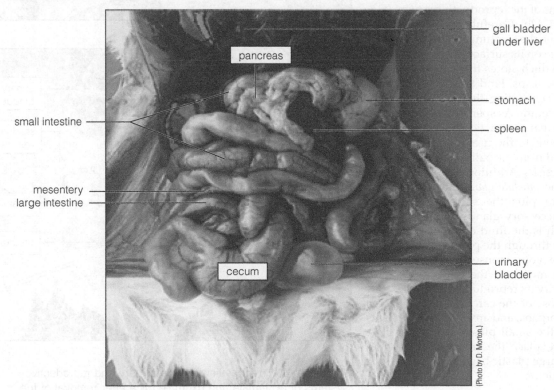

Figure 21-41 Ventral view of organs of abdominopelvic cavity (2×).

Stretch out the length of the small intestine and measure it.

Length of small intestine = _____ cm

Now measure the mouse from the tip of the nose to the base of the tail.

Length of head and body of mouse = _____ cm

How many times the length of the mouse is its small intestine? _____ times

Dispose of the stomach and intestines in the small plastic bag.

7. The *urinary bladder* stores urine delivered via two *ureters* from the paired *kidneys* (Figure 21-42). Find the kidneys, which are located on the dorsal body wall under the serosal membrane, which in this location is called the *peritoneum*. With a dissecting needle, tear away the serosal membrane above one kidney. Each kidney is like a thick C, with the indentation pointing toward the midline of the body. The ureter arises from the center of this concave surface. Trace the ureter to the urinary bladder. At the lower end of the urinary bladder find the *urethra*, the duct that conveys urine out of the body when the bladder contracts.

8. If your mouse is male, skip to step 9. If you have a female mouse, examine the organs of the reproductive system (Figure 21-42a). Find the *body of the uterus* immediately behind the urinary bladder. Trace the body of the uterus upward until it branches into two *uterine horns*, which pass under the ureters and extend almost to the kidneys. Here each uterine horn joins a short *oviduct*, which expands to help form a sac surrounding the *ovary*. The paired ovaries are the source of female sex hormones and eggs. If the mouse is pregnant, the uterine horns will be quite large. Follow the uterus in the other direction and see where it joins the *vagina*, which in turn opens into the vulva.

Skip step 9 and go on to step 10.

9. If you have a male mouse, examine the organs of the reproductive system (Figure 21-42b). Carefully use your scissors to open the scrotum and expose a *testis*. Locate on its surface the coiled *epididymis*, which stores the sperm produced by the testis. Find the *vas deferens*, the tube connecting the epididymis with the urethra. As sperm are propelled by muscular contractions through the vas deferens to the urethra, secretions are added from the paired, glandular *seminal vesicles*. Additional fluid is secreted by the *prostate gland* into the urethra. Sperm plus the secretions of these sex accessory glands comprise *semen*, which is the fluid ejaculated from the body through the penis.

10. Trade your mouse to another group for a mouse of the opposite sex and identify its reproductive organs.

11. Dispose of the carcass, cut-out tissues and organs, and any soiled cotton balls into the small plastic bag. After closing it, place the small plastic bag in the large plastic bag provided by your instructor.

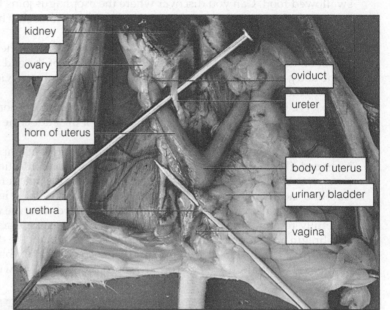

a

b

(Photos by D. Morton.)

Figure 21-42 Ventral views of the urinary and reproductive systems of (**a**) female and (**b**) male mice after removal of the digestive system (3×).

_____ 1. The study of a plant's structure is
  (a) physiology
  (b) morphology
  (c) taxonomy
  (d) botany

_____ 2. A plant with two seed leaves is
  (a) a monocotyledon
  (b) a dicotyledon
  (c) exemplified by corn
  (d) a dihybrid

_____ 3. A taproot system lacks
  (a) lateral roots
  (b) a taproot
  (c) both of the above
  (d) none of the above

_____ 4. Which structure is not part of the shoot system?
  (a) stems
  (b) leaves
  (c) lateral roots
  (d) axillary buds

_____ 5. An axillary bud
  (a) is found along internodes
  (b) produces new roots
  (c) is the structure from which branches and flowers arise
  (d) is the same as a terminal bud

_____ 6. The endodermis
  (a) is the outer covering of the root
  (b) is part of the vascular tissue
  (c) contains the Casparian strip, which regulates the movement of substances
  (d) is none of the above

_____ 7. Meristems are
  (a) located at the tips of stems
  (b) located at the tips of roots
  (c) regions of active growth
  (d) all of the above

_____ 8. To determine the age of a woody twig, one counts the
  (a) nodes
  (b) leaf scars
  (c) lenticels
  (d) regions between sets of terminal bud scale scars

_____ 9. The midrib of a leaf
  (a) contains the midvein
  (b) contains only xylem
  (c) is part of the spongy mesophyll
  (d) contains only phloem

_____ 10. The bundle sheath in a monocot leaf
  (a) is filled with intercellular space
  (b) is the location of stomata
  (c) is covered with a cuticle to prevent water loss
  (d) functions in a manner somewhat similar to the root endodermis

_____ 11. Histology is the study of
  (a) cells
  (b) organelles
  (c) tissues
  (d) organisms

_____ 12. A collection of similarly specialized cells and any extracellular material they secrete and maintain describes
  (a) an organ
  (b) a system
  (c) a tissue
  (d) organelles

_____ 13. Organs strung together functionally and usually structurally form
  (a) organs
  (b) systems
  (c) tissues
  (d) organelles

_____ 14. _____ are constructed of all four basic tissue types.
  (a) organs
  (b) systems
  (c) cells
  (d) organelles

_____ 15. An epithelial tissue formed by more than one layer of cells and with columnlike cells at the surface is called
  (a) simple squamous
  (b) stratified squamous
  (c) simple columnar
  (d) stratified columnar

_____ 16. The middle layer of the skin
  (a) is called the dermis
  (b) is primarily connective tissue
  (c) contains collagen fibers
  (d) is all of the above

_____ 17. To which subtype of muscle tissue does a fiber with cross striations and many peripherally located nuclei belong?
  (a) skeletal
  (b) cardiac
  (c) smooth
  (d) none of the above

_____ 18. In which tissue would you look for cells that function in point-to-point communication?
  (a) connective
  (b) epithelial
  (c) muscle
  (d) nervous

_____ 19. The ventral body cavities include
  (a) the thoracic cavity
  (b) the cranial cavity
  (c) the abdominopelvic cavity
  (d) both a and c

_____ 20. The cranial cavity contains
  (a) the lungs and heart
  (b) the spinal cord
  (c) the brain
  (d) both b and c

Name _____ Section Number _____

EXERCISE 21

Plant and Animal Organization

POST-LAB QUESTIONS

Introduction

1. In the correct order from smallest to largest, list the levels of organization present in most animals.

21.1 External Structure of the Flowering Plant

2. What type of root system do you see on the dandelion at the right?

(0.25×).

21.3 The Shoot System: Stems

3. On the figure below, identify structures **a**, **b**, and **c**.

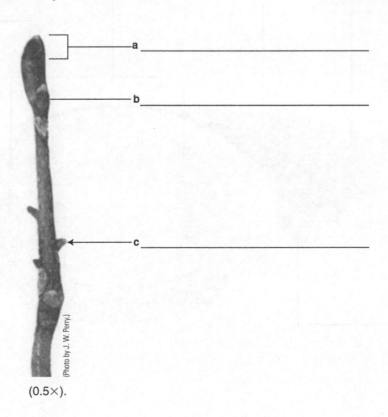

a _____

b _____

c _____

(0.5×).

4. Identify the structures labeled **a** and **b** on the figure below.

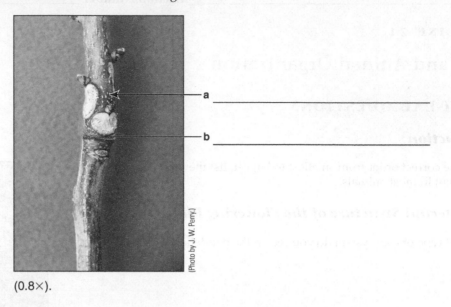

a _____

b _____

(Photo by J. W. Perry.)

(0.8×).

5. What feature(s) would you use to determine the age of a woody twig?

6. Label the diagram using the following terms: bark; heartwood; sapwood; secondary phloem; vascular cambium.

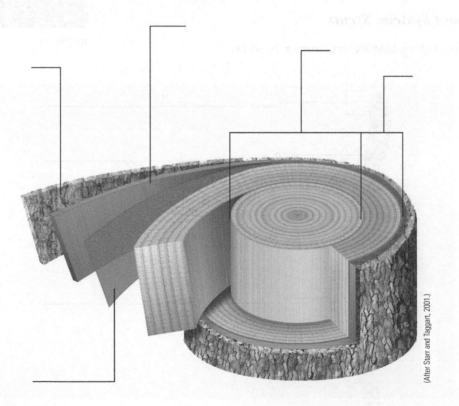

(After Starr and Taggart, 2001.)

7. The photo shows the microscopic appearance of maple wood. Using your knowledge of the woody stem section, identify cell type **a** and the "line" of cells at **b**. (Note: The outside of the tree from which this section was taken is shown in the photo that accompanies question 8.)

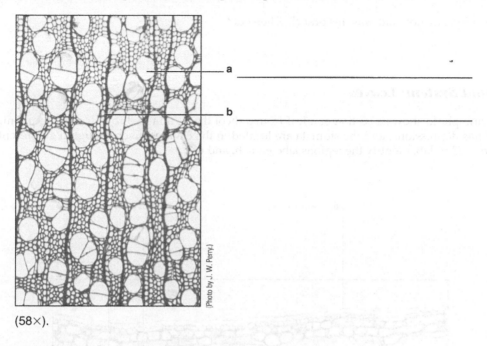

(58×).

a _____

b _____

8. A section cut from an ash branch is shown below.

a.   Identify region **a.**

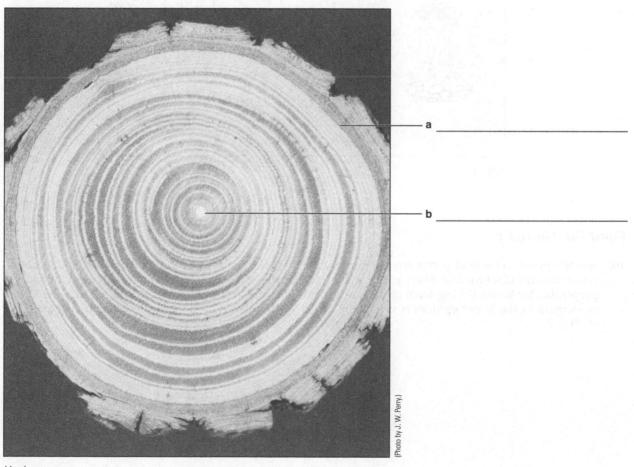

(1×).

a _____

b _____

b.   Which meristem is located at **b**?

c.   Within ±3 years, how old was this branch when cut?

21.4 The Shoot System: Leaves

9.   The following photo shows a section of a leaf from a dicot that is adapted to a dry environment. Its lower epidermis has depressions, and the stomata are located in the cavities. Even though it's different from the leaf you studied in lab, identify the regions labeled **a, b,** and **c.**

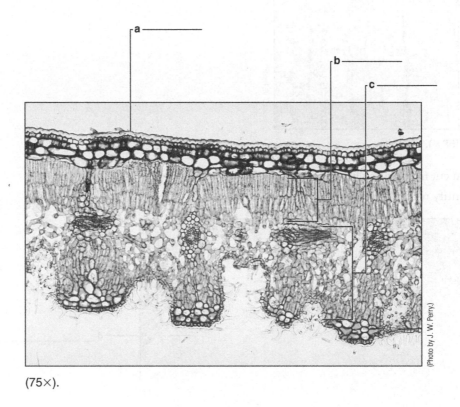

(Photo by J. W. Perry.)

(75×).

Food for Thought

10.   A major problem for land plants is water conservation. Most water is lost through stomata due to evaporation at the surface of the leaf. Many plants, including lilac (the leaf section you examined), orient their leaves perpendicular to the drying force of the sun's rays. What did you observe about the relative abundance of stomata in the lower epidermis versus the upper epidermis? Why do you think this distribution has evolved?

21.5 Tissues

11. Describe the main structural characteristics of the four basic tissues.

    a. epithelial tissue

    b. connective tissue

    c. muscle tissue

    d. nervous tissue

12. Describe the main functions of the four basic tissues.

    a. epithelial tissue

    b. connective tissue

    c. muscle tissue

    d. nervous tissue

13. Identify the following tissues.

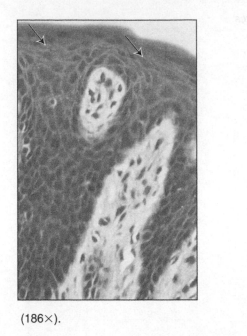

(186×).

a _____

space within nephron tubule

(Photos by D. Morton.)

(480×).

b _____

21.6 Observation: Mouse Dissection

14. Identify the following structures.

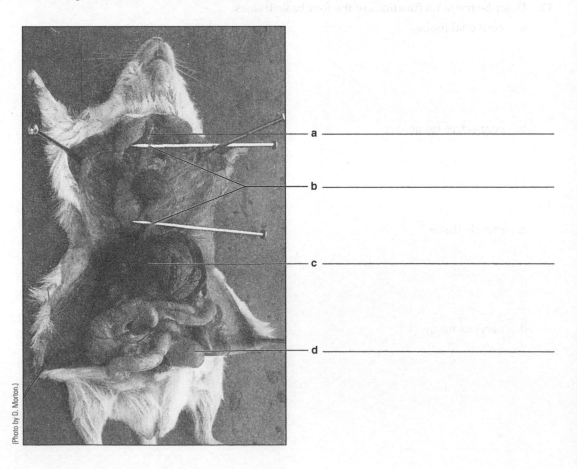

(Photo by D. Morton.)

a _____

b _____

c _____

d _____

Food for Thought

15. Bone is a subtype of connective tissue and bones are organs. How are the two related yet different from each other? Can you think of another tissue/organ pair that potentially creates a similar confusing situation?

16. Search the phrase "tissue engineering" on the World Wide Web. List two sites and briefly summarize their content.

    http://

    http://

Dissection of the Fetal Pig

After completing this exercise, you will be able to

1. define *fetus, digitigrades locomotion, plantigrade locomotion, biped, antagonistic muscles, diaphragm, thoracic cavity, abdominopelvic cavity, exocrine gland, endocrine gland, digestive tract, blood vessels, arteries, veins, capillaries, pulmonary and systemic circuits of circulatory system, portal vein, urea, peritoneum, urine, urinary bladder, homologous, ovulation, semen, inguinal hernia, vasectomy, nephron,* and *meningitis;*

2. locate and describe the external features of a fetal pig;

3. determine the sex of a fetal pig;

4. describe the function of the umbilical cord;

5. define the origin, insertion, and action of a skeletal muscle;

6. identify some of the major skeletal muscles and muscle groups of a mammal;

7. locate the organs of the digestive, respiratory, and circulatory systems in a fetal pig;

8. describe and give the functions of the organs of the digestive, respiratory, and circulatory systems in a living mammal;

9. explain the importance of the digestive, respiratory, and circulatory systems to a living mammal;

10. trace the pathway of ingested food through the digestive tract;

11. trace the pathway of oxygen and carbon dioxide into and out of the lungs of a mammal;

12. identify the major blood vessels of a fetal pig;

13. locate, name; and describe the functions of the chambers of the heart;

14. name the internal structures of the heart;

15. locate the organs of the urinary, reproductive, and nervous systems in a fetal pig;

16. describe and give the functions of the organs of the urinary, reproductive, and nervous systems;

17. explain the importance of the urinary, reproductive, and nervous systems to a living mammal;

18. locate, name, and describe the function of the internal structures of the kidney;

19. list in order the three meninges;

20. explain how the spinal nerves are connected to the spinal cord;

21. identify and give the general functions of the structures of the brain and spinal cord.

INTRODUCTION

In this exercise, you will examine in some detail the external and internal anatomy of a fetal (from **fetus,** an unborn mammal) pig. As the pig is a mammal, many aspects of its structural and functional organization are identical with those of other mammals, including humans. Thus, a study of the fetal pig is in a very real sense a study of ourselves.

The specimens you will use were purchased from a biological supply house, which obtained them from a plant where pregnant sows are slaughtered for food. On average, a sow produces 7–12 offspring per litter. The period of development in the uterus (*gestation period*) is approximately 112–115 days. Generally, lab specimens are approximately 20–30 cm (8–12 in.) long, and their age is between 100 days and nearly full term.

At the slaughterhouse, fetuses are quickly removed from the sow, then cooled and embalmed with preservative, which is injected through one of the umbilical arteries. Following this, the arterial and venous systems are injected under pressure with latex or a rubberlike compound. Red latex is injected into the arteries through the umbilical artery. Then an incision is made in the throat and blue latex is injected into the jugular and other veins.

While dissecting the fetal pig, keep several points in mind. First, be aware that *to dissect* does not mean "to cut up," but rather primarily "to expose to view." Thus, proceed carefully and never cut or move more than is necessary to expose a given part. Second, for each structure or organ that you identify in the pig, ask yourself if an equivalent one is present in your body. If so, where is it located, and is its function similar to that in the fetal pig? Finally, pay particular attention to the spatial relationships of organs, glands, and other structures as you expose them. Carefully identify each structure and determine its organ system, its relationship to that organ system, and its general function. Then determine how it relates to the other organ systems in the body. By proceeding in this manner, you will greatly enhance your understanding of the structure and function of the mammalian body.

To understand the dissection directions, you need to be familiar with the terms used in virtually all anatomical work. Spend a few minutes relating each of these terms to the fetal pig body and to your body as well. Figure 22-1 will aid you in this activity.

Caution

Preserved specimens are kept in preservative solutions. Use latex gloves whenever you handle a specimen. Wash any part of your body exposed to this solution with copious amounts of water. If preservative solution is splashed into your eyes, wash them with the safety eyewash bottle for 15 minutes. If you wear contact lenses during a dissection, wear safety goggles.

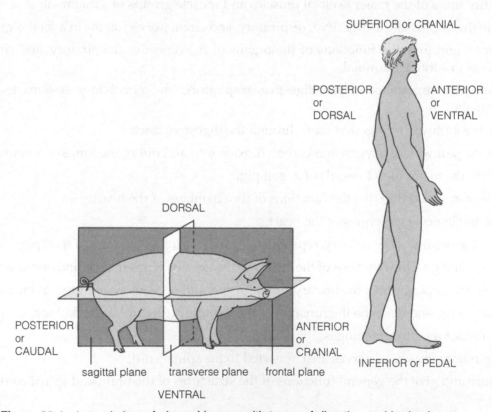

Figure 22-1 Lateral view of pig and human with terms of direction and body planes.

22.1   External Anatomy of the Fetal Pig *(10–15 min.)*

Before you begin your examination of the internal structure of the fetal pig, you will examine the external features of its body. This gives you an opportunity to compare the body of the pig with other mammals. Remember, it is the external surface of an organism that has the greatest amount of contact with the environment. Thus, the greatest differences between two organisms may be their external features rather than their internal features.

MATERIALS

Per student pair:

- one preserved fetal pig injected with red and blue latex
- plastic bag to store fetal pig
- dissection pan

Per lab room:

- liquid waste disposal bottle
- box of nametags (tags may be provided with fetal pigs)
- permanent marking pens or pencils
- boxes of different sizes of latex or vinyl gloves
- box of safety goggles

PROCEDURE

A. Preparing the Fetal Pig for Examination

1. The fluid used to preserve fetal pigs may be irritating to hands and eyes. Therefore, wear protective gloves during dissection and put on protective goggles if you use contact lenses.
2. Obtain a fetal pig and place it on a dissection pan lined with paper towels. If the plastic bag supplied with your pig contains any excess preservative, pour the liquid into the waste bottle provided by your instructor, and save the bag. As you will be using the same fetal pig for several days, you should place it in the plastic bag at the end of each day's exercise so it doesn't dry out.
3. For easy identification, tie a nametag to a hindleg, the bag, or both, *according to the instructor's wishes.* Use a marking pen or pencil to fill in the tag.

B. Body Regions and Their Features

1. Identify the four regions of the fetal pig body: the large, compact **head;** the **neck;** the **trunk** with four **limbs** (or appendages); and the *tail* (Figure 22-2).

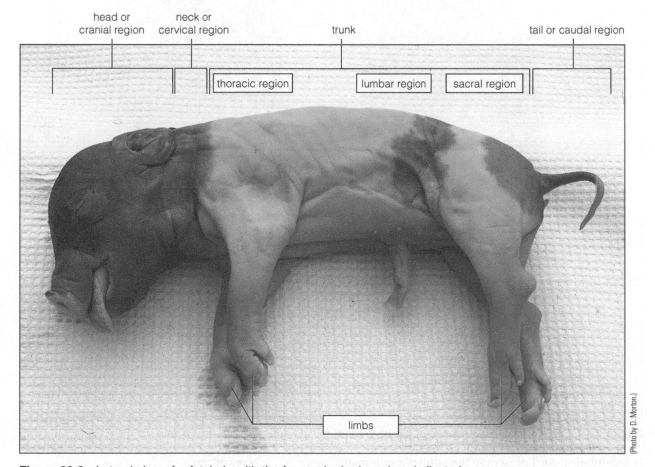

Figure 22-2 Lateral view of a fetal pig with the four major body regions indicated.

2. Examine the head in more detail (Figure 22-3) and identify the **eyes** with **upper** and **lower lids,** the **external ears,** the **mouth,** and the characteristic **nose** or *snout.* Note the position of the **nostrils** or *external nares* on the snout. Feel the texture of the snout. It is composed of bone, cartilage, and other tough connective tissue and as such allows the pig to root and push soil and debris in its search for food.

3. Open the pig's mouth and note the **tongue** with its covering of **papillae,** which contain *taste buds.* Papillae are especially concentrated and prominent along the posterior edges and tip of the tongue. Also notice if any *baby teeth* are present. Like humans, pigs are omnivores; that is, they eat both animal and plant matter.

4. Note that the trunk is divided further into the **thoracic region, lumbar region,** and **sacral region.** These regions, along with the cervical region in the neck, also describe the corresponding regions of the vertebral column or spine.

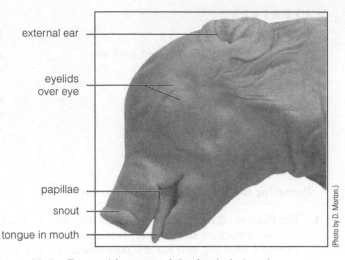

Figure 22-3 External features of the fetal pig head.

5. Place the pig on its back (dorsal surface) and examine its *abdomen* (belly). The most prominent feature of the underside (ventral surface) of the fetal pig is the **umbilical cord** seen near the posterior end of the abdomen (see Figure 22-4).

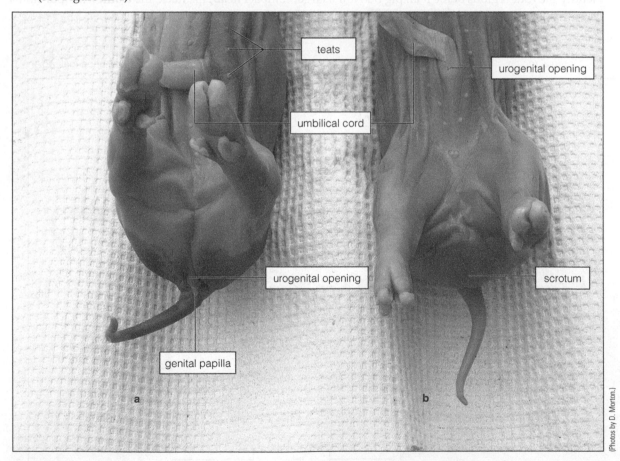

Figure 22-4 Ventral view of lower half of the body of (**a**) female and (**b**) male fetal pigs.

Is an umbilical cord present in the adult pig or human? _____ (yes or no)

During its development, the fetus was connected to the placenta on the uterine wall of its mother's reproductive system via the umbilical cord. The cord contains two *arteries* (red), a large *vein* (blue), and a fourth vessel, usually collapsed, the *allantoic duct.* The blood in the umbilical vein carries nutrients and oxygen from the mother to the fetus, and blood in the umbilical arteries carries waste materials and carbon dioxide from the fetus to the mother.

6. Note on the ventral surface of the pig the pairs of **nipples** or *teats* (Figure 22-4). Both male and female pigs may have from five to eight pairs of these structures situated in two parallel rows on the **thoracic region** (chest) and **abdominal region** of the body. Finally, locate the **anus,** the posterior opening of the digestive tract. The anus is situated immediately under (ventral to) the tail.

7. Locate and identify the following: the **wrist, lower forelimb, elbow, upper forelimb, shoulder, ankle, shank, knee, thigh,** and **hip** (Figure 22-5).

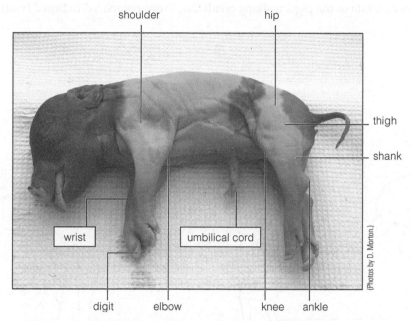

Figure 22-5 Lateral view of external features of the fetal pig.

8. Examine the toes. The first *toe,* or *digit,* which corresponds to your big toe or thumb, is absent in both fore-limbs and hindlimbs of the pig. Furthermore, the second and fifth digits are reduced in size, and the middle two digits, the third and fourth, are flattened or *hoofed.*

Pigs and other hoofed animals walk with the weight of their body borne on the tips of the digits. This type of walking is referred to as **digitigrade locomotion.** By contrast, humans use the entire foot for walking and have **plantigrade locomotion.** Compare the structure of your hands and feet with the foot of a pig.

C. Determining the Sex of Your Fetal Pig

1. Find the *urogenital opening* immediately ventral to the anus in females (Figure 22-4a) and just behind the umbilical cord in males (Figure 22-4b). A small fleshy *genital papilla* projects from the urogenital opening of female fetal pigs.

All fetal pigs have a common urogenital opening shared by the urinary system and the genital (reproductive) system. This situation persists in adult male pigs and humans. In adult female pigs and humans, however, there are separate openings to the urinary and reproductive systems.

2. In your or another group's male fetal pig, find a swelling on the posterior portion of the abdomen between the upper ends of the hindlimbs. The swelling is the **scrotum,** which contains the sperm producing *testes,* a pair of small, oval structures that are part of the male reproductive system. These are generally easy to locate in older fetuses because during late development they descend into the scrotum. Identify the *penis,* a large, tubular structure immediately under the skin just posterior to the urogenital opening.

22.2   Muscular System *(About 40–60 min.)*

The contractions of *skeletal muscle* organs enable the body to move. Through the contractions of skeletal muscles, you maintain your posture and can thus wink your eye, wave at a friend, or tap your foot in time to the beat of music. In this section, you will examine the structure of skeletal muscles and learn to identify some of them and describe their functions. The movement or movements produced by a particular skeletal muscle is called its **action.**

Skeletal muscles are attached to the various bones of the skeleton by tough strips of fibrous connective tissue called *tendons.* The three parts of a typical muscle are the **origin** (the end attached to the less mobile portion of the skeleton), the **insertion** (the end attached to the portion of the skeleton that likely moves when the muscle contracts), and the **belly** (the middle portion between the points of attachment).

Realize that the attachments of the muscles can only be understood with reference to the skeleton and its movable joints. Although we won't examine the skeleton in detail, it will be necessary to refer to its various parts (Figure 22-6). The skeletal muscles and their attachments to the skeleton of pigs and humans are remarkably similar. Most differences relate to the pig's walking on all four limbs compared to **biped** (walk on the hindlimbs only) humans.

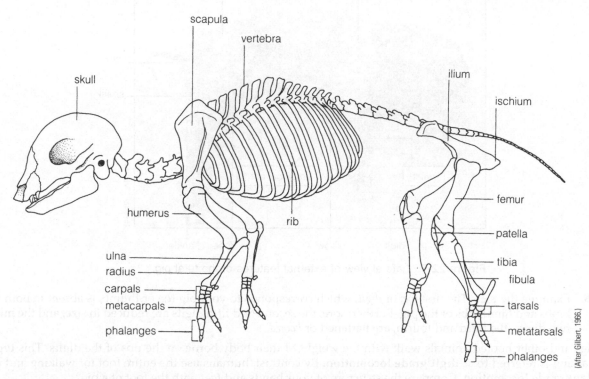

Figure 22-6   Skeleton of the fetal pig. The hyoid bone, located in the upper neck, and the sternum, to which many of the ribs are attached midventrally, are not included in this illustration.

A skeletal muscle that moves an appendage one way usually has one or more opposing muscles that move it in the opposite direction. Muscles with such opposite actions are called **antagonistic muscles.** For example, the biceps brachii is responsible for bending (flexing) the lower forelimb (in the pig) or the forearm (in humans) at the elbow, and the triceps brachii (its antagonist) straightens or extends them. The forearm can be held in any position by the simultaneous contraction of both of these opposing muscles.

MATERIALS

Per student pair:

- one preserved fetal pig injected with red and blue latex
- plastic bag to store fetal pig
- dissection pan
- one dissecting kit including the following: scalpel, blunt probe, dissecting needle, scissors, and forceps
- 4 large rubber bands *or* 2 pieces of string, each 60 cm long

Per lab room:

- boxes of different sizes of latex or vinyl gloves
- box of safety goggles

PROCEDURE

A. Preparing the Fetal Pig for Dissection of the Skeletal Muscles

1. Place your specimen on its dorsal side in a dissection pan lined with paper towels.
2. Tie one end of a string to the left forelimb at the wrist, pass the string underneath the dissection pan, and then tie the other end of the string to the right wrist so that the legs are spread apart under tension. Repeat with the hindlimbs, using the second piece of string. If rubber bands are available, tie two rubber bands together to make them longer. Then loop one end of the band around the right forelimb close to the foot. Bring the rubber band under the dissecting pan and loop it around the other forelimb to anchor the feet securely. Repeat this procedure with another set of rubber bands and anchor the hindlimbs to the pan.
3. For the following dissection, refer to Figure 22-7. The numbers in the drawing refer to the incisions to be made in the dissection.

Using a scalpel, make an incision on the ventral side of the pig at the base of the neck (1). *Make sure you cut only through the skin and not the underlying tissues.* With your scissors, continue the incision posteriorly to the umbilical cord, and then around the cord on the left side (the pig's left side) to the region between the hindlegs. From this point, carry on the cut along the medial surface of the left hindleg toward the foot (2) and around the pig's ankle (3). Make a similar cut from the midventral incision in the thorax down the medial surface of the left forelimb (4) and around the pig's wrist (5).

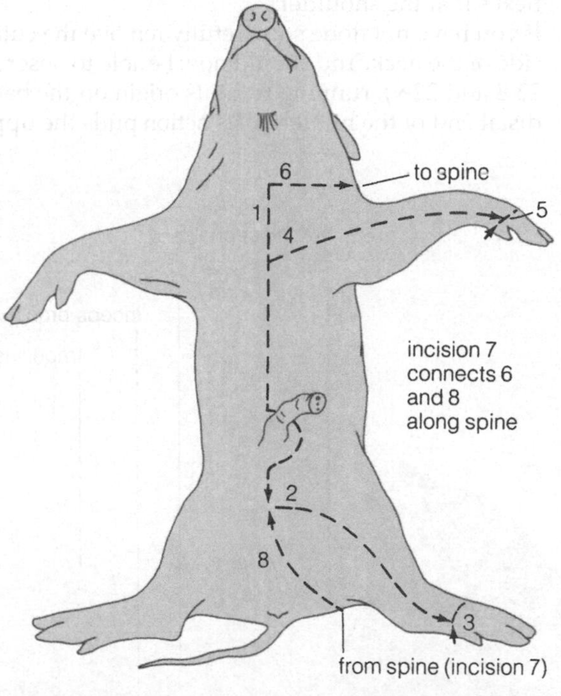

Figure 22-7 Cuts needed to expose muscles.

4. Return to the original incision on the ventral surface of the neck and extend it around the left side of the pig's neck dorsally toward the spine (6). To do this, you'll need to remove the rubber bands or string and place the pig on its right side in the dissecting pan. When you reach the spine, continue the incision posteriorly along the spine to the base of the tail (7). Finally, connect the posterior ends of incisions 7 and 1 (8). After completing the incisions, place the pig on its back and secure the legs with rubber bands or string as you did earlier.
5. To skin your specimen, grasp the cut edge of the skin at the base of the throat and begin easing the skin loose from the underlying tissues. Use your blunt probe between the skin and underlying connective tissues, working slowly until you have removed the skin from the ventral portion of the trunk and the limbs. Then, turn your pig on its right side and remove the skin along the lateral (left side) and dorsal surfaces to the vertebral region (skin over the spine).
6. In the region of the neck, shoulder, and trunk, you may notice a layer of light brown muscle fibers adhering to the skin. These fibers comprise the *cutaneous maximus* muscle, which is responsible for the twitching of the skin that rids the pig of insects and other irritants. Humans don't have this layer of muscle.
7. After your specimen is skinned, the muscles will not appear as clearly defined as in the illustrations. This is because adipose tissue (fat) and a layer of relatively loose connective tissue (*superficial fascia*) covers a denser, tougher layer of *deep fascia*. This deep fascia connects one muscle to another and maintains them in their proper position relative to one another. It will be necessary to break through the deep fascia as you proceed. Remember, however, that you are working with a fetus and as such its structures are immature and can be easily torn. Therefore, proceed with care.
8. As you attempt to identify the various muscles, you may find that the boundary of a given muscle is easily seen, whereas in others it seems to blend with those around it. In order to define the limits of a muscle, use a blunt probe to tease away the overlying adipose and connective tissues until you can see the direction of the muscle fibers. Look for changes in the direction of the muscle fibers and attempt to slip the blunt probe or flat edge of your scalpel handle between the two separate layers at this point. If the two layers separate readily from one another, you have located the boundary between two different muscles.

Now proceed with the exercise and attempt to identify some of the major muscles in the fetal pig *as directed by your instructor.* Obviously, not all of the skeletal muscles have been included, but only those that are relatively easy to identify and that will illustrate the general principles of skeletal muscular action.

1. The **latissimus dorsi** (Figure 22-8) is a broad muscle wrapped around the sides of the thoracic region and chest. Carefully pick away the adipose tissue and loose connective tissue from the sides of the chest until you can see its fibers. The origin of this muscle is the **lumbodorsal fascia** and nearby bones (Figure 22-8). It inserts by a tendon into the medial side of the proximal end of the humerus (the major bone of the upper forelimb). The action of the latissimus dorsi is to move the upper forelimb dorsally and posteriorly (or extend it at the shoulder joint).

2. The **trapezius** (Figure 22-8) is a broad muscle anterior to the latissimus dorsi. Its origin is from the base of the skull to the tenth thoracic vertebrae, and it inserts on the spine of the scapula (the shoulder blade). Its action is to draw or pull the shoulder medially (or elevates it).

3. The **deltoid** (Figure 22-8) is a relatively broad muscle that covers the shoulder region. It originates from the scapula and inserts on the humerus. Its action raises and pulls the upper forelimb anteriorly (or abducts and flexes it at the shoulder).

4. If you have not done so, carefully remove the cutaneous muscle and jellylike connective tissues covering the side of the neck. You should now be able to observe a broad, flat strip of muscle, the **brachiocephalic** (Figures 22-8 and 22-9), running from its origin on the back of the skull to its insertion on the anterior surface of the distal end of the humerus. Its action pulls the upper forelimb anteriorly (or flexes it at the shoulder).

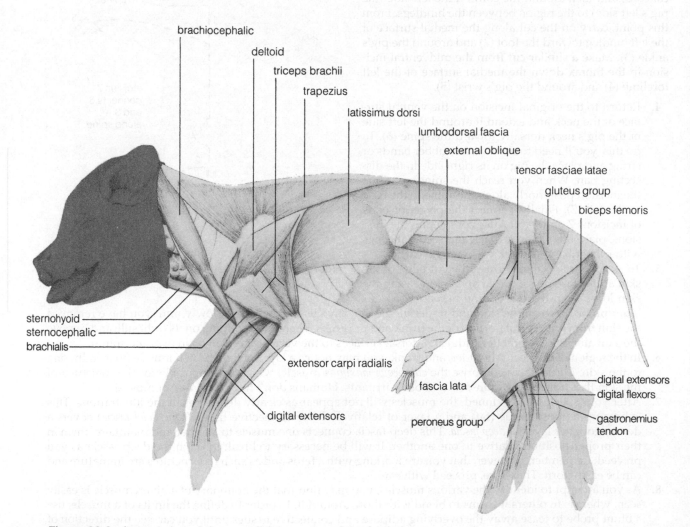

Figure 22-8 Lateral view of the superficial muscles of the fetal pig.

C. Muscles of the Forelimb

1. First find the **triceps brachii** (Figures 22-8 and 22-9), a large muscle that virtually covers the entire outer and posterior surface of the upper forelimb. It originates from the scapula and proximal third of the humerus

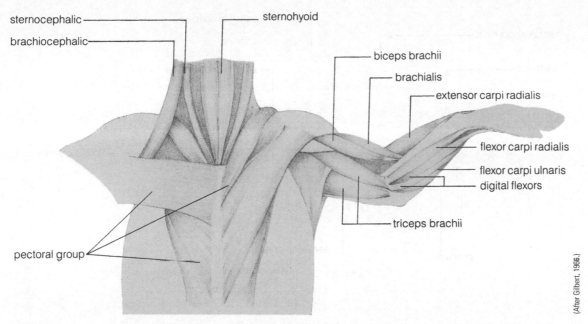

sternocephalic
brachiocephalic
sternohyoid
biceps brachii
brachialis
extensor carpi radialis
flexor carpi radialis
flexor carpi ulnaris
digital flexors
triceps brachii
pectoral group

(After Gilbert, 1966.)

Figure 22-9 Muscles of the ventral thoracic region and forelimb. Some of the superficial muscles of the pig's left side have been removed.

and inserts on the proximal end of the ulna (one of the two bones of the lower forelimb). Its action is to extend the lower forelimb at the elbow.

2. To locate the **biceps brachii** (Figure 22-9), untie your pig, place it on its dorsal side, and again secure the legs with string or rubber bands. Look for a rather small, spindle-shaped muscle covering the anterior surface of the humerus. This muscle originates on the scapula and inserts on the radius (the other bone of the lower forelimb) and ulna. Its action is to flex the lower forelimb and to act antagonistically with the triceps. In order to see it clearly, you'll need to cut through the overlying muscle (the **brachialis;** see Figure 22-9) and pull the cut edges back. The brachialis is a small muscle that also flexes the elbow. Its origin is the proximal humerus, and it inserts on the ulna.

3. There are numerous other muscles in the lower forelimb, most of which are concerned with extending or flexing the foot and digits (see Figures 22-8 and 22-9).

D. Muscles of the Throat and Chest

1. Now locate the **sternocephalic** muscle (Figures 22-8, 22-9), a long muscle lying ventral to the brachiocephalic. Its origin is the sternum, and it inserts into the lateral portion of the skull just behind the external ear known as the mastoid process. Its action draws the snout toward the chest (or flexes the head).

2. Next find the **sternohyoid** muscle (Figures 22-8, 22-9), which consists of two long flat strips running from the sternum (origin) to the hyoid (insertion). When this muscle contracts, it retracts and depresses the base of the tongue and the larynx, as, for example, when swallowing.

3. Note the large **pectoral group** of muscles (Figure 22-9) that originate on the ventral side of the sternum and insert on the humerus. The pectoral group moves the upper forelimb medially toward the chest (or adducts it at the shoulder). If you wish, you may carefully cut the belly of the superficial pectoral and bend it back to examine the underlying muscles more closely.

E. Muscles of the Abdominal Region

The major lateral abdominal muscles consist of the outer **external oblique,** the **internal oblique** (the midlayer), and the **transversus** (the inner layer). To locate these muscles, turn your specimen on its side. Now find the fibers of the external oblique (Figures 22-8 and 22-10) and observe that they run diagonally, so that their ventral ends are posterior to their dorsal ends. Carefully cut a "window" approximately 1 cm square in the external oblique to expose a portion of the internal oblique. (Remember that the muscles of the fetus are extremely thin. Take care not to cut into the body cavity at this time, as you will release a large amount of fluid.) Notice that the fibers of the internal oblique run at nearly right angles to the direction of those of the external oblique. Now, using the same careful technique as previously, remove a small portion of the internal oblique and attempt to reveal the fibers

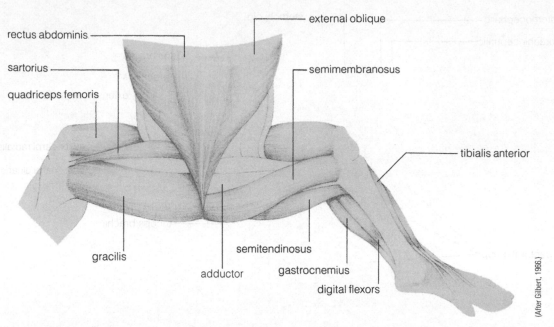

rectus abdominis
external oblique
sartorius
quadriceps femoris
semimembranosus
tibialis anterior
gracilis
semitendinosus
adductor
gastrocnemius
digital flexors

(After Gilbert, 1966.)

Figure 22-10 Ventral view of muscles of the abdominal region and hindlimb. Some superficial muscles on the pig's left side have been removed.

of the transversus, which as the name suggests run transversely across the abdomen. Collectively, the action of these abdominal muscles, together with the **rectus abdominus** (Figure 22-10), is to flex the trunk and compress the abdominal viscera (for example, to aid expiration or defecation).

F. Muscles of the Hip and Thigh

1. Begin your dissection of the muscles of the hip and thigh by locating the **biceps femoris** (Figure 22-8). This conspicuous superficial muscle covers most of the caudal half of the lateral surface of the thigh. It originates from the ischium of the pelvic girdle and inserts on the lateral fascia attached to the femur and tibia near the knee. Its action pulls the thigh posteriorly and laterally (or extends and abducts it at the hip).
2. Next, locate the **tensor fasciae latae** (Figure 22-8), the most anterior of the thigh muscles. Its origin is the ilium of the pelvic girdle and this short, triangular-shaped muscle continues down the anterior and lateral surface of the thigh as a sheet of connective tissue, the *fascia lata,* which attaches to the patellar ligament. Its actions are to tense the fascia lata, draw the thigh anteriorly (or flex it at the hip), and extend the shank at the knee.
3. Now carefully free the tensor fasciae latae at its insertion, but leave the origin and its medial portion intact. Cut the biceps femoris at its insertion near the tibia, and peel it back to expose the deeper muscles of the thigh. Identify the *vastus lateralis,* the most lateral of the four quadriceps femoris muscles, and the **semitendinosus** (Figures 22-10 and 22-11). Carefully expose the lateral portion of the **semimembranosus,** which is situated under the semitendinosus (Figure 22-11). The biceps femoris, semitendinosus, and semimembranosus are collectively called the **hamstring muscles.** In general, the hamstrings originate on the ischium of the pelvic girdle and insert around the back of the knee. Their collective actions are to pull the thigh posteriorly (or extend it at the hip) and flex the shank at the knee.
4. To locate other muscles of the thigh, place your pig on its dorsal side and tie back the legs with string or rubber bands. Referring to Figures 22-10 and 22-11, locate the *vastus medialis*—the most medial of the quadriceps femoris muscles, the **sartorius, the gracilis,** and the medial side of the **semimembranosus.** The sartorius and gracilis originate from the bones of the pelvic girdle and insert on the medial side of the tibia. These muscles draw the thigh medially (or adduct it at the hip) and the sartorius helps pull the thigh anteriorly (or flex it at the hip).
5. Separate the anterior edge of the vastus medialis and find between it and the vastus lateralis muscle a third quadriceps femoris muscle, the *rectus femoris.* A much-reduced fourth quadriceps femoris muscle, the *vastus intermedius,* lies beneath the rectus femoris. Collectively, the **quadriceps femoris** muscles originate on the upper surface of the femur and on the ilium of the pelvic girdle (rectus femoris) and insert via the patellar ligament onto the anterior surface of the tibia. Their general actions are to pull the thigh anteriorly (or flex it at the hip) and extend the shank at the knee. If you wish, you may carefully cut the belly of the sartorius and gracilis and bend the bellies back to examine the underlying muscles more closely.

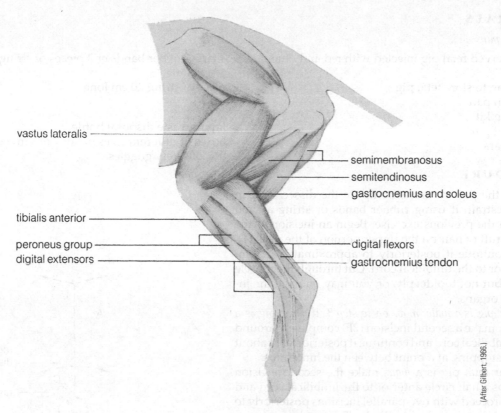

vastus lateralis

semimembranosus

semitendinosus

gastrocnemius and soleus

tibialis anterior

peroneus group

digital extensors

digital flexors

gastrocnemius tendon

(After Gilbert, 1966.)

Figure 22-11   Lateral view of the muscles of the hindlimb.

G. Muscles of the Hindlimb

1. The **gastrocnemius** (Figures 22-10 and 22-11) and underlying *soleus* (Figure 22-11) originate, respectively, from the distal end of the femur and the head of the fibula. The Achilles tendon inserts them both to the calcaneus (heel bone). These muscles extend the foot at the ankle, and the gastrocnemius helps flex the shank at the knee.
2. Other muscles of the lower hindlimb flex the foot (*tibialis anterior*, Figures 22-10 and 22-11; *peroneus group*, Figures 22-8 and 22-11), while others flex and extend the digits (*digital flexors*, Figures 22-8 and 22-11; *digital extensors*, Figures 22-8 and 22-11). Spend a few minutes examining the lower hindlimb and attempt to identify several of these muscles.

22.3   Ventral Body Cavities *(10 min.)*

The ventral body cavities of your fetal pig contain most of the organs of the digestive, respiratory, and urogenital systems, as well as the heart and major vessels of the circulatory system. Remember that *dissecting* does not primarily mean "cutting up" but rather "exposing to view." Thus, proceed carefully, as the internal organs are fragile. Do not remove any structure unless directed to do so. Work closely with your partner, making sure that each organ and structure is fully identified and studied by both of you before you proceed to the next step.

Caution

Specimens are kept in preservative solutions. Use latex gloves whenever you handle a specimen. Wash any part of your body exposed to this solution with copious amounts of water. If preservative solution is splashed into your eyes, wash them with the safety eyewash bottle for 15 minutes. If you wear contact lenses during a dissection, wear safety goggles.

For the following dissection, use Figure 22-12 as a guide for making the various incisions. The numbers in the figure correspond to the numbers in the following directions.

MATERIALS

Per student pair:

- one preserved fetal pig injected with red and blue latex
- plastic bag to store fetal pig
- dissection pan
- dissecting kit
- dissecting pins
- bone shears

- 4 large rubber bands *or* 2 pieces of string, each 60 cm long
- piece of string 20 cm long

Per lab room:

- liquid waste disposal bottle
- boxes of different sizes of latex or vinyl gloves
- box of safety goggles

PROCEDURE

1. Place the pig, ventral side up, in the dissection pan and restrain it using rubber bands or string as you did in the previous exercise. Begin an incision at the small tuft of hair on the upper portion of the neck (1), and continue it posteriorly to approximately 1.5 cm anterior to the umbilical cord. Cut through the muscle layer but not too deeply, or you may damage the internal organs.

2. *If your pig is a male, move on to step 3.* If your pig is a *female,* make a second incision (2F) completely around the umbilical cord and continue it posteriorly for about 3 cm, stopping at a point between the hindlimbs.

3. If your fetal pig is a *male,* make the second incision (2M) as a half circle anterior to the umbilical cord and then proceed with two parallel incisions posteriorly to a region between the hindlimbs. The two incisions are necessary to avoid cutting the *penis,* which lies under the skin just posterior to the umbilical cord. The incisions made in the region of the *scrotum* should be made carefully so as not to damage the testes, which are lying just under the skin.

4. Carefully deepen incisions 1 and 2 until the body cavity is exposed. In order to make lateral flaps of the muscle tissue, which can be folded out of the way, make a third (3) and fourth (4) incision as illustrated in Figure 22-12. Now, carefully open the body cavity. If it is filled with fluid, pour the fluid into the waste container provided (not into the sink!) and carefully rinse out the cavity with a little water.

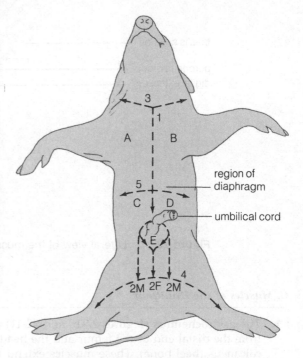

Figure 22-12 Ventral view of the fetal pig, with the position of incisions indicated. Specific incisions for male and female specimens are marked M and F, respectively.

5. Use your fingers to locate the lower margin of the *rib cage.* Just below it, make a fifth (5) incision laterally in both directions from the first incision (1). In this region is the **diaphragm,** a skeletal muscular sheet connected to the body wall and separating the two major ventral body cavities: an anterior **thoracic cavity** and a posterior **abdominopelvic cavity.** Use your scalpel to free the diaphragm from the body wall (do not remove it, however).

6. Carefully peel back flaps A, B, C, and D (see Figure 22-12) and pin them beneath your pig. It may be necessary to cut through the ventral part of the rib cage with a pair of bone shears to separate body wall flaps A and B. Do so carefully, so as not to damage the heart and lungs, which are located in the thoracic cavity.

7. To free the umbilical cord and the flesh immediately surrounding it (flap E), first locate the umbilical vein—a dark, tubular structure extending from the umbilical cord forward (anteriorly) to the liver. Then tie small pieces of string around the vein in two places (approximately 1.5 cm apart) and with your scissors cut through the vein between the pieces of string. Now pull back the umbilical cord to a position between the hindlegs and pin the flesh surrounding it to the body. The pieces of string around the umbilical vein will aid in identifying this structure during the section on the circulatory system.

8. When the body cavities are fully exposed, carefully remove any excess red or blue latex that may be present. (This occurs when some veins and arteries burst when injected with latex.) Remove large pieces with forceps. Remove your pig from the dissection pan and rinse out any smaller pieces of latex in a sink equipped to screen out debris.

9. *If you are continuing with the dissection, move on to the next section.* Otherwise, place your pig back in the plastic bag. Tie it shut to prevent your pig from drying out. Dispose of any paper towels that contain preservative *as directed by your instructor.* Clean the tray, dissecting tools, and the laboratory table.

The digestive system of a vertebrate consists of the **digestive tract** or *alimentary canal* (mouth, oral cavity, pharynx, esophagus, stomach, small intestine, large intestine or colon, cecum, rectum, and anus) and associated structures and glands (salivary glands, gall bladder, liver, and pancreas). In addition to these digestive system organs, you will locate and identify the thymus and thyroid, two endocrine system glands. The digestive system digests the complex molecules in food, absorbs the useful end products of digestion along with water and most everything else that the body adds to digesting food, and processes indigestible remnants for defecation.

MATERIALS

Per student pair:

- one preserved fetal pig injected with red and blue latex
- plastic bag to store fetal pig
- dissection pan
- dissecting kit
- dissecting pins
- bone shears

- 4 large rubber bands *or* 2 pieces of string, each 60 cm long
- dissection microscope
- meter stick (optional)

Per lab room:

- liquid waste disposal bottle
- boxes of different sizes of latex or vinyl gloves
- box of safety goggles

PROCEDURE

A. Mouth

1. Place your pig in the dissection pan. Observe the area between the **lips** and **gums;** this is called the **vestibule.** The larger area behind the gums is referred to as the **oral cavity.**
2. Carefully cut through the corners of the mouth and back toward the ears with bone shears until the lower jaw can be dropped and the oral cavity exposed (Figure 22-13).
3. If teeth are present, carefully extract a **tooth** and examine it. A tooth consists of the *crown,* the *neck* (surrounded by the gum), and the *root* (embedded in the jawbone). If your specimen does not have exposed teeth, cut into the gums and look for developing teeth.
4. Feel the roof of the oral cavity and determine the position of the **hard palate** and **soft palate** (Figure 22-13). What is the difference between the two regions?

_____

_____

_____

_____

_____

5. Posterior to the soft palate is the **pharynx** (Figure 22-13). In humans the portion of the pharynx just posterior to the oral cavity is also referred to as the *throat.* Note that unlike humans, the pig does not

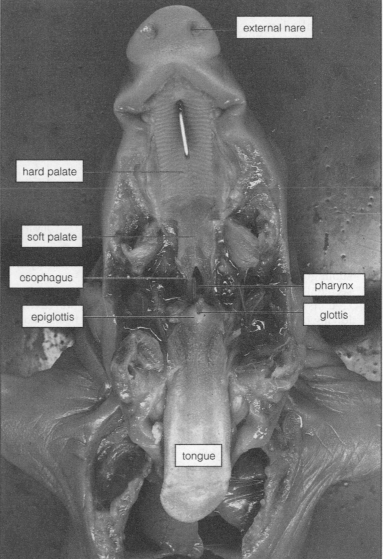

Figure 22-13 Structures of the oral cavity and pharynx.

(Photo by D. Morton.)

have a fingerlike piece of tissue, the **uvula,** projecting from the posterior region of the soft palate. Confirm its presence in humans by looking into the throat of your lab partner.

6. Carefully close the pig's jaws. Now, in the neck locate the **trachea** (Figure 22-15), a tube that is kept open throughout its length by a series of cartilaginous supports. Although the trachea is actually a part of the respiratory system, its identification aids in finding the **esophagus** (Figure 22-15), which lies behind it on its dorsal surface. Carefully slit the esophagus and insert a blunt probe into it and run it back toward the mouth. Open the mouth and note where the probe emerges. This is the opening of the esophagus (Figure 22-13).

If you would run your probe posteriorly through the esophagus, where would it emerge?

_____

7. Continue your study of the pharynx by locating the opening to the *larynx,* the **glottis.** It can be identified by the presence of a small white cartilaginous flap, the **epiglottis,** on its ventral surface (Figure 22-13). The epiglottis covers the glottis when a mammal swallows.

B. Salivary Glands

1. Place your pig on its right side and, proceeding from the base of the ear, carefully cut through the skin to the corner of the eye, then ventrally toward the chin, and, finally, continue the incision posteriorly toward the forelimb. Carefully remove the skin.

2. Tease away the muscle tissue below the ear to reveal a large, relatively dark, triangular **parotid gland.** This salivary gland extends from the edge of the ear posteriorly to halfway down the neck (Figure 22-14). Note that the parotid appears to be composed of many small nodules compared to the fibrous large masseter muscle lying underneath and anterior to it.

3. Cut through the middle of the parotid gland to expose the somewhat lobed **mandibular gland** (Figure 22-14). Do not confuse this second salivary gland with the small, oval lymph nodes present in the head and neck region.

4. The third salivary gland is the **sublingual gland.** It is located under the tongue and there is not enough time to dissect it today. The fluids secreted by the mandibular and sublingual glands are more viscous than that secreted by the parotid. Collectively, the secretions by the three salivary glands maintain the oral cavity in a moist condition, ease the mixing and swallowing of food, and contain enzymes that begin the breakdown of starch to sugars.

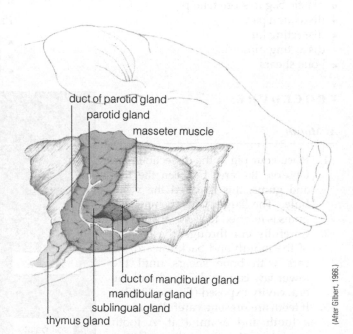

(After Gilbert, 1966.)

Figure 22-14 Lateral view of the fetal pig head with salivary and thymus glands exposed.

C. Thymus and Thyroid Glands

1. Work from the ventral side of your pig with the legs secured by string or rubber bands and the body wall flaps pinned to the sides of the body.

2. Identify the **thymus gland,** a whitish structure that is divided into two lobes. It extends from the neck where it covers the trachea and larynx to upper thoracic cavity where it partially covers the anterior portion of the heart (Figures 22-14, 22-15). The thymus plays important roles in the development and maintenance of the body's immune system.

3. Immediately beneath the thymus in the neck region find the **thyroid gland,** a small reddish, oval mass with a relatively solid consistency (Figure 22-15). Thyroid hormones function in the regulation of metabolism, growth, and development.

D. Liver, Gallbladder, and Pancreas

1. Identify the brownish colored **liver** (Figure 22-15), which is largest organ of the abdominopelvic cavity. Count and carefully determine the extent of all of its four lobes. The liver has many functions, including secreting *bile.*

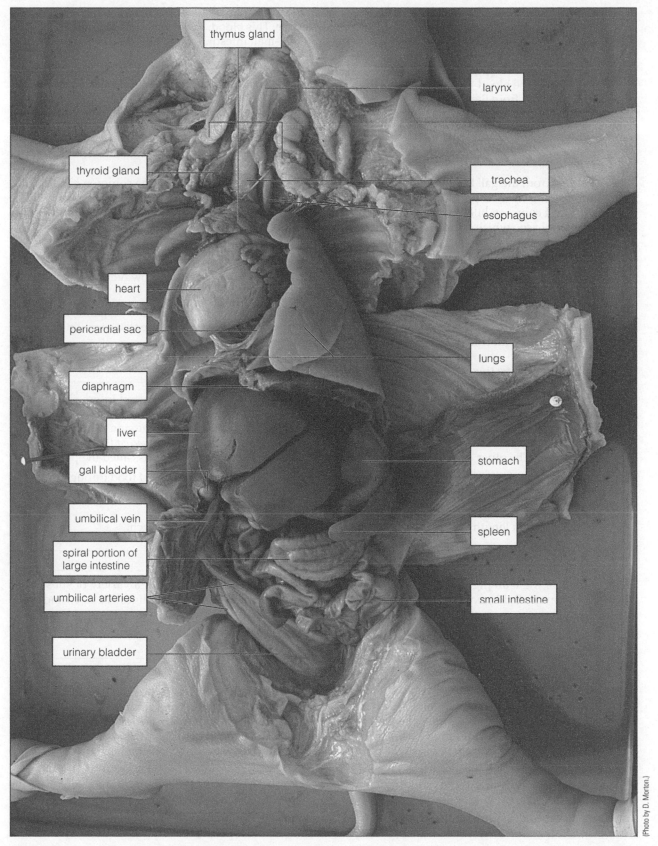

thymus gland

larynx

thyroid gland

trachea

esophagus

heart

pericardial sac

lungs

diaphragm

liver

stomach

gall bladder

umbilical vein

spleen

spiral portion of
large intestine

umbilical arteries

small intestine

urinary bladder

(Photo by D. Morton.)

Figure 22-15 Ventral view of the general internal anatomy of the fetal pig.

The liver also plays a very important role in maintaining the stable composition of the blood. The nutrients from a digested meal are absorbed into the blood of the small intestine. This blood, which contains high concentrations of sugars like glucose and amino acids, is transported from the small intestine to the liver (via the hepatic portal vein), and there the excess glucose is converted to glycogen for storage. If the liver has stored a full capacity of glycogen, it converts the glucose into fat, which is stored in other parts of the body. The liver also removes excess amino acids from the blood by converting them to carbohydrates and fats. During this process, an amino group ($-NH_2$) is removed from the amino acid and converted into ammonia (NH_3). Ammonia is a very toxic substance, and the liver combines it with carbon dioxide to form urea. The urea, which is less toxic than ammonia, is then eliminated from the body in urine.

2. Lift the right central lobe of the liver to expose the **gall bladder.** This saclike organ stores the bile secreted by the liver.

The *cystic duct* from the gall bladder unites with the *hepatic duct* from the liver to form the common *bile duct.* The latter empties into the first portion of the *small intestine* (Figure 22-15). *If your instructor wishes,* attempt to locate the hepatic and common bile ducts and trace them from the liver to the small intestine. Be careful not to injure the *hepatic portal vein,* which parallels these ducts.

3. Carefully move the small intestine and locate the **pancreas,** an elongated globular mass lying between the *stomach* and small intestine. The *pancreatic duct* carries digestive enzymes and other substances produced by the pancreas to the duodenum. (Do not attempt to find the pancreatic duct, however, as it is too small be dissected satisfactorily.)

The pancreas is both an **exocrine gland** (whose secretions are released into a duct) and an **endocrine gland** (whose hormones are released into the blood). The endocrine portion of the pancreas secretes insulin and other hormones involved with controlling the levels of glucose in the blood of mammals.

E. Stomach, Small Intestine, Large Intestine or Colon, Rectum, and Anus

1. Earlier in this section, you made a small slit in the esophagus. Return to this incision and insert a blunt probe through the slit in the esophagus only this time posteriorly until its tip enters the **stomach,** a bean-shaped organ dorsal and to the left of the liver (Figure 22-15). Push the lobes of the liver to one side to fully expose the stomach. Note that the esophagus penetrates the diaphragm before joining the *cardiac end* (near the heart) of the stomach. The other end of the stomach, which empties into the small intestine, is called the pyloric end.

Two muscular rings, the *cardiac sphincter* and the *pyloric sphincter,* control the movement of food through the stomach. Feel for these sphincters by gently squeezing these rings of smooth muscle tissue situated at the entrance and exit of the stomach between your index finger and thumb.

2. Cut the stomach lengthwise with your scissors. Describe any contents of the stomach.

_____

_____

The contents of the fetal pig's digestive tract are called *meconium* and are composed of a variety of substances, including amniotic fluid swallowed by the fetus, epithelial cells sloughed off from the digestive tract, and hair.

3. Clean out the stomach and note the folds or *rugae* on its internal surface. What role might the rugae play in digestion?

_____

_____

4. The **small intestine** (Figure 22-15) is divided into three regions: the *duodenum,* the *jejunum,* and the *ileum.* The first portion, the duodenum, leaves the pyloric end of the stomach and runs along the edge of the pancreas. The junctions of the duodenum and ileum with the jejunum cannot easily be distinguished.
5. A thin membrane, the **mesentery,** holds the coils of the small intestine together. Cut the mesentery from the dorsal body wall and from between the coils of the small intestine. Uncoil the small intestine.

Measure the length with a meter stick and record it: _____ cm
   A rule of thumb is that the small intestine in both pigs and humans is about five times the length of the individual.

6. Using your scissors, cut a 0.5-cm section of the intestine, slit it lengthwise, and place it in a clear, shallow dish filled with water. Now examine it using a dissection microscope.

How does the inner surface appear?

_____

_____

_____

Locate the *villi*. Most of the nutrients provided by the digestive process are absorbed by these small projections from the wall of the small intestine.

7. Locate the juncture of the **large intestine** (Figure 22-15), or *colon*, and the ileum. This may be more difficult in a pig than in a human because in the former there is not such a noticeable difference in the size of the small and large intestines. However, this juncture is marked by the presence of a blind pouch, the *cecum*, which in the pig is relatively large. In humans, the cecum is very short and bears a small fingerlike projection known as the *appendix*.
8. As with other junctures in the digestive tract, the region where the ileum joins the large intestine is the site of a sphincter of smooth muscle, sometimes called the *ileocecal valve*. Feel for it by rolling the junction between your index finger and thumb.
9. The coiled large intestine stretches from the cecum to the straight **rectum,** which opens to the outside at the **anus.** The anus is the site of the final muscle in the alimentary canal, the *anal sphincter.* Locate the rectum, anus, and anal sphincter, but do not dissect these structures at this time. You may, however, *at the direction of your instructor,* remove a piece of the colon and examine it with a dissecting microscope.

How does the colon's internal surface compare with that of the small intestine?

_____

_____

_____

10. Review the digestive system by tracing the pathway of an ingested indigestible fiber in and out of the body.
11. *If you are continuing with the dissection, move on to the next section.* Otherwise, place your pig and any excised organs back in the plastic bag. Tie it shut to prevent your pig from drying out. Dispose of any paper towels that contain preservative *as directed by your instructor.* Clean the tray, dissecting tools, and the laboratory table.

22.5 Respiratory System *(10 min.)*

The respiratory system of a mammal consists of various organs and structures associated with the exchange of gasses between the internal and external environment. Air rich in oxygen is inhaled into the air sacs of the lungs. Oxygen diffuses from this air into the capillaries of the lungs while carbon dioxide moves in the other direction. Carbon dioxide–rich air is then exhaled from the body.

MATERIALS

Per student pair:

- one preserved fetal pig injected with red and blue latex
- plastic bag to store fetal pig
- dissection pan
- dissecting kit
- dissecting pins
- 4 large rubber bands *or* 2 pieces of string, each 60 cm long

Per lab room:

- liquid waste disposal bottle
- boxes of different sizes of latex or vinyl gloves
- box of safety goggles

PROCEDURE

A. Nose

1. Remove your pig from the dissection pan. In the pig and other mammals, molecules of air enter the body through the *nostrils* and pass through a pair of **nasal cavities** dorsal to the hard palate and into the nasal portion of the pharynx or *nasopharynx* (Figure 22-13). Examine the nostrils and hard and soft palates, and then carefully cut the soft palate longitudinally to examine the nasopharynx of your specimen.
2. From the nasopharynx, air passes through the *glottis* into the **larynx** (Figures 22-15 and 22-14a and b). In humans, the front of the larynx is often referred to as the *Adam's apple* or *voice box.*

a

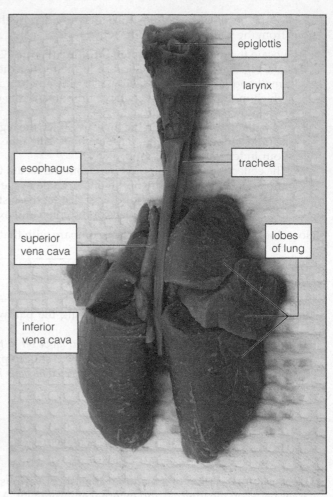

b

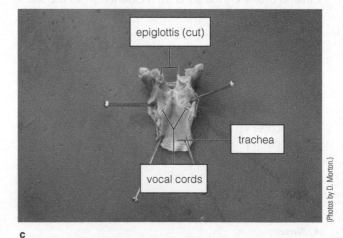

(Photos by D. Morton.)

c

Figure 22-16 (a) Ventral and (b) dorsal views of the respiratory system of the fetal pig. In (c), the larynx has been slit open to show the vocal cords.

B. Trachea, Bronchial Tubes, and Lungs

1. Place the pig, ventral side up, in the dissection pan and restrain it using rubber bands or string as you did in the first section of this exercise. Pin back the body flaps to the sides of the body.
2. Slit the larynx longitudinally to expose the *vocal cords* (Figure 22-16c).
3. Locate the trachea (Figure 22-16). The trachea extends from the larynx and divides into two major branches, the **bronchi** (singular, *bronchus*), to the lungs. Note again the series of cartilaginous structures that prevent the trachea from collapsing. These apparent rings of cartilage are actually incomplete on their dorsal side.

4. Note that the thoracic cavity is divided into two **pleural cavities,** which contain the lungs, and a **pericardial sac** that is located between them. Inside this sac is the *pericardial cavity* and the *heart.* Carefully examine the lungs and note the thin, transparent **pleural membrane** that lines the inner surface of the thoracic cavity and the outer surface of the lungs. The right lung consists of four lobes and the left of two or three. Are the lungs of the fetal pig filled with air? _____ (yes or no)

5. Carefully push the *heart* to one side and gently tease away some of the lung tissue to expose the bronchi. Notice that the bronchi divide into smaller and smaller branches.

These branches continue to divide and branch into finer and finer structures, eventually ending as microscopic air sacs called **alveoli** (singular, *alveolus*). The thin walls of the alveoli are well supplied with blood capillaries, and it is here that the exchange of carbon dioxide and oxygen occurs after birth.

Where does this exchange occur in the fetus? _____

6. Now relocate the diaphragm and note its position in relation to the lungs. Contraction of this skeletal muscle in part results in inhalation.

7. Complete your study of the respiratory system of the fetal pig by tracing the pathway of a carbon dioxide molecule from an alveolus to the nostrils.

8. *If you are continuing with the dissection, move on to the next section.* Otherwise, place your pig and any excised organs back in the plastic bag. Tie it shut to prevent your pig from drying out. Dispose of any paper towels that contain preservative *as directed by your instructor.* Clean the tray, dissecting tools, and the laboratory table.

22.6 Blood Vessels and the Surface Anatomy of the Heart *(15–20 min.)*

The mammalian body contains a vast network of **blood vessels** that transport blood—water, oxygen, carbon dioxide, nutrients, metabolic wastes, hormones, and other substances—to and from every living cell. In mammals, blood is pumped through the arteries, arterioles, and capillaries by a muscular four-chambered **heart.** Thus, **arteries,** their branches, and their final branches, arterioles, are vessels that carry blood away from the heart to the capillaries of organs and their tissues. Venules drain the tissues and organs and deliver blood into **veins,** which transport it back to the heart. The exchange of substances between the blood and tissues occurs across the thin walls of **capillaries.**

The circulatory system is divided into two circuits: the **pulmonary circuit,** which involves blood flow to and from the lungs, and the **systemic circuit,** which is concerned with the flow of blood to and from the rest of the body. In this section, you will study these two circuits and examine how the heart directs the flow of blood through them both in a fetus and in an adult.

MATERIALS

Per student pair:

- one preserved fetal pig injected with red and blue latex
- plastic bag to store fetal pig
- dissection pan
- dissecting kit
- dissecting pins
- 4 large rubber bands *or* 2 pieces of string, each 60 cm long

Per lab room:

- liquid waste disposal bottle
- boxes of different sizes of latex or vinyl gloves
- box of safety goggles

PROCEDURE

A. Pulmonary Circuit and Surface Anatomy of the Heart

1. Place the pig, ventral side up, in the dissection pan and restrain it using rubber bands or string as you did in the first section of this exercise. Pin back the body flaps to the sides of the body.

2. If it is not already torn, open the pericardial sac. Similar to the situation in the pleural cavities, the inside of the pericardial sac and the outside of the heart is lined by the **pericardial membrane.**

3. Identify the four chambers of the heart (Figure 22-17)—the **right** and **left atria** (singular, *atrium*), and the larger **right** and **left ventricles.** On the surface of the heart locate the **coronary vessels** lying in the diagonal groove between the two ventricles (Figure 22-17). The coronary arteries and their branches supply blood directly to the heart. (The heart is a muscle and as such has the same requirements as any other organ.) When these vessels become severely occluded, a heart attack may occur. It is the coronary arteries and their branches that are replaced or "bypassed" in coronary bypass surgery.

4. Gently push the heart to the left and identify two relatively large blue veins (Figure 22-17), called the **superior vena cava** and the **inferior vena cava** in humans and the *anterior vena cava* and the *posterior vena cava* in pigs.

After birth, oxygen-poor (or carbon dioxide–rich) blood from all of the body except the lungs and heart returns from the systemic circuit to the right atrium of the heart through these large veins. Trace the inferior vena cava a short distance from the heart.

Through what structure does the inferior vena cava pass? _____

5. The blood that enters the right atrium passes to the right ventricle and then is forced into the pulmonary circuit as the heart contracts. Identify the **pulmonary trunk,** which lies between the left and right atria and extends dorsally and to the left (Figure 22-17a). It branches to form the **left** and **right pulmonary arteries** (Figure 22-17b).

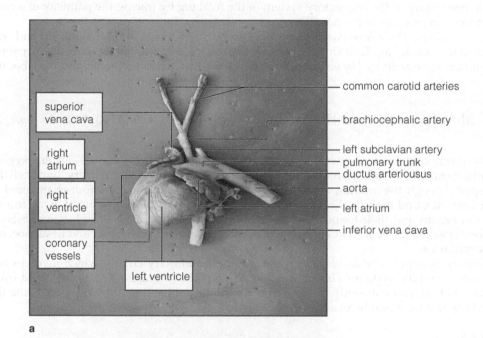

a

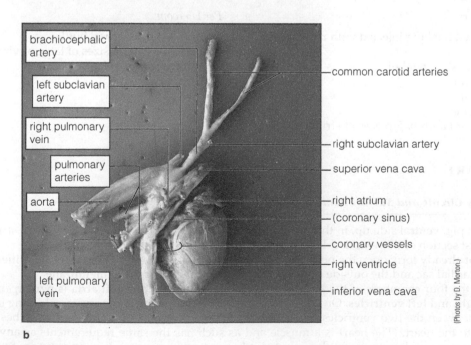

b

(Photos by D. Morton.)

Figure 22-17 (a) Ventral and (b) dorsal views of a fetal pig heart.

Do these arteries contain red or blue latex? _____

After birth, do these arteries carry oxygen-rich or oxygen-poor blood? _____

Carefully move the heart aside and follow the pulmonary arteries to the lungs.

6. After birth, once the blood is oxygenated (and the carbon dioxide removed) in the lungs, it returns to the left atrium of the heart via the **left** or **right pulmonary veins**, which complete the pulmonary circuit. Carefully move the lungs and heart aside and locate these large vessels (Figure 22-17b).

7. From the left atrium of the adult, the oxygenated blood passes to the left ventricle. Blood is forced into the **aorta** as the heart contracts, starting its trip through the systemic circuit. Locate the aorta (Figures 22-16a and b), which leads dorsally out of the left ventricle of the heart. Note that its base is partially covered by the pulmonary trunk coming from the right ventricle.

8. Examine how blood circulation is different in fetal mammals (Figure 22-18). In the fetus, most of the pulmonary circuit is bypassed twice. First, most blood from the right ventricle enters the aorta directly from the pulmonary trunk through the **ductus arteriosus**, a large but short vessel connecting the pulmonary trunk directly to the aorta. The second bypass occurs when most of the blood delivered to the right atrium, mostly from the posterior portions of the body by the inferior vena cava, passes directly into the left atrium via a temporary opening in the wall separating the right and left atria (*foramen ovale*). Thus, this blood entirely bypasses the pulmonary circuit. Identify the ductus arteriosus (Figures 22-17 and 22-18).

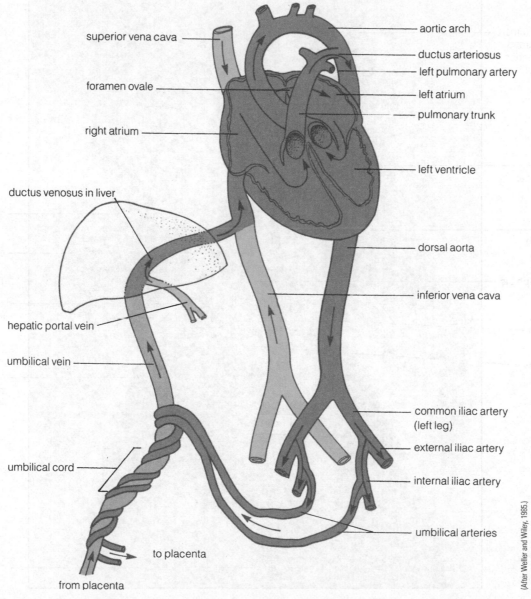

(After Weller and Wiley, 1985.)

Figure 22-18 Diagram of the circulatory system of a fetal mammal. Arrows indicate the flow of blood. Pink represents fully oxygenated blood. The blue indicates oxygen-depleted blood.

Why is it not necessary for large quantities of blood to enter the pulmonary system of a fetus?

_____

_____

With the first breath of the newborn, the ductus arteriosus contracts and the foramen ovale closes. Circulation through the pulmonary circuit is increased dramatically. Then, during the eight weeks following birth, the ductus arteriosus forms a fibrous strand of connective tissue, the *ligamentum arteriosum*, and the foramen ovale permanently fuses shut.

B. Systemic Circuit—Major Arteries and Veins Anterior to the Heart

1. The systemic circuit begins with the aorta. This large vessel leads anteriorly out of the left ventricle of the heart and makes a sharp turn to the left (the so-called **aortic arch,** Figure 22-18) and proceeds posteriorly through the body as the **dorsal aorta** (Figure 22-19a). All of the major arteries of the body arise from these two regions (the aortic arch and dorsal aorta) of the aorta.

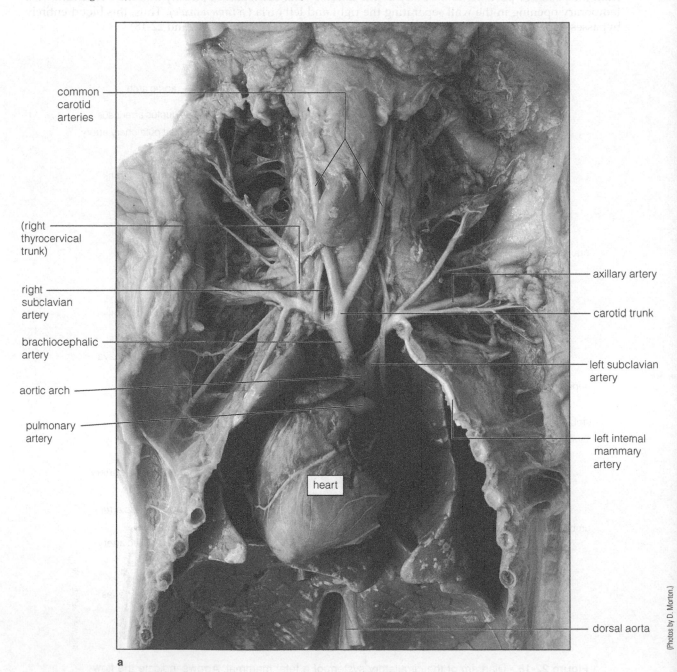

(Photos by D. Morton.)

Figure 22-19 Ventral views of (**a**) arteries and (**b**) veins anterior to the heart of a fetal pig.

2. Locate the first visible vessel, the **brachiocephalic artery** (Figure 22-19a), to branch from the aortic arch. The first vessels to branch from the aorta are the coronary arteries; these cannot be seen without dissecting the heart. The brachiocephalic artery branches to give rise to the **right subclavian artery** (Figure 22-19a), going to the right forelimb, and the **carotid trunk** (Figure 22-19a), whose branches course anteriorly through the neck and head. Trace the right subclavian artery and its branches through the shoulder region to the right forelimb. As it passes through the shoulder region, the name of the right subclavian changes to the *axillary artery* and then to the *brachial artery* when it enters the upper forelimb.

3. Return to the aorta and locate the second visible vessel to branch from the aortic arch, the **left subclavian artery** (Figure 22-19a). The left subclavian artery and its branches pass through the shoulder and left forelimb or arm in the same manner as the right subclavian artery, described above. As you trace the course of the left subclavian, notice that some of its branches feed the muscles of the chest and back.

4. Return to the right forelimb and locate the venous system that passes through this appendage. Because the veins are relatively thin-walled, this may be very difficult. Also, some of them may not be injected with blue latex and will appear a brownish color. If possible, follow the *brachial vein* to the *axillary* and the **subclavian vein** (Figure 22-19b) until the latter becomes the **brachiocephalic vein** (Figure 22-19b). It should be relatively easy to follow the brachiocephalic to its juncture with the previously identified superior vena cava (Figure 22-19a; it forms a prominent V), which returns blood to the right atrium of the heart.

5. In order to examine the arterial system that serves the throat and head, locate the carotid trunk (Figure 22-19a; a branch of the brachiocephalic artery; see above). This short branch of the brachiocephalic artery immediately splits into the **left** and **right common carotid arteries.** Each of these vessels divides into the *internal* and *external common carotid arteries.* Remove the thymus and thyroid glands and considerable muscle tissue in the throat to locate the anterior portions of the common carotid arteries.

As you locate and trace the carotid arteries, look for a white "fiber" that parallels them. This is the *vagus nerve.*

6. On either side of the neck are the major veins that drain the head and throat region. The **internal** and **external jugular veins** (Figure 22-19b) join the subclavian veins (from the forelimbs) to form the previously identified brachiocephalic vein. The latter leads into the superior vena cava.

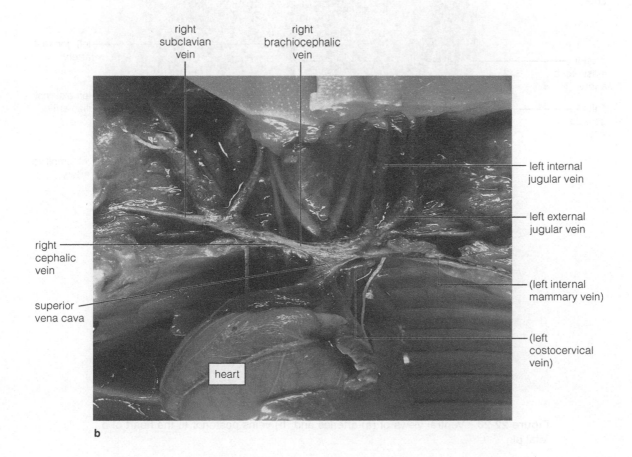

1. The posterior extension of the aortic arch is the dorsal aorta. As the name implies, the dorsal aorta lies near the dorsal body wall running parallel to the spine. From this large vessel arise all of the arterial branches that feed the organs, glands, and muscles of the abdominal region and the muscles of the hindlimbs and tail.

2. Follow the dorsal aorta posteriorly. Carefully move the liver and stomach of the pig and use a dissection needle to scrape away the sheet of tissue that connects the dorsal aorta to the pig's back. Locate the **celiac artery** (Figure 22-20a), whose branches deliver blood to the stomach, spleen, and liver. Continue to follow the dorsal aorta posteriorly and locate the **superior mesenteric artery** (Figure 22-20a). This vessel, just posterior to the celiac artery, branches to the pancreas and duodenum of the small intestine.

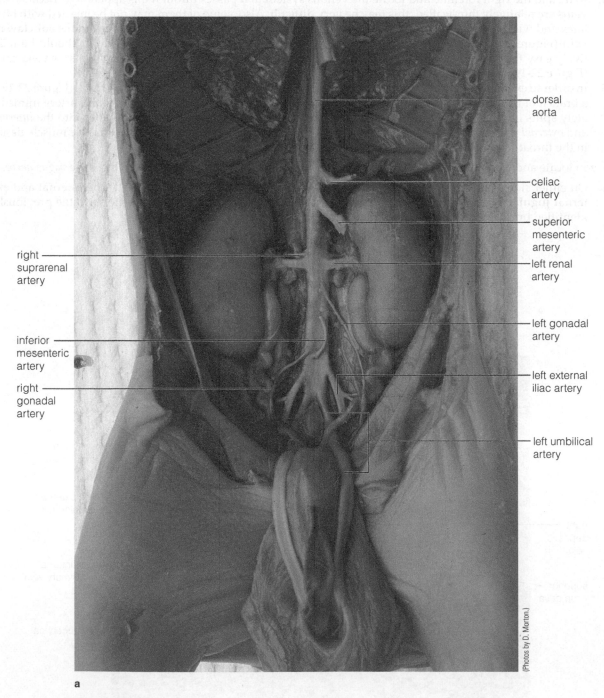

a

Figure 22-20 Ventral views of (**a**) arteries and (**b**) veins posterior to the heart of a fetal pig.

3. Posterior to the superior mesenteric artery are the **renal arteries** (Figure 22-20a), relatively short vessels that connect the dorsal aorta and the kidneys. At this time, it's easy to locate the **renal veins** (Figure 22-20b) which drain blood from the kidneys to the inferior vena cava.

4. As you follow the dorsal aorta posteriorly beyond the kidneys, the **external iliac arteries** (Figure 22-18) branch, one into each hindlimb. Each leg is also drained with a major vein, the **common iliac vein** (Figure 22-20a), which joins the inferior vena cava.

5. Follow the branches of the dorsal aorta into the tail region, being careful not to cut the two intervening branches. The small extension toward the tail region is called the *sacral artery* as it leaves the dorsal aorta and the *caudal artery* when it enters the tail.

6. Just anterior to the sacral artery, the **internal iliac arteries** (Figure 22-18) branch from the dorsal aorta. These enlarge and form the two **umbilical arteries** (Figure 22-18), which run through the umbilical cord to the placenta. Cut the umbilical cord transversely and note the arrangement of the umbilical arteries within it. Consider the composition of the blood as it travels through the umbilical arteries to the placenta. Is it rich in oxygen or carbon dioxide? _____

7. Locate the two pieces of vein that you tied with string in Section 22.1. This is the **umbilical vein** (Figure 22-18), through which blood rich in nutrients and oxygen flows from the placenta of the mother back to the fetus. Locate the umbilical vein in the umbilical cord and follow it anteriorly toward the liver. When the umbilical vein reaches the liver, it becomes the **ductus venosus** (Figure 22-18), which continues anteriorly within the substance of the liver and joins the inferior vena cava. The umbilical arteries, the umbilical veins, and ductus venosus become modified into ligaments following the birth of the fetus.

What is the relationship between the navel and the umbilical cord?

_____

_____

_____

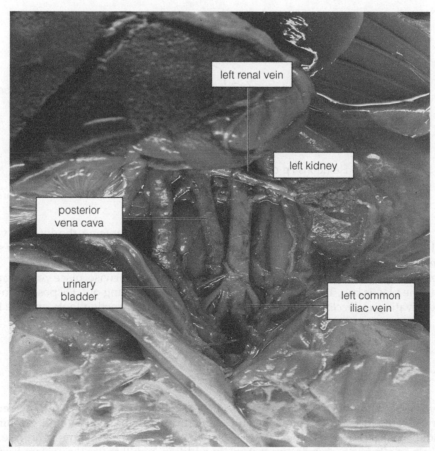

b

After birth, the *hepatic portal system*, which consists of a network of veins, collects all of the blood from the lower digestive tract and associated organs (stomach, small intestine, pancreas, and spleen) and carries it to the liver via the *hepatic portal vein*. In general, a **portal vein** is one that collects blood from the capillaries of one organ and transfers it to the capillaries of another organ. After birth, the *hepatic veins* drain the blood of the liver. These join the inferior vena cava just anterior to the point where the ductus venosus joins the inferior vena cava (Figure 22-21).

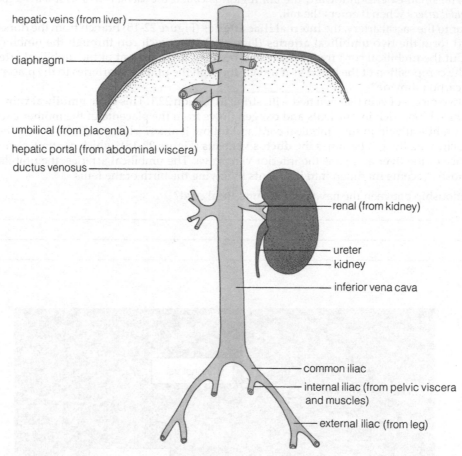

hepatic veins (from liver)

diaphragm

umbilical (from placenta)
hepatic portal (from abdominal viscera)
ductus venosus

renal (from kidney)

ureter
kidney

inferior vena cava

common iliac

internal iliac (from pelvic viscera and muscles)

external iliac (from leg)

Figure 22-21 Relationship of umbilical vein, hepatic portal vein, and hepatic veins to the inferior vena cava in a fetal pig.

8. Complete your dissection of the systemic circulatory system by tracing the inferior vena cava from the abdominal cavity through the **diaphragm** and to the thoracic cavity, where it joins the superior vena cava before entering the right atrium of the heart.

9. *If you are continuing with the dissection, move on to the next section.* Otherwise, place your pig and any excised organs back in the plastic bag. Tie it shut to prevent your pig from drying out. Dispose of any paper towels that contain preservative *as directed by your instructor.* Clean the tray, dissecting tools, and the laboratory table.

22.7   Urogenital System *(30 min.)*

In this section you will complete your study of the internal anatomy of the fetal pig. In previous sections, you dissected the muscular, digestive, respiratory, and circulatory systems. Now you'll examine the system that removes metabolic wastes from the bloodstream (*urinary system*) and the system responsible for the production of new individuals or offspring (*reproductive system*). In addition, you'll study the system largely responsible for the integration and the control of all the other organs and systems (*nervous system*).

The urinary and reproductive systems are traditionally studied together (as the *urogenital system*) because they share several anatomical features.

Specimens are kept in preservative solutions. Use latex gloves whenever you handle a specimen. Wash any part of your body exposed to this solution with copious amounts of water. If preservative solution is splashed into your eyes, wash them with the safety eyewash bottle for 15 minutes. If you wear contact lenses during a dissection, wear safety goggles.

MATERIALS

Per student pair:

- one preserved fetal pig injected with red and blue latex
- plastic bag to store fetal pig
- dissection pan
- dissecting kit
- dissecting pins
- 4 large rubber bands *or* 2 pieces of string, each 60 cm long

Per lab room:

- liquid waste disposal bottle
- boxes of different sizes of latex or vinyl gloves
- box of safety goggles

PROCEDURE

A. Urinary System

Like humans, the pig is a terrestrial organism and, as such, must conserve water. At the same time, metabolic wastes must be continuously removed from the blood. Furthermore, the composition of the blood must be constantly monitored and adjusted so that the cells of the body are bathed in a fluid of constant composition.

Much of the potentially poisonous waste occurs in the form of **urea** and results from the metabolism of amino acids in the liver. Urea is filtered from the bloodstream in the kidneys, which also regulate water and salt balance.

1. Place your pig on its back in the dissection pan and use string or rubber bands to secure the legs, as you did in the preceding exercises. Pin the lateral body-wall flaps to the dorsal side of your specimen and pull the umbilical cord and surrounding tissue back between the hindlimbs.
2. Find the paired **kidneys** in the lumbar region of the body cavity pressed against the dorsal body wall (Figure 22-22). They are covered by the **peritoneum,** the smooth, rather shiny membrane that lines the abdominopelvic cavity. (You may have already removed much of this during the dissection of the circulatory

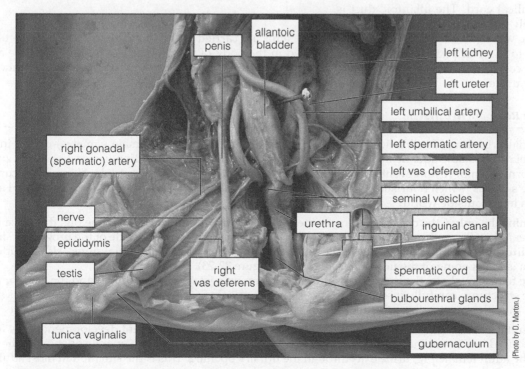

Figure 22-22 Ventral view of the male urogenital system of the fetal pig.

system in the preceding exercise.) The main function of the kidneys is the production of **urine,** a fluid containing urea and other waste substances dissolved in water.

3. To expose the right kidney, carefully lift up the abdominal organs and move them anteriorly and to the left. Using a dissecting needle, carefully scrape away the peritoneum so that the kidney bean–shaped kidney and the **ureter**—the duct that connects it to the bladder—are easily seen. Note the central depression in the surface of the kidney. This is the *hilus,* the region where the ureter leaves, the *renal vein* leaves, and the *renal artery* enters the kidney.

4. Carefully follow the ureter from the hilus to the **allantoic bladder.** Then lift the bladder and find the **urethra.** The latter is the structure through which urine passes from the bladder to the outside of the animal. In the male, the urethra is very long and passes through the *penis* to the outside of the body (Figure 22-22). Notice that the urethra passes posteriorly for a distance of approximately 2 cm and then turns sharply anteriorly and ventrally before entering the penis. In the female fetal pig, the urethra is short and passes posteriorly to join with the *vagina* to form the *urogenital sinus* (Figure 22-23).

5. Locate the **allantoic duct,** which leads from the allantoic bladder into the umbilical cord. The allantoic duct is largely a vestigial structure, for most

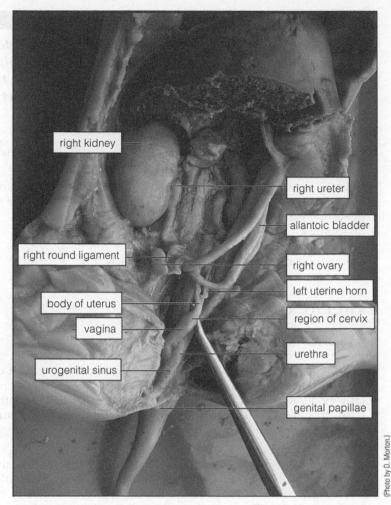

(Photo by D. Morton.)

Figure 22-23 Ventral view of the female urogenital system of the fetal pig.

of the wastes produced in the kidneys of the fetus are carried in the bloodstream through the umbilical vein to the placenta, where they are removed in the body of the mother. Following birth, the allantoic duct collapses, and the allantoic bladder is incorporated into the **urinary bladder,** which functions to store urine.

B. Female Reproductive System

In terrestrial organisms, fertilization (the fusion of the nuclei of male and female gametes) occurs internally, where a relatively stable aquaticlike environment is maintained. Once fertilization has occurred, the zygote divides to form an embryo and eventually a fetus. In mammals, all of this growth and development occurs within the female's uterus, which nourishes the developing offspring until it can pass through the birth canal and exist on its own in the outside world.

1. Examine the **vulva,** the collective term for the external genitalia of the female. In the pig, the vulva includes the *genital papilla* on the outside of the body, the *labia* or lips found on either side of the urogenital sinus, and the **clitoris,** a small body of erectile tissue on the ventral portion of the urogenital sinus. Also included in the vulva is the opening of the urogenital sinus itself (Figure 22-23).

2. In the female fetal pig, the **urogenital sinus** is the common passage for the urethra and the **vagina.** To locate these structures, carefully insert your scissors into the opening of the urogenital sinus and cut this structure from the side. Locate where the ducts of the vagina and urethra enter to form the urogenital sinus.

The urogenital sinus is not present in the adult female pig. During the subsequent development of the fetus, the sinus is reduced in size until the vagina and the urethra each develop their own, separate opening to the outside. Thus, in the adult female pig, urine exits through the urinary opening. This is the situation in most adult female mammals.

How does this compare with the structure of the reproductive system of most male mammals?

_____

_____

_____

3. Again, locate the clitoris. This small, rounded region on the inner ventral surface of the urogenital sinus is **homologous** (that is, similar in developmental origin) with the part of the male penis. In the male, the tissues of the penis develop around and enclose the urethra, while in the female, the urethra opens posteriorly to the clitoris.

4. Follow the urogenital sinus anteriorly and identify the thick-walled muscular vagina, which is continuous with the **uterus**. In the pig, the uterus consists of three structures or regions: the **cervix** at the entrance to the uterus, the **uterine body,** and the two **uterine horns** (Figure 22-23). Note that the uterine horns unite to form the body of the uterus. The pig has a *bicornuate uterus,* in which the fetuses develop within the uterine horns. In the human female there are no uterine horns, and the fetus develops within the body of the *simplex uterus.*

5. From the uterine horns, follow the **oviducts** to the **ovaries** (female gonads). The ovaries are small, yellowish kidney bean–shaped structures that lie just posterior to the kidneys. They are the sites of egg production and the source of the female sex hormones, estrogen and progesterone.

All of the eggs that a female produces during her lifetime are present in the ovaries at birth. After puberty, eggs will mature, rupture from the surface of the ovaries, and enter the oviducts. This process is called **ovulation.**

If viable sperm are present in the upper third of an oviduct when it contains eggs, fertilization may occur. In this case, the fertilized egg or zygote will develop into an embryo and pass down the oviduct to become implanted in the wall of the uterine horn. In the human female, however, it becomes implanted in the uterine body.

6. Identify the membranous broad and round ligaments. The **broad ligament,** which originates from the **dorsal body wall** (Figure 22-23), supports the ovaries, oviducts, and uterine horns. The **round ligament**, which also supports the ovaries, extends from the lateral wall and crosses the broad ligament diagonally.

C. Male Reproductive System

The male reproductive system of a mammal consists of organs and structures that primarily function in the production of sperm, their transit to the base of the penis during sexual excitement, and their subsequent ejaculation in **semen**—sperm plus the secretions of the *sex accessory glands.*

1. Locate the **testes** (male gonads; Figure 22-22), the site of **sperm** production and source of *testosterone,* the male sex hormone. In older fetuses they are located in the *scrotum,* but in younger fetuses they can be found anywhere between the kidneys and the scrotum.

For viable sperm to be produced in adult males, the testes must be situated outside of the abdominopelvic cavity, where body temperatures are slightly lower than within. Thus, during normal development, the testes undergo a posterior migration, or descent, into the scrotum.

2. Locate the scrotum. Make a midline incision through this structure, cutting through the muscle tissue. Pull out the two elongated bulbous structures covered with a transparent membrane. This membrane is the *tunica vaginalis* and is actually an outpocketing of the abdominal wall. Notice the tough white cord that connects the posterior end of the testes to the inner face of the sac (Figure 22-22). This cord, the **gubernaculum,** is homologous to the round ligament in the female reproductive system. It grows more slowly than the surrounding tissues and thus aids in pulling the testes posteriorly into the scrotal sacs.

3. Cut through the tunica vaginalis to expose a single testis and find the **epididymis.** This is a tightly coiled tube that lies along one side of the testis. Sperm produced in the testes are stored in the epididymides until ejaculation.

4. Note the slender, elongated **spermatic cord** that emerges anteriorly from each testis (Figure 22-22). Gently pull the cord and note that it passes through an opening, the **inguinal canal,** which is actually an opening from the abdominopelvic cavity into the scrotum. It is through this opening that the testes descend during their migration into the scrotum.

Some human males develop an **inguinal hernia,** a condition in which part of the intestine drops through the inguinal canal into the scrotum. Pigs and other four-legged mammals (hint) do not develop inguinal hernias. Why do you think this is so?

_____

_____

_____

5. Using the tips of your scissors, slit open the spermatic cord attached to the dissected testis. Note that it contains the *sperm duct* or **vas deferens** (plural, *vasa deferentia*), the spermatic vein, the **spermatic artery,** and the spermatic nerve. It is the vas deferens that is severed when a human male has a **vasectomy.** Follow the vas deferens to the base of the bladder, where it loops up and over the ureter and then continues posteriorly to enter the urethra.

6. Expose the full length of the **penis** and its juncture with the urethra. To find the latter, make an incision with your scalpel through the muscles in the midventral line between the hindlegs. When the cut is deep enough, they will lie flat. Carefully remove the muscle tissue and pubic bone on each side of the cut until the urethra is exposed. With a blunt probe, tear the connective tissue connecting the urethra to the rectum, which lies dorsal to it.

7. Locate a pair of small glands, the **seminal vesicles,** on the dorsal surface of the urethra where the two vasa deferentia enter. Situated between the bases of the seminal vesicles is the **prostate gland.** The other sex accessory glands are the **bulbourethral glands,** two elongated structures lying on either side of the juncture of the penis and urethra.

The seminal vesicles, the prostate gland, and the bulbourethral glands all secrete fluids that, together with sperm, form semen, which is ejaculated during sexual intercourse. In addition to sperm, semen is mostly water, sugar, and other molecules that provide a supportive environment for the sperm.

8. *If you are continuing with the dissection, move on to the next section.* Otherwise, place your pig back in the plastic bag. Tie it shut to prevent your pig from drying out. Dispose of any paper towels that contain preservative *as directed by your instructor.* Clean the tray, dissecting tools, and the laboratory table.

22.8 Kidney *(About 10 min.)*

In the first part of this exercise, you located the kidneys, a pair of kidney bean–shaped structures lying on either side of the spine in the lumbar region of the body. Although the kidneys are situated below the diaphragm, they are actually located outside of the peritoneum, the membrane that lines the abdominal cavity. During this procedure, you will examine the internal anatomy of the kidney, including its functional unit, the **nephron.**

MATERIALS

Per student pair:

- one preserved fetal pig injected with red and blue latex
- plastic bag to store fetal pig
- dissection pan
- dissecting kit
- dissecting pins
- 4 large rubber bands *or* 2 pieces of string, each 60 cm long
- compound light microscope
- prepared section of kidney

Per lab room:

- liquid waste disposal bottle
- boxes of different sizes of latex or vinyl gloves
- box of safety goggles

PROCEDURE

A. Anatomy

Each kidney consists of numerous nephrons, which function to filter the blood and start the process of urine formation. A system of ducts completes urine formation and transports it to a space, the *renal pelvis*, which is drained by the ureter. The ureter transports the urine to the urinary bladder.

1. Return to the right kidney and attempt to identify the *adrenal gland.* Look for a tiny, cream colored, comma-shaped body located on the medial, anterior side of the kidney. This is another of the body's endocrine organs. The paired adrenal glands secrete several hormones, including adrenaline (epinephrine). Free the right kidney by severing the renal vein, renal artery, and ureter. Remove the kidney from the body cavity and place it on a paper towel with the central depression to the right.

2. With your scalpel, carefully cut the kidney in half lengthwise, as you would separate the two halves of a peanut. Examine the cut surface of one of the halves and locate the three major regions of the kidney—the

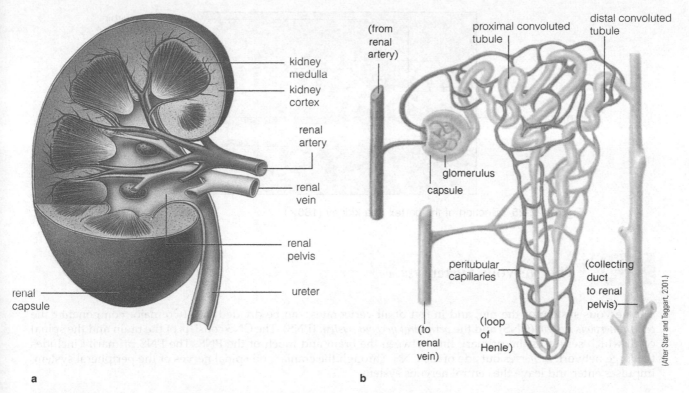

a

b

(After Starr and Taggart, 2001.)

Figure 22-24 (a) Longitudinal section of human kidney. (b) A nephron, the functional unit of the kidney.

outer **cortex**, the inner **medulla**, and the **renal pelvis** (Figure 22-24a). The cortex and medulla contain different portions of the nephrons and their associated blood vessels.

B. Microscopic Structure of the Nephrons

Each **nephron** is a little tube or tubule that is closed at one end. The closed end is expanded and collapsed upon itself much like a deflated basketball pushed in by your fist, only the inside space is connected to the space within a tube attached to the other side. The expanded, collapsed end of the nephron is called the *capsule* and in the kidney the space within its inner wall (where your fist would be) contains a network of capillaries, the *glomerulus* (plural, glomeruli; Figure 22-24b). The combination of the capsule and the glomerulus is called a *renal corpuscle.*

Urine formation begins with the ultrafiltration of blood across the capillary and inner capsule walls into the space within the capsule. After ultrafiltration, the filtrate travels through the rest of the nephron—*proximal convoluted tubule,* the *loop of Henle,* and the *distal convoluted tubule*). Much of the water, ions, sugars, and other useful substances are reabsorbed into the blood in the *peritubular capillaries.* While the reabsorption process recaptures many useful molecules, substances such as ammonia, potassium, and hydrogen ions are secreted to join the urea in the forming urine. Thus, the nephron carries out its excretory and salt balance functions in three steps: filtration, reabsorption, and tubular secretion.

Although all of the activities of the nephron are extremely vital to the health of a mammal, the importance of the reabsorption function is easily illustrated with a few numbers. The renal corpuscles of human kidneys produce approximately 180 L (approximately 180 quarts) of filtrate each day. About 99% of this ultrafiltrate is reabsorbed as water, primarily by the nephrons. If they were not so efficient, we would have to drink constantly just to replenish the fluid lost.

Examine Figure 22-24b, which shows a nephron, its named portions, and associated blood vessels.

1. Examine a prepared section of the kidney with your compound microscope. Identify the *cortex* and *medulla.* In the cortex locate *renal corpuscles, glomeruli,* and *capsules* (Figure 22-25).
2. Identify cross sections of the tubular portion of nephrons.
3. *If you are continuing with the dissection, move on to the next section.* Otherwise, place your pig and any excised organs back in the plastic bag. Tie it shut to prevent your pig from drying out. Dispose of any paper towels that contain preservative *as directed by your instructor.* Clean the tray, dissecting tools, and the laboratory table.

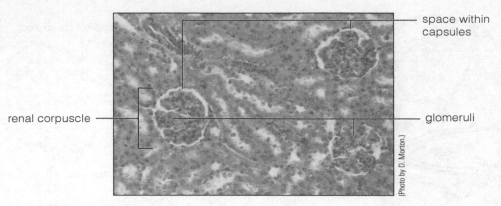

Figure 22-25   Section of the cortex of a kidney (186×).

renal corpuscle

space within capsules

glomeruli

(Photo by D. Morton.)

The Nervous System *(20 min.)*

The nervous system of the pig, and in fact of all vertebrates, can be divided into two major components: the *central nervous system* (CNS) and the *peripheral nervous system* (PNS). The CNS consists of the brain and the spinal cord, which serves as the primary link between the brain and much of the PNS. The PNS primarily includes the large network of *nerves* outside of the CNS. Through the cranial and spinal nerves of the peripheral system, impulses enter and leave the central nervous system.

MATERIALS

Per student group (4):

- one preserved fetal pig injected with red and blue latex
- plastic bag to store fetal pig
- dissection pan
- dissecting kit
- dissecting pins
- 4 large rubber bands *or* 2 pieces of string, each 60 cm long
- dissection microscope

Per lab room:

- liquid waste disposal bottle
- boxes of different sizes of latex or vinyl gloves
- box of safety goggles

PROCEDURE

As it is very time-consuming to expose the full length of the spinal cord, work in groups of four. One pair in the group exposes the anterior portion of the spinal cord and the other pair its posterior region.

A. Spinal Cord

1. Turn the body-wall flaps of the fetal pig inward and place your specimen ventral-side down on the dissection pan. Proceed carefully with this portion of the dissection, as the nervous tissues of a fetus are extremely delicate and can be easily destroyed.
2. Carefully remove a strip of muscle about 1.5 cm wide from the base of the neck posteriorly along the spinal column to the tail. This will expose the **spines** and the **vertebral arches** of the vertebrae (Figure 22-26). With a sharp scalpel, gradually cut away the spines and the neural arches of the vertebrae until the **spinal cord** is exposed (Figure 22-27).
3. Note the enlargements of the spinal cord at the level of the forelimbs and hind-limbs. These are the **cervical** and **lumbar enlargements,** respectively; they result from the presence of a large number of nerve cells, or *neurons,* supplying the appendages in these regions.
4. At the anterior end of the body, the spinal cord widens to become the **medulla oblongata,** the most posterior portion of the brain. At its caudal end, the spinal cord narrows to a relatively thin strand of tissue called the *filum terminale.*
5. Surrounding the spinal cord and the brain are a set of three membranes, the **meninges** (Figure 22-26). The outermost layer, the **dura mater,** is the most apparent and adheres to the underside of the cranial and spinal bones. The dura mater is a tough, fibrous connective tissue sheath that you cut through to expose the spinal

cord. The middle layer, the **anachnoid,** is very delicate and collapses when the spinal canal is opened. The innermost layer, the **pia mater,** adheres closely to the surface of the spinal cord and brain. If you can't identify the outer and inner meninges, attempt to locate them when you dissect the brain.

In certain severe viral or bacterial infections, the meninges around the spinal cord and/or brain can become inflamed. This serious condition is known as **meningitis.**

6. Note the origin of the **spinal nerves** on either side of the spinal cord (see Figures 22-26 and 22-27 to determine the relationships of the spinal nerves to the spinal cord). Each spinal nerve is composed of a **dorsal** and a **ventral root** (Figure 22-26). The dorsal root, carrying sensory impulses into the spinal cord, can be easily identified by the presence of a distinct swelling called the **dorsal root ganglion.** The ventral root, which carries motor impulses from the cord to some type of effector (a skeletal muscle, for example), has no ganglion and is not as easily seen as the dorsal root in a dorsal view.

7. Remove a short cross section (0.5 cm long) of the spinal cord and examine it with a dissection microscope. Observe the cut end and note the prominent H-shaped area in the center (Figure 22-26). This is the **gray matter,** which is composed of the cell bodies of neurons, their unmyelinated processes, and accessory cells. The **white matter** around the "H" is primarily made up of myelinated (surrounded by a white, fatty insulating material) neuron processes that conduct messages to and from the brain.

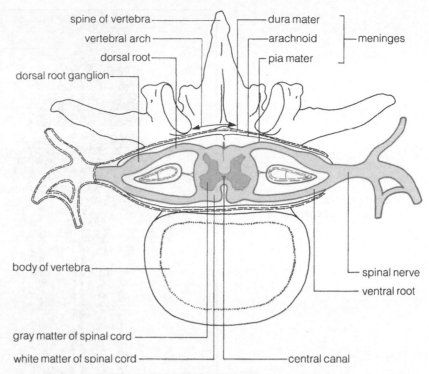

Figure 22-26   Cross section of a vertebra and spinal cord of a mammal.

B. Brain

1. Using your scalpel and scissors, make a longitudinal cut through the skin and muscle tissue of the dorsal portion of the head, beginning at the base of the snout and ending at the base of the skull. From the anterior portion of this incision, make a transverse cut to the angle of the jaws and another transverse incision from the base of the skull to a level just ventral to the ears. Peel back the skin and muscle to expose the skull.

2. Penetrate the skull with the lower blade of your scissors and make a cut along its middorsal line. Do not cut too deeply or you will damage the brain. Now make two cuts, about 2 cm apart, at right angles to the first cut. With forceps, carefully break off pieces of the skull until the entire dorsal and lateral areas of the brain are exposed (Figure 22-26).

3. If you were not able to identify the meninges covering the spinal cord, locate the dura mater lining the inside of the skull and the pia mater on the surface of the brain. As with the spinal cord, the arachnoid layer (middle layer) will not be apparent.

4. Cut the spinal cord at the base of the brain and carefully remove it from the skull. The brain is composed of the right and left *cerebral hemispheres* (collectively, the largest part of the brain, the **cerebrum;** Figure 22-27), separated by a prominent *longitudinal groove;* a smaller mass posterior to the cerebrum, the **cerebellum;** and the medulla oblongata or, more simply, the *medulla,* poking out from under the cerebellum. The **pons** lies just anterior to the medulla on the ventral side of the brain.

In general, the more posterior portions of the brain direct most involuntary, unconscious, and mechanical processes. For example, the cerebellum unconsciously controls posture and contains motor programs (like computer programs) for many complex body movements. The cerebrum is responsible for such activities as reasoning, memory, conscious thought, language, and sensory decoding—activities that are generally associated with intelligence.

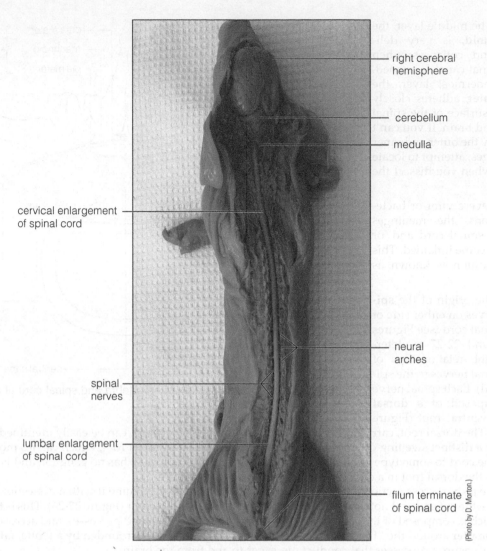

right cerebral hemisphere

cerebellum

medulla

cervical enlargement of spinal cord

neural arches

spinal nerves

lumbar enlargement of spinal cord

filum terminate of spinal cord

(Photo by D. Morton.)

Figure 22-27 Dorsal view of the brain, spinal cord, and spinal nerves of a fetal pig.

5. Place your pig and any excised organs back in the plastic bag. Tie it shut to prevent your pig from drying out. Dispose of any paper towels that contain preservative *as directed by your instructor*. Clean the tray, dissecting tools, and the laboratory table.

_____ 1. A fetus is
  (a) a newborn pig
  (b) a newborn human
  (c) an unborn mammal
  (d) all of the above

_____ 2. *To dissect* means primarily
  (a) to cut open
  (b) to remove all internal organs
  (c) to expose to view
  (d) all of the above

_____ 3. When the directions for a fetal pig dissection refer to the left, they are referring to
  (a) your left
  (b) the pig's left
  (c) the pig's right
  (d) both a and c

_____ 4. In a fetal pig, *dorsal* and *ventral* refer respectively to
  (a) the head and tail regions of the body
  (b) the tail and the head regions of the body
  (c) the upper (back) portion and the lower (underside) portion of the body
  (d) the lower (underside) portion and the upper (back) portion of the body

_____ 5. The umbilical cord functions to
  (a) carry waste products in the blood from the fetus to the mother
  (b) carry waste products in the blood from the mother to the fetus
  (c) carry oxygen in the blood from the mother to the fetus
  (d) do both a and c

_____ 6. The female *fetal* pig is similar to the male *fetal* pig in that its body
  (a) has separate openings for the urinary system and the reproductive system
  (b) has a common opening for the urinary system and the reproductive system
  (c) has an opening for the urinary system but none for the reproductive system
  (d) has an opening for the reproductive system but none for the excretory system

_____ 7. The two *major* body cavities of a fetal pig are the
  (a) thoracic and pleural
  (b) thoracic and pericardial
  (c) abdominopelvic and thoracic
  (d) abdominopelvic and pericardial

_____ 8. The diaphragm is a sheetlike skeletal muscle that separates the
  (a) thoracic and pleural cavities
  (b) thoracic and pericardial cavities
  (c) thoracic and abdominopelvic cavities
  (d) pleural and pericardial cavities

_____ 9. The digestive system is concerned with
  (a) blood circulation
  (b) digestion and the absorption of nutrients
  (c) reproduction
  (d) excretion of urine

_____ 10. The liver functions to
  (a) produce bile
  (b) pump blood
  (c) form urea
  (d) do both a and c

_____ 11. As a general rule, the small intestine of a pig or human is
  (a) about 60 cm long
  (b) about 1.5 m long
  (c) about as long as the individual is tall (or long in the case of the pig)
  (d) about five times the height of the individual

_____ 12. The cardiac, pyloric, anal, and ileocecal sphincters are all part of the
  (a) digestive tract
  (b) respiratory tract
  (c) circulatory system
  (d) muscular system

_____ 13. In humans, the front of the larynx is commonly referred to as
  (a) the voice box
  (b) the Adam's apple
  (c) the food pipe
  (d) both a and b

_____ 14. The microscopic air sacs or alveoli are the sites where blood
(a) picks up oxygen
(b) gives up carbon dioxide
(c) gives up oxygen
(d) both a and b

_____ 15. A vein is a blood vessel that always carries
(a) blood toward the heart
(b) blood away from the heart
(c) oxygen-rich blood
(d) oxygen-poor blood

_____ 16. The urogenital system refers to the
(a) urinary and reproductive systems
(b) urinary and excretory systems
(c) reproductive system
(d) external genitalia

_____ 17. The ureters drain urine into the
(a) renal pelvis
(b) cecum
(c) urinary bladder
(d) small intestine

_____ 18. The clitoris of the female and a portion of the penis of the male are homologous structures. This means they have a similar
(a) function
(b) structure
(c) developmental origin
(d) origin and structure

_____ 19. The testes of a male differ from the ovaries of a female in that the testes
(a) develop in the body cavity and migrate to a position outside of the body cavity
(b) require a slightly higher temperature than that of the body to produce viable gametes
(c) produce zygotes
(d) do both a and b

_____ 20. When a human male has a vasectomy, the operation involves
(a) removal of the male gonads or testes
(b) removal of the urethra
(c) the severing of the vas deferens
(d) removal of the prostate gland

_____ 21. Semen contains
(a) sperm
(b) the secretions of sex accessory glands
(c) eggs
(d) both a and b

_____ 22. The functional unit of the kidney is the
(a) renal pelvis
(b) ureter
(c) cortex
(d) nephron

_____ 23. The central nervous system of a mammal includes the
(a) brain
(b) spinal cord
(c) brain and spinal cord
(d) brain, spinal cord, and every major nerve in the body

_____ 24. The brain is surrounded by a set of membranes called the
(a) pleural membranes
(b) peritoneum
(c) pericardial membranes
(d) meninges

_____ 25. The largest part of the brain of a mammal is the
(a) cerebrum
(b) cerebellum
(c) pons
(d) medulla oblongata

Name _____ Section Number _____

EXERCISE 22

Dissection of the Fetal Pig

POST-LAB QUESTIONS

22.1 External Anatomy of the Fetal Pig

1. Identify the external features of a fetal pig noted on the drawing.

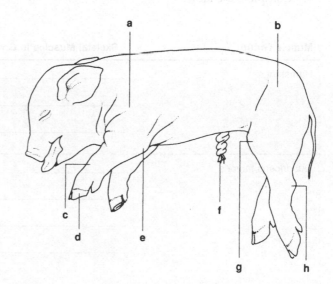

a _____

b _____

c _____

d _____

e _____

f _____

g _____

h _____

2. What is the function of the umbilical cord?

3. Briefly describe how your feet differ from those of the pig.

4. Using external features, briefly describe how you can determine the difference between a male and a female fetal pig.

22.2 Muscular System

5. Define in general terms the *origin, insertion,* and *action* of a skeletal muscle organ.

6. What does the phrase *antagonistic muscles* mean?

7. Complete the table.

Muscle Group	Skeletal Muscles in Group	General Actions of Group
Hamstrings	_____ _____ _____	
Quadriceps femoris	_____ _____ _____ _____	

Food for Thought

8. The right and left sternocleidomastoid muscles of humans originate on the upper surfaces of breast bone and collar bones, pass along both sides of the neck, and insert just behind the ears on the mastoid processes. Describe the action that occurs if both sides contract together.

9. Explain the basic difference between digitigrade and plantigrade locomotion.

10. Search the World Wide Web for sites that describe human body movements. List two sites and briefly summarize their contents.

    http://

    http://

22.3 Ventral Body Cavities

11. Describe the location of the two *major* ventral body cavities.

22.4 Digestive System

12. Describe the difference between the digestive system and the digestive tract.

13. List in order the organs through which food and other substances pass in their journey into, through, and out the digestive tract.

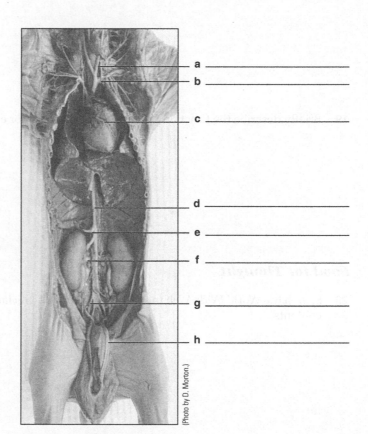

22.5 Respiratory System

14. Describe the major similarities and differences in the location, structure, and function of the trachea and esophagus.

a _____

b _____

c _____

d _____

e _____

f _____

g _____

h _____

(Photo by D. Morton.)

22.6 Blood Vessels and the Surface Anatomy of the Heart

15. Identify the structures in the photo on the right.

16. What is the main difference between the pulmonary and the systemic circuits of the circulatory system?

17. What is the foramen ovale? What is its fate after birth?

18. With regard to blood circulation, what is the difference between an artery and a vein?

19. Briefly describe the function of a portal vein system (for example, the hepatic portal vein).

Food for Thought

20. Search the World Wide Web for sites that describe artificial hearts. List two sites and briefly summarize their contents.

http://

http://

22.7 Urogenital System

21. What is the urogenital system?

22. Briefly describe the functions of the kidney, ureters, bladder, and urethra in the adult male pig.

23. What is the vulva?

24. How does the uterus of female pigs and humans differ? Include in your discussion the site of embryo implantation.

25. What is the inguinal canal in males and how does it form during development?

26. Identify the structures in the photo on the right.

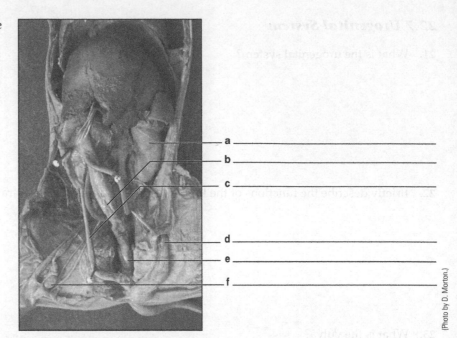

a _____

b _____

c _____

d _____

e _____

f _____

22.8 Kidney

27. Name the functional unit of the kidney. Briefly describe how it operates.

22.9 The Nervous System

28. Describe the basic organization of the nervous system of a mammal.

29. List the three meninges in order from the surface of the brain to the inside of the skull.

Food for Thought

30. Search the World Wide Web for sites that describe kidney transplants and dialysis treatments. List two sites and briefly summarize their contents.

http://

http://

Human Sensations, Reflexes, Reactions, and the Structure and Function of Sensory Organs

OBJECTIVES

After completing this exercise, you will be able to

1. define *consciousness, sensory neurons, receptors, stimulus, motor neurons, effectors, somatic motor neurons, autonomic motor neurons, interneurons, chemical synapse, integration, proprioception, sensations, modality, free neuron endings, encapsulated neuron endings—Meissner's and Pacinian corpuscles, projection, phantom pain, adaptation, reflex, reflex arc, stretch reflexes, patella reflex, muscle spindle, pupillary reflex, swallowing reflex, reaction, visual acuity, myopia, hyperopia, astigmatism, near point of eye,* and *presbyopia;*

2. describe the flow of information through the nervous system;

3. state the nature and function of sensations;

4. describe a stretch reflex;

5. describe the papillary reflex;

6. distinguish between a reflex and a reaction;

7. measure visual reaction time;

8. explain the differences among photoreceptors, mechanoreceptors, and chemoreceptors;

9. recognize and describe the structures of the external eye, eyeball, and ear;

10. determine visual acuity and the presence or absence of astigmatism;

11. describe the various reflexes important to proper vision;

12. recognize the retina, organ of Corti, taste buds, and olfactory epithelium;

13. explain briefly the function of rods and cones;

14. describe afterimages;

15. explain how we sense the source of a sound;

16. discuss the similarities and differences between the senses of taste and smell.

INTRODUCTION

How do you interact with the external environment? To answer this question, you first have to be able to analyze your interactions. This means you have to be conscious. **Consciousness** is the state of being aware of the things around you, your responses, and your own thoughts. Being conscious allows you to learn, to remember, and to feel and show emotion. Second, you have to understand the flow of information through the nervous system (Figure 23-1).

Sensory neurons carry messages from **receptors** to the spinal cord and brain, which comprise the central nervous system (CNS). Receptors are located both within the body and on its surface. Receptors within the body receive information from the internal environment, while those on the surface of the body receive information from the external environment. Each piece of information received by a receptor is called a **stimulus** (plural, *stimuli*).

Motor neurons carry messages from the CNS to **effectors**. Effectors are muscles or glands that respond to stimuli. **Somatic motor neurons** control skeletal muscles, and **autonomic motor neurons** control smooth muscles, cardiac muscle, and glands.

In the CNS, a sensory neuron can directly stimulate a motor neuron across a *chemical synapse;* more frequently, though, one or more **interneurons** (*association neurons*) connect the sensory and motor neurons.

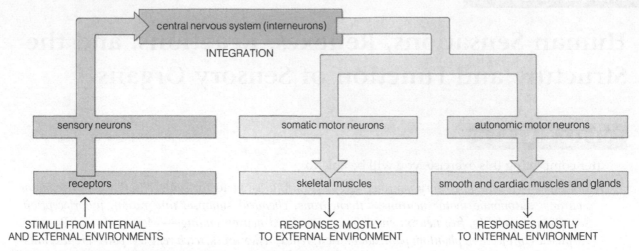

Figure 23-1 Flow of information through the nervous system.

A **chemical synapse** is a junction between two neurons (or between a neuron and an effector) that are separated by a small gap. A chemical transmitter substance released from the first neuron diffuses across the gap and then binds to (and produces changes in) the receiving cell.

The function of the interneurons is **integration.** At this level, integration is the processing of messages received from receptors via sensory neurons and the activation of the appropriate motor neurons, if any, to initiate responses by effectors. Your conscious mind is located in the cerebral cortex of the brain and is aware of (and indeed plays a part in) some of this activity.

<table>
<tr><td>23.1</td><td>Sensations (50 min.)</td></tr>
</table>

A receptor is the smallest part of a sense organ (such as skin) that can respond to a stimulus. The receptor is linked to the CNS by a single sensory neuron. Our bodies have receptors for light, sound waves, chemicals, heat, cold, tissue damage, and mechanical displacement. Senses for which we have sensations include sight, hearing, taste, smell, pain, touch, pressure, temperature, vibration, equilibrium, and **proprioception** (knowledge of the position and movement of the various body parts). **Sensations** are that portion of the sensory input to the CNS that is perceived by the conscious mind. There are also a number of complex sensations such as thirst, hunger, and nausea.

Most sensations inform the conscious mind about the state of the external environment. Sensations from the internal environment inform the conscious mind about problems such as dehydration. If you are thirsty, you will make a conscious decision to find and drink water.

Receptors and the sensations they produce have three characteristics: *modality, projection,* and *adaptation.* These characteristics can be easily demonstrated by investigating the skin's receptors.

MATERIALS

Per student pair:

- compound microscope
- prepared slide of mammalian skin stained with hematoxylin and eosin
- felt-tip, nonpermanent pen
- bristle
- dissecting needle
- scientific calculator
- 2 blunt probes in a 250-mL beaker of ice water
- 2 blunt probes in a 250-mL beaker of hot tap water (the hot water will have to be changed every 5 minutes)

- ice bag
- camel-hair brush
- reflex hammer
- three 1000-mL beakers containing
  - ice water
  - 45°C water
  - room-temperature water
- tissue paper

Per lab room:

- demonstration slide of a Pacinian corpuscle

PROCEDURE

A. Modality

Modality is the particular sensation that results from the stimulation of a particular receptor. For example, the modalities of taste—bitter, salty, sour, and sweet—are associated with four different types of taste buds. However, although every receptor has evolved to be most sensitive to one type of stimulus, modality actually depends on where in the brain the sensory neurons from the receptor (or the interneurons to which they connect) terminate. Modality cannot be encoded in the messages carried by sensory neurons, because every impulse in that message is identical. The only information neurons transmit is the absence or presence of a stimulus and (if one is present) its intensity—low stimulus-intensities produce a low frequency of impulses and high stimulus-intensities produce a high frequency of impulses.

1. Examine a prepared section of skin (Figure 23-2). Two categories of receptors are present: **free neuron endings** and **encapsulated neuron endings.**

Free neuron endings are almost impossible to see in typically stained sections, but note their distribution in Figure 23-2. Stimulating different free neuron endings produces sensations of pain, crude touch, and perhaps cold and hot. Encapsulated neuron endings consist of neuron endings surrounded by a connective tissue capsule.

Find **Meissner's corpuscles** in the *dermal papillae* (Figure 23-2). Meissner's corpuscles are receptors for fine touch and low-frequency vibration. Now look for **Pacinian corpuscles** between the dermis and hypodermis. Pacinian corpuscles look like a cut onion and are receptors for pressure and high-frequency vibration. Not all skin sections will contain a Pacinian corpuscle. If you cannot locate one, look at the demonstration slide.

2. With a felt-tip, nonpermanent ink pen, draw a 25-cell, 0.5-cm grid (Figure 23-3) on the inside of your lab partner's forearm, just above the wrist.
3. You are now the investigator, and your lab partner is the subject. At this point, ask your lab partner to close his or her eyes. Using a bristle, touch the center of each box in the grid. If the bristle bends, you are pressing too hard. Ask your lab partner to announce when the touch is felt. Do not count responses given when you remove the bristle. Just count those that coincide with the initial touch. Mark each positive response with a T in the upper left-hand corner of the corresponding box in Figure 23-3.

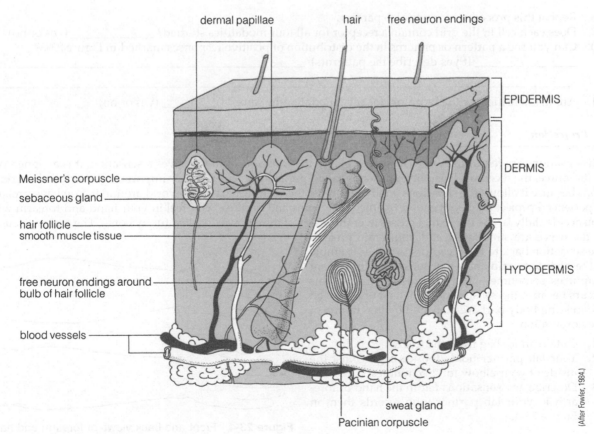

Figure 23-2 Section of skin.

Human Sensations, Reflexes, Reactions, and the Structure and Function of Sensory Organs **523**

4. Repeat the above with a clean dissecting needle. This time, if you feel a prick, mark P for "pain" in the upper right-hand corner of the corresponding box in Figure 23-3.

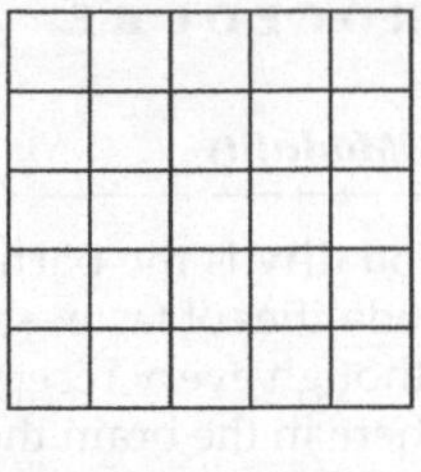

Figure 23-3 Grid for testing skin stimuli and recording modality data.

Caution

Do not press; simply let the tip of the dissecting needle rest on the surface of the skin.

5. Repeat the above with a chilled blunt probe. Before using the blunt probe, dry it with tissue paper. The blunt probe will warm up over time, so switch it with the second chilled blunt probe every five trials. This time, mark each positive response with a C for "cold" in the lower left-hand corner of the corresponding box in Figure 23-3.
6. Repeat the above with a heated blunt probe. Before using the blunt probe, dry it with tissue paper. Use the two blunt probes alternately every five trials. This time, mark each positive response with an H for "hot" in the lower right-hand corner of the corresponding box in Figure 23-3.
7. Record the total number of positive responses for each stimulus in Table 23-1. Calculate the density of receptors for each modality (multiply the number of positive responses for each stimulus by 4) and record them in the third column of Table 23-1.

TABLE 23-1 Positive Identifications to Stimuli Applied to 25 Cells in a 0.25-cm^2 Patch of Skin

Stimulus	Number of Responses	Density (Responses/cm^2)
Touch		
Pain		
Cold		
Hot		

8. Repeat this procedure for your lab partner.
9. Does each cell in the grid contain a receptor for all four modalities studied? ____________ (yes or no)
10. Can you see a pattern or patterns in the distribution of positive responses marked in Figure 23-3? ______________ If yes describe the pattern(s): ____

____

11. Are the densities for the receptors for each modality the same? __________ (yes or no)

B. Projection

All sensations are felt in the brain. However, before the conscious mind receives a sensation, it is assigned back to its source, the receptor. This phenomenon is called **projection.** This is a very important characteristic of sensations because it allows the conscious mind to perceive the body as part of the world around it. You have probably experienced projection. A common example is the "pins and needles" you feel in your hand and forearm when you accidentally jar the nerve that passes over the inside of the elbow (so-called funny bone). The sensory neurons in the nerve are stimulated, and your brain projects the sensation back to the receptors. Another example is the **phantom pain** and other sensations that recent amputees sometimes "feel" in missing limbs. This occurs because the sensory neurons that once served the missing body part are activated by the trauma of the amputation.

1. Obtain an ice bag from the freezer.
2. Your lab partner holds the ice bag against the inside of your elbow for 2–5 minutes.
3. Describe any sensations felt in the hand or forearm to your lab partner who records them in Figure 23-4.

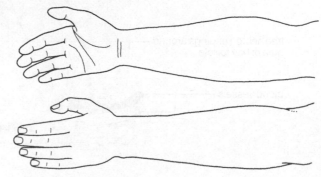

Figure 23-4 Front and back views of forearm and hand for recording projection data.

4. While the ice bag is applied to your elbow, your lab partner checks you for any loss of sensation by gently stroking your forearm and hand with a camel-hair brush. Sensations may also be felt after the ice bag is removed.
5. If no results are obtained, try tapping the inside of the elbow with the reflex hammer.
6. Similarly test your lab partner.
7. What can you conclude about projection and the receptors on the surface of the hand and forearm?

_____

_____

_____

C. Adaptation

The intensity of the signal produced by a receptor depends in part on the strength of the stimulus and in part on the degree to which the receptor was stimulated before the stimulus. Receptors undergo **adaptation** to a constant stimulus over time. For example, when you first enter a dark room after being in bright light, you cannot see. After a while, your photoreceptors adapt to the new light conditions, and your vision improves.

1. Partially fill each of three 1000-mL beakers with ice water, water at room temperature, and water at 45°C.
2. Place one hand in the ice water and one in the warm water. After 1 minute, place both hands simultaneously in the water at room temperature.
3. Describe the sensation of temperature in each hand to your lab partner, who records these descriptions in Table 23-2.

TABLE 23-2 Sensations Felt When Preadapted Hands Are Placed in Room-Temperature Water

Relative Temperature of Preadaption	Result
Cold	
Warm	

4. What can you conclude about the skin receptors for temperature and their capacity for adaptation?

_____

_____

What about the ability of other kinds of receptors to adapt? Use your own experiences for smell, touch, and pain to answer this question. (Hints for touch: Can you feel your clothes? How about when you first get dressed after a shower?)

_____

_____

_____

_____

_____

23.2   Reflexes (15 min.)

A **reflex** is an involuntary response to the reception of a stimulus. A **reflex arc** consists of the nervous system components activated during the reflex. The simplest reflex arc consists of a receptor, sensory neuron, motor neuron, and effector. Involuntary means that your conscious mind does not decide the response to the stimulus. However, the conscious mind may be aware after the fact that the reflex has taken place. Reflexes of which we are not aware occur most often in the internal environment (for example, reflexes involved in adjustments of blood pressure).

MATERIALS

Per student pair:

- reflex hammer
- penlight

PROCEDURE

A. Stretch Reflexes

Stretch reflexes are the simplest type of reflex because the interneurons are not directly involved (Figure 23-5). The sensory neuron connects directly with the motor neuron in the spinal cord. Stretch reflexes are important in controlling balance and complex skeletal muscular movements such as walking. Physicians often test these reflexes during physicals to check for spinal nerve damage. You've probably experienced one of these tests, the **patella reflex.** In this test, the receptor is the **muscle spindle** in the quadriceps femoris muscle of the front of the thigh, which is attached through its tendon and the patellar ligament to the top of the front surface of the tibia. The tibia is the larger of the two lower leg bones. The patella (kneecap) is embedded in the middle of the combined tendon/ligament. The muscle spindle detects any stretching of the muscle. The effector is the muscle itself.

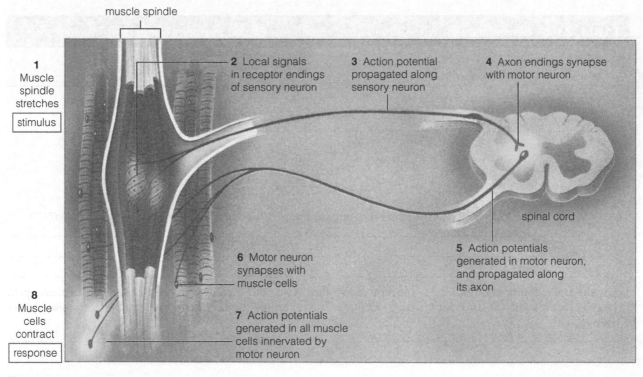

Figure 23-5 The stretch reflex.

1. Sit on a clean lab bench and shut your eyes.
2. Your lab partner taps the patella ligament with a reflex hammer (Figure 23-6). Describe the response.

_____

_____

If you have trouble producing a response, distract yourself by counting backward from 10. During this count, your lab partner again taps the patella ligament.

3. Even with your eyes shut, are you aware of the stimulus and the response? _____ (yes or no) This is because of pressure receptors that sense the tap and because of proprioceptors that sense movement of the leg.
4. Stretch reflexes are *somatic reflexes* because they involve somatic motor neurons and skeletal muscles. Can you willfully inhibit a stretch reflex? _____ (yes or no)
5. Similarly test your lab partner.

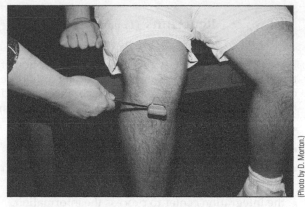

Figure 23-6 Area to tap to produce patella reflex.

B. Pupillary Reflex

1. Shine the penlight into your lab partner's eyes. Does the size of the pupil (the diameter of the opening into the eye that is surrounded by the pigmented iris) get larger or smaller? _____
2. Now turn off the penlight. Does the size of the pupil get larger or smaller? _____
3. Repeat steps 1 and 2. Which is faster, constriction of the iris (which makes the pupil smaller) or dilation of the iris (which makes the pupil larger)? _____
4. Are you aware of the pupil's changing diameter? _____ (yes or no)
5. The **pupillary reflex** is an autonomic reflex because it involves an autonomic motor neuron and, in this case, smooth muscle. Can you willfully inhibit the pupillary reflex? _____ (yes or no)
6. Similarly test your lab partner.

C. Complex Reflexes

Complex reflexes involve many reflex arcs and interneurons. A good example is swallowing. The stimulus in the **swallowing reflex** is the movement of saliva, food, or drink into the posterior oral cavity. The response is swallowing.

1. Cup your hand around your neck and swallow. Feel the complex skeletal muscular movements involved in swallowing. Do you consciously control all these muscles? _____ (yes or no)
2. Is it possible to swallow several times in quick succession? _____ (yes or no)
3. Explain this result. It has something to do with the stimulus.

_____

_____

_____

_____

4. What part of swallowing does your conscious mind control, and what part is a reflex?

_____

_____

_____

A **reaction** is a voluntary response to the reception of a stimulus. *Voluntary* means that your conscious mind initiates the reaction. An example is swatting a fly once it has landed in an accessible spot. Because neurons must carry the sensory message to the cerebral cortex and the message to react back to the motor neuron, a reaction takes more time than a reflex. *Reaction time* is the sum of the time it takes for

- the stimulus to reach the receptive unit,
- the receptor to process the message,
- a sensory neuron to carry the message to the integration center,
- the integration center to process the information,
- a motor neuron to carry the response to the effector, and
- the effector to respond.

Visual reaction time can easily be measured with a reaction-time ruler. This device makes use of the principle of progressive acceleration of a falling object.

MATERIALS

Per student pair:

- Reaction Time Kit (Carolina Biological Supply Company)
- chair or stool
- scientific calculator

PROCEDURE

The following instructions are modified from the *Reaction Time Kit Instructions* booklet.

1. Sit on a chair or stool (Figure 23-7).
2. Your lab partner stands facing you and holds the *release end* of the reaction-time ruler with the thumb and forefinger of the dominant hand, at eye level or higher.
3. Position the thumb and forefinger of your dominant hand around the *thumb line* on the ruler. The space between the thumb and forefinger should be about 1 inch.
4. Tell your lab partner when you are ready to be tested.
5. Any time during the next 10 seconds, your lab partner lets go of the ruler.
6. Catch the ruler between the thumb and forefinger as soon as it starts to fall. The line under your thumb represents visual reaction time in milliseconds.
7. Read the reaction time from the ruler out loud, and your lab partner records the data in Table 23-3.
8. Repeat steps 1–7 ten times and calculate the average reaction time from the ten trials.
9. Similarly test your lab partner.
10. The reaction times of most of the ten trials should be similar, but perhaps the first few or one at random may be relatively different from the others. If this is true for your own or your lab partner's data, suggest some reasons for this variability.

_____

_____

_____

_____

_____

11. If opportunity and interest allow, the *Reaction Time Kit Instructions* booklet has a number of suggestions for other experiments that you can easily do with the reaction-time ruler.

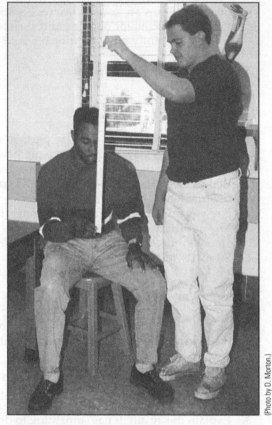

Figure 23-7 Two students measuring visual reaction time.

(Photo by D. Morton.)

TABLE 23-3	Reaction-Time Data	
Trial	Subject 1	Subject 2
1		
2		
3		
4		
5		
6		
7		
8		
9		
10		
Total		
Average (Total/10)		

23.4 The Eyes and Vision (50 min.)

What we know about the world around us is the direct result of the activity of our sensory organs and structures on the surface of our bodies, especially our heads. These sensory organs and structures contain receptors that receive information about the form and amount of light, atmospheric sound waves, movements of our heads, chemicals that enter our mouths and nostrils, and other stimuli. They contain specialized neurons and sometimes accessory cells that process this information and transform it into bioelectrical energy. This bioelectrical energy produces impulses in sensory neurons that extend from the receptors to particular parts of our brains. Here the impulses are interpreted as a particular sensation. The frequency of the impulses is interpreted as intensity. Finally, this information is projected to our conscious minds. Being aware of the nature and limitations of our senses is crucial to understanding ourselves.

Of all our senses, vision is the most developed. The retina contains **photoreceptors** (receptors sensitive not only to light intensity but also to particular groups of wavelengths of light, which the brain interprets as color). You will discover that the eye is similar to a camera in structure in that our eyes use a lens to focus, and the retina is the "camera film" our brain uses to "make" images. Actually a digital camera is an even better analogy because the array of microscopic detectors that replace "camera film" functions more like the retina.

MATERIALS

Per student:

- compound light microscope
- section of eye with retina and optic nerve

Per student pair:

- 2 sheets of white paper, 1 blank and 1 with dime-sized black dot; black and colored construction paper; a white paper dot cut-out and cut-outs of different colored paper shapes (heart, star, etc.)
- mirror
- preserved sheep eye
- dissecting scissors
- blunt probe or dissecting needle
- forceps
- scalpel

- measuring tape
- penlight

Per student group (4):

- eye model

Per lab room:

- liquid waste disposal bottle
- boxes of different sizes of latex or vinyl gloves
- box of safety goggles
- Snellen chart
- astigmatism chart
- place to stand with a taped line 10 ft away and a red taped line 20 ft away

PROCEDURE

A. External Eye Structures

The surface of the eyes is kept moist by *lacrimal fluid* (tears) secreted by the *lacrimal gland,* which is located under the skin above and just lateral to the center of the visible portion of the eyeball (Figure 23-8). Its ducts open under the lateral third of the upper *eyelid*. This fluid is swept across the visible surface of the eyeball when we blink. Excess fluid drains into the openings (*lacrimal puncta*) of ducts located at the tip of small elevations (*lacrimal papillae*) situated on the medial rims of the eyelids. These canals join to the *nasolacrimal sac,* which connects via the *nasolacrimal duct* to the *nasal cavity*.

1. *Wash your hands thoroughly.* Consult Figure 23-8 and examine one of your eyes with a mirror. Identify the **eyelids, eyelashes, lacrimal papillae, lacrimal punta,** and **lacrimal caruncle,** a reddish body located at the inner corner of each eye. It contains lubricating glands. From your own experience, suggest an additional function for eyelids and eyelashes.

_____

_____

_____

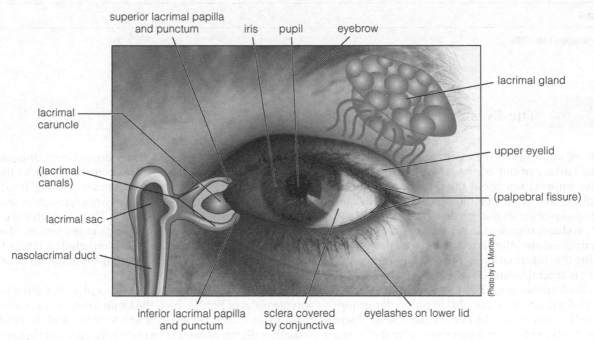

Figure 23-8 Human eye.

2. Which eyelid moves when you open and close your eyes?

_____

3. Why do your eyes water during a cold?

_____

_____

4. Open your eyes wide. Note that only a small part of the eyeball can be seen through the fissure in the skin located between the eyelids. The "white of the eye" is the **sclera.** It is actually covered by an unseen transparent epithelial membrane, the **conjunctiva,** which is continuous with the epithelium lining the insides of the eyelids. Identify the equally transparent cornea, a dome of sclera at the center of the visible eyeball. Through it you can see the pigmented **iris,** which functions as a diaphragm that opens (dilates) and closes (constricts) the hole or **pupil** through which light passes into the eyeball.

B. Structure and Function of the Eyeball

Many human eye models are available. Figure 23-9 shows two such models.

1. Examine a model of the eye and, consulting Figure 23-9, identify the following features:
   - **Outer eye muscles**—skeletal muscles that move the eyeball.
   - Sclera and cornea.
   - **Choroid**—the black-pigmented middle layer of the eye. This layer is continuous with the blue, brown, or other-colored iris, which surrounds the pupil. Just behind the iris and also part of the choroid is the ringlike **ciliary body**. Both the iris and ciliary body have *inner eye muscles* composed of smooth muscle tissue.

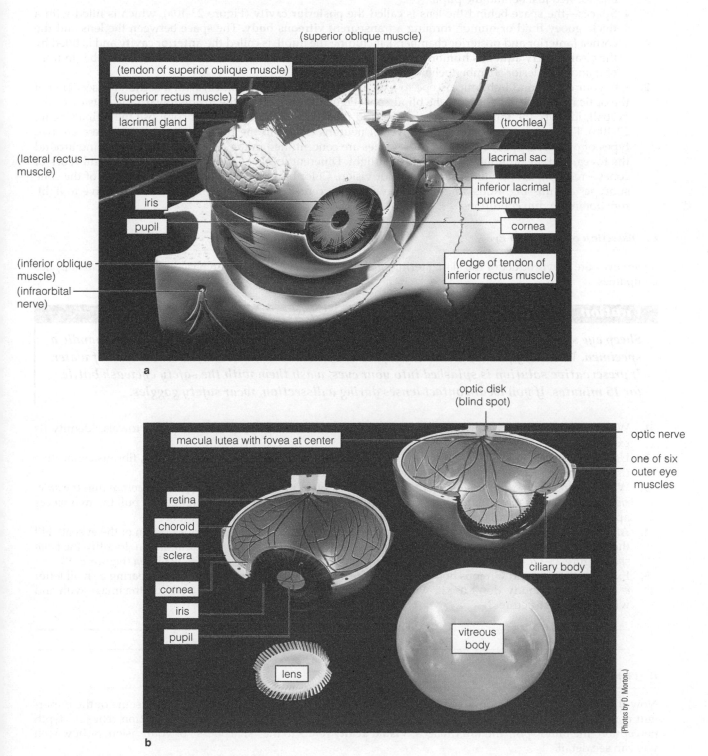

Figure 23-9 Typical eye models, with labels.

- **Retina**—the inner layer of the eye, which lines the back two-thirds of the eyeball. The retina contains photoreceptors. At the optical center of the retina is a slight depression called the **fovea,** which contains the greatest concentration of photoreceptors for color and is surrounded by a yellowish spot, the **macula lutea.** Neuron pathways from the photoreceptors converge on the optical center to form the root of the optic nerve (sometimes called the optic disk). As there are no photoreceptors in this region, its presence creates a hole in the image perceived by the brain and therefore it is also called the **blind spot.**
- **Optic nerve**—connects the retinal neuron pathways to the brain. Note that its outer layer is continuous with the sclera.
- **Lens**—a transparent, convex structure suspended by threadlike ligaments connected to the ciliary body and located just behind the pupil.
- **Spaces**—the space behind the lens is called the **posterior cavity** (Figure 23-10a), which is filled with a thick, gooey fluid or humor, forming a transparent **vitreous body.** The space between the lens and the cornea (anterior and posterior chambers), including the pupil, is called the **anterior cavity** and is filled by the clear, watery **aqueous humor.** The pupil divides the anterior cavity into an anterior chamber in front of it and a posterior chamber behind.

2. Use your compound light microscope to look at a prepared slide (or slides) with a section (or sections) of the optic nerve and retina. Note the blind spot (Figure 23-10a). From the inside toward the outside of the eyeball, identify the three layers of neurons and the pigment epithelium that compose the retina (Figure 23-10b). Photoreceptors comprise the layer of neurons closest to the pigment epithelium. There are two types of photoreceptors, **cones** and **rods.** Cones are concentrated in the center of the retina, in and around the fovea, and are stimulated only by bright light. Different levels of stimulation of the three subtypes of cones—red, green, and blue—allow for color vision. Color blindness results when one or two of the cone subtypes are missing. Rods are found away from the center of the retina and are more sensitive to light, functioning in dim conditions.

C. Dissection of the Sheep Eye

Sheep eyes are a byproduct of the slaughtering of sheep for food. They are purchased from biological supplies companies.

> **Caution**
>
> *Sheep eye specimens are kept in preservative solutions. Use latex gloves whenever you handle a specimen. Wash any part of your body exposed to this solution with copious amounts of water. If preservative solution is splashed into your eyes, wash them with the safety eyewash bottle for 15 minutes. If you wear contact lenses during a dissection, wear safety goggles.*

1. Work in pairs. Obtain a preserved sheep eye and place it on a stack of several paper towels. Identify its external features (Figure 23-11a).
2. Use dissecting scissors to dissect out the *outer eye muscles* and expose the tough, white, fibrous connective tissue of the sclera (Figure 23-11b). In life, these muscles move the eyeball.
3. With a scalpel, make an incision about ¼ inch in back of and parallel to the edge of the *cornea* into the *anterior cavity*. Use scissors to extend the incision completely around the eyeball and gently pull the two pieces apart. Identify the internal features of the eyeball illustrated in Figures 23-12a and b.
4. Allow the *lens* and *vitreous body* of the posterior cavity to flow out of the posterior portion of the eyeball. Fill the posterior portion of the eyeball with water so as to flatten the retina for examination. Identify the *blind spot* (Figure 23-12c) and, depending on the state of preservation, perhaps the macula lutea (Figure 23-9).
5. Examine the lens and vitreous body. One after the other, place them on an index card bearing a small letter *e*. If the lens is cloudy, don't use it. What is the effect of the lens and vitreous body on the image with and without these structures?

_____

_____

D. Vision

Now that you know the anatomy and function of the structures of the eye, let's investigate some of the important concepts related to the sense of vision—visual acuity, the lens and pupil accommodation reflexes, depth perception, the blind spot, and afterimages. **Visual acuity** refers to the "sharpness" of your vision, or how well you can see detail.

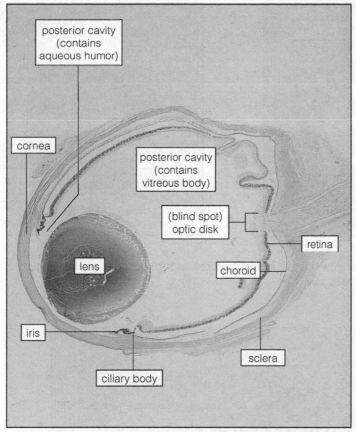

posterior cavity
(contains
aqueous humor)

cornea

posterior cavity
(contains
vitreous body)

(blind spot)
optic disk

retina

choroid

lens

iris

sclera

ciliary body

a

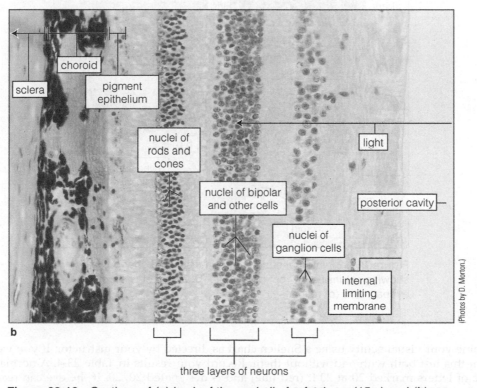

choroid

sclera

pigment
epithelium

nuclei of
rods and
cones

nuclei of bipolar
and other cells

light

nuclei of
ganglion cells

posterior cavity

internal
limiting
membrane

(Photos by D. Morton.)

b

three layers of neurons

Figure 23-10 Sections of (**a**) back of the eyeball of a fetal eye (15×) and (**b**) a higher magnification view of the retina of a monkey (300×). The pigment epithelium and three layers of neurons comprise the retina.

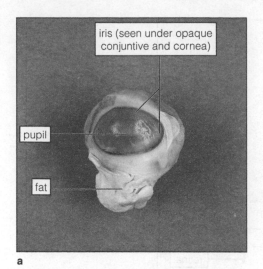

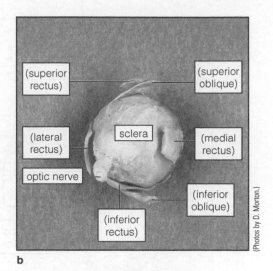

Figure 23-11 (a) Anterior view of the external features and (b) posterior view with dissected muscles of a preserved sheep eye.

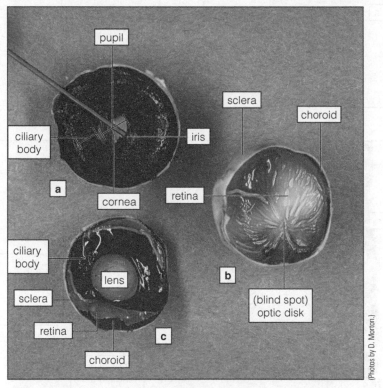

Figure 23-12 Internal structures of a dissected sheep eye showing (a) inside of front of eyeball, (b) inside of back of eyeball, and (c) retina.

1. Determine your visual acuity using a Snellen chart as directed by your instructor. If you wear eyeglasses, perform this test both with and without them. Record your results in Table 23-4. A normal eye can read the line of letters marked 20 at 20 ft (red line) and is designated 20/20. If the eye can read only the letter marked 200, it is designated 20/200, and so on. Such an eye has **myopia** and is nearsighted. It is also possible for an eye to be farsighted (**hyperopia**). These conditions are usually the result of an elongated (myopia) or shortened (hyperopia) eyeball.

2. Visual acuity also can be reduced by **astigmatism**. Astigmatism is caused when one of the transparent surfaces (that is, the cornea or lens) of the eye is not uniformly curved in all planes. Check for astigmatism

TABLE 23-4 Visual Acuity

Eye	With Corrective Glasses or Contacts	Without Corrective Lenses
Right		
Left		

by viewing a series of radiating lines on an astigmatism chart from a distance of 10 ft. To astigmatic individuals, some lines will look different, appearing sharper, thicker, or darker than other lines. Determine whether your right and/or left eye is astigmatic. If you wear eyeglasses, perform this test with them and then without them. Note the numbers of any lines that appear sharper, thicker, or darker in Table 23-5. If you wear glasses, detect the correction for astigmatism in your eyeglasses by looking at the chart after rotating the lenses 90°.

TABLE 23-5 Astigmatism

Eye	With Corrective Glasses or Contacts	Without Corrective Lenses
Right		
Left		

3. Your eyes accommodate focusing on objects at varying distances from the eye. Light rays are virtually parallel from an object 20 feet and farther away. Light rays from closer objects diverge, and these divergent rays must be bent at a sharper angle to focus them on the retina. **Lens accommodation** is accomplished by contracting the ciliary muscles to take tension off the lens, thus allowing it to bulge and become more convex.

Have your lab partner measure the distance from your eyes to the nearest point at which you can sharply focus on an object such as the words in this lab manual. Record this value in Table 23-6. Repeat this procedure with and without glasses, if you wear them. This distance is called the **near point** and is a relative measure of the ability of the lens to accommodate for close vision. Lens accommodation requires an elastic lens. As an individual grows older, the lens becomes more inelastic, and to focus on an object it must be held farther and farther from the eye. However, the object then appears smaller and smaller. This condition is called **presbyopia.** To get letters large enough to read, older people use bifocals (or similar eyeglasses) or extra big print.

TABLE 23-6 Near Point

Eye	With Corrective Glasses or Contacts	Without Corrective Lenses
Right		
Left		

4. To demonstrate pupil accommodation for changes in light intensity, shine a penlight intermittently into your lab partner's eye and note any changes in the diameter of the pupil. These responses are due to the **photopupil reflex.** Describe these changes.

_____

_____

_____

Which occurs more rapidly, constriction or dilation (opening)? _____

What advantage might this difference confer in changing light conditions?

_____

_____

_____

Note that when your lab partner switches focus from a distant to a nearby object, the pupils also constrict (**pupil accommodation for distance reflex**). This is done to block out the most divergent (most difficult to focus) rays of light.

5. Observe that when your lab partner switches focus from a distant to a nearby object, the eyeballs converge (**eye convergence reflex**). Convergence allows for binocular vision (overlapped images), which is necessary for depth perception. Gently press one eyeball out of line while viewing an object.

How many objects do you see? _____

6. Hold Figure 23-13 about 20 inches from your right eye. Close your left eye and place your right eye directly over the dot. Stare at the dot while bringing the page closer to the eye until the × disappears. At this point, the image of the × is falling on the blind spot of the retina.

7. Place a sheet of white paper with a black spot about the size of a dime onto another sheet of blank white paper. Under a bright light, stare at the black spot for 20 seconds, trying to move your head and eyes as little as possible. Slide off the top sheet, continuing to look at the blank sheet. Describe the **afterimage**. _____

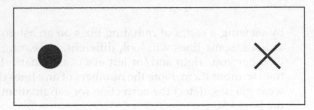

Figure 23-13 Demonstration of blind spot.

_____

Repeat this test, substituting a top sheet of black construction paper with a white dot on it. Describe the afterimage.

_____

_____

Bits of bright red, blue, green, and yellow paper can also be used in this activity. Just as the afterimage of black is a more intense white and vice versa, color vision has afterimages that are essentially the "opposite" color of the original color. To demonstrate this, place a red heart on a green sheet of construction paper; then place this combination over the sheet of blank white paper. Repeat the experiment and describe the results.

_____

_____

23.5   The Ears and Hearing *(20 min.)*

The ear contains several groups of mechanoreceptors (receptors for mechanical energy). Depending on their specific location, their stimulation is interpreted by the brain as the sense of hearing or the sense of equilibrium.

MATERIALS

Per student:

- compound light microscope
- section of cochlea

Per student pair:

- mug-sized container stuffed to overflowing with cotton wool

Per student group (4):

- ear model

PROCEDURE

A. Structure and Function of the Ear

Except for the visible flaps (*auricles*) located on either side of the head, the paired ears are embedded in the temporal bones of the skull.

1. Examine a model of the ear and, consulting Figure 23-14, identify the following features:
   - **Outer ear**—includes **auricles** and **outer ear canal,** which funnel sound waves in air to the **tympanic membrane.**

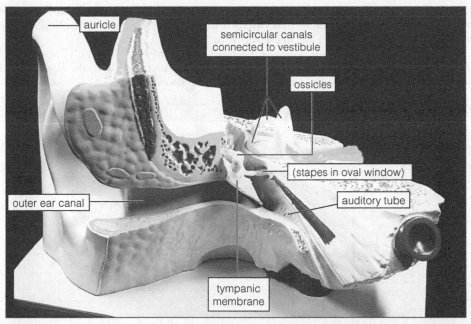

a

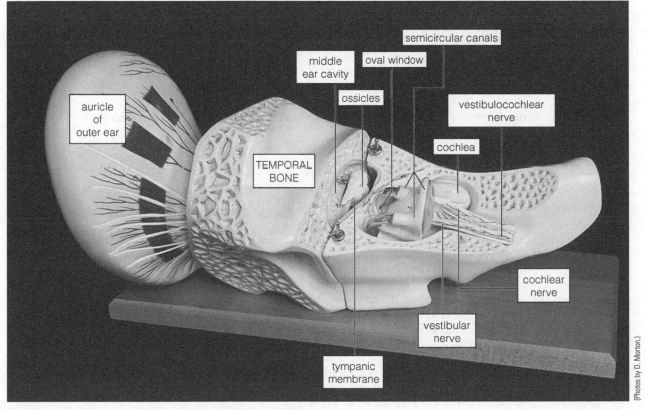

b

Figure 23-14 Typical ear models, with labels.

(Photos by D. Morton.)

- **Middle ear**—contains inner ear **ossicles** (three small bones shaped like a hammer, anvil, and stirrup), which conduct vibrations in bone tissue from the tympanic membrane to the **oval window** of the cochlea. The middle ear space is connected to the upper pharynx (space behind the nasal and oral cavities) by the **auditory tube,** which functions to equalize air pressure on both sides of the tympanic membrane.
- **Inner ear**—composed of the snail shell–like cochlea, vestibule, and semicircular canals. The **cochlea** contains fluid-filled spaces that conduct vibrations from the end plate of the stirrup-shaped ossicle, which is lodged in the **oval window,** to mechanoreceptors for hearing in the **organ of Corti** (or *spiral organ*;

Figure 23-15). A membrane-covered *round window* dissipates old vibrations from the cochlea into the air in the middle ear space. The **semicircular canals** and **vestibule** contain fluid-filled sacs with structures containing mechanoreceptors that aid in equilibrium.

2. Observe a section of the cochlea with your compound light microscope. Consult Figure 23-15 and identify the *scala vestibuli*, the *vestibular membrane*, the *cochlear duct*, the *organ of Corti*, the *basilar membrane*, and the *scala tympani*. The arrow indicates the direction followed by sound waves.

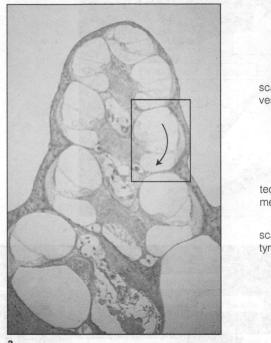

a

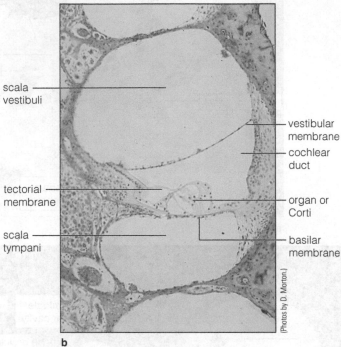

scala vestibuli

vestibular membrane

cochlear duct

tectorial membrane

organ or Corti

scala tympani

basilar membrane

(Photos by D. Morton.)

b

Figure 23-15 Cross section of the cochlea. The arrow shows direction taken by sound waves (25×). The boxed area in (**a**) is magnified in (**b**) 107×.

B. Sense of Direction of Sound

Why do we have two ears? Two ears enable our brains to perceive the direction of sound waves.

1. To demonstrate this, work in pairs. One of you sits on a stool with eyes closed. The other circles the stool quietly, stops, and calls the sitter's name in a high-pitched voice. (This might seem funny, but high-pitched sound does not penetrate bone as well as low-pitched sound.) Each time the sitter's name is called, the sitter tries to point to the caller. If a straight line from the sitter's finger "hits" the caller, place a check mark after trial 1 in the second column of Table 23-7. Repeat this test for a total of 10 trials.

2. Now cover one ear with a mug-sized container stuffed to overflowing with cotton wool and repeat step 1 ten more times. Use the third column of Table 23-7 to record "hits."

3. Total the number of check marks in both columns.

4. Consider the time of arrival and the relative intensity of the waves at each ear, and suggest two ideas as to how the brain determines the direction of the sound source.

TABLE 23-7	Sense of Sound Direction	
Trial	**Ears Uncovered**	**One Ear Covered**
1		
2		
3		
4		
5		
6		
7		
8		
9		
10		
Total		

Chemoreceptors (receptors for chemicals) are located in the taste buds of the tongue and in the olfactory epithelium of the nasal cavity.

MATERIALS

Per student:

- compound light microscope
- section of rabbit tongue
- section of nasal cavity

Per student group (4):

- box of tissues
- permanent marker
- four 250-mL Erlenmeyer flasks, one each with
    10% sucrose solution marked SW
    1% acetic acid solution marked SR
    5% NaCl solution marked SL
    0.5% quinine sulfate solution or tonic water
    marked BT
- 3 dishes containing cubes of a different vegetable
  or fruit (e.g., apple, onion, and potato)

Per lab room:

- box of applicator sticks
- container of small disposable beakers
- box of paper cups
- source of drinking water

PROCEDURE

A. Location of Chemoreceptors

Four types of taste buds function to detect sugars, acid, salt, and bitter substances in food and drink. A much larger number of different types of chemoreceptors in the olfactory epithelium detect a larger number of more volatile substances.

1. Examine a section of rabbit tongue with your compound light microscope. Look at the outer edge of the organ between the projections (foliate papillae) and identify the **taste buds** (Figure 23-16a). They consist of neuron endings surrounded by *taste cells* (chemoreceptors) with *microvilli* projecting through a *taste pore* and supportive epithelial cells.
2. Likewise, examine a section of the nasal cavity. Identify the **olfactory epithelium** lining the roof of the nasal cavity (Figure 23-17). Like taste buds, the olfactory epithelium contains neuron endings, chemoreceptors, and supportive epithelial cells.

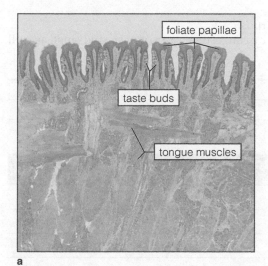

a

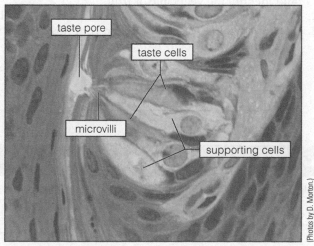

b

(Photos by D. Morton.)

Figure 23-16   (a) Foliate papillae in rabbit tongue (25×) and (b) magnified taste bud (1000×).

B. Distribution of Taste Buds on the Tongue

1. Set up a personal station for yourself with four new applicator sticks, a box of tissues, a cup of drinking water, and four new small disposable beakers. Label them with a permanent marker: SW for sweet, SR for sour, SL for salty, and BT for bitter. Half-fill each beaker with solution from the similarly marked flask.

2. Dry your tongue *with a clean tissue and, using an unused applicator stick*, have your lab partner apply a dab of sweet solution to the tip of the tongue. Do not close your mouth. Repeat using first the sides and then the back of the tongue. Where do you taste the sweetness? Record the location in Figure 23-18, using the code SW for sweet. Dispose of the tissue and used applicator stick in the trash or as otherwise directed by your instructor.

3. Repeat this procedure using the sour, salty, and bitter solutions, being sure to rinse the mouth after each solution and to redry the tongue. On the tongue outline in Figure 23-18, record the location of each solution tested: SL for salty, SR for sour, and BT for bitter.

4. Can you taste salt, bitter, sweet, and sour with your nostrils pinched closed? (yes or no) _____

C. Interaction Between Taste and Smell

On an everyday level, we tend to mix up the sense of taste and smell. For example, how many times have you heard a statement like, "This pizza tastes great"? How much of this "great taste" is actually taste and how much of it is smell?

1. Rinse your mouth. Shut your eyes, pinch your nostrils closed, and have your lab partner place a small piece of vegetable or fruit on your tongue. Attempt to identify it. Record its identity or, if you can't identify it, put a question mark in Table 23-8. With your nostrils still pinched closed, chew the piece thoroughly. Again, attempt to identify it. Record this and subsequent results in Table 23-8. Open your nostrils and identify the vegetable or fruit.

2. Repeat step 1 twice, using two more different vegetables or fruits.

3. In a strictly biological sense, when your nostrils are blocked by congestion, can you taste and smell as well as when they are open? Explain your answers.

_____

_____

_____

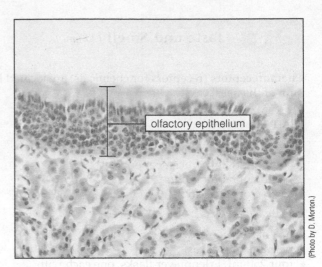

Figure 23-17 Olfactory epithelium (250×).

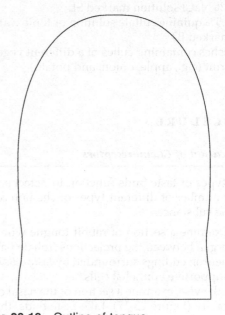

Figure 23-18 Outline of tongue.

TABLE 23-8	Identity of Vegetable/Fruit		
Trial	Placed on Tongue	After Chewing	Nostrils Open
1			
2			
3			

_____ 1. Neurons that carry messages from receptors to the CNS are
(a) sensory
(b) motor
(c) interneurons
(d) autonomic

_____ 2. Neurons that carry messages from the CNS to effectors are
(a) sensory
(b) motor
(c) interneurons
(d) both a and b

_____ 3. Neurons that carry messages within the CNS are
(a) sensory
(b) motor
(c) interneurons
(d) autonomic

_____ 4. Knowledge of the position and movement of the various body parts is
(a) modality
(b) projection
(c) adaptation
(d) proprioception

_____ 5. Skin contains
(a) free neuron endings
(b) encapsulated neuron endings
(c) no nervous tissue
(d) both a and b

_____ 6. Which characteristic of receptors does phantom pain illustrate?
(a) modality
(b) projection
(c) adaptation
(d) proprioception

_____ 7. A simple reflex arc is made up of a receptor and
(a) a sensory neuron
(b) a motor neuron
(c) an effector
(d) all of the above

_____ 8. A stretch reflex is
(a) somatic
(b) autonomic
(c) both a and b
(d) none of the above

_____ 9. A pupillary reflex is
(a) somatic
(b) autonomic
(c) both a and b
(d) none of the above

_____ 10. A reaction is
(a) a reflex
(b) involuntary
(c) voluntary
(d) both a and b

_____ 11. The cornea is part of the
  (a) sclera
  (b) choroid
  (c) retina
  (d) lens

_____ 12. Parts of the choroid form
  (a) the lens
  (b) the ciliary body
  (c) the iris
  (d) both b and c

_____ 13. Photoreceptors are found in the
  (a) lens
  (b) ciliary body
  (c) iris
  (d) retina

_____ 14. The space between the lens and iris is
  (a) the posterior cavity
  (b) the posterior chamber
  (c) the anterior chamber
  (d) none of the above

_____ 15. Nearsightedness is also called
  (a) myopia
  (b) hyperopia
  (c) presbyopia
  (d) none of the above

_____ 16. Ossicles are found in
  (a) the outer ear
  (b) the middle ear
  (c) the inner ear
  (d) both a and c

_____ 17. Receptors found in the ear are
  (a) chemoreceptors
  (b) mechanoreceptors
  (c) photoreceptors
  (d) none of the above

_____ 18. Taste buds detect
  (a) light
  (b) atmospheric sound waves
  (c) chemicals in food and drink
  (d) movements of the head

_____ 19. The olfactory epithelium is found in the
  (a) ear
  (b) nasal cavity
  (c) eye
  (d) tongue

_____ 20. The most developed sense in humans is
  (a) hearing
  (b) vision
  (c) taste
  (d) smell

Name _____ Section Number _____

EXERCISE 23

Human Sensations, Reflexes, Reactions, and the Structure and Function of Sensory Organs

POST-LAB QUESTIONS

Introduction

1. Where in the brain does your consciousness reside?

23.1 Sensations

2. In your own words, define the following terms:
   a. modality

   b. projection

   c. adaptation

3. Identify the structure indicated in these photos.

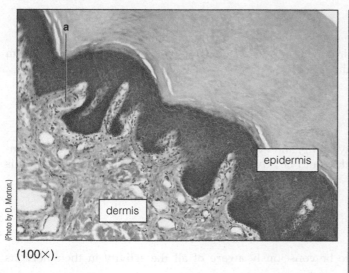

(Photo by D. Morton.)

epidermis

dermis

(100×).

a _____

b _____

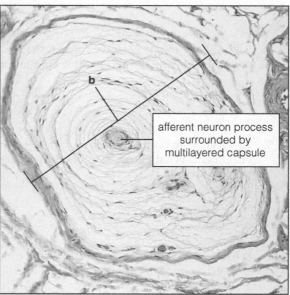

b

afferent neuron process surrounded by multilayered capsule

(Photo by D. Morton.)

(110×).

23.2 Reflexes

4. Diagram the basic steps of a simple reflex arc like a stretch reflex.

5. How does the patella reflex differ from the pupillary reflex?

23.3 Reactions

6. Indicate whether the following actions are due to reactions or reflexes by putting a check in the correct column in the table.

Action	Reaction	Reflex
A baby wetting a diaper		
Braking a car to avoid an accident		
Withdrawing your hand from a hot stove surface		
Sneezing		
Waving to a friend across the street		

Food for Thought

7. What are the advantages and disadvantages to an organism of receptor adaptation?

8. All animals do not perceive the external environment in exactly the same way. List some examples from your own knowledge and readings in the textbook.

9. To survive, an animal needs all its receptors working, and even then it cannot fully sense the external environment. What extra receptors do you think would be advantageous to the survival of humans in this modern world?

10. Why is it advantageous for organisms *not* to be consciously aware of all the activity in their nervous systems?

23.4 The Eyes and Vision

11. Define the following terms:
    a. myopia

    b. hyperopia

    c. presbyopia

12. How does uncorrected myopia or hyperopia affect the near point?

13. What types and subtypes of photoreceptors are present in the retina? Briefly describe their function.

14. What is an afterimage? How is it formed?

15. List the eye structures that light passes through as it travels from outside the body to the retina.

23.5 The Ears and Hearing

16. List the ear structures that sound waves and resulting vibrations pass through as they travel from outside the body to the organ of Corti.

23.6 Taste and Smell

17. Why do humans and other animals have two nostrils?

18. When your nasal passages are blocked due to a heavy cold, food seems to lose its flavor. For example, pizza tastes like salt. Explain why this is so.

Food for Thought

19. In what part of the eye are cataracts located?

20. What is the relationship between the fluid in the anterior cavity and the disease glaucoma?

Human Skeletal and Muscular Systems

After completing this exercise, you will be able to

1. define *bones, ligaments, joints, skeletal muscles, tendons, muscle tone, posture, sutures, synovial joints, diaphysis, epiphyses, compact bone, spongy bone, marrow cavity, lever;*

2. identify the major bones of the human skeleton;

3. describe the structure of a typical bone;

4. define *origin, insertion,* and *action* as these terms apply to skeletal muscles and their tendons;

5. distinguish between isometric and isotonic contractions of skeletal muscles;

6. give everyday and anatomical examples of the three classes of levers;

7. present a simple biomechanical analysis of walking.

INTRODUCTION

The skeletal system and muscular system are often considered together to stress their close structural and functional ties. These two systems are referred to as the musculoskeletal system. They determine the basic shape of your body, support your other systems, and provide the means by which you move in the external environment.

Bones are the main organs of the skeletal system. They are primarily bone tissue, although all four basic tissue types are present. The places in the body where two or more bones are connected are called **joints.** The joints you are most familiar with are the shoulder, elbow, wrist, hip, knee, and ankle. However, there are many others. Around many joints, bones are held together by straplike structures called **ligaments.** Ligaments are primarily dense connective tissue that is more or less elastic. Elastic ligaments around mobile joints stretch to allow movement.

Skeletal muscles are the main organs of the muscular system and are composed primarily of skeletal muscle tissue. Skeletal muscles are connected to bones by dense fibrous connective tissue structures called **tendons.** Tendons are inelastic, so all of the force of skeletal muscle contraction is transferred to the skeleton.

When a skeletal muscle contracts, movement may or may not occur. If the skeletal muscle is allowed to shorten, the bone moves, and in doing so it moves some body part. On the other hand, if the skeletal muscle does not shorten, the tension in that muscle and in its tendons increases. All skeletal muscles exhibit tension or **muscle tone** except when you are asleep. This tension maintains **posture**—the ability to hold the body erect and to keep the position of its parts, all against the pull of gravity.

The organs of the skeletal and muscular systems have other functions. Bones protect internal organs (for example, the skull protects the brain, eyes, and ears). Bones also store minerals and produce blood cells in the bone marrow. When body temperature drops below a certain level, skeletal muscles produce heat by shivering.

24.1 Adult Human Skeleton *(About 45 min.)*

An articulated human skeleton is prepared by joining the degreased and bleached bones of an individual so that many of the bones can be moved as they were in life. Often, plastic casts of the original bones are used.

Although bone tissue predominates, fresh bones are composed of all four basic tissue types. However, when they are prepared for study, their organic portion is lost. These bones consist only of the mineral portion of bone tissue. Original details remain, but the bones are brittle. *Therefore, you must handle bones gently. Use a pipe cleaner to point out details and never use a pencil or pen because it is very difficult to remove marks.*

MATERIALS

Per student:

- pipe cleaner
- compound microscope, lens paper, a bottle of lens-cleaning solution, a lint-free cloth

■ prepared slides of
   a synovial joint
   a ground cross section of compact bone (optional)
   a cross section of a skeletal muscle

Per student group:

■ articulated adult human skeleton (natural bone or plastic)
■ femur
■ femur that has been sawed into two halves lengthwise

Per lab room:

■ labeled chart and illustrations of the adult human skeleton

PROCEDURE

A. Identification of Some Bones

There are 206 separate bones in the adult human skeleton. Using the labeled chart and illustrations of the human skeleton, identify the following bones on the articulated human skeleton and label them in Figure 24-1.

1. Axial skeleton
   ■ **skull** (28 separate bones, including middle ear bones) (Figure 24-2)
   ■ **vertebrae** (singular, *vertebra;* 26 separate bones, including the **sacrum,** which is composed of five fused vertebrae, and the **coccyx,** which is usually composed of four fused vertebrae) (Figure 24-3)
   ■ **ribs** (12 pairs of ribs for a total of 24 separate bones) (Figure 24-4)
   ■ **sternum** (three fused bones) (Figure 24-4)
   ■ **hyoid** (only bone that does not form a joint with another bone) (Figure 24-4)
2. Appendicular skeleton (these are all paired bones found on the right and left sides of the body)

pectoral girdle (shoulder) (Figure 24-5)
   ■ **scapula**
   ■ **clavicle**

arm (Figures 24-5, 24-6)
   ■ **humerus**
   ■ **radius**
   ■ **ulna**
   ■ **carpals** (8)
   ■ **metacarpals** (5)
   ■ **phalanges** (14)

pelvic girdle (hip) (Figure 24-7)
   ■ **coxal bone** (three fused bones— **pubis, ischium, ilium**)

leg (Figure 24-8)
   ■ **femur**
   ■ **tibia**
   ■ **fibula**
   ■ **patella**
   ■ **tarsals** (7)
   ■ **metatarsals** (5)
   ■ **phalanges** (14)

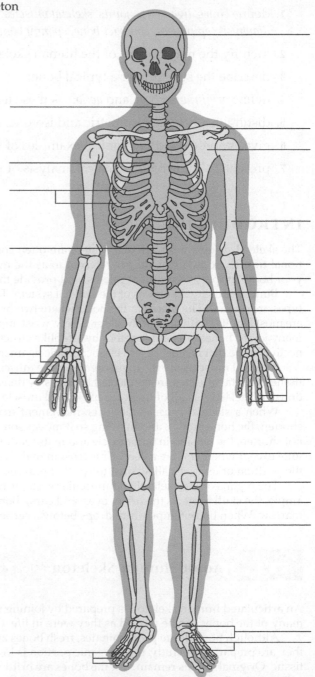

Figure 24-1 Label this front view of adult human skeleton (axial portion shaded gray and appendicular portion colored yellow).
Labels: bones listed in section A, wrist, elbow, shoulder, hip, knee, ankle

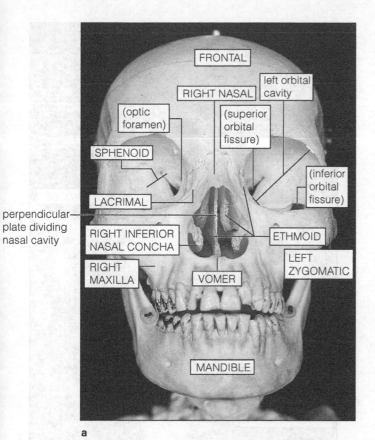

a

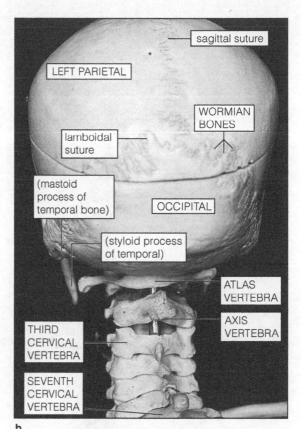

b

Figure 24-2 (**a**) Front, (**b**) back, and (**c**) side views of the skull. The names of bones are capitalized in this and subsequent figures to clearly separate them from other features.

c

(Photos by D. Morton.)

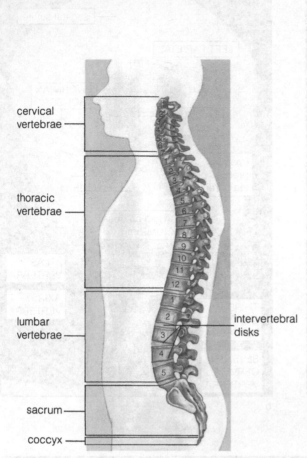

Figure 24-3 Side of the vertebral column.
(After Starr, 1991.)

cervical vertebrae

thoracic vertebrae

lumbar vertebrae

intervertebral disks

sacrum

coccyx

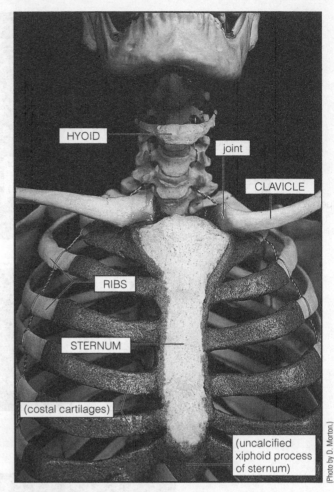

Figure 24-4 Anterior view of skeleton of upper trunk.

HYOID

joint

CLAVICLE

RIBS

STERNUM

(costal cartilages)

(uncalcified xiphoid process of sternum)

(Photo by D. Morton.)

B. Joints

The degree of movements allowed at different joints ranges from none to freely movable (Figure 24-9). Examples of immovable joints are the **sutures** that connect the bones of the roof of the skull of young adults. Joints that allow the freest movement are **synovial joints,** such as the ones listed in Table 24-1.

1. Examine a prepared section of a synovial joint with your compound microscope. Identify the structures labeled in Figure 24-9. The *fibrous capsule* of synovial joints is lined by a *synovial membrane,* which secretes lubricating *synovial fluid.* The fibrous capsule and ligaments function to stabilize synovial joints. Ligaments can be located outside and inside the capsule, and they may be thickenings of its wall.
2. Identify the synovial joints listed in Figure 24-1 and list the adjacent bones that form them in Table 24-1.

TABLE 24-1 Bones That Form the Major Synovial Joints

Joint	Adjacent Bones
Wrist	
Elbow	
Shoulder	
Hip	
Knee	
Ankle	

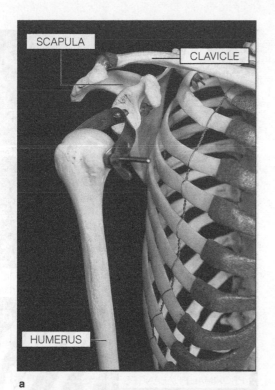

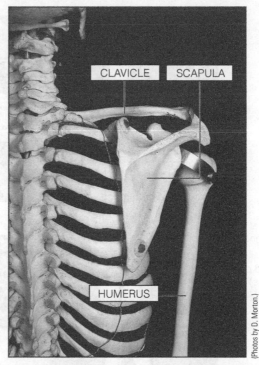

a

b

(Photos by D. Morton.)

Figure 24-5 (**a**) Anterior and (**b**) posterior views of pectoral girdle and shoulder.

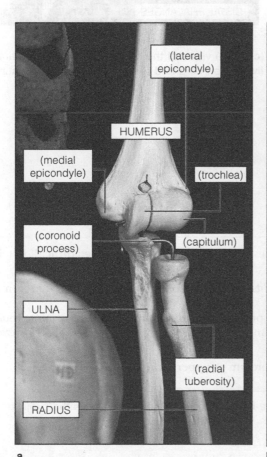

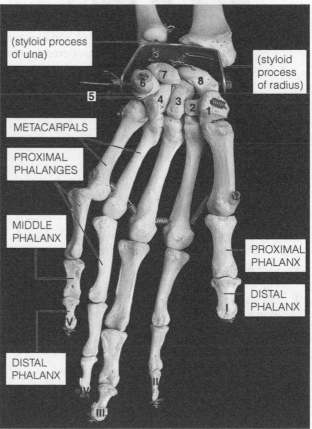

a

b

Figure 24-6 Anterior (**a** and **b**) and posterior (**c** and **d**) views of the bones of the arm and hand. The eight carpal bones in b are numbered and the five digits in b and d are labeled with Roman numerals. (Photos by D. Morton.) *Continues.*

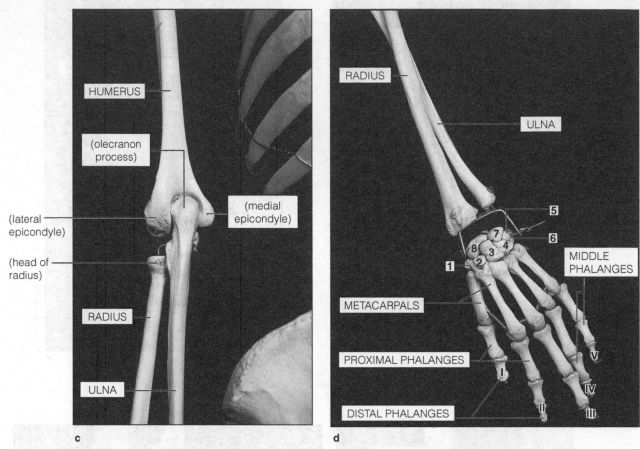

Figure 24-6 Anterior (**a** and **b**) and posterior (**c** and **d**) views of the bones of the arm and hand. The eight carpal bones in b are numbered and the five digits in b and d are labeled with Roman numerals. (Photos by D. Morton.) *Continued.*

C. Surface Features

There are many places on your body surface where bones can be felt. However, it is often difficult to tell specifically which bone you are feeling. Some are easy.

1. Feel the bone supporting your lower jaw, the *mandible*. This is the only bone of the skull that forms a synovial joint with another skull bone.
2. Let's try a harder example, the piece of bone that projects from the point of the elbow joint. Touch it and alternately extend and flex the forearm, increasing and decreasing the angle between the forearm and upper arm, respectively.

Which part of the arm does the projection move with, forearm or upper arm?

_____

While still touching this projection, alternately turn the hand palm down and up. Does the projection move? _____ (yes or no)

Which bone belongs to this projection? _____

In general, to identify a portion of a bone near a joint, move the body parts adjacent to the joint while touching the bone.

3. Identify the bones that have the surface features listed in Table 24-2.

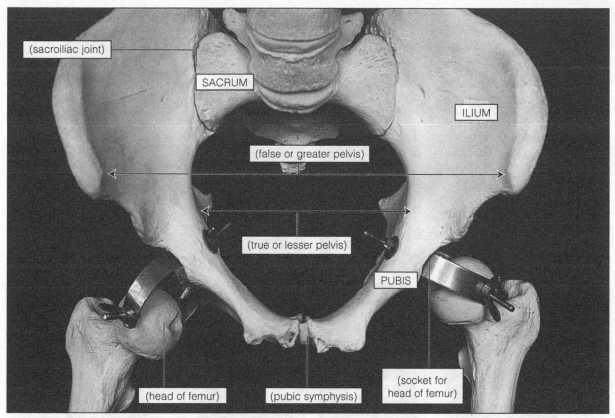

(sacroiliac joint)

SACRUM

ILIUM

(false or greater pelvis)

(true or lesser pelvis)

PUBIS

(head of femur)

(pubic symphysis)

(socket for head of femur)

a

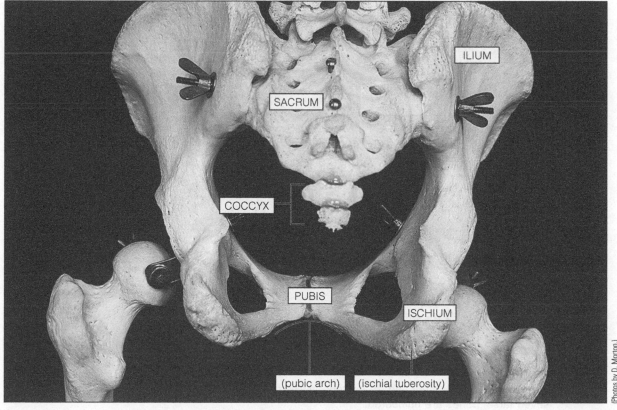

ILIUM

SACRUM

COCCYX

PUBIS

ISCHIUM

(pubic arch)

(ischial tuberosity)

(Photos by D. Morton.)

b

Figure 24-7   (a) Anterior and (b) posterior views of pelvic girdle and hip.

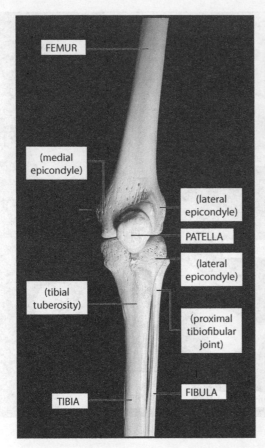

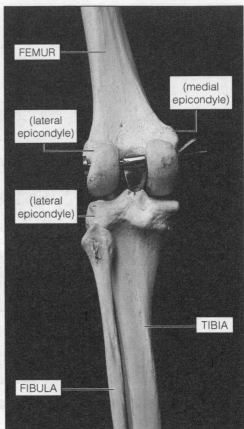

a

b

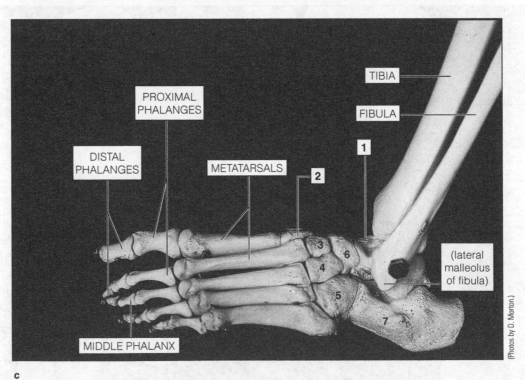

c

Figure 24-8 (**a**) Anterior and (**b**) posterior views of knee and lateral view of foot. (**c**) The seven tarsal bones are numbered.

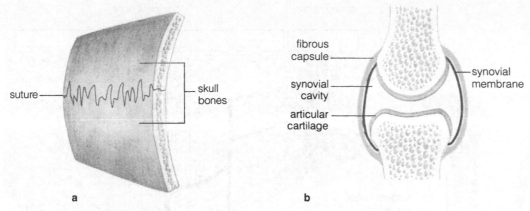

Figure 24-9 Diagrams of (**a**) a suture and (**b**) a synovial joint.

suture — skull bones

fibrous capsule — synovial membrane
synovial cavity
articular cartilage

a

b

TABLE 24-2 Surface Features and Bones	
Surface Feature	**Bone**
Knuckles	
Bump next to the wrist and on the same side of the upper appendage as the little finger	
Smaller bump next to the wrist and on the same side of the upper appendage as the thumb	
Bump next to and outside the ankle	
Bump next to and inside the ankle	

D. Structure of a Bone

1. Look at a femur, the longest bone of the skeleton (Figure 24-10). It consists of a shaft, or **diaphysis,** with two knobby ends, or **epiphyses** (singular, *epiphysis*). One end has a narrow neck and a round head.

Which bone does the femur join? _____

To which bone of the skeleton does the other end join? _____

Note the other surface features on the femur, such as projections of various sizes and lines. These surface features are attachment sites for tendons and ligaments.

Are there small tunnels opening onto the surface of the femur? _____ (yes or no)

In life, these nutrient canals serve as routes for blood vessels and nerves.

2. Examine a femur that has been sawed in half lengthwise. There are two kinds of adult bone tissue: **compact bone** and **spongy bone.** Compact bone is solid and dense and is found on the surface of the femur. Spongy bone is latticelike and is found on the inside of the femur, primarily in the epiphyses and surrounding the **marrow cavity.**

Which kind of bone tissue looks denser? _____

Comparing pieces of equal size, which kind of bone tissue looks lighter? _____

3. *Optional.* Instructions for the study of a transverse section of ground compact bone are located on page 460 in Exercise 21.

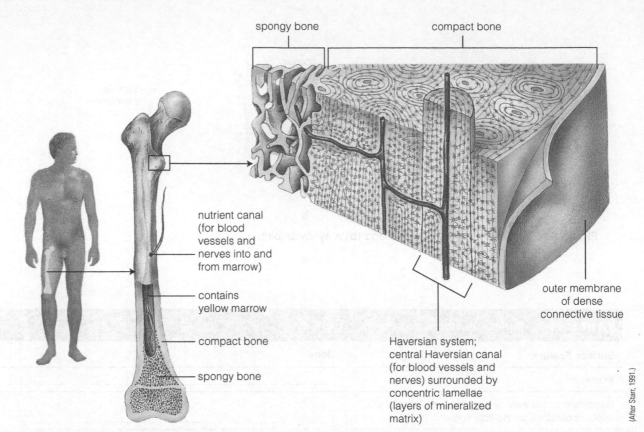

spongy bone   compact bone

nutrient canal
(for blood
vessels and
nerves into and
from marrow)

contains
yellow marrow

compact bone

spongy bone

Haversian system;
central Haversian canal
(for blood vessels and
nerves) surrounded by
concentric lamellae
(layers of mineralized
matrix)

outer membrane
of dense
connective tissue

(After Starr, 1991.)

Figure 24-10   Structure of the femur.

E. Structure of a Skeletal Muscle

Like bones, skeletal muscles are composed of all four basic tissue types. Skeletal muscles are mostly skeletal muscle tissue with the individual skeletal muscle fibers arranged parallel to the axis along which the muscle shortens when contracting. A substantial amount of connective tissue surrounds the fibers and connects them to the tendons.

Use your compound microscope to examine a section of a skeletal muscle. Look for fibers, fiber bundles, the more or less loose connective tissue located between fibers and between bundles of fibers, and the fibrous connective tissue that surrounds the entire organ (Figure 24-11).

24.2   Leverage and Movement *(45 min.)*

Much of the skeletal system is a system of levers, in which each bone is a lever and the joints are fulcrums. During a typical movement, one end of a skeletal muscle, the **origin,** remains stationary. The other end, the **insertion,** moves along with the bone and surrounding body part. The movement produced by the contraction is the **action** of the skeletal muscle. Most insertions are close to their joints, and the advantage gained by this is that the muscle has to shorten a small distance to produce a large movement of the corresponding body part.

MATERIALS

Per student pair:

- pair of scissors
- toggle switch mounted on a board
  (Alternatively, you can use any light switches present in the room.)

- pair of forceps
- pencil
- textbook

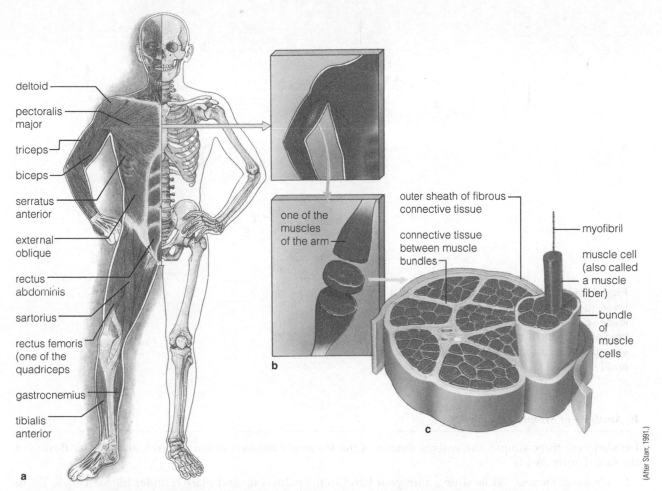

deltoid

pectoralis major

triceps

biceps

serratus anterior

external oblique

rectus abdominis

sartorius

rectus femoris (one of the quadriceps

gastrocnemius

tibialis anterior

a

one of the muscles of the arm

b

outer sheath of fibrous connective tissue

connective tissue between muscle bundles

myofibril

muscle cell (also called a muscle fiber)

bundle of muscle cells

c

(After Starr, 1991.)

Figure 24-11 (a) Some of the major skeletal muscles of the human muscular system. (b) Closer look at the structure of a skeletal muscle organ. (c) Transverse section of a skeletal muscle organ.

PROCEDURE

A. Classes of Levers

Levers are simple machines. When a pulling force or effort is applied to a lever, it moves about its fulcrum, overcoming a resistance or moving a load.

1. There are three classes of levers (Figure 24-12):
    - Class I. The fulcrum is located between the effort and the load.
    - Class II. The load is located between the fulcrum and the effort.
    - Class III. The effort is located between the fulcrum and the load. Class III levers are the most common in the skeletal system.

Test your understanding of the three classes of levers by examining the objects listed in the following table and then matching them with the appropriate class of lever.

2. To remember the relative position of the fulcrum, load, and effort for each class of lever, use this mnemonic (memory device): "1, 2, 3; F, L, E." For example, because 2 has the same relative position as L in Figure 24-12b, the mnemonic tells you that a class II lever has the Load in the center, the Effort at one end, and the Fulcrum at the other. Try it for the other two classes of levers.

Lever	Object
Class I. _____	a. scissors
Class II. _____	b. toggle switch or light switch
Class III. _____	c. forceps

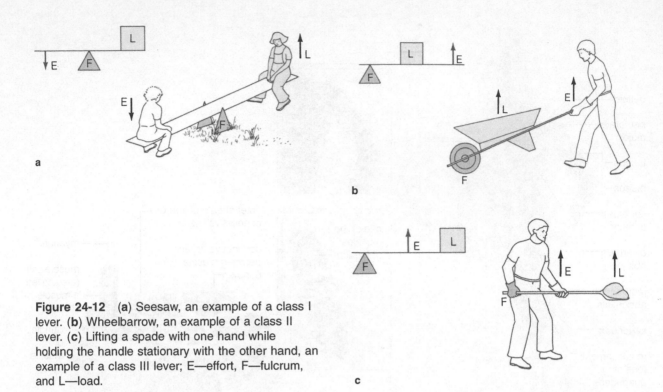

Figure 24-12 (a) Seesaw, an example of a class I lever. (b) Wheelbarrow, an example of a class II lever. (c) Lifting a spade with one hand while holding the handle stationary with the other hand, an example of a class III lever; E—effort, F—fulcrum, and L—load.

B. Analysis of Simple Movements

Let's analyze three simple movements: flexion of the forearm, extension of the forearm, and plantar flexion of the foot (Figure 24-13).

1. *Flexion of forearm*. While sitting, turn your hand so the palm is up and place it under the lab bench. Try to flex the forearm (decrease the angle between the forearm and upper arm). Because the skeletal muscle that is attempting to flex the forearm cannot shorten, the tension in it will increase. A contraction of a skeletal muscle in which tension increases but no movement results is called an **isometric contraction.** Feel with your other hand the front surface of the upper arm. The large tense muscle is the biceps brachii. Its origin is the scapula, and its insertion is the radius.

Which joint is the fulcrum? _____

Now place a pencil in the palm of your hand and flex the forearm. A contraction of a skeletal muscle that results in movement is called an **isotonic contraction.** There is no increase in tension during the movement. Feel

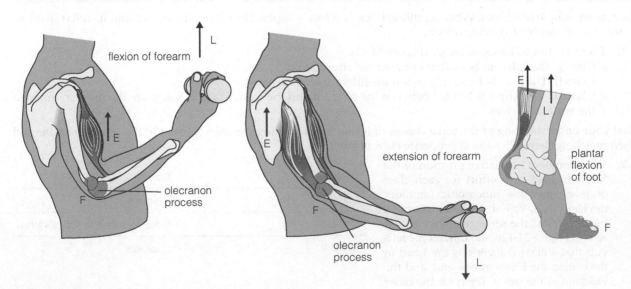

Figure 24-13 Some simple actions of skeletal muscles.

the tension in the biceps brachii as you make this movement. Repeat this procedure, but replace the pencil with a textbook. Both the pencil and the book are adding to the load being lifted, the forearm.

In which case—lifting the pencil or the textbook—was the tension in the biceps brachii the greatest?

_____

When you lift any object, the tension in the muscle must equal the weight of that object before movement can occur. Therefore, normal movements have an isometric phase followed by an isotonic phase.

Where is the pulling force applied? (insertion, origin, or both the insertion and origin)

_____

Even simple movements require the coordination of a group of muscles. For example, the origin does not move because other skeletal muscles hold the scapula stationary.

What class of lever (I, II, or III) is illustrated by the preceding example? _____

2. *Extension of forearm.* Place the hand, still palm up, on the top of the lab bench and try to extend the forearm (increase the angle between the forearm and the upper arm). Feel for a tense muscle on the back surface of the upper arm. This is the triceps brachii. The origin of the triceps brachii is the scapula and the upper humerus; its insertion is the *olecranon process* of the ulna. The fulcrum is the same as the previous example, except that it has shifted position relative to the effort and the load.

What class of lever is illustrated by this movement? _____

Extension of the forearm is the opposite movement to flexion of the forearm. Hold the textbook, palm still up, halfway between full flexion and full extension. Feel the tension in the biceps brachii and triceps brachii. Repeat this procedure without the book.

Is the tension in the biceps brachii greater with or without the book? _____

Is the tension in the triceps brachii greater with or without the book? _____

The state of contraction of a group of skeletal muscles has to be coordinated to accomplish a particular movement or element of posture. Both the tendons of the biceps brachii and the triceps brachii are pulling on their insertions on the bones of the forearm to keep the forearm stationary. Other muscles are keeping the shoulder stationary.

3. *Plantar flexion of foot.* You need to stand up to do this movement. A lab partner should stand behind you and watch that you do not fall during this procedure. With one hand on the lab bench to steady your balance, stand on the tips of your toes. With your other hand, feel one of the very large tense muscles on the back of each calf. This is the gastrocnemius. The origin of the gastrocnemius is the femur, and its insertion is a tarsal—the calcaneus or heel bone. The fulcrum is the metatarsal-phalangeal joints, and the weight is the weight of the body transmitted through the tibia.

What class of lever does this movement exemplify? _____

24.3   Walking (20 min.)

Walking is a complex activity that requires many movements and the coordinated contractions of several groups of skeletal muscles. For each leg, walking involves two phases, which together make up the *step cycle*. The *stance phase* occurs when the leg bears weight, and the *swing phase* occurs when the leg is in the air.

MATERIALS

Per lab room:

■ safe place to walk

PROCEDURE

1. *Follow your instructor's directions as to where to walk safely.* Walk a few normal steps, concentrating on one leg.

What part of the foot (toe or heel) strikes the ground first? _____

What part of your foot leaves the ground last? _____

Does it leave passively, or does it push off? _____

2. Now put your hands on your hips and concentrate on what your pelvic girdle is doing while you walk. First take short strides and then long ones.

Does the pelvic girdle rotate more during short or long strides? _____

Rotation of the pelvic girdle can be demonstrated in a different way. Find a lab partner of about equal height. Walk right next to each other but out of step, that is, with opposite feet leading. First take short steps and then long ones. What happens?

_____

_____

This sideways movement is called *lateral displacement*. Incidentally, females in general have to rotate their pelvic girdles a little more than males for a given length of stride. This is due to differences in the proportions of the female and male pelvic girdles.

3. *Vertical displacement* also occurs during walking. From the side, observe two individuals of equal height walking out of step and next to each other.

Do their heads remain at the same level, or do they bob up and down? _____

PRE-LAB QUESTIONS

_____ 1. Ligaments connect
(a) bones to bones
(b) skeletal muscles to bones
(c) tendons to bones
(d) skeletal muscles to tendons

_____ 2. Tendons connect
(a) bones to bones
(b) skeletal muscles to bones
(c) ligaments to bones
(d) skeletal muscles to tendons

_____ 3. Which bone is part of the axial skeleton?
(a) clavicle
(b) radius
(c) coxal bone
(d) sternum

_____ 4. The two kinds of bone tissue are
(a) compact and loose
(b) compact and spongy
(c) dense and spongy
(d) loose and dense

_____ 5. There are _____ classes of levers.
(a) two
(b) three
(c) four
(d) more than four

_____ 6. The class of lever in which the effort is
located between the fulcrum and the load
is called
(a) class I
(b) class II
(c) class III
(d) class IV

_____ 7. The end of the skeletal muscle that
remains stationary during a movement is
(a) the action
(b) the origin
(c) the insertion
(d) none of the above

_____ 8. In an isotonic contraction of a skeletal
muscle,
(a) the tension in the muscle increases
(b) movement occurs
(c) no movement occurs
(d) both a and c occur

_____ 9. In an isometric contraction of a skeletal
muscle,
(a) the tension in the muscle increases
(b) movement occurs
(c) no movement occurs
(d) both a and c occur

_____ 10. The step cycle of walking consists of
(a) a stance phase
(b) a swing phase
(c) both a and b
(d) none of the above

EXERCISE 24

Human Skeletal and Muscular Systems

POST-LAB QUESTIONS

24.1 Adult Human Skeleton

1. Match the following bones to their location in the body.

Bone	Location
_____ radius	a. pectoral girdle
_____ coxal bone	b. leg
_____ ribs	c. axial skeleton
_____ scapula	d. arm
_____ fibula	e. pelvic girdle

2. Label this photo of a femur that has been sawed in half lengthwise.

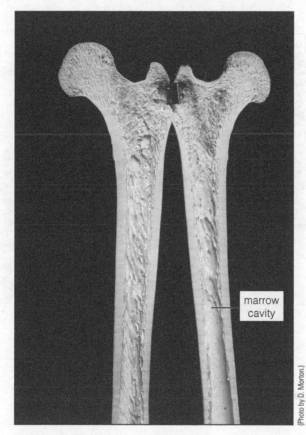

marrow cavity

(Photo by D. Morton.)

Labels: compact bone, spongy bone

3. Identify the bones indicated in this photo.

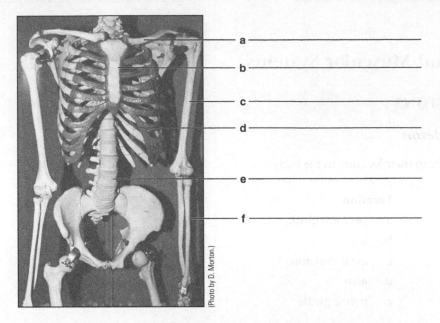

a _____

b _____

c _____

d _____

e _____

f _____

(Photo by D. Morton.)

4. Label the fibrous capsule and synovial membrane of this joint.

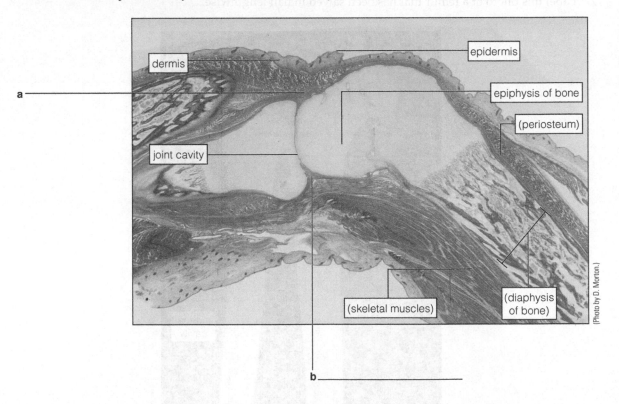

dermis

epidermis

a _____

epiphysis of bone

(periosteum)

joint cavity

(diaphysis of bone)

(skeletal muscles)

(Photo by D. Morton.)

b _____

24.2 Leverage and Movement

5. Define the following terms:

a. the *insertion* of a skeletal muscle

b. the *origin* of a skeletal muscle

c. the *action* of a skeletal muscle

6. Explain the difference between isometric and isotonic contractions. How are both important to normal body movements?

7. Draw and label the structures of a typical skeletal muscle organ.

24.3 Walking

8. In your own words, describe one step in the walking cycle.

Food for Thought

9. The skeletal muscle that flexes (bends) the forearm after pronation (palm down position as in a pull-up) is the brachialis. Its origin is the humerus, and the insertion is the upper front of the ulna. Identify the class of lever involved and explain why you made this choice.

10. Search the World Wide Web for sites that describe diseases of bones (for example, osteoporosis) and of skeletal muscles (such as muscular dystrophy). List two sites and briefly summarize their contents.

    http://

    http://

Human Blood, Circulation, and Respiration

After completing this exercise, you will be able to

1. define *blood vessels* (different types), *blood, heart, pulmonary circuit, systemic circuit, homeostasis, plasma, hematocrit, agglutination, blood pressure, elastic membranes, valves, sinoatrial node, acetylcholine, epinephrine, ventilation, inhalation, exhalation, breathing, cohesion, tidal volume, forced inhalation volume, forced exhalation volume, residual volume, vital capacity, chemoreceptor;*

2. identify and give the characteristics and functions of the different types of blood cells;

3. explain the ABO and Rh blood group systems;

4. describe how to perform a hematocrit and blood typing;

5. give the structure and function of the different types of blood vessels;

6. name the four chambers and four valves of the heart and describe the route blood takes through them;

7. describe how the heart contracts;

8. explain how the heart is controlled;

9. list the skeletal muscles used in breathing and give the specific function of each;

10. trace the flow of air through the organs and structures of the respiratory system;

11. explain how air moves in and out of the lungs during respiration in humans;

12. explain how air moves in and out of the lungs during respiration in frogs;

13. distinguish among negative pressure inhalation, positive pressure exhalation, and positive pressure inhalation;

14. describe the relationship between vital capacity and lung volumes and the interrelationships among lung volumes;

15. explain the importance of CO_2 concentration in the blood and other body fluids to the control of respiration.

INTRODUCTION

In multicellular organisms, the circulatory and the respiratory systems work together and play a central role in the maintenance of **homeostasis**—a stable environment. Homeostatic mechanisms throughout the body function to keep the physical and chemical properties of the **blood** within physiological limits. By means of the quickly circulation blood, the effects on homeostatic mechanisms are spread first to the interstitial fluid and then to the intracellular fluid throughout the body. Thus, circulation—the bulk transport of fluid around the body—connects the specialized cells of multicellular organisms even though they are separated physically. A circulatory system is a necessary step in the evolution of complex, larger organisms. Because diffusion works well only over short distances, animals about the size of earthworms and larger have circulatory systems that move dissolved gases around the body; animals a little larger have respiratory systems. For example, the clam and crayfish move water across gills to facilitate the body's gain of O_2 and loss of CO_2.

In humans and other air-breathing vertebrates, O_2 uptake and CO_2 elimination occur by diffusion across the moistened thin membranes of millions of alveoli (singular, *alveolus*) and their surrounding capillaries located in the lungs. These animals are protected from excessive water loss via evaporation from the very large, moist respiratory surface by having the lungs positioned inside the body.

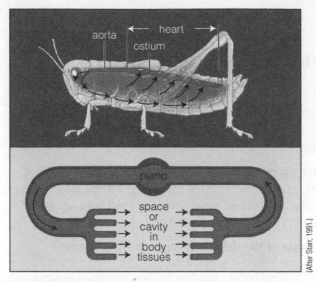

Figure 25-1 Open circulatory system.

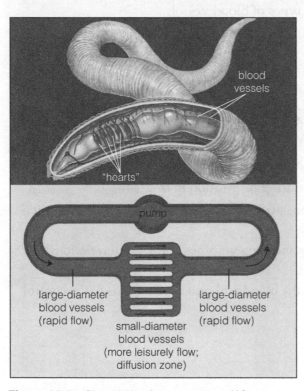

Figure 25-2 Closed circulatory system. (After Starr, 1991.)

(After Starr, 1991.)

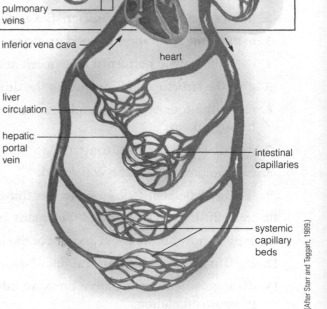

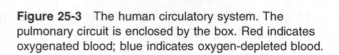

(After Starr and Taggart, 1989.)

Figure 25-3 The human circulatory system. The pulmonary circuit is enclosed by the box. Red indicates oxygenated blood; blue indicates oxygen-depleted blood.

The main function of the rest of the respiratory system is ventilation—the exchange of gases between the lungs and the atmosphere. The movement of gases in (**inhalation**) and out (**exhalation**) of the respiratory system requires the rhythmical contraction of skeletal muscles (**breathing**).

Most coelomates (animals with a true body cavity) have a circulatory system with a branching network of pipes (**blood vessels**) that contain fluid (**blood**). Blood is pumped through the blood vessels by one or more *hearts*. Some invertebrates, like the clam, crayfish, and insect, have an *open circulatory system,* in which the blood percolates directly through the body tissues (Figure 25-1). Vertebrates and some invertebrates, like the earthworm, have a *closed circulatory system,* in which the blood flows solely within the blood vessels (Figure 25-2).

Humans and other mammals and birds have a single four-chambered heart with two completely separate networks of blood vessels, the pulmonary and systemic circuits (Figure 25-3). The **heart** receives blood from both circuits and pumps it back into them, alternating the blood between each one. The pattern of blood vessels is the same in each circuit (Table 25-1).

The **pulmonary circuit** carries oxygen-depleted blood to the capillary beds of the lungs, where oxygen is loaded and where excess carbon dioxide is unloaded. Pulmonary veins drain the oxygen-rich blood back to the heart. The **systemic circuit** takes the oxygen-rich blood from the heart and conveys it to the rest of the body's capillary beds, where oxygen is unloaded and excess carbon dioxide is picked up. Systemic veins drain the oxygen-depleted blood back to the heart.

In a few cases, this pattern of blood flow from

heart → arteries → arterioles → capillaries → venules → veins → heart

is interrupted by a **portal vein,** which connects two sets of capillary beds. The most prominent example is the *hepatic portal vein,* which transports blood from capillary beds in the intestines, stomach, and spleen to beds of large capillaries in the liver.

The circulatory system plays a central role in the maintenance of **homeostasis**—a stable internal environment. Homeostatic mechanisms throughout the body function to keep the physical and chemical properties of the blood within physiological limits. By means of the fast-circulating blood, the effects of homeostatic mechanisms are spread first to the interstitial fluid and then to the intracellular fluid throughout the body.

25.1	Blood (30 min. or more)

Human blood is about 45% cells by volume, although it is only slightly thicker than water. Blood cells are suspended in a straw-colored fluid called **plasma** (about 55% of total blood volume). Plasma is mostly water but contains many dissolved substances, including gases, nutrients, wastes, ions, hormones, enzymes, antibodies, and other proteins.

MATERIALS

Per student:
- compound microscope, lens paper, bottle of lens-cleaning solution (optional), lint-free cloth (optional), dropper bottle of immersion oil (optional)
- prepared slide of a Wright- or Giemsa-stained smear of human blood

Per lab room:
- eosinophil on demonstration (compound microscope)

- basophil on demonstration (compound microscope)
- dropper bottle(s) of aseptic or simulated blood
- dropper bottles of anti-A, anti-B, anti-D (Rh) sera
- box of depression or stirring sticks
- glass slides and wax pencil or plastic blood typing trays

Per student group (4):
- plain capillary tubes

TABLE 25-1 Types of Blood Vessels and Their Major Functions

Types	Functions
Elastic arteries	1. Receive blood from heart
	2. Deliver blood to more numerous muscular arteries
	3. Maintain blood pressure between contractions of the heart
Muscular arteries	1. Deliver blood to more numerous and smaller arterioles
	2. Regulate blood flow to organs
Arterioles	1. Deliver blood to more numerous and smaller capillaries
	2. Regulate peripheral resistance
	3. Precapillary sphincters of smooth muscle regulate blood flow into particular capillary beds
Capillaries	1. Exchange dissolved gases, nutrients, wastes, etc. with fluid surrounding cells (interstitial fluid)
	2. Form interstitial fluid
Venules	1. Drain blood into fewer and larger veins
	2. Serve as a blood reservoir
Veins	1. Drain blood into fewer and larger veins, and finally back to heart
	2. Serve as a blood reservoir

TABLE 25-2   Characteristics of Formed Elements of Blood

Cell or Fragment	Number/mm³ in Peripheral Blood	Percent of Leukocytes	Size (μm)
Erythrocytes	4.5–5.5 million	—	7 by 2
Platelets	250,000–300,000	—	2–5
Neutrophils	3000–6750	65	10–12
Eosinophils	100–360	3	10–12
Basophils	25–90	1	8–10
Lymphocytes	1000–2700	25	5–8
Monocytes	150–750	6	9–15

- microhematocrit centrifuge
- ruler or hematocrit reader
- blood kits (optional)

Per lab section:
- blood waste disposal jar

PROCEDURE

A. Formed Elements (Cells and Platelets) of Blood

The abundance and size of the various blood cells are presented in Table 25-2. Table 25-3 lists the major functions of the blood cells.

1. Use your compound microscope to examine the prepared slide of a Wright-stained smear of blood with medium power. Note the numerous, small pink-stained red blood cells, or **erythrocytes** (Figure 25-4a). Each erythrocyte is a biconcave disk without a nucleus. Scattered among the erythrocytes are a much smaller number of blue/purple-stained cells. These are white blood cells, or **leukocytes.** Center a leukocyte and rotate the nosepiece to the high-dry objective.

What part of the cell, nucleus or cytoplasm, is stained blue/purple? _____

2. Using high power, preferably the oil-immersion objective, move the slide slowly and look for cell fragments between the erythrocytes and leukocytes. They usually have one small blue-stained granule in them and are often clumped together. These cell fragments are **platelets,** or thrombocytes (Figure 25-4).

TABLE 25-3   Functions of Formed Elements of Blood

Cell or Fragment	Functions
Erythrocytes	Contain hemoglobin, which transports oxygen, and carbonic anhydrase, which promotes transport of carbon dioxide by the blood
Platelets	Source of substances that aid in blood clotting
Neutrophils	Leave the blood early in an inflammation to become phagocytes (cells that eat bacteria and debris)
Eosinophils	Phagocytosis of antigen-antibody complexes; numbers are elevated during allergic reactions
Basophils	Granules contain a substance (histamine) that makes blood vessels leaky and a substance (heparin) that inhibits blood clotting
Lymphocytes	Perform many functions central to immunity
Monocytes	Leave the blood to form phagocytic cells called macrophages

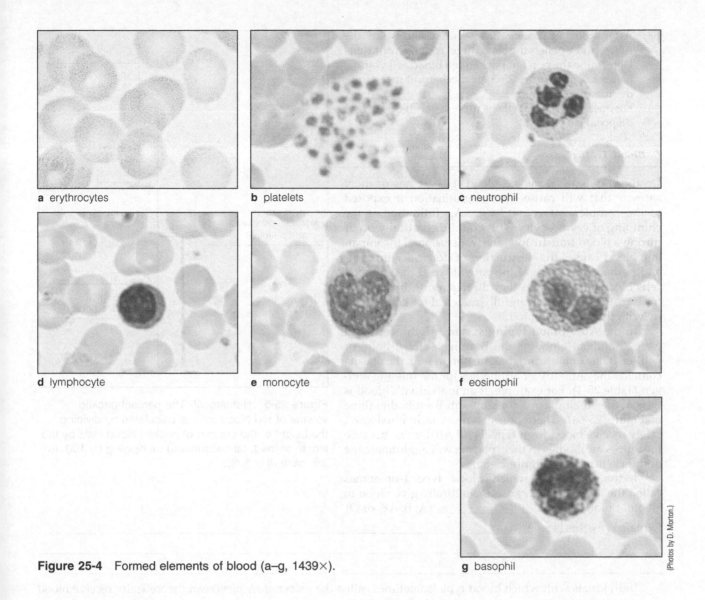

a erythrocytes   b platelets   c neutrophil

d lymphocyte   e monocyte   f eosinophil

g basophil

(Photos by D. Morton.)

Figure 25-4 Formed elements of blood (a–g, 1439×).

3. Locate at least three of the five leukocytes—**neutrophils, lymphocytes,** and **monocytes** (Figures 25-4c, d, and e). Search for them with the high-dry objective. When you find one, center and examine it. If your microscope has an oil-immersion objective (and with the permission of the instructor), use it to look at each blood cell. You may also find **eosinophils** and **basophils** (Figures 25-4f and g), which are normally the rarest leukocyte types. If you have not found an eosinophil or a basophil by the time you have identified the three common leukocyte types, look at one or both of the demonstrations of these cells that have been set up by your instructor.

The three types of leukocytes with the suffix -*phil* (for *philic*, meaning "to like") have large *specific granules.* The prefix in their names refers to the staining characteristics of the specific granules: *neutro-* for neutral (that is, little staining by either of the two dyes in typical blood stains—eosin and methylene blue); *eosino-* because the specific granules stain with the pink dye eosin; and *baso-* because the specific granules stain with the *basic* dye methylene blue.

B. Hematocrit

If you have donated blood, you've probably had your **hematocrit** or percent packed red blood cell volume taken. A lower than normal hematocrit is one indicator of anemia—a lower than normal hemoglobin concentration.

1. Work in groups of four. Fill a plain capillary tube to the mark with aseptic or simulated blood and seal it with clay. Fresh blood requires a capillary tube with the inside surface coated with the anticoagulant heparin to prevent clotting.
2. Place the sealed capillary tube in a microhematocrit centrifuge along with those from other groups.

3. After your instructor has secured and spun the tubes, and opened the centrifuge, recover your tube. Determine the percent packed red blood cell volume using a ruler (Figure 25-5) or mechanical measuring device and record it: _____ %

4. Dispose of your capillary tube in the blood waste disposal jar.

C. Blood Typing

On the surfaces of your red blood cells are one or more *antigens* that will cause their agglutination if exposed to the complementary *antibodies*. **Agglutination** is the clumping of erythrocytes. This could theoretically occur during a blood transfusion. The transfusion of incompatible blood causes the destruction of donor erythrocytes and perhaps the death of the patient when the clumped cells block blood vessels. For these reasons, blood used for transfusion is very carefully matched for compatibility with the patient's blood.

In the **ABO blood typing system,** erythrocytes can have A antigen alone or the B antigen alone, both, or neither. If one or both antigens are not present, the plasma contains the antibody or antibodies for the missing antigen (Table 25-4). For example, if an individual's blood is type A, then their plasma contains anti-B antibodies (blue in Figure 25-6a). Therefore, a type A individual can't safely receive blood from type B and AB donors, because the anti-A antibodies in their plasma will agglutinate the donor erythrocytes (Figure 25-6b).

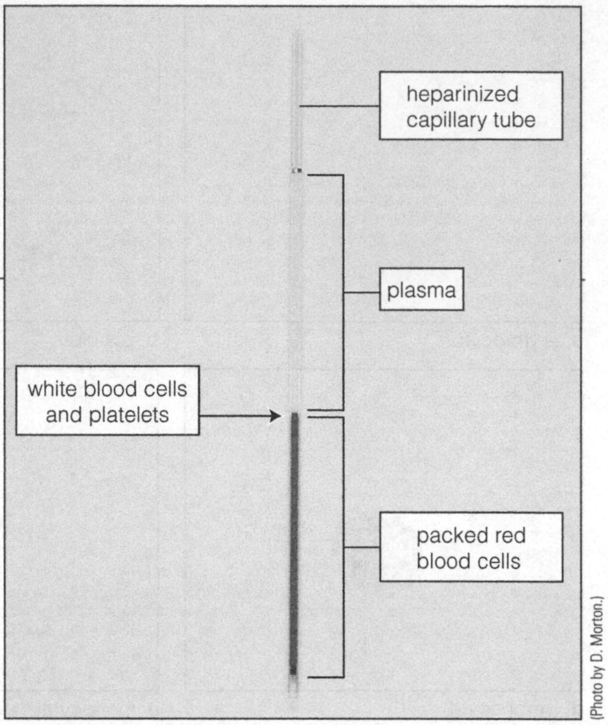

Individuals with which blood type (sometimes called the *universal donor*) can theoretically give blood to all other blood types? _____ (A, B, AB, or O)

Explain why this is so. _____
_____
_____

Individuals with which blood type (sometimes called the *universal recipient*) can theoretically receive blood from any other type? _____ (A, B, AB, or O)

Explain why this is so. _____
_____
_____

Usually a standard blood typing procedure includes a test for the Rh factor or D antigen. For example, people with A+ blood have both the A and D antigens on the surface of their erythrocytes. An A− individual has only the A antigen. About 86% of the population in the United States is Rh+. However, Rh− individuals don't

Figure 25-5 Hematocrit. The percent packed volume of red blood cells is calculated by dividing the height of the column of packed blood cells by the length of the total column and multiplying by 100. In this case, it is 53%.

labels: heparinized capillary tube; plasma; white blood cells and platelets; packed red blood cells

(Photo by D. Morton.)

TABLE 25-4 ABO Blood Types			
Blood Types	Antigens Present on Erythrocytes	Antibodies Present in Plasma	Plasma Agglutinates
A	A	Anti-B	B and AB
B	B	Anti-A	A and B
AB	A and B	None	None
O	None	Anti-A and B	A, B, and AB

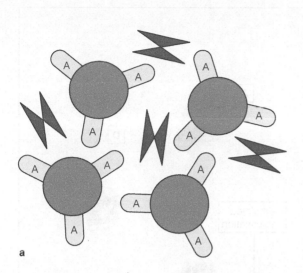

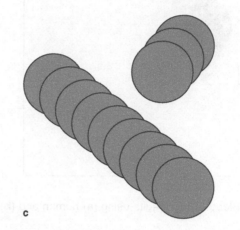

Figure 25-6 (a) Red blood cells (RBCs) from a type A individual. The RBCs have A antigen on their surfaces and anti-B antibody (blue) in the plasma. (b) If type B RBCs are introduced into the circulation, agglutination of these cells will occur. (c) The round surfaces of the RBCs have more antigen, so they tend to stack up in a "slipped stack of poker chips" pattern.

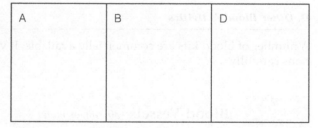

have the anti-D antibody unless they have been exposed to Rh+ erythrocytes. This could happen during a transfusion of Rh+ blood or during the birth of an Rh+ child if not prevented by injecting antibody (RhoGAM) during the pregnancy and shortly after birth. The injected antibody ties up any D antigen and prevents the mother's body from making anti-D antibody, which would otherwise attack a subsequent Rh+ fetus.

A	B	D

Figure 25-7 An ABO/D grid made with a glass slide.

1. Work in groups of four. Gather a microscope slide, several mixing sticks, and a wax pencil. Alternatively, use a plastic blood typing tray. If you use the latter, skip step 2.
2. With the wax pencil, divide each microscope slide into thirds with two lines perpendicular to the long axis of the slide (see Figure 25-7). Label the upper left-hand corner of the left-hand box or depression with an A. Similarly, label the next box B and the last box D.
3. Place one drop of aseptic or simulated blood in the middle of each box on the slide or each depression in the tray. Then place one drop of anti-A serum next to the box A blood cells, one drop of anti-B serum next to the box B blood cells, and one drop of anti-D serum next to box D blood cells. To mix the drops, stir each set with an unused end of a mixing stick, or rock the tray back and forth.
4. If the antigen is present, the erythrocytes in that mixture will clump or the mixture will be cloudier. If it isn't present, the mixture will not change (Figure 25-8).

Record the blood type. _____ (A+, A−, B+, B−, AB+, AB−, or O+ or O−)

5. Your instructor may ask you to type several samples of aseptic or simulated blood. If so instructed, repeat steps 1–4 above.

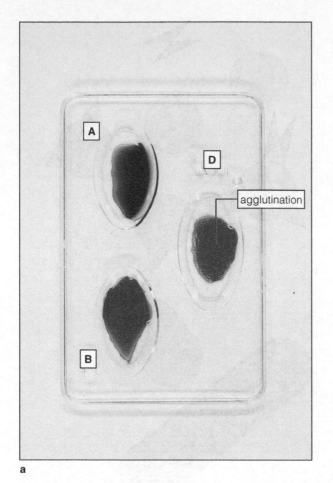

a

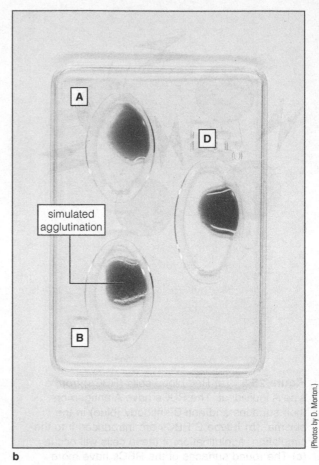

b

(Photos by D. Morton.)

Figure 25-8 Blood typing results using (**a**) human and (**b**) simulated blood.

D. Other Blood Activities

A number of blood kits are commercially available. If you are asked to use one of these kits, follow the instructions carefully.

25.2   Blood Vessels *(30 min. or more)*

The basic structure of blood vessels is shown in Figure 25-9.

MATERIALS

Per student:

- compound microscope
- prepared slide of a companion artery and vein, c.s.
- prepared slide of a whole mount of mesentery

Per student pair:

- fish net
- small fish (3–4 cm long) in an aquarium
- 3 × 7 cm piece of absorbent cotton
- petri dish
- coverslip
- dissecting needle

Per student group (4):

- container of anesthetic dissolved in dechlorinated water
- squeeze bottle of dechlorinated water

Per lab room:

- safe area to run in place
- several meter sticks taped vertically to the walls
- clock with a second hand

PROCEDURE

A. Arteries

Each contraction of the heart pumps blood into the space within the arteries. The rate of flow of blood out of the heart and into the arteries per minute is called *cardiac output* (CO). The arterial space is fairly constant, and it is somewhat difficult for blood to flow through the blood vessels, especially out of the arterioles. This resistance to the flow of blood is called *peripheral resistance* (PR). As cardiac output or peripheral resistance, or both, increase, more blood has to fit into the arterial space, which increases the force that the blood exerts on the walls of the arteries. This force is called **blood pressure** (BP). Blood pressure is directly proportional to the product of cardiac output and peripheral resistance (BP ∝ CO × PR).

Pressure in blood vessels is highest in the arteries leaving the heart, gradually decreasing the farther a vessel is from the heart (Table 25-5). Blood or any other liquid or gas always flows from high to low pressure. So accordingly, blood flows through the circulatory system down this pressure gradient.

Would you expect blood flow to be more rapid in arteries or veins? _____

Explain your answer. (*Hint:* look at the pressure differences between arteries, capillaries, and veins in Table 25-5.)

_____

_____

_____

The walls of the largest arteries contain many **elastic membranes,** which are stretched during contraction (*systole*) of the heart. When the heart is relaxing (*diastole*), these membranes rebound and squeeze the blood, maintaining blood pressure and flow.

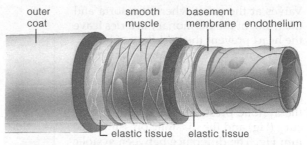

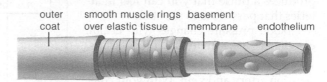

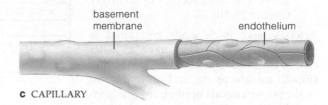

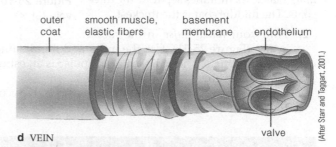

(After Starr and Taggart, 2001.)

Figure 25-9 Structure of blood vessels.

Locations	Blood Pressures (mm Hg)
Right atrium of heart	5/0 (systolic/diastolic)
Right ventricle of heart	25/5
Pulmonary arteries	20
Arterioles and capillaries of lung	20–10
Pulmonary veins	10
Left atrium of heart	10/0
Left ventricle of heart	120/10
Brachial artery	120/80
Arterioles	100–50
Capillaries	50–20
Veins	20–0

TABLE 25-5 Blood Pressure in Different Parts of the Circulatory System of a Young Man at Rest

Valves at the point where the aorta and the trunk of the pulmonary arteries leave the heart prevent the backflow of blood.

When a physician takes your blood pressure, it's usually of the brachial artery of the upper arm and with the body at rest. A blood pressure of 120/80 means the systolic pressure is 120 mm of mercury (Hg) and the diastolic pressure is 80 mm Hg. The difference between systolic and diastolic pressures (*pulse pressure*) produces a pulse that you can feel in arteries that pass close to the skin.

Blood pressure changes with health, emotional state, activity, and other factors.

1. Get a prepared slide of a companion artery and vein. With your compound microscope, locate and examine the cross section of an artery (Figure 25-10).

Arteries have thick walls compared to other blood vessels. They have an outer coat of connective tissue, a middle coat of smooth muscle tissue, and an inner coat of simple squamous epithelium (*endothelium*). Elastic membranes separate the three coats. The middle coat is the thickest.

2. Find your *radial pulse* in the radial artery (Figure 25-11). Use the index and middle fingers of your other hand. A pulse occurs every time the heart contracts. The strength of the pulse is an estimate of the difference between the systolic and diastolic blood pressure.

3. Sit down and count the number of pulses in 15 seconds. To calculate your heart rate, multiply by 4. Record these numbers in Table 25-6.

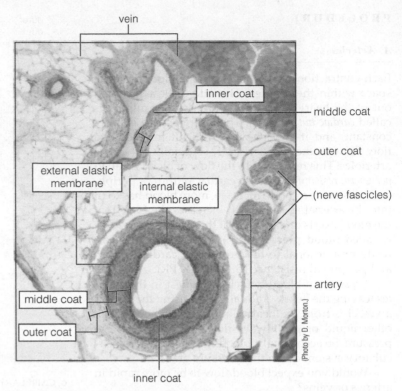

(Photo by D. Morton.)

Figure 25-10 Photomicrograph of cross section of an artery and vein (35×).

TABLE 25-6 Heart Rate Under Different Conditions

Condition	15 Second Counts	Heart Rate (beats per minute)
Sitting at rest	_____ × 4 =	
Holding breath	_____ × 4 =	
Running in place	_____ × 4 =	

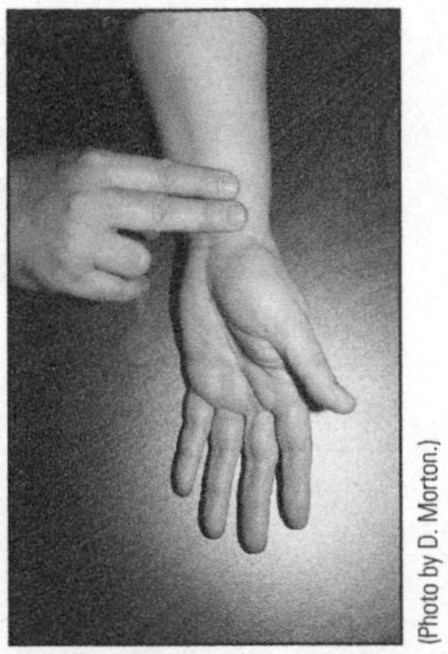

Caution

Do not do the following procedures if you have any medical problems with your lungs or heart. All subjects should be seated except where otherwise indicated and should stop immediately if they feel faint.

Figure 25-11 Feeling the radial pulse.

(Photo by D. Morton.)

4. Hold your breath. After 10 seconds have passed, count the number of pulses, then calculate your heart rate as in step 3. Record these numbers in Table 25-6.

Compared to your resting heart rate, does your heart rate increase, decrease, or remain the same? _____

When you hold your breath, you decrease the return of blood to the heart. This reduces pulse pressure. Homeostatic mechanisms increase heart rate to compensate for reduced blood pressure.

5. Now run in place for 2 minutes in the area designated by your lab instructor. Immediately after sitting down, count the number of pulses, calculate your heart rate, and record these numbers.

After running in place, does your heart rate increase, decrease, or remain the same? _____ Explain these results. _____

_____

_____

_____

B. Capillaries

Capillaries have a very thin wall that consists of endothelium.

1. Obtain a prepared slide of a whole mount of mesentery and examine it with your compound microscope. Look for groups of blood vessels running through the connective tissue (Figure 25-12). The smallest vessels, which branch and join with each other, are capillaries.

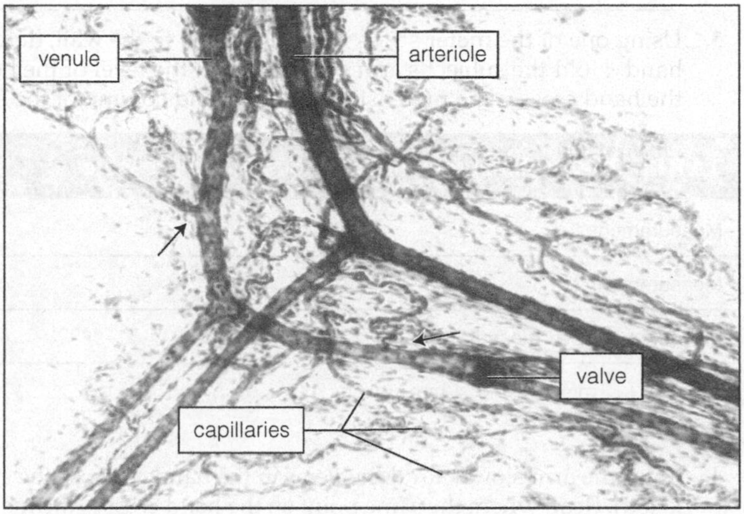

Can you see red blood cells inside the capillaries? _____ (yes, no)

2. Use the net to catch a small fish from the aquarium and place it in the anesthetic fluid. Treat the fish gently, and it will not be harmed by this procedure.
3. After the fish turns belly up, wrap its body in cotton made soaking wet with dechlorinated water. Place the fish in half a petri dish so that the tail is in the center.
4. Using dechlorinated water, make a wet mount of the posterior two-thirds of the fish's tail and examine it with the low-power, medium-power, and high-dry objectives of the compound microscope. Use the lowest illumination that still allows you to see the blood flowing in the vessels. If necessary, you can temporarily close the condenser iris diaphragm to create more contrast. The smallest blood vessels are capillaries.

Figure 25-12 Arteriole, venule, and capillary bed in a whole mount of mesentery. Arrows indicate capillaries joining venule (100×).

Can you see red blood cells moving through them? _____ (yes or no)

What other vessels can you identify?

_____

_____

Is the blood flowing at the same speed in all of the capillaries? _____(yes or no)

Describe blood flow in the fish's tail.

_____

_____

_____

_____

5. Return the fish to the aquarium, wash the half petri dish, and squeeze out the cotton in the sink before dropping it in the trash can.

C. Veins

For the blood to return to the heart after passing through capillary beds below the heart, it must overcome the force of gravity. Veins have **valves** to prevent the backflow of blood away from the heart. Primarily muscular and breathing movements move blood from one segment between valves to another.

1. Again look at a prepared slide of a companion artery and vein. Find and examine the cross section of a vein (Figure 25-12).

The vein has thinner walls and a larger lumen compared to its companion artery. Veins have an outer coat of connective tissue, a middle coat of smooth muscle tissue, and an inner coat of endothelium. Elastic membranes may be present. The outer coat is the thickest layer of the three coats. Compared to arteries, the walls of veins are more disorganized.

2. Work in pairs. Notice the veins as the subject's arm hangs down at the side of the body. You can easily see the veins because they are full of blood. This is usually best seen on the back of the hand. Now raise the arm above the head. Describe and explain any changes that take place.

_____

_____

_____

3. Using one of the meter sticks vertically taped to the wall, determine the venous pressure in the veins of the hand. Hold the subject's arm straight out at the level of the heart. Record the reading in millimeters where the hand crosses the meter stick in the second column of the second row in Table 25-7 (measurement 1).

TABLE 25-7 Measurement of Venous Blood Pressure

Measurement 1	_____ mm
Measurement 2	_____ mm
Measurement 2 − measurement 1	_____ mm H_2O
_____ mm H_2O 0.074 mm Hg/mm H_2O	_____ mm Hg

4. Raise the arm slowly until the veins in the hand collapse (be sure that most of the muscles in the arm are relaxed). Read where the same point on the hand crosses the meter stick and record it in millimeters in the second column of the third row of Table 25-7 (measurement 2). The difference between the two readings gives you the venous pressure expressed in millimeters of water. Calculate the venous pressure in millimeters of mercury (mm Hg) by multiplying by 0.074 mm Hg/mm H_2O.

How does the venous pressure compare to arterial pressures?

_____

_____

5. Look at the veins of the subject's forearm and hand. The swellings that occur at various intervals are valves. Figure 25-12 shows a smaller valve in a venule. Choose a section between two swellings that doesn't have any side branches. Place one finger on the swelling away from the heart and with another finger press the blood forward (toward the heart) beyond the next swelling.

Does the vein fill up with blood again? _____ (yes or no)

Now remove the finger and observe what happens. Try this again, but press blood in the opposite direction. Discuss your observations.

_____

_____

_____

_____

25.3   The Heart (20 min.)

Normally the heart beats over 100,000 times a day, pumping the blood around the circulatory system (Figure 25-13).

MATERIALS

Per lab group bench:

■ human heart model

Per lab section:

■ demonstration of the effects of acetylcholine and epinephrine on the heart of a doubly pithed frog (no brain or spinal cord) kept moist with amphibian Ringer's solution (balanced salts solution)

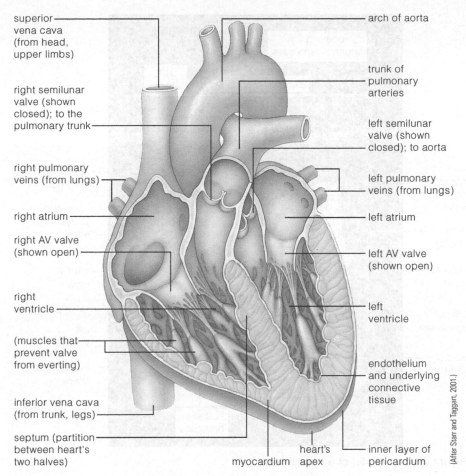

superior
vena cava
(from head,
upper limbs)

arch of aorta

right semilunar
valve (shown
closed); to the
pulmonary trunk

trunk of
pulmonary
arteries

right pulmonary
veins (from lungs)

left semilunar
valve (shown
closed); to aorta

right atrium

left pulmonary
veins (from lungs)

right AV valve
(shown open)

left atrium

right
ventricle

left AV valve
(shown open)

(muscles that
prevent valve
from everting)

left
ventricle

inferior vena cava
(from trunk, legs)

endothelium
and underlying
connective
tissue

septum (partition
between heart's
two halves)

heart's
apex

inner layer of
pericardium

myocardium

(After Starr and Taggart, 2001.)

Figure 25-13   Ventral view of a sectioned human heart.

PROCEDURE

A. Human Heart Model

Examine a model of the human heart and identify and trace the flow through its four chambers and four valves (Figure 25-14).

1. The **right atrium** (Figures 25-13 and 25-14) receives blood from the systemic circuit via the *superior vena cava*, the *inferior vena cava*, and the *coronary sinus* (which drains blood from capillary beds in the heart itself).
2. When the right atrium contracts, blood is pushed through the **right AV valve** (*atrioventricular* or *tricuspid valve*) into the **right ventricle**.
3. Contraction of the right ventricle forces this blood into the trunk of the *pulmonary arteries to* start its journey through the pulmonary circuit.
4. The **left atrium** receives blood from pulmonary circuit via the *pulmonary veins*.
5. Now, blood is forced through the **left AV valve** (*bicuspid valve*) into the **left ventricle** by the contraction of the left atrium.
6. Contraction of the left ventricle pushes blood into the *aorta* to begin its travel through the systemic circuit. The *aortic semilunar valves* prevent the backflow of blood into the left ventricle.

B. Demonstration of Frog Heart Function

1. Your instructor has set up a demonstration of a frog heart in place in the opened thorax of a frog. Although the frog has a three-chambered heart—two atria and one ventricle—the heart's function and control are essentially the same as those of humans. The nervous system of the frog has been destroyed, so it does not feel pain or control heart action.
    Is the heart beating? _____ (yes or no)

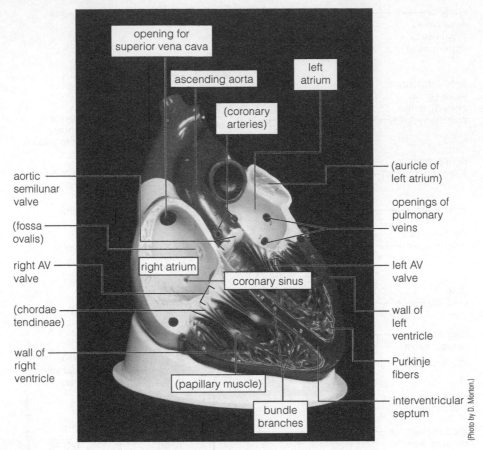

Figure 25-14   Model of a human heart.

opening for
superior vena cava

ascending aorta

left
atrium

(coronary
arteries)

aortic
semilunar
valve

(fossa
ovalis)

right AV
valve

right atrium

(chordae
tendineae)

wall of
right
ventricle

(papillary muscle)

coronary sinus

bundle
branches

(auricle of
left atrium)

openings of
pulmonary
veins

left AV
valve

wall of
left
ventricle

Purkinje
fibers

interventricular
septum

(Photo by D. Morton.)

2. Carefully observe the beating heart.

Does the entire heart muscle contract simultaneously or do some parts contract before others?_____
_____

Is there a pattern in the way each contraction sweeps across the heart or is it totally random as to when a particular part contracts? _____

If you observe carefully, you can see the order in which the chambers contract. Record your observations.
_____
_____
_____

The primary pacemaker of the heart, the **sinoatrial node,** is located in the right atrium. It functions independently of the nervous system, firing rhythmically. Each time the sinoatrial node fires, it initiates a message to contract. This message spreads over the atria. Then special heart cells amplify and conduct the message throughout the ventricle.

3. Count how many times the heart contracts in 15 seconds, record it in Table 25-8 as the control 1 rate, then calculate the heart rate.

TABLE 25-8   Effect of Acetylcholine and Epinephrine on the Heart Rate of a Frog

Conditions	15-sec Counts	Heart Rate (beats per minute)
Control 1	_____ × 4 =	
After acetylcholine	_____ × 4 =	
Control 2	_____ × 4 =	
After epinephrine	_____ × 4 =	

4. Note when your instructor places several drops of an acetylcholine solution on the heart. After a minute passes, repeat step 3, recording the count in row 2.

5. Watch your instructor thoroughly flush the thoracic cavity with balanced salt solution, wait 3 minutes, then determine the control 2 heart rate.

Now observe as your instructor places several drops of an epinephrine solution on the heart. After 1 minute, repeat step 3, recording the count in row 4.

6. Plot your results on Figure 25-15.
7. Describe the effects of acetylcholine and epinephrine on the heart rate.

_____

_____

_____

_____

In an intact frog, the heart rate is modified by input from the central nervous system. The heart rate is affected by the amount of **acetylcholine** and **norepinephrine** secreted by neurons around the sinoatrial node. During restful activities, acetylcholine slows the heart rate and thus acts as a brake on the sinoatrial node. Pain, strong emotions, the anticipation of exercise, and the fight-or-flight response can all increase the secretion of norepinephrine. Norepinephrine, which has a molecular structure and action similar to epinephrine, speeds the heart rate and thus acts as an accelerator on the sinoatrial node and can override the parasympathetic brake.

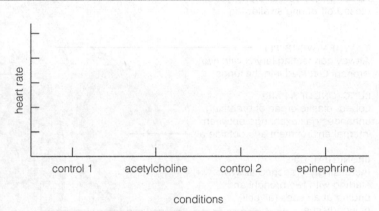

Figure 25-15 Bar graph for plotting heart rate data.

25.4   Breathing *(40 min.)*

Before we continue with this exercise, let's review some anatomical terms. The trunk of the body is divided into an upper *thorax,* which is supported by the rib cage and contains the *thoracic cavity,* and a lower *abdomen.* The thoracic and abdominal cavities are separated by a partition of skeletal muscle called the *diaphragm.* The thoracic cavity contains two *pleural cavities,* which contain the lungs and the *pericardial cavity* surrounding the heart.

The muscles of breathing and their roles in inhalation and exhalation are listed in Table 25-9. The external and internal intercostal muscles are located between the ribs. Contraction of the diaphragm increases the size of the thoracic cavity by lowering its floor. Relaxation of the diaphragm allows it to spring back to its original position. When they are contracted, the abdominal muscles can squeeze the internal organs, pushing up the diaphragm to decrease the size of the thoracic cavity.

MATERIALS

Per student:

- 2 pieces of paper (each 14 × 21.5 cm—half of a sheet of notebook paper)

Per group (2):

- metric tape measure
- large caliper with linear scale (for example, Collyer pelvimeter)

Per group (4):

- functional model of lung
- models of the organs and structures of the respiratory system
- prepared slide of a section of mammalian lung

Per lab room:

- frogs in terrarium or video of breathing frog
- clock with second hand

PROCEDURE

A. Ventilatory Ducts and Lungs

The respiratory system consists of the lungs and the ducts that shuttle air between the atmosphere and the lungs. Its organs and structures from the outside in are the two **external nares** (nostrils), the **nasal cavity** (divided into

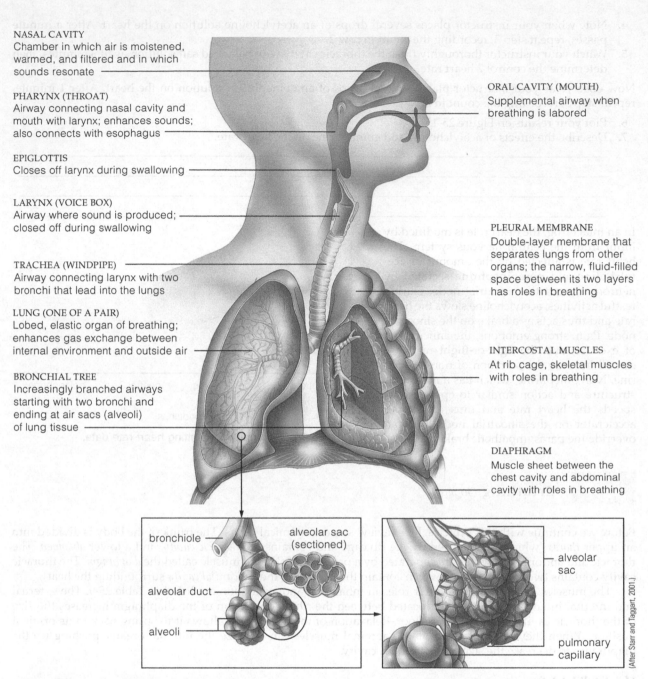

NASAL CAVITY
Chamber in which air is moistened, warmed, and filtered and in which sounds resonate

PHARYNX (THROAT)
Airway connecting nasal cavity and mouth with larynx; enhances sounds; also connects with esophagus

EPIGLOTTIS
Closes off larynx during swallowing

LARYNX (VOICE BOX)
Airway where sound is produced; closed off during swallowing

TRACHEA (WINDPIPE)
Airway connecting larynx with two bronchi that lead into the lungs

LUNG (ONE OF A PAIR)
Lobed, elastic organ of breathing; enhances gas exchange between internal environment and outside air

BRONCHIAL TREE
Increasingly branched airways starting with two bronchi and ending at air sacs (alveoli) of lung tissue

ORAL CAVITY (MOUTH)
Supplemental airway when breathing is labored

PLEURAL MEMBRANE
Double-layer membrane that separates lungs from other organs; the narrow, fluid-filled space between its two layers has roles in breathing

INTERCOSTAL MUSCLES
At rib cage, skeletal muscles with roles in breathing

DIAPHRAGM
Muscle sheet between the chest cavity and abdominal cavity with roles in breathing

bronchiole

alveolar sac (sectioned)

alveolar duct

alveoli

alveolar sac

pulmonary capillary

(After Starr and Taggart, 2001.)

Figure 25-16 Human respiratory system.

TABLE 25-9   Muscles Used to Breathe

Stage of Respiratory Cycle	Muscles				
	External Intercostals	Internal Intercostals	Diaphragm	Neck and Shoulder	Abdominal
Restful inhalation	Relaxed	Relaxed	Contracted	Relaxed	Relaxed
Forced inhalation	Contracted	Relaxed	Contracted	Contracted	Relaxed
Restful exhalation	Relaxed	Relaxed	Relaxed	Relaxed	Relaxed
Forced exhalation	Relaxed	Contracted	Relaxed	Relaxed	Contracted

right and left sides by a **nasal septum** or partition), the **pharynx** (shared with the digestive tract), the **glottis** (and its cover the **epiglottis**), the **larynx** (voice box), the **trachea** (windpipe), two **bronchi** (singular, bronchus), and its branches, which terminally open into myriad **alveoli** (singular, alveolus) in the **lungs.** Along with respiration, sound production is a major function of the respiratory tract.

1. Look at the various models of respiratory tract organs and structures and identify as many of the preceding boldfaced terms as possible.
2. Get your compound light microscope and examine a lung section at high power; identify the alveoli (Figure 25-17).

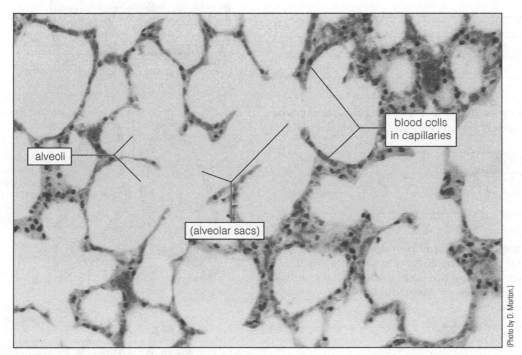

Figure 25-17   Alveoli in the lung (500×).

B. Ventilation

All flow occurs down a pressure gradient. When you let go of an untied inflated balloon, it flies away, propelled by the jet of air flowing out of it. The air flows out because the pressure is higher inside the balloon. The high pressure inside the balloon is maintained by the energy stored in its stretched elastic wall.

When the thoracic cavity expands during inspiration, first the pressure in the pleural sacs decreases, and then the pressure within the lungs decreases. Because the pressure outside the body is now higher than that in the lungs, and assuming the connecting *ventilatory ducts* (trachea and so on) are not blocked, air flows into the lungs (Figure 25-18). This is called **negative pressure inhalation.**

The opposite occurs during expiration. The size of the thorax and pleural sacs decreases, the pressure in the lungs increases, and air flows out of the body down its concentration gradient. This is called **positive pressure exhalation.**

The pressure in the pleural sacs is actually always below atmospheric pressure, which means the lungs are always partially inflated after birth. Thus, a hole in a pleural sac or lung will result in a collapsed lung. **Cohesion** (sticking together) of the wet pleural membranes lining the outsides of the lungs and insides of the body walls of the pleural cavities aids inhalation. Exhalation depends in part on the *elastic recoil* (like letting go of a stretched rubber band) of lung tissue.

1. Work in groups of four. Look at the functional lung model. The "Y" tube is analogous to the ventilatory ducts. The balloons represent the lungs. The space within the transparent chamber represents the thoracic spaces and its rubber floor (rubber "diaphragm"), the muscular diaphragm.
2. Pull down the rubber diaphragm. Describe what happens to the balloons.

As you pull down the rubber diaphragm, does the volume of the space in the container increase or decrease?

As the volume changes, does the pressure in the container increase or decrease?

_____

As the balloons inflate, does the volume of air in the balloons increase or decrease?

_____

Why do the balloons inflate?

_____

_____

_____

_____

_____

3. Push up on the rubber diaphragm. Describe what happens to the balloons and why it takes place.

_____

_____

_____

_____

_____

4. Pull the rubber diaphragm down and push it up several times in succession to simulate breathing.
5. Pucker up your lips and inhale. As you inhale, place one piece of paper directly over your lips. What occurs?

_____

_____

The negative pressure created in your lungs by the contraction of the muscles of inhalation causes this suction.

6. Fold the narrow ends of the two pieces of paper to produce 2- to 3-cm flaps. Open the flaps and use them as handles. Hold a piece of paper with each hand and touch the papers' flat surfaces together in front of you (Figure 25-19). Pull them apart.

Now, thoroughly wet both pieces of paper with water and again touch their flat surfaces together in front of you. Pull them apart. What difference did the water make?

_____

_____

7. Some vertebrates, such as the frog, inhale by pushing air into the lungs. This is the positive pressure inhalation. Observe a frog out of water or watch a video of a breathing frog.

The frog inhales by sucking in air through the nostrils by lowering the floor of the mouth. Valves in the nostrils are then closed and the floor of the mouth raised, thus increasing the pressure and forcing the air into the lungs. The upper portion of the ventilatory duct can be closed to keep the air in the lungs. Exhalation occurs by elastic recoil of the lungs with the ventilatory duct open. In the frog, both inhalation and exhalation are the result of positive pressure.

What is the frog's respiratory rate (breaths per minute)? Count and record how many times the frog lowers and raises the floor of the mouth (one breath) in 3 minutes. _____ breaths/3 minutes

Divide by 3 to calculate the average respiratory rate and record it: _____ breaths/minute

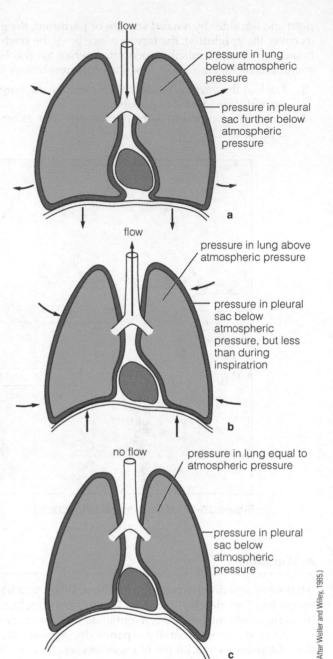

Figure 25-18 Changes in the thoracic cavity during (**a**) negative pressure inhalation and (**b**) positive pressure exhalation, and corresponding movements of air. (**c**) The thoracic cavity at the end of an expiration.

(After Weller and Wiley, 1985.)

Figure 25-19 Use of two pieces of paper to demonstrate cohesion.

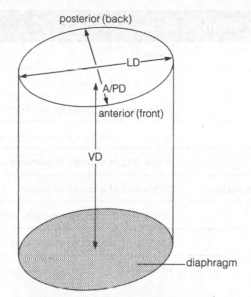

Figure 25-20 Thoracic diameters: LD, lateral diameter; A/PD, anterioposterior diameter; and VD, vertical diameter.

Respiration in the frog is supplemented by gas exchange across the moist skin. Also, as they are ectotherms (do not maintain a high body temperature using physiological means), frogs generally have a lower metabolic rate and, therefore, a lesser demand for O_2 than a mammal of the same size.

C. Breathing Movements

1. Place your hands on your abdomen and take three deep breaths—three inspirations followed by three expirations. What do you feel during each inspiration?

    _____

    each expiration?

    _____

    _____

2. Place your hands on your chest and repeat step 1. What do you feel during each inspiration?

    _____

    _____

    each expiration?

    _____

    _____

D. Measurements of the Thorax

The size of the thorax can be described by three so-called diameters: the lateral diameter (LD), the anterioposterior diameter (A/PD), and the vertical diameter (VD; Figure 25-20). The vertical diameter is the only one that can't be measured easily.

Make the following observations and record them in Table 25-10.

1. Work in pairs. Take turns measuring the circumference of each other's chest with a tape measure at two levels, under the armpits (axillae, C_{AX}) and at the lower tip of the sternum (xiphoid process, C_{XP}) for the conditions listed in Table 25-10. While the measurements are being taken, it's extremely important not to tense muscles other than those used for respiration. For example, do not raise the arms.

Subject	Condition	C_{AX}	C_{XP}	A/PD	LD
TABLE 25-10 Chest Measurements (cm)					
You	At the end of a restful inhalation				
	At the end of a restful (passive) exhalation				
	At the end of a forced (maximum) inhalation				
	At the end of a forced (maximum) exhalation				
Your lab partner	At the end of a restful inhalation				
	At the end of a restful (passive) expiration				
	At the end of a forced (maximum) inhalation				
	At the end of a forced (maximum) exhalation				

2. With calipers, also measure the A/PD and LD at the nipple line for these same conditions. The distance between the tips of the calipers is read off the scale in centimeters.

3. About two-thirds of the air inhaled during a restful inhalation is due to contraction of the diaphragm. Interpret the data in Table 25-10 and in your own words describe changes in the size of the thorax during

  a. a restful inhalation:

  C_{AX} _____

  C_{XP} _____

  A/PD _____

  LD _____

  b. a passive exhalation:

  C_{AX} _____

  C_{XP} _____

  A/PD _____

  LD _____

  c. a forced inhalation:

  C_{AX} _____

  C_{XP} _____

  A/PD _____

  LD _____

  d. a forced exhalation:

  C_{AX} _____

  C_{XP} _____

  A/PD _____

  LD _____

Does the size of the thorax change significantly during a restful inhalation or a passive exhalation? _____ (yes or no)

How does the shape of the thorax change during a forced inhalation?

_____

_____

_____

_____

How does the shape of the thorax change during a subsequent forced exhalation?

_____

_____

_____

_____

25.5   Spirometry *(20 min.)*

Air in the lungs is divided into four mutually ex-clusive volumes (Figure 25-21): tidal volume (TV), forced inhalation volume (FIV), forced exhalation volume (FEV), and residual volume (RV).

Tidal volume is the volume of air inhaled or exhaled during breathing. It normally varies from a minimum at rest to a maximum during strenuous exercise.

Forced inhalation volume is the volume of air you can voluntarily inhale after inhalation of the tidal volume. **Forced exhalation volume** is the volume of air you can voluntarily exhale after an exhalation of the tidal volume. FIV and FEV both decrease as TV increases.

Residual volume is the volume of air that cannot be exhaled from the lungs. That is, normal lungs are always partially inflated.

There are four capacities derived from the four volumes:

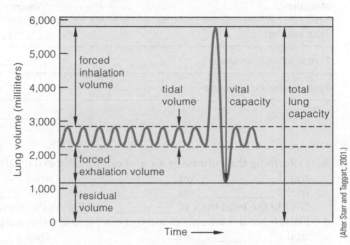

Figure 25-21   Spirogram shows the defined lung volumes and capacities.

(After Starr and Taggart, 2001.)

$$\text{inspiratory capacity (IC)} = \text{TV} + \text{FIV}$$

$$\text{functional residual capacity (FRC)} = \text{FEV} + \text{RV}$$

$$\text{vital capacity (VC)} = \text{TV} + \text{FIV} + \text{FEV}$$

$$\text{total lung capacity (TLC)} = \text{total of all four lung volumes}$$

All the lung volumes except the residual volume can be measured or calculated from measurements obtained using a simple spirometer or lung volume bag. A more sophisticated recording spirometer plots respiration over time (a spirogram). Figure 25-20 shows a spirogram and the relationships of the lung volumes and capacities.

Caution

Always use a sterile mouthpiece and do not inhale air from either a simple spirometer or a lung volume bag.

MATERIALS

Per group (4):

- noseclip (optional)
- simple spirometer or lung volume bags

PROCEDURE

1. Work in groups of four. Sit quietly and breathe restfully. Use a noseclip or hold your nose. After you feel comfortable, start counting as you inhale. After the fourth inhalation, exhale normally into the spirometer

or lung volume bag. Read the volume indicated by the spirometer or squeeze the air to the end of the lung volume bag and read the volume from the wall of the bag. Record the volume below (trial 1). Reset the spirometer or squeeze the air out of the lung volume bag. Repeat this procedure two more times (trials 2 and 3) and calculate the total and average tidal volume at rest.

trial 1 _____ mL   trial 3 _____ mL

trial 2 _____ mL   total = _____ mL

Divide the total by 3 = _____ mL to calculate the average TV at rest. Record the TV in Table 25-11.

TABLE 25-11 Vital Capacity and Lung Volumes at Rest (mL)

Measure	Volume (mL)
Tidal volume	_____
Forced inhalation volume	_____
Forced exhalation volume	_____
Vital capacity	_____

2. Determine the volume of air you can forcibly exhale after a restful inhalation (average of three trials).
   trial 1 _____ mL   trial 3 _____ mL
   trial 2 _____ mL   total = _____ mL
   Divide the total by 3 = _____ mL. This is the average sum of FEV and TV at rest.
3. Determine the volume of air you can forcibly exhale after a forceful inhalation (average of three trials).
   trial 1 _____ mL   trial 3 _____ mL
   trial 2 _____ mL   total = _____ mL
   Divide the total by 3 = _____ mL to calculate the average VC. Record VC in Table 25-11.
4. Calculate the FEV at rest by subtracting step 1's result from step 2's result.
   _____ mL (step 2) − _____ mL (step 1) = _____ mL.
   Record FEV in Table 25-11.
5. Calculate the FIV at rest by subtracting step 2's result from step 3's result.
   _____ mL (step 3) − _____ mL (step 2) = _____ mL.
   Record the FIV in Table 25-11.
6. Does **vital capacity** change as tidal volume increases or decreases? _____ (yes or no)
7. Measure and record your height in centimeters. _____ cm

Write your vital capacity/height on the board—your name isn't necessary.

Plot the vital capacity of each student in your lab section on the following graph.

Is there a relationship between vital capacity and height? If so, describe it mathematically using a graphing calculator or with words.

_____

_____

_____

25.6 Control of Respiration (20 min.)

The control of respiration, both the rate and depth of breathing, is very complex. Simply stated, **chemoreceptors** (receptors for chemicals such as O_2, CO_2, and hydrogen ions, or H^+), stretch receptors in the ventilatory ducts, and centers in the brain stem (part of the brain that connects to the spinal cord) control respiration.

Your own experience has taught you that respiration is to some extent under the control of the conscious mind. We can decide to stop breathing or to breathe more rapidly and deeply. However, the unconscious mind can override voluntary control. The classic example of this is the inability to hold one's breath for more than a few minutes. In this section you will make further pertinent observations related to the hypothesis that CO_2 concentration in the blood and other body fluids is the most important stimulus for the control of respiration.

MATERIALS

Per student pair:
- small mirror

Per lab room:
- a safe place to exercise
- clock with second hand

PROCEDURE

Caution

Do not do the following activities if you have any medical problems with your lungs or heart. All subjects should stop immediately if they feel faint.

1. Work in pairs. Write two predictions in Table 25-12. In prediction 1 forecast the effects on respiratory rate and the ability to hold one's breath after hyperventilation (overventilating the lungs by forced deep breathing). (For example, if I hyperventilate, then. . . .) In prediction 2, forecast the effects on respiratory rate and the ability to hold one's breath after exercise.
2. Sit down and after you feel comfortable, have your lab partner count the number of times you breathe in 3 minutes. If it's difficult to see you breathe, have your partner place a small mirror under your nose. Record the number of breaths: _____ breaths

Divide this number by 3 to calculate the average respiratory rate *at rest* and write it in the second column of Table 25-12.

3. Determine how many minutes you can hold your breath after a restful inhalation and record the time in the third column of Table 25-12.
4. Now breathe as deeply and as rapidly as possible (*hyperventilate*). Try to take at least 10 breaths but stop as soon as you can answer the following question. (In any case, do not continue for more than 20 breaths.) As times goes on, does it become easier or more difficult to continue rapid deep breathing?

_____

TABLE 25-12 Respiratory Data			
Prediction 1:			
Prediction 2:			
	Condition	**Respiratory Rate (breaths/minute)**	**Breath Holding (minutes)**
You	At rest		
	After hyperventilation		
	After exercise		
Your lab partner	At rest		
	After hyperventilation		
	After exercise		
Conclusions:			

Immediately have your lab partner count and record the number of breaths you take in the next 3 minutes: _____ breaths

5. Now determine how many minutes you can hold your breath *immediately* after hyperventilating. Record this time in the third column of Table 25-12.

6. Divide the number in step 4 by 3 to calculate the average respiratory rate *after hyperventilation* and write it in the second column of Table 25-12.

Does hyperventilation increase, decrease, or have no effect on the CO_2 concentration of the blood?
_____

7. When fully recovered from step 2, carefully run in place for 2 minutes in the area designated by your lab instructor. Immediately after sitting down, again have your lab partner count how many breaths you take in 3 minutes and record it: _____ breaths

Now determine how long you can hold your breath *immediately* after 2 minutes of exercise and record it in the third column of Table 25-12.

As you did earlier, calculate the average respiratory rate *after exercise* and record it in the second column of Table 25-12.

Does running in place increase or decrease the CO_2 concentration of the blood? _____

What causes the CO_2 concentration to change while you are running in place?
_____
_____
_____

8. Switch roles with your lab partner and repeat steps 1–7, recording this data in Table 25-12.
9. Use Table 25-12 to summarize your results:
_____
_____
_____
_____
_____

Write a conclusion as to whether your results supported your predictions.

PRE-LAB QUESTIONS

_____ 1. The number of circuits in the circulatory systems of humans is
  (a) one
  (b) two
  (c) three
  (d) four

_____ 2. Blood contains
  (a) dissolved gases
  (b) dissolved nutrients
  (c) dissolved hormones
  (d) all of the above

_____ 3. Red blood cells are
  (a) erythrocytes
  (b) leukocytes
  (c) platelets
  (d) all of the above

_____ 4. The most common leukocyte in the blood is
  (a) a lymphocyte
  (b) an eosinophil
  (c) a basophil
  (d) a neutrophil

_____ 5. The cellular fragments in the blood that function in blood clotting are
  (a) erythrocytes
  (b) leukocytes
  (c) platelets
  (d) none of the above

_____ 6. Blood vessels that return blood from capillaries back to the heart are
  (a) arteries
  (b) veins
  (c) portal veins
  (d) arterioles

_____ 7. Blood vessels that connect networks of capillary beds are
  (a) arteries
  (b) veins
  (c) portal veins
  (d) both b and c

_____ 8. From which chamber of the heart does the right ventricle receive blood?
  (a) right atrium
  (b) left atrium
  (c) left ventricle
  (d) none of the above

_____ 9. How many chambers does the frog heart have?
  (a) one
  (b) two
  (c) three
  (d) four

_____ 10. The primary pacemaker of the heart is the
  (a) bicuspid valve
  (b) tricuspid valve
  (c) aorta
  (d) sinoatrial node

_____ 11. Which muscles may contract during inspiration?
   (a) external intercostals
   (b) internal intercostals
   (c) abdominal
   (d) both b and c

_____ 12. Which muscles contract during a restful expiration?
   (a) external intercostals
   (b) internal intercostals
   (c) diaphragm
   (d) none of the above

_____ 13. Which muscles may contract during a more forceful expiration?
   (a) external intercostals
   (b) diaphragm
   (c) abdominal
   (d) both b and c

_____ 14. An untied inflated balloon flies because
   (a) the pressure is higher inside than outside the balloon
   (b) the pressure is lower inside than outside the balloon
   (c) air flows down its pressure gradient
   (d) both a and c occur

_____ 15. Human ventilation is
   (a) negative pressure inhalation
   (b) positive pressure inhalation
   (c) negative pressure exhalation
   (d) both b and c

_____ 16. Frog ventilation is
   (a) negative pressure inhalation
   (b) positive pressure inhalation
   (c) positive pressure exhalation
   (d) both b and c

_____ 17. Vital capacity is always equal to
   (a) tidal volume
   (b) forced inhalation volume
   (c) forced exhalation volume
   (d) a + b + c

_____ 18. An instrument that measures lung volumes is a
   (a) caliper
   (b) spirometer
   (c) barometer
   (d) stethoscope

_____ 19. Respiration is controlled by
   (a) chemoreceptors
   (b) stretch receptors
   (c) centers in the brain stem
   (d) all of the above

_____ 20. The most important stimulus in the control of respiration is the concentration in the blood and other body fluids of
   (a) oxygen (O_2)
   (b) carbon dioxide (CO_2)
   (c) hydrogen ions (H^+)
   (d) nitrogen (N_2)

EXERCISE 25

Human Blood, Circulation, and Respiration

POST-LAB QUESTIONS

25.1 Blood

1. Name and give the staining characteristics and functions of the three leukocytes with specific granules.
   a.

   b.

   c.

2. Describe the shape, content, and function of an erythrocyte.

3. Why can't you give type A blood to a type B patient?

25.2 Blood Vessels

4. Describe how to take someone's hematocrit.

5. Identify the indicated blood vessel.

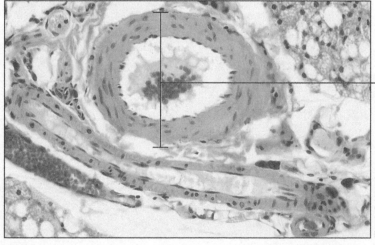

(Photo by D. Morton.)

(250×).

6. Pretend you are an erythrocyte in the right atrium of the heart. Describe one trip through the human circulatory system, ending back where you started.

25.3 The Heart

7. Explain how the heart of a double-pithed frog can continue to contract in an organized manner after the nervous system is destroyed.

8. How is the heart rate controlled by the nervous system in an intact organism?

Food for Thought

9. Considering its relationship with the other systems of the body, what is the importance of the circulatory system?

10. Search the World Wide Web for sites about arteriosclerosis. List two sites and briefly summarize their contents.

    http://

    http://

25.4 Breathing

11. Which skeletal muscles are contracted during
    a. restful inhalation?

    b. forced inhalation?

    c. restful exhalation?

    d. forced exhalation?

12. How does the size of the thorax change during
    a. inhalation?

    b. exhalation?

13. How does the potential volume of the pleural sacs change during
    a. inhalation?

    b. exhalation?

14. Define these terms:
    a. negative pressure inhalation

    b. positive pressure exhalation

15. How does breathing in a human differ from that in a frog?

25.5 Spirometry

16. Explain the relationship among vital capacity, tidal volume, forced inhalation volume, and forced exhalation volume.

25.6 Control of Respiration

17. What substance is the most important stimulus in the control of respiration? How is its production linked to changes in metabolic rate, such as occur during exercise?

Food for Thought

18. Explain why hyperventilation can prolong the time you can hold your breath. Can this be dangerous (for example, hyperventilation followed by swimming under water)?

19. Search the World Wide Web for sites about emphysema. List two sites below and briefly summarize their contents.

    http://

    http://

Animal Development

After completing this exercise, you will be able to

1. define *sexual reproduction, fertilization, gametes, gonads, ovum, sperm, dioecious, zygote, monoecious, asexual reproduction, parthenogenesis, yolk, acrosome, blastodisc, vegetal pole, animal pole, gametogenesis, testis, ovary, seminiferous tubules, interstitial cells, testosterone, Sertoli cells, follicle, ovulation, estrogen, progesterone, corpus luteum, external fertilization, external development, internal fertilization, internal development, uterus, blastomere, morula, blastula, blastocoel, gray crescent, blastoderm, blastocyst, inner cell mass, trophoblast, archenteron, blastopore, primitive streak, notochord, neural tube, somites, amniotic cavity, embryonic disk, implantation, placenta, umbilical cord,* and *fetus;*

2. draw a mammalian sperm and label it;

3. describe the structure of chicken and frog ova, and tell how they differ from a typical mammalian ovum;

4. recognize interstitial cells, seminiferous tubules, Sertoli cells, and sperm in a prepared section of a mammalian testis;

5. recognize follicles, primary oocytes, primordial and secondary follicles, and a corpus luteum in a prepared slide of a mammalian ovary;

6. describe spermatogenesis and oogenesis in mammals;

7. describe the events and consequences of sperm penetration and fertilization;

8. list and define the five stages of development;

9. compare and contrast cleavage in the sea star, frog, chicken, and human;

10. explain differences in cleavage according to (a) the amount and distribution of yolk in the ovum and (b) the evolution of mammals;

11. describe gastrulation, the formation of the primary germ layers—extoderm, mesoderm, endoderm, and their derivatives;

12. list the four extraembryonic membranes;

13. give the functions of the extraembryonic membranes in birds and mammals.

INTRODUCTION

Most animal species reproduce sexually. **Sexual reproduction** usually involves the fusion of the nuclei of two gametes, called the ovum (plural, *ova*) and sperm. This fusion is referred to as **fertilization.**

Gametes are produced by *meiosis* in reproductive organs called **gonads,** usually in individuals of two separate sexes. A female's gametes are **ova,** while those of a male are **sperm.** Species with male and female gonads in separate individuals are said to be **dioecious.**

Each gamete is haploid, and fertilization creates a new diploid cell, the **zygote,** whose combination of genes is unlike those of either parent. It is also very unlikely that the genes of one zygote will be identical to those of any other zygote, even those derived from the same parents. What two Mendelian principles largely account for this?

_____

_____

_____

This variation is an advantage to a species in a changing and unpredictable environment. Why?

_____

_____

_____

Some animals (earthworms and snails, for example) produce both ova and sperm in the same individual, but self-fertilization is rare, occurring only in parasites with constant and predictable environments (tapeworms, for example). A species with both male and female reproductive organs in the same individual is referred to as **monoecious** or _hermaphroditic._

Some animals reproduce asexually. **Asexual reproduction** is the production of new individuals by any mechanism that does not involve gametes (budding in sponges, for example). Also, the ova of many animals, either naturally or in the laboratory, are capable of development without fertilization. This is called **parthenogenesis.**

Note: Depending on the timing of your lab and to prepare for the observation of a live frog zygote, your instructor may demonstrate the fertilization of frog eggs before you start this exercise.

26.1   Gametes _(35 min.)_

Most sperm have at least one flagellum. Sperm are specialized for motility and contribute little more than their chromosomes to the zygote.

Ova are specialized for storing nutrients, and they contain the molecules and organelles needed to fuel, direct, and maintain the early development of the embryo. Nutrients are stored as **yolk** in the cytoplasm of the ovum. Consequently, mammalian ova are larger than body cells and in some species reach a diameter of 0.2 mm. In the frog, additional yolk increases the diameter of the ovum to 2 mm, and in the chicken it reaches about 3 cm.

MATERIALS

Per student:

- compound microscope, lens paper, bottle of lens-cleaning solution (optional), lint-free cloth (optional), dropper bottle of immersion oil (optional)
- prepared slide with a whole mount of bull sperm
- glass microscope slide, a coverslip, and a dissecting needle
- one-piece plastic dropping pipet

Per student pair:

- unfertilized hen's egg
- several paper towels
- dissection pan
- Syracuse dish

- 2 camel-hair brushes
- dissection microscope

Per lab group (table or bench):

- a model of a frog ovum (optional)

Per lab section:

- live frog sperm in pond water
- live frog ova in pond water
- pond water

Per lab room:

- phase-contrast compound light microscope (optional)

PROCEDURE

A. Sperm

1. With your compound microscope, study a prepared slide of bull sperm and draw several sperm in Figure 26-1 as seen with the high-dry objective. Each sperm has three major segments: the _head, midpiece,_ and _tail._ Label the major segments of one sperm in your drawing. The tail is composed primarily of a single _flagellum._

2. _Skip this step if your compound microscope doesn't have an oil-immersion objective._ Using the oil-immersion objective, find the **acrosome** covering the _nucleus_ in the head of a sperm and _mitochondria_ in its midpiece. The acrosome contains enzymes that aid in the penetration of the egg. Considering its cell size, why does a sperm have a lot of mitochondria?

_____

_____

_____

Figure 26-1   Drawing of bull sperm
(_____ x).

3. Place a drop of frog sperm suspension on a glass microscope slide and make a wet mount. Observe the movement of sperm flagella using either the phase-contrast compound microscope (a microscope that increases the contrast of transparent specimens) or your compound microscope with the iris diaphragm partially closed to increase contrast. Describe what you see.

_____

_____

_____

B. Ova

1. *Chicken egg.*
   (a) Obtain an unfertilized chicken egg (Figure 26-2). Crack it open as you would in the kitchen and carefully spill the contents into a hollow made from paper towels placed in a dissection pan.
   (b) Only the yolk is the ovum. Look on its surface for the **blastodisc,** a small white spot just under the cell membrane. This area is free of yolk and contains the nucleus. This is where fertilization would have occurred had sperm been present in the hen's oviduct. The walls of the oviduct secrete the *albumin* (egg white).
   (c) Examine the *shell* and the two *shell membranes*. The shell membranes are fused except in the region of the air space at the blunt end of the egg. Note the two shock absorber–like *chalazae* (singular, *chalaza*), which suspend the ovum between the ends of the egg. They are made of thickened albumin and may help rotate the ovum to keep the blastodisc always on top of the yolk.

2. *Frog egg.*
   (a) Gently place a live frog egg in a Syracuse dish half-filled with pond water and examine it with a dissection microscope. Use two camel-hair brushes to transfer the egg. *Do not let the egg dry out*. The ovum is enclosed in a protective jelly membrane.

Which surface (light side or dark side) of the ovum floats up in the pond water? _____

   (b) Look at a model of a frog ovum. Ova from different species of animals vary in the amount and distribution of yolk. Frog ova have a moderate amount of yolk that is concentrated in the lower half of the ovum (Figure 26-3). This half of the ovum is called the **vegetal pole.** The nucleus is located in the upper, yolk-free half, the **animal pole.** Note that the animal pole is black. This is because it contains pigment granules. Why does a frog ovum have less yolk than a bird ovum?

_____

_____

Suggest one or more possible functions for the black pigment in the animal pole. (*Hint:* One function is the same as that for the pigment in your skin that increases when exposed to sunlight.)

_____

_____

_____

3. *Human egg.* Figure 26-8 shows a human egg as it would appear in the upper oviduct. The egg is surrounded by a membrane, the *zona pellucida*, and a capsule of follicle cells. Why do most mammalian ova have very little yolk?

_____

_____

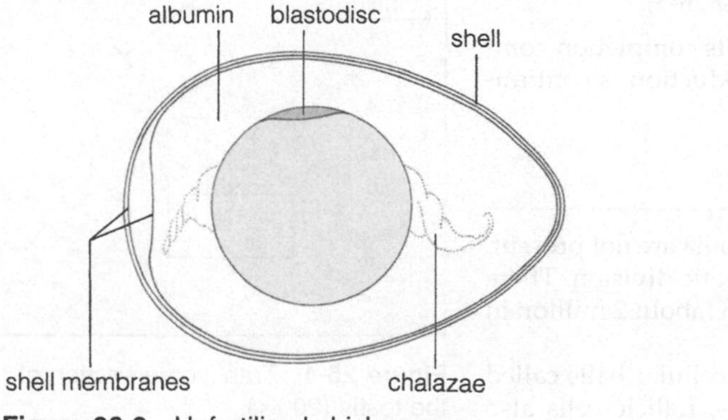

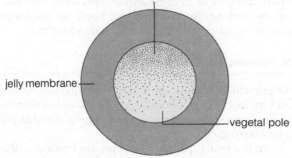

Figure 26-2 Unfertilized chicken egg.

Figure 26-3 Frog egg.

Gamete formation (**gametogenesis**) is called **oogenesis** in the female and **spermatogenesis** in the male. The general scheme and terminology of gametogenesis are summarized in Table 26-1. You studied the events of meiosis that bring about the haploid number of chromosomes present in gametes in Exercise 8. Here we examine the gamete formation in the gonads of mammals. The gonads are paired organs and are called **testes** (singular, *testis*) in males and **ovaries** in females.

TABLE 26-1 Gametogenesis

Condition of Cell	Type of Cell	
	Male	Female
Mitotically active	Spermatogonium	Oogonium
Before meiosis I	Primary spermatocyte	Primary oocyte
Before meiosis II	Secondary spermatocyte	Secondary oocyte and first polar body
After meiosis II	Spermatid	Ovum and three polar bodies
After differentiation	Sperm	—

MATERIALS

Per student:

- compound microscope
- prepared slides of a section of
    adult mammalian testis stained with iron
    hematoxylin

    adult mammalian ovary with follicles
    adult mammalian ovary with a corpus luteum

PROCEDURE

A. Mammalian Spermatogenesis

1. With your compound microscope, look at a prepared slide of the testis (Figure 26-4). Most of the interior of a testis is filled with coiled **seminiferous tubules,** which coil. Transverse and oblique sections are present in your slide. Look for glandular **interstitial cells** between the seminiferous tubules. Interstitial cells secrete the male sex hormone **testosterone.**

2. Center a cross section of a seminiferous tubule in the field of view and increase the magnification until you see only a portion of the tubule's wall. In the wall of the seminiferous tubule, identify as many stages of spermatogenesis as possible, as well as **Sertoli cells,** which function to nurture the developing sperm (Figure 26-5).

Spermatogenesis in most animal species is seasonal, its completion coinciding with mating. In humans, however, sperm production is continuous from puberty throughout a male's lifetime.

B. Mammalian Oogenesis

After birth, in humans and most other mammals, oogonia are not present, and meiosis is suspended in prophase of the first meiotic division. There is an excess supply of primary oocytes present at birth (about 2 million in a newborn girl).

In the ovary, primary oocytes are located within cellular balls called **follicles.** Besides supporting the developing oocytes, follicle cells also

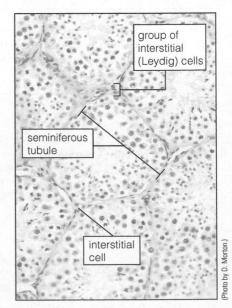

Figure 26-4 Transverse section of the testis (200×).

secrete the female sex hormones, **estrogen** and **progesterone.** A typical mammalian ovary contains a number of stages of follicular development (Figure 26-6).

A primary follicle whose thin walls are one cell thick initially surrounds each primary oocyte. Most follicles degenerate, and in humans only about 300,000 primary oocytes remain at puberty. The majority of these oocytes will also degenerate. With each turn of the female cycle, several follicles begin to mature, and one (rarely two or more) of each batch of oocytes finally bursts from the surface of the ovary and is swept into the oviduct. The release of oocytes from the ovary is called **ovulation** (Figure 26-7).

Around the time of ovulation, the first meiotic division is completed, a polar body is split off, and the secondary oocyte enters but does not complete the second meiotic division. The zona pellucida and a capsule of follicle cells (Figure 26-8) surround an ovulated secondary oocyte. A sperm must first penetrate this barrier before penetration of the egg membrane and fertilization can occur.

Follicle cells left behind in the ovary after ovulation develop collectively into a large roundish structure called the **corpus luteum** (the name means "yellow body" and refers to its color in a live ovary). The corpus luteum continues to secrete female sex hormones, especially progesterone.

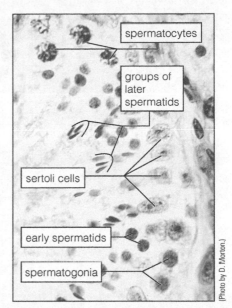

Figure 26-5 Spermatogenesis (800×).

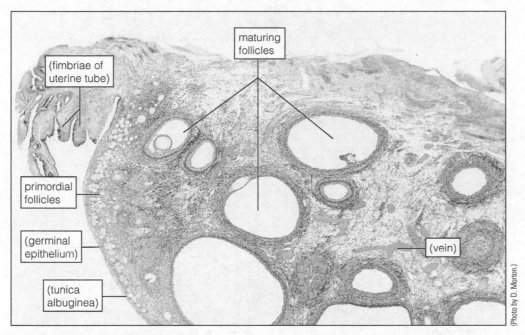

Figure 26-6 Mammalian ovary (25×).

1. With the compound microscope, examine a prepared slide of a section of a mammalian ovary. Look for primordial follicles (Figure 26-9a).

Both the follicles and their primary oocytes are small compared with maturing follicles and their primary oocytes. Groups of primordial follicles tend to occur between maturing follicles or between maturing follicles and the ovarian wall. Note that the wall of the primordial follicle is composed of a single layer of smaller cells.

2. Now look for maturing follicles.

In maturing follicles, the size of the cells in the follicle wall and of the primary oocyte itself increases. Also, the follicle cells divide, causing the wall to become first two cells thick and then multilayered. Then a space appears between the follicle cells (Figure 26-9b). This fluid-filled space increases in size until the primary oocyte and the follicle cells immediately around the primary oocyte are suspended in it (Figure 26-9c). The mass of cells is connected to the rest of the wall by a narrow stalk of follicle cells. Just prior to ovulation, the follicle reaches its maximum size and bulges from the surface of the ovary.

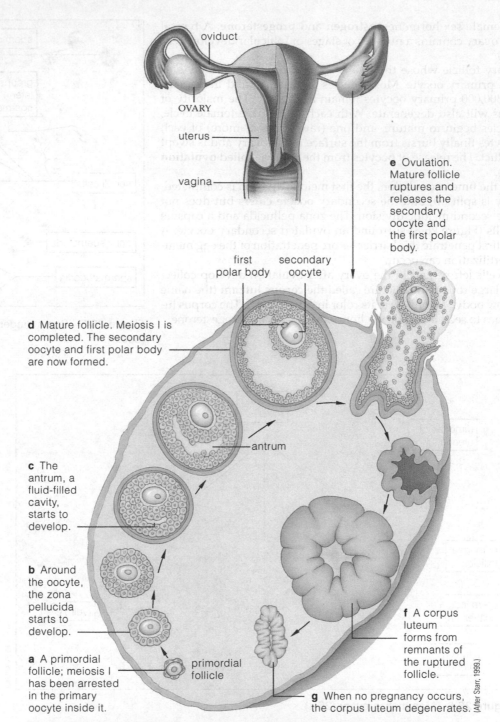

d Mature follicle. Meiosis I is completed. The secondary oocyte and first polar body are now formed.

c The antrum, a fluid-filled cavity, starts to develop.

b Around the oocyte, the zona pellucida starts to develop.

a A primordial follicle; meiosis I has been arrested in the primary oocyte inside it.

oviduct

OVARY

uterus

vagina

e Ovulation. Mature follicle ruptures and releases the secondary oocyte and the first polar body.

first polar body

secondary oocyte

antrum

primordial follicle

f A corpus luteum forms from remnants of the ruptured follicle.

g When no pregnancy occurs, the corpus luteum degenerates.

(After Starr, 1999.)

Figure 26-7 Cyclic events in a human ovary, drawn as if sliced lengthwise through its midsection. Events in the ovarian cycle proceed from the growth and maturation of follicles, through ovulation (rupturing of a mature follicle with a concurrent release of a secondary oocyte). For illustrative purposes, the formation of these structures is drawn as if they move clockwise around the ovary. In reality, their maturation occurs at the same site.

3. Replace the slide on the microscope with one of a mammalian ovary with a corpus luteum. Locate a corpus luteum (Figure 26-10).

If fertilization and implantation of the embryo in the uterus do not occur, the corpus luteum degenerates and is replaced by scar tissue. This scar is now called the *corpus albicans* (Figure 26-11).

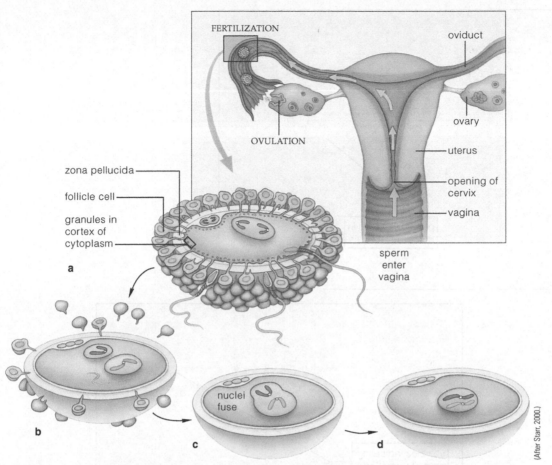

Figure 26-8 Fertilization. (**a**) Many sperm surround a secondary oocyte. Acrosomal enzymes clear a path through the zona pellucida. (**b**) When a sperm does penetrate the secondary oocyte, granules in the egg cortex release substances that make the zona pellucida impenetrable to other sperm. Penetration also stimulates meiosis II of the oocyte's nucleus. (**c**) The sperm's tail degenerates and its nucleus enlarges and fuses with the oocyte nucleus. (**d**) With fusion, fertilization is over. The zygote has formed.

26.3   Sperm Penetration and Fertilization in the Frog *(20 min.)*

Fertilization and subsequent development may be internal or external to the female's body. **External fertilization** with **external development** is common in invertebrates, fish, and amphibians. Why is external fertilization generally limited to aquatic animals?

_____

_____

_____

    Land animals have **internal fertilization,** with either external development in the shell (true of most reptiles and birds) or **internal development** in the mother's **uterus** (true of most mammals). The uterus is the organ of the female reproductive system where most mammalian embryos develop until birth.

    At the beginning of the lab, or just before, your instructor induced ovulation in a female frog by injecting pituitary gland extract into the abdominopelvic cavity. Sperm obtained from the shredded testes of a male frog were then used to fertilize these ova.

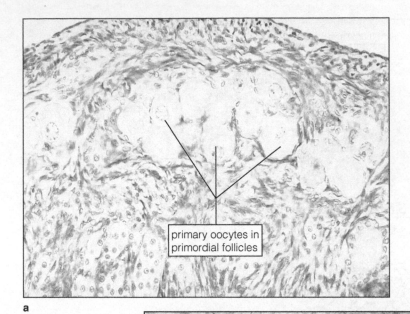

a

primary oocytes in
primordial follicles

vesicles indicating
start of formation
of fluid-filled space

primary oocyte

b

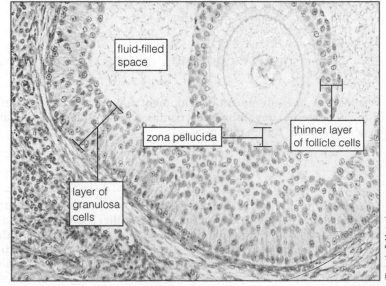

fluid-filled
space

zona pellucida

thinner layer
of follicle cells

layer of
granulosa
cells

c

(Photos by D. Morton.)

Figure 26-9 **(a)** Group of primordial
follicles (450×), **(b)** maturing follicles
(450×); and **(c)** portion of a mature
secondary follicle (400×).

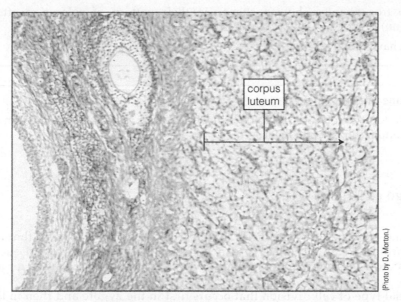

Figure 26-10 Corpus luteum (100×).

(Photo by D. Morton.)

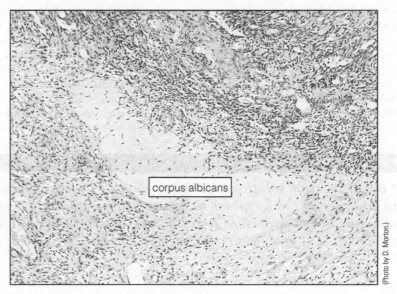

Figure 26-11 Corpus albicans (100×).

(Photo by D. Morton.)

MATERIALS

Per student pair:

- Syracuse dish
- 2 camel-hair brushes
- dissection microscope

Per lab group (table or bench):

- model of a frog zygote (optional)

Per lab section:

- live frog zygotes in pond water
- pond water
- film or videotape of sperm penetration and fertilization in the frog (optional)

PROCEDURE

1. Using two camel-hair brushes, gently transfer a live frog zygote that has not yet begun to cleave to a Syracuse dish half-filled with pond water and examine it with a dissecting microscope. *Do not let the zygote dry out.*

Which surface of the zygote (light or dark side) floats up in the pond water? _____

Compared to ova, do frog zygotes show a change in their pattern of coloration after fertilization? _____ (yes, no)

If so, describe what you have observed.

_____

_____

Is a fertilization membrane present? _____ (yes, no)

2. If you can't observe fertilization directly, study a model of a frog zygote or watch a film or videotape on sperm penetration and fertilization in the frog.

26.4　　**Cleavage** (*25 min. or more*)

Animal development has six stages (Figure 26-12). In this exercise you will pick up the story with cleavage and continue on through organ formation (organogenesis). During growth and tissue specialization, the sixth stage, organs grow in size and acquire the specialized functions necessary for an independent life.

Cleavage is a special type of cell division that occurs first in the zygote and then in the cells formed by successive cleavages, the **blastomeres.** Unlike typical cell division, there is no intervening period of cytoplasmic growth between mitotic divisions. Thus, the blastomeres become smaller and smaller. After a number of cleavages, the blastomeres form a solid cluster of cells called the **morula.** The formation of a hollow ball of cells, called the **blastula** in invertebrates and amphibians, marks the end of cleavage. Because there is no cytoplasmic growth, the size of the blastula is only slightly larger than that of the zygote.

In many organisms whose ova have little yolk (the sea star and human, for example), cleavage is *complete* and nearly *equal*, resulting in separate blastomeres that are all about the same size. In amphibians like the frog and other animals with moderate amounts of yolk, cleavage is complete but unequal. There is so much yolk in the ova of many animals (most fish, reptiles, birds, and the two mammals that lay eggs—the platypus and spiny anteater) that complete cleavage is impossible. This type of cleavage is incomplete and is called *discoidal* (disk like).

Caution

Specimens are kept in preservative solutions. Use latex gloves whenever you handle a specimen. Wash any part of your body exposed to this solution with copious amounts of water. If preservative solution is splashed into your eyes, wash them with the safety eyewash bottle for 15 minutes. If you wear contact lenses during a dissection, wear safety goggles.

MATERIALS

Per student:

- compound microscope, lens paper, bottle of lens-cleaning solution (optional), lint-free cloth (optional), dropper bottle of immersion oil (optional)
- prepared slide of a whole mount of sea star development through gastrulation

Per student pair:

- 2 Syracuse dishes
- 2 blue camel-hair brushes (optional)
- 2 red camel-hair brushes
- dissection microscope

Per lab section:

- frog embryos in pond water from eggs fertilized 1 hr before lab (optional)

Per lab room:

- preserved specimens of two-, four-, and eight-cell cleavage stages; morulae (32-cell cleavage stage); and blastulae of frog in easily accessible screw-top containers
- source of dH_2O
- pond water (optional)
- models of early frog development
- models of human development
- boxes of different sizes of latex or vinyl gloves
- box of safety goggles

PROCEDURE

A. Cleavage in the Sea Star

With your compound microscope, observe a prepared slide with whole mounts of early sea star embryos. Find a zygote; two-, four-, and eight-cell cleavage stages; a morula; and a blastula; and draw them in Figure 26-13. Be sure to adjust the fine-focus knob to see the three-dimensional aspects of these stages. The blastula is a hollow ball of flagellated blastomeres surrounding a cavity called the **blastocoel.** Label the blastomeres and blastocoel in your drawing of the sea star blastula.

What is the orientation (parallel or perpendicular) relative to each other of the first two cleavage planes?

_____

What is the orientation (parallel or perpendicular) of the third cleavage plane compared to the first cleavage plane? _____

B. Cleavage in the Frog

1. *Optional.* If your instructor fertilized frog eggs 1 hour before lab, you'll be able to watch the first two cleavage divisions during this laboratory. At room temperature about an hour after addition of the sperm to the eggs, a region of less pigmented cytoplasm, the **gray crescent,** appears opposite the site of sperm penetration (Figure 26-14). The first cleavage division occurs about 2 hours after fertilization, the second a half hour later, and the third after another 2 hours.

Gently place several live frog zygotes in a Syracuse dish half-filled with pond water and examine them with a

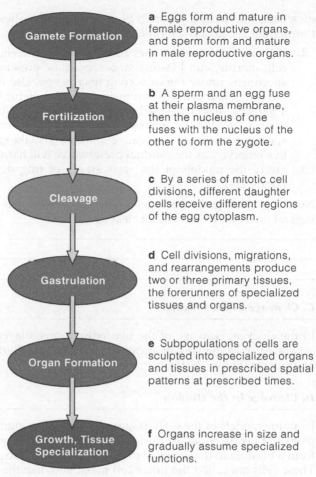

a Eggs form and mature in female reproductive organs, and sperm form and mature in male reproductive organs.

b A sperm and an egg fuse at their plasma membrane, then the nucleus of one fuses with the nucleus of the other to form the zygote.

c By a series of mitotic cell divisions, different daughter cells receive different regions of the egg cytoplasm.

d Cell divisions, migrations, and rearrangements produce two or three primary tissues, the forerunners of specialized tissues and organs.

e Subpopulations of cells are sculpted into specialized organs and tissues in prescribed spatial patterns at prescribed times.

f Organs increase in size and gradually assume specialized functions.

(After Starr and Taggart, 2001.)

Figure 26-12 Generalized scheme and control of animal development.

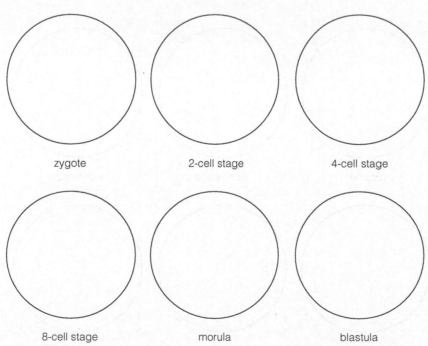

zygote 2-cell stage 4-cell stage

8-cell stage morula blastula

Figure 26-13 Drawings of early sea star developmental stages (_____×).
Labels: blastomeres, blastocoel

dissection microscope. Use two blue camel-hair brushes to transfer the zygote. *Do not let the zygote dry out.*

2. If living embryos are unavailable, and for the eight-cell, morula, and blastula stages, examine preserved specimens under the dissecting microscope. Use two red camel-hair brushes to transfer each stage in turn to a Syracuse dish half-filled with distilled water. When done, return the specimen to its container. The embryo specimens are in a preservative solution, so make sure that you *do not* use the *red* brushes to manipulate the live embryos, as the residual preservative will harm them.

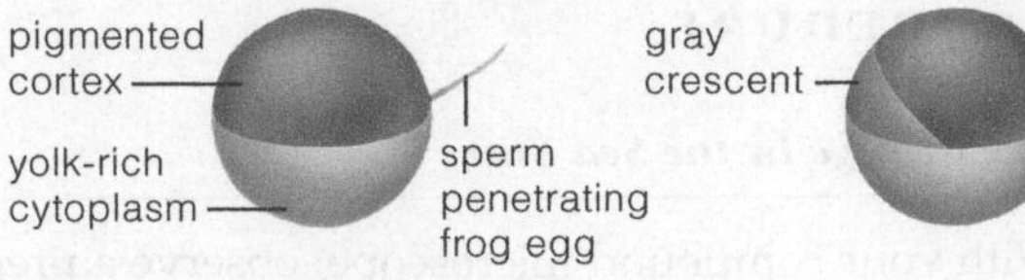

Figure 26-14 Formation of the gray crescent. (After Starr and Taggart, 2001.)

3. Study the models of the early stages of frog development. Draw the stages of frog development in Figure 26-15.

Note that cleavage in the vegetal pole lags behind that of the animal pole. Why? (*Hint:* What is present in the vegetal pole that would hinder cleavage?)

_____

_____

_____

C. Cleavage in the Chicken

In the chicken, cleavage of the blastodisc forms a layer of cells called the *blastoderm*, which in time becomes separated from the yolk by a cavity, the blastocoel. Further development of the embryo will occur only in the blastoderm (Figure 26-16).

D. Cleavage in the Human

Examine models of the early stages of human development. Because the ovum has little yolk, cleavage is complete and nearly equal. At the end of cleavage, a hollow ball of cells is formed. However, this **blastocyst** differs from a blastula in that a group of cells aggregate at one pole of the inner surface of the blastocoel (Figure 26-17). These cells are called the **inner cell mass,** and, like the blastoderm of the chicken, further development of the embryo will proceed only here. The remaining cells that surround the blastocoel are called the **trophoblast.** This stage of embryonic development coincides with implantation in the uterus.

Current evolutionary thought is that modern reptiles, birds, and mammals evolved from earlier reptilian ancestors that had external development and an ovum rich in yolk. Internal development in mammals linked

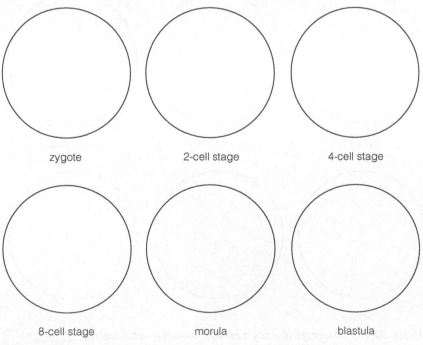

zygote   2-cell stage   4-cell stage

8-cell stage   morula   blastula

Figure 26-15 Drawings of early frog developmental stages (_____×).

nourishment of the embryo directly to the mother and removed the need for large yolky eggs. Excessive yolk would have been a biological liability; therefore, by natural selection, mammals have developed ova with little yolk. However, gastrulation and subsequent developmental events in reptiles, birds, and mammals are remarkably similar. This theme will be expanded in the discussion of extraembryonic membranes later in this exercise.

26.5   Gastrulation *(25 min. or more)*

Gastrulation is the first time embryonic cells undergo typical cell division with periods of growth between divisions. These cells start the process of specialization and undertake migrations from one embryonic location to another. Groups of cells undergo programmed cell death (*apoptosis*). These processes produce, in most animals, a body comprised of three **primary germ layers** or tissues and a body with a front and back end, which can be divided in half along its longitudinal axis. In animals having at least the organ level of organization, the primary germ layers are called **ectoderm, mesoderm**, and **endoderm** and give rise to the four adult basic tissue types (Table 26-2).

TABLE 26-2   Tissue Derivatives of the Primary Germ Layers

Layer	Derivatives
Ectoderm	Nervous tissues, epidermis (keratinized stratified squamous epithelium) and the structures it forms
Mesoderm	Muscle tissues, connective tissues, and epithelia of the urinary and reproductive systems
Endoderm	Epithelial lining of most of the digestive tract and respiratory tract, and associated glands (for example, liver)

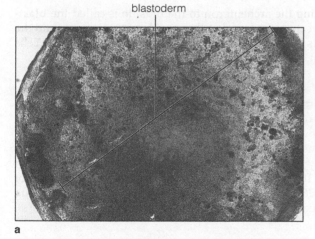

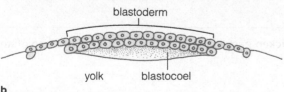

Figure 26-16 Late cleavage in the chicken. **(a)** Surface view (42×). (Photo courtesy Biodisc, Inc.) **(b)** Transverse section along diameter indicated in **(a)**.

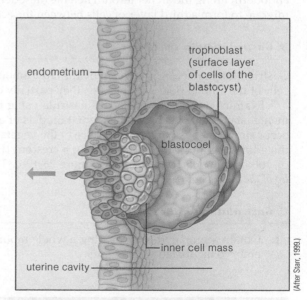

(After Starr, 1999.)

Figure 26-17 Section through a human blastocyst as it attaches to the inner lining (endometrium) of the uterus and starts implantation by burrowing into it (day 6–7 after fertilization).

MATERIALS

Per student:

- compound microscope
- prepared slide of a whole mount of sea star development through gastrulation
- prepared slides with whole mounts of chicken embryos at 18 and 24 hr incubation

Per student pair:

- 2 Syracuse dishes
- 2 blue camel-hair brushes (optional)
- 2 red camel-hair brushes
- dissection microscope
- fertile hen's eggs incubated for 18 and 24 hr (optional)

Per student group (bench or table):

- models of human gastrulation

Per lab section:

- frog embryos (early gastrulae) in pond water from eggs fertilized 21 hr before lab (optional)
- frog embryos (late gastrulae) in pond water from eggs fertilized 36 hr before lab (optional)
- pond water

Per lab room:

- preserved specimens of early and late gastrulae of frog in easily accessible screw-top containers
- a source of dH$_2$O
- pond water (optional)
- models of early and late gastrulae of frog
- boxes of different sizes of latex or vinyl gloves
- box of safety goggles

PROCEDURE

A. Gastrulation in the Sea Star

The sea star has complete and nearly equal cleavage, resulting in a blastula with one layer of flagellated blastomeres that are slightly elongated at the *vegetal pole* (Figure 26-18a).

Re-examine the prepared slide bearing whole mounts of the early stages of sea star development. Find an early gastrula (Figure 26-18b). As gastrulation starts, the vegetal pole flattens and folds in like a pocket to create a new cavity, the **archenteron** ("ancient gut").

Find a late gastrula (Figure 26-18c). The hole connecting the archenteron to the outside is called the **blastopore.** Gastrulation initially forms two layers of cells—the ectoderm covering the outside of the gastrula and the endoderm lining the archenteron. Then the mesoderm buds off from the inner tip of the archenteron, and its cells migrate to form a third layer of cells between the ectoderm and endoderm.

B. Gastrulation in the Frog

Gastrulation in the frog is affected by the large amount of yolk in the vegetal hemisphere. Because the pigmented cells of the animal pole divide faster, they partially overgrow the yolk-laden cells of the vegetal pole.

Examine a living or preserved gastrula using the same protocol previously described for earlier developmental stages, as well as supplemental models of an early and a late gastrula. Note that gastrulation does not occur simultaneously over the surface of the vegetal pole. It starts at a point just under what will be the anus of the adult frog and continues, forming a crescent (Figure 26-19a) that will close to form a circle around a plug of yolk-laden cells, the *yolk plug* (Figure 26-19b). The initial point of the folding-in is referred to as the *dorsal lip of the blastopore.*

C. Gastrulation in the Chicken

1. Obtain a prepared slide bearing a whole mount of a chicken gastrula (18 hours incubation) and one of an embryo incubated for 24 hours.

Caution

Do not use the high-power objectives when examining whole mounts of thick material. Its use will break the coverslip.

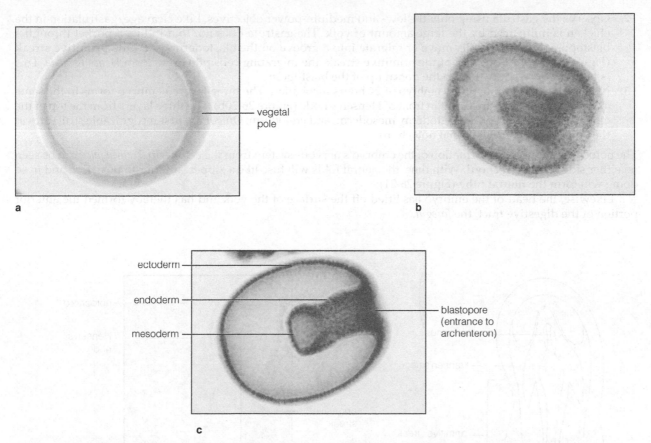

Figure 26-18 Gastrulation in the sea star (320×). (**a**) Late blastula, (**b**) early gastrula, and (**c**) late gastrula. (Photos by D. Morton.)

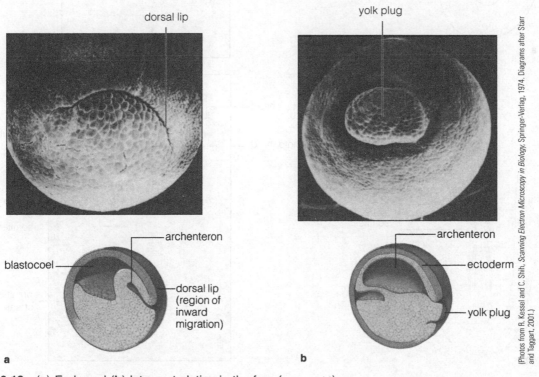

(Photos from R. Kessel and C. Shih, *Scanning Electron Microscopy in Biology*, Springer-Verlag, 1974. Diagrams after Starr and Taggart, 2001.)

Figure 26-19 (**a**) Early and (**b**) late gastrulation in the frog (_____×).

2. Observe the gastrula using only the low- and medium-power objectives. Like cleavage, gastrulation in the chicken is influenced by the large amount of yolk. The gastrula does not fold in like a pocket through a blastopore, but rather cells move or migrate into a groove on the blastoderm called the **primitive streak** (Figure 26-20a). At one end of the primitive streak, the migrating cells pile up to form *Hensen's node*. This is thought to be equivalent to the dorsal lip of the blastopore.

3. Now examine the slide of the embryo of 24 hours incubation. The three-layered embryo forms in the same axis as the primitive streak but in front of Hensen's node (Figure 26-20b). The three layers from the top of the embryo toward the yolk are ectoderm, mesoderm, and endoderm. One of the first recognizable structures in the embryo is the mesodermal **notochord.**

The notochord induces the formation of the embryo's nervous system from the ectoderm. *Neural folds* can be seen on either side of the notochord. With time, the neural folds will fuse like a zipper, anterior to posterior, and in so doing will form the **neural tube** (Figure 26-21).

Likewise, the head of the embryo has lifted off the surface of the yolk and has thereby formed the anterior portion of the digestive tract, the *foregut.*

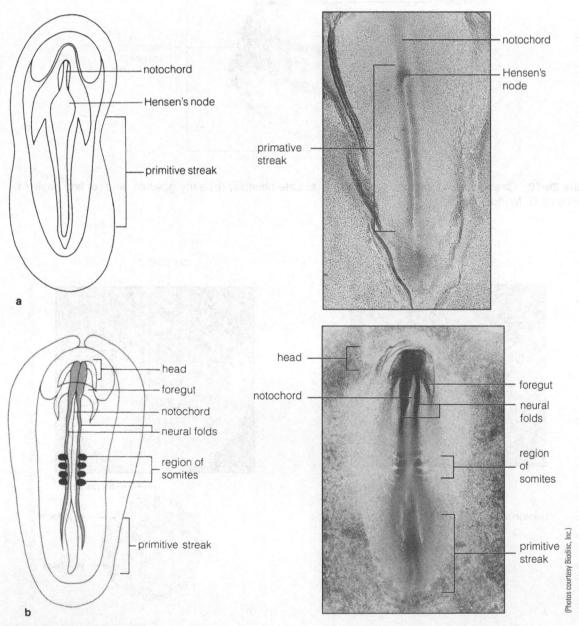

Figure 26-20 Gastrulation in the chicken. (**a**) gastrulation (18 hr of incubation) (28×); (**b**) early embryo (24 hours of incubation) (16×). The embryos have been removed from the surface of the yolk.

4. Note the two rows of **somites** on either side of the notochord. These are segmental condensations of mesoderm that will later develop into the skeleton, the skeletal muscles of the trunk, and the dermis of the skin. Count the number of pairs of somites present in your embryo. Record this number:

_____

5. *Optional.* If your instructor has fertile hen's eggs incubated for 18 and 24 hours, examine them under the dissection microscope.

D. Gastrulation in the Human

Examine the models of human gastrulation. Human gastrulation follows much the same scenario as that of the chicken. However, before gastrulation occurs, a cavity forms between the inner cell mass and the trophoblast. This is the **amniotic cavity** of the *amnion* (Figure 26-22). Also, cells from the inner cell mass grow downward, along the inner surface of the trophoblast, fuse, and form the **yolk sac cavity.** Between the amniotic and yolk sac cavities is the embryonic disk. The **embryonic disk** in mammals is the equivalent of the blastoderm in the chicken. Thus, gastrulation commences in the embryonic disk with the formation of a primitive streak.

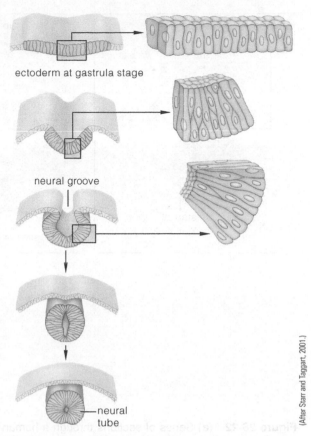

ectoderm at gastrula stage

neural groove

neural tube

(After Starr and Taggart, 2001.)

Figure 26-21 Formation of neural tube.

26.6 Extraembryonic Membranes
(20 min. or more)

The **amnion, yolk sac, chorion,** and **allantois** comprise the four **extraembryonic membranes** of birds and mammals. **Implantation** of the embryo in the uterus occurs at the same time as the formation of the amnion and yolk sac, about 8 days after fertilization of the ovum in the upper oviduct. The outer layer of the trophoblast erodes away the maternal tissues so that the blastocyst can sink into the wall of the uterus. The inner layer of the trophoblast forms the chorion. By this time, the ectoderm of the amnion and chorion as well as the endoderm of the yolk sac are coated with mesoderm derived from the inner cell mass.

MATERIALS

Per student pair:

- dissection microscope (optional)
- fertile hen's eggs incubated for 5 days (optional)

Per student group (bench or table):

- preserved pig fetus and placenta with injected vessels
- models of human intrauterine development

PROCEDURE

A. The Extraembryonic Membranes of the Chicken (Optional)

Examine a hen's egg incubated for 5 days. The amnion in the chicken forms from four folds of ectoderm and mesoderm: one in front of the developing embryo, one behind it, and one on each side. These in time fuse over the developing embryo's back. The outer wall of the fused folds becomes the chorion; the inner wall becomes the amnion. The chorion forms a sac that surrounds the developing embryo and the other extraembryonic membranes. The amnion forms the fluid-filled *amniotic cavity*, in which the developing embryo is suspended (Figure 26-23).

The yolk sac is formed when endoderm, accompanied by mesoderm containing a rich network of blood vessels, spreads over the yolk. The yolk sac serves as the digestive organ for the developing embryo. The endodermal cells secrete enzymes that digest the yolk. The end products of digestion diffuse into the blood vessels and are carried into the developing embryo.

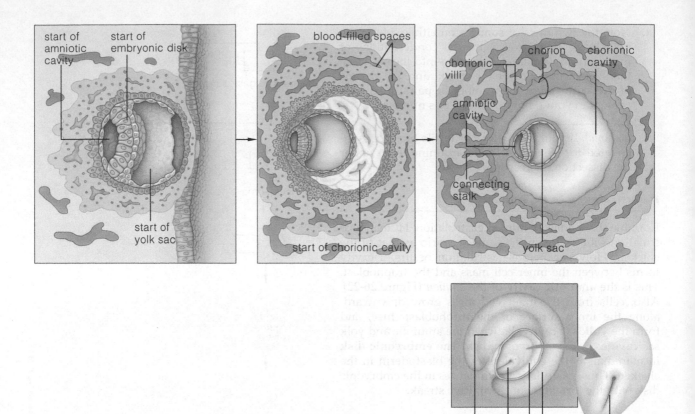

Figure 26-22 (a) Series of sections through a human embryo showing development just after implantation (days 10–14 after fertilization). (b) Formation of the primitive streak (day 15 after fertilization).

(After Starr, 1999.)

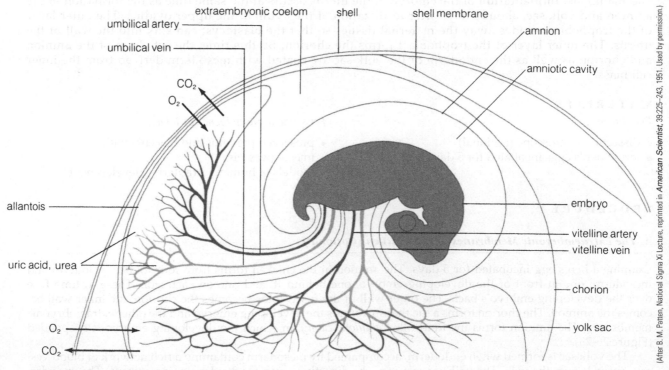

Figure 26-23 Extraembryonic membranes of the chicken.

(After B. M. Patten, National Sigma Xi Lecture, reprinted in *American Scientist*, 39:225–243, 1951. Used by permission.)

The floor of the hindgut folds out like a pocket to form the allantois, which is lined on the inside by endoderm and covered on the outside by mesoderm. The mesoderm forms a rich network of blood vessels. The cavity of the allantois functions as a dump for excretory wastes. Also the allantois and the yolk sac function as embryonic respiratory organs.

B. The Extraembryonic Membranes in Mammals

The formation of the extraembryonic membranes in mammals is quite similar to that of the chicken. One difference is the earlier formation of the amnion and allantois. Because there is little yolk in the zygote of most mammals, the yolk sac is generally smaller.

After implantation, the trophoblast and the maternal tissues of the uterus start to form the **placenta.** When complete, the placenta brings the blood of the mother and the embryo very close to each other but does not allow them to mix. Diffusion across this thin barrier allows the placenta to function as the digestive, respiratory, and excretory organs of the developing embryo.

The mesoderm of the allantois forms the **umbilical cord** and the *umbilical arteries* and *umbilical vein* contained therein. The mesoderm of the allantois also directs the formation of connecting blood vessels in the placenta. By the time the embryo's circulatory system is established, a circuit of vessels to and from the placenta is complete (Figure 26-24).

1. Look at the models of intrauterine (within the uterus) development in humans and identify the structures labeled in Figure 26-25. Once the organs and basic body shape of an embryo are established, the embryo is called a **fetus.** This transition occurs about one-third of the way through the time spent in the uterus. This time spent in the uterus is called the *gestation period.*
2. Examine the preserved pig fetus and its placenta (Figure 26-26). The umbilical arteries have been injected with different colored latex to help you identify the maternal and fetal blood vessels. Identify the fetus, umbilical cord, amnion, fetal vessels, and maternal vessels.

26.7   Organ Formation (20 min. or more)

Organ formation is the fifth stage of development. As the name suggests, it is during this period that the developing animal's organs and adult form are achieved.

MATERIALS

Per student:

- compound microscope
- prepared slides with whole mounts of chicken embryos at 48 and 72 hr of incubation

Per student group:

- fertile hen's eggs incubated for 48 and 72 hr and longer (optional)
- dissection microscope

PROCEDURE

1. Under the low-power lens of your compound microscope, examine a slide of a whole mount of a chicken embryo incubated for 48 hours (Figure 26-27).

At this time, development of the front end of the neural tube has produced the *forebrain, midbrain,* and *hindbrain.* Growth of the brain has caused it to bend over on itself. Note that the head of the embryo has turned to the right. Pairs of primitive eyes and ears are forming on both sides of the brain. Locate the right *eye* and *ear,* respectively, on the sides of the forebrain and the hindbrain.

Find the primitive *heart,* which is the first portion of the circulatory system to develop, bulging to the embryo's right. Three pairs of *aortic arches* are present as well as other blood vessels.

Has the number of somites increased compared with those of the chicken embryo incubated for 24 hours? _____ (yes or no)

Note the lower edge of the *amnion.* By this point in development, the folds that form the amnion and chorion have enclosed the front half of the embryo.

2. Now examine the slide of a chicken embryo incubated for 72 hours with the dissection microscope (Figure 26-28). Observe that by this stage the whole embryo is lying on its right side, and the rate of development is accelerating. Note a number of new features—another *pair of aortic arches,* the *wing buds, hindlimb*

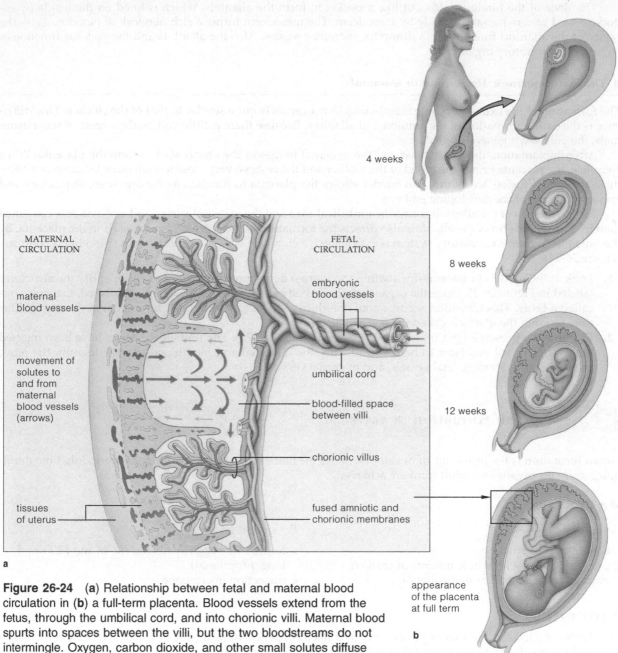

4 weeks

8 weeks

12 weeks

MATERNAL CIRCULATION

FETAL CIRCULATION

maternal blood vessels

embryonic blood vessels

movement of solutes to and from maternal blood vessels (arrows)

umbilical cord

blood-filled space between villi

chorionic villus

tissues of uterus

fused amniotic and chorionic membranes

a

appearance of the placenta at full term

b

Figure 26-24 (a) Relationship between fetal and maternal blood circulation in (b) a full-term placenta. Blood vessels extend from the fetus, through the umbilical cord, and into chorionic villi. Maternal blood spurts into spaces between the villi, but the two bloodstreams do not intermingle. Oxygen, carbon dioxide, and other small solutes diffuse across the placental membrane surface. (After Starr, 2000.)

buds, and a *tail bud*. Also, the heart has established contact with the *vitelline arteries* and *vein*, which connect the embryo with its source of nourishment, the yolk in the yolk sac.

3. *Optional.* If your instructor has fertile hen's eggs incubated for 48 and 72 hours, examine them under the dissection microscope.

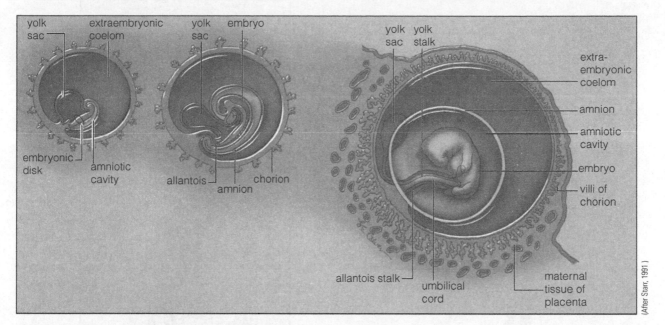

Figure 26-25   Human intrauterine development.

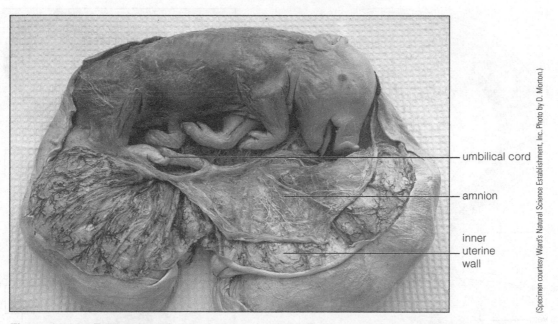

Figure 26-26   Extraembryonic membranes of the pig. Fetal vessels have been injected with yellow latex. The maternal vessels have been injected with red latex for arteries and blue latex for veins.

Animal Development

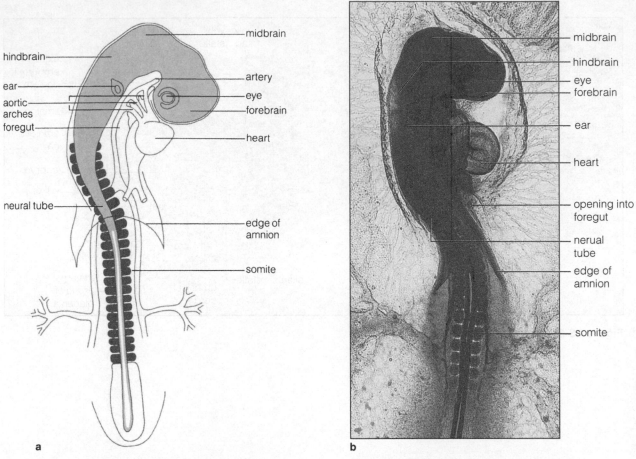

Figure 26-27   Development of the chicken after 48 hours of incubation (15×).

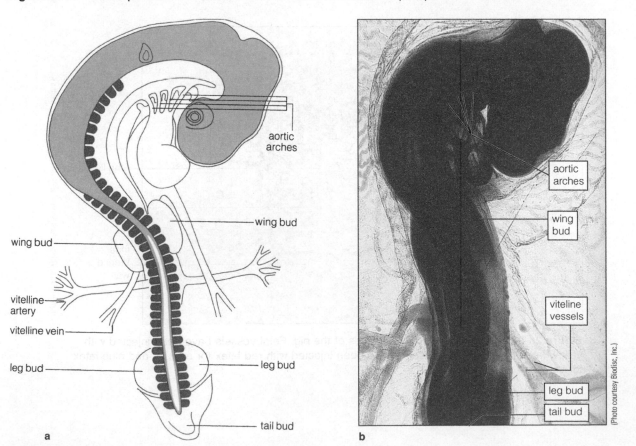

Figure 26-28   Development of the chicken after 72 hours of incubation (12×).

_____ **1.** Most animals reproduce by
(a) asexual means
(b) sexual means
(c) both of these means
(d) none of these means

_____ **2.** Most animals
(a) are monoecious
(b) are dioecious
(c) have two sexes
(d) are both b and c

_____ **3.** Sperm
(a) are male gametes
(b) are female gametes
(c) contain yolk
(d) are produced in the ovaries

_____ **4.** Ova are
(a) male gametes
(b) female gametes
(c) specialized for motility
(d) both a and c

_____ **5.** The formation of gametes in the gonads is called
(a) gametogenesis
(b) spermatogenesis
(c) oogenesis
(d) none of the above

_____ **6.** One primary oocyte will form
(a) one ovum
(b) four ova
(c) up to three polar bodies
(d) both a and c

_____ **7.** Which structure is found in a section of the testis?
(a) secondary spermatocytes
(b) sperm
(c) Sertoli cells
(d) all of the above

_____ **8.** Oocytes are found in an ovary in
(a) seminiferous tubules
(b) follicles
(c) corpora lutea
(d) none of the above

_____ **9.** Most mammals have
(a) internal fertilization
(b) external fertilization
(c) internal development
(d) both a and c

_____ **10.** Which process results in a zygote?
(a) meiosis
(b) mitosis
(c) fertilization
(d) none of the above

_____ 11. The number of circuits in the circulatory systems of humans is
  (a) one
  (b) two
  (c) three
  (d) four

_____ 12. Blood contains
  (a) dissolved gases
  (b) dissolved nutrients
  (c) dissolved hormones
  (d) all of the above

_____ 13. Red blood cells are
  (a) erythrocytes
  (b) leukocytes
  (c) platelets
  (d) all of the above

_____ 14. The most common leukocyte in the blood is
  (a) a lymphocyte
  (b) an eosinophil
  (c) a basophil
  (d) a neutrophil

_____ 15. The cellular fragments in the blood that function in blood clotting are
  (a) erythrocytes
  (b) leukocytes
  (c) platelets
  (d) none of the above

_____ 16. Blood vessels that return blood from capillaries back to the heart are
  (a) arteries
  (b) veins
  (c) portal veins
  (d) arterioles

_____ 17. Blood vessels that connect networks of capillary beds are
  (a) arteries
  (b) veins
  (c) portal veins
  (d) both b and c

_____ 18. From which chamber of the heart does the right ventricle receive blood?
  (a) right atrium
  (b) left atrium
  (c) left ventricle
  (d) none of the above

_____ 19. How many chambers does the frog heart have?
  (a) one
  (b) two
  (c) three
  (d) four

_____ 20. The primary pacemaker of the heart is the
  (a) bicuspid valve
  (b) tricuspid valve
  (c) aorta
  (d) sinoatrial node

EXERCISE 26

Animal Development

POST-LAB QUESTIONS

26.1 Gametes

1. Define and characterize the following:

   a. gametes

   b. gonads

   c. gametogenesis

2. Describe the similarities and differences between sperm and ova.

3. Compare chicken, frog, and mammalian ova as to size and the amount of yolk present.

26.2 Gametogenesis

4. Why do you think four sperm cells are produced as a result of gametogenesis, but only one ovum?

5. Why is meiosis a necessary part of gametogenesis?

6. Are oogonia present in adult human females? _____(yes or no)
   Explain your answer.

26.3 Sperm Penetration and Fertilization in the Frog

7. Read about the gray crescent in your textbook. How does its location relate to the site of sperm penetration?

Food for Thought

8. What substances are contained in the acrosome of a sperm? What is the function of these substances during sperm penetration?

9. As various newspaper articles, books, and movies suggest, the cloning of human beings—producing new individuals from activated somatic cells, perhaps followed by uterine implant—is now a distinct possibility. Can you suggest any biological advantages or disadvantages to having the earth populated with clones of a few of the best examples of our species?

10. Search the World Wide Web for sites about identical twins. List two sites and briefly summarize their contents.

    http://

    http://

26.4 Cleavage

11. Define *cleavage* and describe how it differs from typical cell division.

12. Identify the following sea star developmental stages (185×). (Photos by D. Morton.)

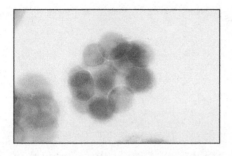

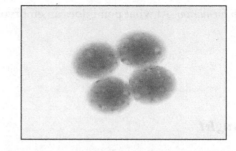

a. _____

b. _____

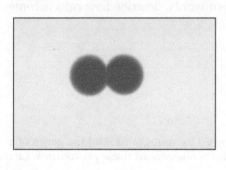

c. _____

d. _____

13. How do the amount and distribution of yolk in animal zygotes affect cleavage?

26.5 Gastrulation

14. In what ways is gastrulation in humans more like that of the chicken than that of the sea star?

15. List the primary germ layers formed by gastrulation. What tissues will they form in the adult?

26.6 Extraembryonic Membranes

16. List the extraembryonic membranes of a chicken embryo and give their functions.

17. Briefly describe the formation of the placenta in mammals. What are its functions?

26.7 Organ Formation

18. Define *organ formation*. At what point does an embryo become a fetus?

Food for Thought

19. Most body cells have a full set of chromosomes and identical genes. Yet there are many different kinds of body cells (for example, cells of the epidermis, blood cells, skeletal muscle fibers, and neurons). Look up the definition of *cell differentiation* in your textbook and, in your own words, describe how cells become specialized.

20. These days, the progress of intrauterine development is usually followed by sonograms, and the genes of a fetus may be checked by amniocentesis. Search the World Wide Web for sites about these procedures. List one site for each and briefly summarize its contents.

http://

http://

The Natural Arsenal: An Experimental Study of the Relationships Between Plants and Animals

OBJECTIVES

After completing this exercise, you should be able to

1. define *herbivore, plant secondary compound, allelochemical, bioactive, cytotoxic, bioassay;*

2. explain the ecological role of many plant toxins and their potential effects on herbivores, pathogens, or competitors;

3. describe the nature of the chemical "arms race" between plants and herbivores;

4. state scientific hypotheses and predictions, design a simple experiment with proper controls and replication, evaluate the results, and reach conclusions based on the evidence provided by the experiment.

INTRODUCTION

"NO! Don't touch that leaf!" . . . "Don't eat that berry!" Sound like something your parents said to you and your siblings when you were young? Why shouldn't we eat the berries of American yew bushes, or any part of the foxglove plant? Why is one nightshade species called "deadly"? Why have so many of us learned to identify poison ivy ("leaves of three, let it be")? Because they're "poisonous," right?

So why do plants produce chemicals that are harmful to people or other animals? Recall that plants occupy the producer trophic level. They use solar energy to build organic molecules from water and carbon dioxide in the process of photosynthesis, and also incorporate into their tissues mineral nutrients absorbed via their roots. Plants therefore compete with each other for crucial resources and occupy the base of most food webs, with animals and other **herbivore** consumers (herbivores are organisms that eat plants) eating them in order to acquire the nutrients and energy stored in the plant body.

Plants would seem to be at a distinct disadvantage in the predation and competition relationships we just described. The vast majority are rooted in the earth; certainly plants aren't able to evade predators by moving away. Nor can they physically move to more advantageous locations for gathering sunshine or minerals. Nonetheless, they are not defenseless against the vast array of herbivores and competitors. In addition to various physical defenses like thorns and tough tissues, most plants have highly effective chemical weapons.

All higher plants have **secondary compounds.** These are chemicals that do NOT fall into the major chemical groupings of carbohydrates, proteins, fats, or nucleic acids. They occur irregularly, with most secondary compounds occurring in a few kinds of plants but not in others. More importantly, most secondary substances have no known role in the metabolism of the plants in which they occur. Rather, they seem to perform a variety of functions, including that of chemical defense agent.

The secondary chemical compound signals that are active between different species are called **allelochemicals.** Caffeine, nicotine, strychnine, organic cyanide compounds, and curare are just a few of the "weapons" in the chemical arsenal that plants have evolved to defend themselves against animal consumers, plant competitors, and pathogenic (disease-causing) microbes. For example, tannins in the leaves of oaks and other plants combine with leaf proteins and digestive enzymes in an insect's gut, inhibiting protein digestion by the insect. Thus tannins considerably slow the growth of most caterpillars and other herbivores.

Of course, herbivores may counter such toxic effects through the evolutionary modification of their own biochemistry or physiology. (Toxins react with specific cellular components to kill cells or alter growth or development in harmful ways.) Herbivore species that coevolve detoxification mechanisms of secondary chemicals may be able to specialize on plant hosts that are poisonous to most other species.

One familiar example is the Monarch butterfly. The adult female lays eggs on milkweed (*Asclepias*) plants. The eggs hatch into an immature stage called a *larva,* which we know as the "caterpillar." The larvae eat the milkweed leaves and eventually complete their developmental cycle. What makes the Monarch/milkweed relationship so intriguing is that milkweed plants contain potent heart poisons that deter virtually all herbivore feeding. Monarch butterflies have evolved metabolic pathways, however, to detoxify the milkweed poison. In fact, Monarch larvae store the milkweed toxin in their tissues and thereby protect themselves from predators such as birds!

Not all allelochemicals act as toxins that reduce herbivore survival. Some alter the development of insects by interfering with the production of crucial hormones, while others act as repellants that prevent feeding. Still others have antibiotic effects that inhibit infections by pathogens.

In this exercise, you will investigate the harmful effects of some plant compounds on animals and other plants, focusing especially on those bioactive compounds that may be **cytotoxic,** or poisonous to specific cell types. You'll work in groups to design and conduct bioassays to identify potentially **bioactive** compounds, that is, molecules that have a targeted biological effect on survival, growth, or reproduction. A **bioassay** is a method of screening for potentially active substances by exposing a test population of living organisms to the substance in a controlled environment. A bioassay provides an evaluation by measuring the quantity of a substance that results in a defined effective dose.

Bioassays are used for a variety of purposes. They enable researchers to find plant chemicals that show biological activity on selected research organisms. For example, botanists and biochemists use bioassays to screen unstudied plants for potentially useful, naturally occurring drugs. The National Institutes of Health, the National Cancer Institute, and several major U.S. pharmaceutical companies have launched intensive bioassay "chemical prospecting" efforts in tropical rainforests to find new sources of drugs in the fights against cancer, AIDS, heart disease, and many other health problems.

Bioassays are also used by the pharmaceutical industry in other ways. Some pharmaceuticals are cytotoxic agents; chemotherapy agents, for example, are toxic to cancer cells. Certain side effects of chemotherapy are due to the chemical's cytotoxic effect on other body cells as well. Cytotoxic effects are due to interactions with specific biochemical pathways. Therefore, these chemicals are toxic to cells that utilize the specific pathway, but have little to no effect on other cells that don't use the pathway.

While it's unlikely that you will discover an unidentified cancer-fighting chemical or natural herbicide in the plants in your lab, you might very well find cytotoxic agents. It's important to keep in mind that the reason those chemicals exist in the first place is the result of coevolution between plants and their herbivores, competitors, and pathogens. The more we know about what organisms make these chemicals, why they are made, how they are made, and the consequences of the secondary compounds to the parties involved, the more we will know about the evolutionary process and about potential practical uses of such chemicals for our own benefit.

You will use two test organisms in your bioassays. The first, *Artemia,* or brine shrimp, is a small crustacean inhabitant of salty lakes like the Great Salt Lake. *Artemia* are filter feeders that sweep algae and other tiny food particles into their mouths as they swim. They therefore occupy a primary consumer position in aquatic ecosystems. One bioassay will look for effects on *Artemia* cells as a general indication of the presence of cytotoxic compounds in a plant extract.

You may have used the second bioassay organism, Wisconsin Fast Plants™ (*Brassica rapa*), in a previous bioassay (see Exercise 1). Fast Plants™ are small, very fast-growing plants of the cabbage family. A bioassay with Fast Plants™ serves to identify potential herbicides (plant-killing compounds) or plant-growth stimulants by looking at the effect of a plant extract on seed germination and plant growth over 1 day.

27.1   Experimental Design

Work in groups of four to explore the effects of plant secondary compounds on cell survival of *Artemia,* and/or the effects on seed germination and growth of Fast Plants™. Your instructor will direct you to the plant materials that are available for testing.

Develop an hypothesis that you will test. Your hypothesis might relate to differences between extracts of different plant species with respect to effects on *Artemia* cell survival. Or you might wish to test the effect that a single plant extract has on *Artemia* cell survival versus its effect on Fast Plant™ germination and growth.

Remember, there is no *single* correct hypothesis.
The hypothesis we will test is

_____

_____

_____

Design an experiment using the materials available in the lab and the following bioassay procedures to test your hypothesis.
Our experimental design is

_____

_____

_____

_____

What will be your control treatment(s)? What is the purpose of a control treatment?

_____

_____

How will you provide replication to improve the reliability of your results?

_____

_____

Write a prediction for your experiment.

_____

_____

List the possible general outcomes of your experiment. What does each tell you about your prediction and hypothesis?

_____

_____

_____

_____

Develop a work plan to carry out your experiment so that each group member has specific tasks to accomplish and contribute to the overall group effort. For example, if your experiment will test the differences between extracts of two plant species with respect to their ability to harm *Artemia*, two group members might prepare the bioassay for one species, and the other two might run the bioassay for the second plant species. You also might divide up the work of collecting data for the duration of the bioassay.

27.2 Bioassay Procedures

A. Preparation of Plant Extract *(About 20 min.)*

MATERIALS

Per student group:

- 25-mL graduated cylinder
- two 20-mL beakers
- 50-mL beaker
- small mortar and pestle
- spatula
- plant material of your choice
- glass stirring rod

Per lab room:

- several electronic balances calibrated to weigh in 0.1-g increments
- weighing papers or boats
- liquid nitrogen (optional)

PROCEDURE

1. Grind about 1 g of the plant tissue to a fine paste with the mortar and pestle. (If liquid nitrogen is available, carefully pour a small amount over the plant material to instantly freeze it.)

Caution

Liquid nitrogen is extremely cold and will cause cryogenic burns. DO NOT touch the liquid or allow any to splash onto your skin.

If using leaf tissue, try not to include large veins; if using root tissue, wash off all soil very well. Grind the tissue *for at least 5 minutes*. The goal is to rupture virtually all the plant cells to mix their contents in the bioassay liquid. When you think you've ground the tissue adequately, continue grinding for two more minutes.

2. Use the balance to measure out two 0.1-g samples of the paste onto two pieces of weighing paper or two weighing boats.

3. Measure 10 mL of prepared brine into one labeled 20-mL beaker, and 10 mL of distilled water into a second labeled 20-mL beaker. Mix a 0.1-g sample of plant paste into each beaker, being careful to scrape off ALL plant material, and stir to create uniform suspensions. This extract has a concentration of 0.1-g plant material/10 mL, or 10 mg/mL. (Recall that 1000 mg = 1 g.) You will use the brine mixture with brine shrimp bioassays, and the dH_2O mixture for Fast Plant™ bioassays.

B. Artemia Bioassay (About 1 hr)

MATERIALS

Per student group:

- glass microscope slides
- coverslips
- fine-pointed water-resistant marker (Sharpie® or similar) or wax pencil
- wide-mouth plastic graduated transfer pipets
- razor blade or scalpel
- trypan blue stain in dropping bottle
- acetic acid in dropping bottle

- dissecting needle
- compound microscope
- plant extract(s) prepared in brine in part A

Per lab room:

- live *Artemia* in aquarium or beaker
- plant extracts
- paper towels

PROCEDURE

1. Label one microscope slide "Negative control" and a second microscope slide "Positive control."
2. Obtain a number of brine shrimp and brine in the 50-mL beaker.
3. Place a brine shrimp onto each slide, transferring only a little brine. Blot excess brine with a paper towel. The brine shrimp should be surrounded in a small brine droplet.
4. Swirl the trypan blue stain bottle to suspend the stain mixture evenly. Place a couple drops of stain atop the brine shrimp on each slide. The brine shrimp should not be able to swim, but should be covered by the liquid. Blot excess stain with a paper towel.
5. Place the negative control slide on the microscope stage and focus on the brine shrimp at low magnification. Observe the structure of the animal, locating the point where the "tail" emerges behind the main body. (See Figure 27-1.)
6. Using a sharp scalpel, cut about $1/2$ to $2/3$ of the tail off the brine shrimp.
7. Put a coverslip over the severed tail sections. Observe the cut ends of the tail. Amoebocytes, the small roundish cells that ooze out of the tail cavity, are excellent bioassay subjects. They respond to cytotoxic agents in much the same way that human cells do. Healthy cells exclude trypan blue stain and appear colorless, but damaged and dead cells will take up the stain and appear blue (Figure 27-2). Do not confuse clumps of dye crystals with the very round cells.
8. Find a field of view where the amoebocytes aren't being moved by liquid movements. Count the number of cells that are colorless in appearance, and the number that are blue for a total of about 25 cells total. Calculate the percentage of dead cells. This number represents the "background" rate for dead cells. Record your data in Table 27-1.
9. Repeat steps 4–5 for the positive control brine shrimp.
10. Add a drop of acetic acid to the severed tail section for the positive control treatment and mix with a dissecting needle. (Acetic acid is a potent cytotoxin.) Add a coverslip and wait a couple minutes before observing with the microscope.
11. Count unstained and stained cells (about 25 total), and calculate the percentage of dead cells. This number illustrates the effects of a strong cytotoxin on *Artemia* amoebocytes. If you see no stained cells, wait up to 30 minutes for a cytotoxic effect to become visible. Record your data in Table 27-1.
12. Test your plant extract(s) similarly by cutting off a brine shrimp tail in a drop of trypan blue stain on a clean slide, then adding 1–2 drops of plant extract. Add a coverslip, wait a couple minutes, then count stained (dead) and unstained (living) cells (about 25 total). Record your data in Table 27-1.

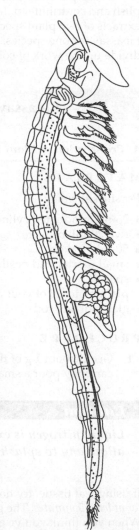

Figure 27-1 *Artemia,* or brine shrimp.

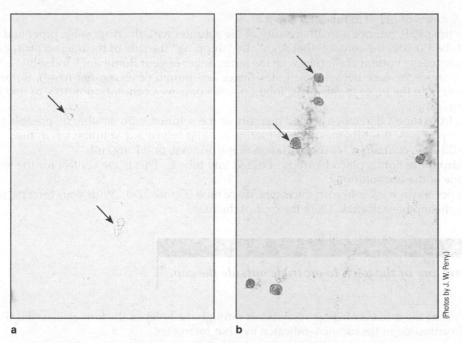

a   b

Figure 27-2 (a) Healthy (unstained) cells, (b) dead (stained) cells (500×).

(Photos by J. W. Perry.)

TABLE 27-1	Effect of Plant Extracts on *Artemia* Amoebocytes				
Treatment	Droplet Order on Slide	Number Unstained (Live) Cells	Number Stained (Dead) Cells	Percentage Dead Cells
Negative control	Brine + trypan blue			
Positive control	Brine + trypan blue + acetic acid			
Plant extract 1	Brine + trypan blue + plant extract 1			
Plant extract 2	Brine + trypan blue + plant extract 2			
Plant extract 3	Brine + trypan blue + plant extract 3			

C. Fast Plant™ Bioassay *(About 30 min.)*

MATERIALS

Per student group:

- fine-pointed water-resistant marker (Sharpie® or similar)
- 35-mm film canister with lid prepunched with 4 holes
- 4 microcentrifuge tubes with caps
- 4 paper-towel wicks

- disposable pipets and bulb
- forceps
- 8 Wisconsin Fast Plants™ seeds
- 15-cm plastic ruler
- dH₂O in dropping bottle
- plant extract(s) prepared in dH₂O in Part A

PROCEDURE

1. Using a fine-pointed, water-resistant marker, label the caps and/or sides of the four microcentrifuge tubes 10.0, 1.0, 0.1, and C (for Control).
2. Place 10 drops of the previously prepared plant extract into the tube labeled 10.0. This tube now contains 10 mg/mL of the test extract. You will now make *serial dilutions* of the solution, producing concentrations of 1.0 mg/mL and 0.1 mg/mL of the test substance:

(a) Add 9 drops of dH$_2$O to tubes 1.0 and 0.1.

(b) From tube 10.0, remove a small quantity of the solution with the disposable pipet and place *one* drop into tube 1.0. Mix the contents thoroughly by "thipping" the side of the microcentrifuge tube (flicking the sides of the bottom of the tube with the index finger of your dominant hand while holding the sides of the top of the tube between the index finger and thumb of your other hand). Return any solution remaining in the pipet to tube 10.0. Tube 1.0 now contains a concentration 10% of that in tube 10.0, or 1.0 mg/mL.

(c) Now, from tube 1.0, remove a small quantity of the solution with another disposable pipet and place *one* drop in tube 0.1. Mix the contents thoroughly and return any solution left in the pipet to tube 1.0. Tube 0.1 now contains a 1% concentration of the original, or 0.1 mg/mL.

3. From the dropping bottle, place 10 drops of dH$_2$O into tube C. This is the control for the experiment, containing none of the test solution.

4. Insert a paper-towel wick into each microcentrifuge tube (Figure 27-3). With your forceps, place two RCBr seeds near the top of each wick. Close the caps of the tubes.

Caution

Do not allow any of the wick to protrude outside the cap.

---

5. Carefully insert the microcentrifuge assay tubes through the holes in the film canister lid (Figure 27-4). Set the experiment aside in the location indicated by your instructor.

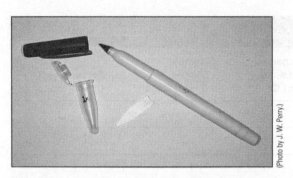

(Photo by J. W. Perry.)

Figure 27-3 One of the four bioassay tubes.

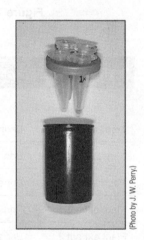

(Photo by J. W. Perry.)

Figure 27-4 Experimental apparatus for Fast Plant™ bioassay.

6. After 24 hours or more, examine your bioassay to see how many and which seeds germinated. Record your results in Table 27-2. Rather than simply recording the response as "germinated" (G) or "did not germinate" (DNG), also indicate the extent of any germination by measuring and recording the length in millimeters (mm) of the roots and shoots of seed 1 (S1) and seed 2 (S2) for each treatment. If you are testing a second plant extract, record those results in Table 27-3.

TABLE 27-2 Effect of on Germination of Fast Plant™ Seeds						
	Germination		**Length of Roots (mm)**		**Length of Shoots (mm)**	
Tube	**Seed 1**	**Seed 2**	**Seed 1**	**Seed 2**	**Seed 1**	**Seed 2**
Control (dH$_2$O)						
0.1 mg/mL extract						
1.0 mg/mL extract						
10.0 mg/mL extract						

TABLE 27-3	Effect of _____ on Germination of Fast Plant™ Seeds					
	Germination		**Length of Roots (mm)**		**Length of Shoots (mm)**	
Tube	Seed 1	Seed 2	Seed 1	Seed 2	Seed 1	Seed 2
Control (dH$_2$O)						
0.1 mg/mL extract						
1.0 mg/mL extract						
10.0 mg/mL extract						

27.3   Analysis of Results

Do your results indicate any cytotoxicity of the plant extract(s) you tested? _____
    If yes, do your results indicate cytotoxicity toward plant cells, animal cells, or both? _____
    Using your experimental results, make a conclusion regarding the effect of the plant extract(s) on your test organism.
_____
_____

    Does your conclusion support your prediction or prove it unsound? Is your hypothesis supported or falsified by your conclusion? Explain.
_____
_____
_____

_____ 1. Plant secondary compounds are chemicals that
    (a) have no known metabolic function in the plant
    (b) are required for enzyme function
    (c) are generally carbohydrates or proteins
    (d) are important in cell division

_____ 2. Allelochemicals are active in preventing
    (a) herbivore feeding
    (b) germination of plant competitors
    (c) infection by pathogens
    (d) all of the above

_____ 3. Chemicals that are poisonous to specific cell types are said to be
    (a) hazardous
    (b) cytotoxic
    (c) allelochemicals
    (d) bioactive

_____ 4. Bioassays
    (a) are performed in a test tube using killed cells
    (b) expose a population of living organisms to a test substance
    (c) are used for chemotherapy
    (d) are none of the above

_____ 5. _Artemia_, brine shrimp,
    (a) are aquatic organisms
    (b) live in salty ecosystems
    (c) are filter-feeding primary consumers
    (d) are all of the above

_____ 6. Plant defenses against herbivore grazing include
    (a) thorns
    (b) plant secondary compounds
    (c) mobility
    (d) both a and b

_____ 7. Trypan blue is
    (a) taken up by living cells
    (b) taken up by dead cells
    (c) taken up by both living and dead cells
    (d) used to identify plant cells

_____ 8. Dead _Artemia_ cells that have been treated with trypan blue appear
    (a) colorless
    (b) blue
    (c) red
    (d) indistinguishable from living treated cells

_____ 9. The dH_2O-only treatment in the Fast Plant™ bioassay provides
    (a) a control
    (b) moisture for seed germination
    (c) dilution of potential cytotoxins
    (d) none of the above

_____ 10. _Artemia_ and Fast Plant™ bioassays can be used to identify
    (a) cytotoxins
    (b) herbicides
    (c) plant growth stimulators
    (d) all of the above

EXERCISE 27

The Natural Arsenal: An Experimental Study of the Relationships Between Plants and Animals

POST-LAB QUESTIONS

Introduction

1. Why have plants evolved secondary compounds?

2. Describe how a pharmaceutical company might use a bioassay to identify potential drugs in rainforest plants.

3. Would a chemical that is cytotoxic to brine shrimp cells necessarily have the same effect on Fast Plant™ cells? Explain.

27.1 Experimental Design

4. How can replication be increased in the *Artemia* bioassay to improve reliability of results?

27.2 Artemia Bioassay

5. What was the purpose of the *Artemia* slide prepared with droplets of
   a. brine and trypan blue only?

   b. brine, trypan blue, and acetic acid?

Fast Plant™ Bioassay

6. If seed germination took place in 100% of plants at the 1.0 mg/mL concentration, and no germination was observed at the 0.1 mg/mL concentration, how would you modify a second experiment to determine the actual effective concentration?

Food for Thought

7. If you found no cytotoxic effects with your plant extract, does this prove the extract is not cytotoxic? Explain.

8. If you found no cytotoxic effects with your plant extract, does this prove that the plant species from which the extract was made has no cytotoxic allelochemicals? Explain.

9. How might you alter the two bioassay procedures to provide more information?
   a. *Artemia* bioassay:

   b. Fast Plant™ bioassay:

10. What other kinds of tests could you perform to provide more information about possible cytotoxicity of a plant extract?

Ecology: Living Organisms in Their Environment

1. define *ecology, population, community, ecosystem, producer, consumer, trophic level, primary consumer, secondary consumer, tertiary consumer, decomposer, parasite, herbivore, carnivore, omnivore, survivorship curve*;

2. construct a food web;

3. determine an organism's trophic level;

4. identify the three basic types of survivorship curves and describe the trends exhibited by each.

INTRODUCTION

Ecology. The word certainly brings to mind groups who advocate for a clean environment, who lobby the government to prevent loss of natural areas, or who protest the policies or actions of other groups. To the biologist, however, this is *not* what ecology as a science is all about. Rather, these activities are environmental concerns that have grown out of ecological studies.

The word *ecology* is derived from the Greek *oikos,* which means "house," so ecology can be thought of as the study of a "home"—be it the earth, a neighborhood, our house, or a pond. Less broadly defined, **ecology** is the study of the interactions between living organisms and their environment. Individuals, whether they are bacteria or humans, do not exist in a vacuum. Rather, they interact with one another in complex ways. As such, ecology is an extremely diverse and complex study. There are so many aspects of ecology that it's difficult to pick one, or even a few, to represent the entire science.

Let's look at the levels of organization considered in an ecological study.

A **population** of organisms is a group of individuals of the same species occupying a given area at a given time. For example, you and your colleagues in the lab at this moment are a population; you are all *Homo sapiens.* Your population consists of humans in biology class right now. The place you occupy, the lab room, is your *habitat.* Of course, you share your habitat with other organisms; unseen bacteria, fungi, and possibly insects are present in the room as well. Maybe plants decorate the lab room; if so, these too, comprise yet another population in your habitat.

A **community** consists of *all* the populations of species that occupy a given area at a given time. To use our simple example, your lab community consists of humans, bacteria, fungi, insects, and plants.

Our community exists in a physical and chemical environment. The combination of the community and its environment comprise an **ecosystem.** Your lab ecosystem consists of humans, bacteria, insects, plants, an atmosphere, lab benches, walls, floors, light, heat, humidity, and so on. As you can see, an ecosystem has both living and nonliving components.

In this exercise, we will introduce some of the principles of ecology. You will study three concepts that demonstrate how ecological studies are performed: food webs, energy flow through ecosystems, and survivorship curves.

28.1   Food Webs *(About 25 min.)*

Let's consider the community of organisms in any environment. Each population of organisms plays a different role in the community, that is, each community has organisms that are producers and organisms that are consumers.

Producers are autotrophic organisms, so most plants and a few microorganisms are producers. They utilize the energy of the sun or chemical energy to synthesize organic (carbon-containing) compounds from inorganic compounds. Photosynthetic producers form the base of nearly all food chains and food webs.

Consumers are heterotrophic organisms that feed on other organisms. In ecological studies, consumers are usually classified into feeding levels (**trophic levels**) by what they eat, as follows:

- **Primary consumers,** known as **herbivores,** are animals that eat plant material. Primary consumers thus consume producers.

- **Secondary consumers** are organisms that eat primary consumers. These organisms are called **carnivores,** animals that consume other animals.
- **Tertiary consumers** consume secondary consumers. They are also called carnivores.
- Sometimes feeding strategies cross trophic levels. **Omnivores** are organisms that eat either producers or consumers. They thus consume at various trophic levels depending on the particular food item eaten.
- **Parasites**—organisms that absorb nutrients from a living host organism for a period of time—might be considered primary, secondary, or tertiary consumers depending on whether they are living on a producer or another consumer.
- **Decomposers** include fungi, most bacteria, and some protistans that break down dead organic material from organisms of *all* trophic levels into smaller molecules. The decomposers then absorb the nutrients and ultimately play a crucial role by recycling the nutrients back into the ecosystem. Decomposers operate at virtually all trophic levels.

How would you classify yourself with respect to your feeding strategy and trophic level?
_____

_____

PROCEDURE

Below you will find a partial listing of species in a forest community and some of their source(s) of energy. In the blanks provided, list the *most specific* trophic level from the descriptions earlier.

- Human (black raspberries, hickory nuts, deer, rabbits) _____
- Black raspberry (sun) _____
- Deer (black raspberries) _____
- Bear (black raspberries, deer) _____
- Coyote (deer, rabbits, black raspberries) _____
- Caterpillar (living hickory leaves) _____
- Bacterial species 1 (living black raspberry leaves) _____
- Bacterial species 2 (dead skin cells of deer) _____
- Bacterial species 3 (dead hickory trees, dead black raspberries, dead humans, dead black bears, dead deer) _____
- Weasels (young rabbits) _____
- Mosquito (blood of living humans, deer, and bears) _____
- Hickory tree (sun) _____
- Fungal species 1 (living black raspberry stems) _____
- Fungal species 2 (dead black raspberries) _____
- Rabbit (black raspberries) _____

Now, in Table 28-1, construct a *food web* by placing the organisms in their appropriate trophic levels. Leave space around the names if more than one organism occupies a trophic level. Then connect lines from the organisms to their energy source(s) to complete the food web.

Example: Secondary consumer (carnivore)   mountain lion
              Primary consumer (herbivore)   deer
              Producer   grass

TABLE 28-1 A Simplified Food Web	
Trophic Level	**Organism(s)**
Decomposer:	
Secondary consumer (carnivore):	
Secondary consumer (parasite):	
Primary consumer (herbivore):	
Primary consumer (parasite):	
Omnivore:	
Producer:	

Suppose a new parasite species that kills black raspberry plants is introduced from another part of the world into this forest community. Speculate on the consequences of loss of black raspberries for the rabbit population.

_____

_____

How might the elimination of black raspberries affect the coyote populations? The weasel population?

_____

_____

What if pesticide spraying unintentionally kills both fungal species? What effect might these species losses have on both the raspberry and rabbit populations?

_____

_____

_____

_____

28.2   Flow of Energy Through an Ecosystem _(About 15 min.)_

Ecologists have learned through careful measurement that the amount of energy captured decreases at each succeeding trophic level relative to the initial sunlight falling on the ecosystem. Of the total amount of sunlight falling on a green plant, the plant captures—at best—1% of this energy. That is, only a small fraction of the light energy falling on the surface of a leaf is converted to the chemical energy of ATP. A primary consumer on average only captures about 10% of the energy stored in a green plant. A secondary consumer usually captures only about 10% of the energy stored in the body of the animal that _it_ eats, and so on.

What percentage of the energy contained in plant foods has been captured by the secondary consumer that feeds on an herbivore (primary consumer)? _____

A tertiary consumer eats the first (herbivore-eating) secondary consumer. Assuming a similar flow of energy, how much of the sun's _original_ energy does this secondary consumer gain? _____

Suppose an omnivore can obtain all the nutritional requirements necessary for life by eating either plant material or animal material. From an energetics standpoint, which route will allow the omnivore to capture the greater amount of the sun's energy? _____

What other routes would enable this organism to obtain equal amounts of the sun's energy? _____

_____

_____

28.3   Survivorship*

Within a population, some individuals die very young while others live into old age. To a large extent, the _pattern_ of survivorship is species dependent. Generally, **survivorship** takes on one of three patterns, which we summarize using **survivorship curves,** graphs that plot the pattern of mortality (death) in a population.

Humans in developed countries with good health-care services are characterized by a Type I curve (see Figure 28-4), in which there is high survivorship until some age, then high mortality. The insurance industry uses this information to determine risk groups. The premiums they charge are based on an individual's risk group, with risk of death (and cost of premiums) highest for the oldest purchasers of insurance.

While survivorship curves for humans are relatively easy to generate, similar information on other species is much more difficult to determine. It can be quite a trick to determine the age of an individual plant or animal, let alone track an entire population over a period of years. However, survivorship rates can be demonstrated in the laboratory using nonliving objects.

In this section, you will study the "populations" of dice and soap bubbles, using them to model real populations and to construct survivorship curves. You will subject these populations to different kinds of stress to determine their effects on survivorship curves.

_____

*Adapted from an exercise courtesy J. Shepherd, Mercer University, personal communication.

MATERIALS

Per student:

■ 15-cm ruler

Per student group (4):

■ bucket containing 50 dice
■ soap bubble solution and wand
■ stopwatch or digital watch

Per lab room:

■ 1–2 survivorship frames (see Figure 28-1)
■ overhead transparency of Figures 28-2 and 28-3
■ 1 set of red, blue, black, and green pens (Sharpie®
  or similar marker)
■ overhead projector
■ projection screen

PROCEDURE

A. Dice Survivorship *(About 30 min.)*

Work in a group of four. One person should be assigned to dump the dice, another to record data, and the other two to count.

Population 1

1. Empty the bucket of 50 dice onto the floor.
2. Assume that individuals who come up as 1's die of heart disease. All others survive.
3. Pick up all the 1's, set them aside (in the cemetery), and count the number of individuals who have survived. Record the number of *survivors* in this generation (generation 1) in the left-hand column of Table 28-2.
4. Return the survivors to the bucket.
5. Dump the survivors onto the floor again and remove the deaths (1's) that occur during this second generation.
6. Count and record the number of survivors.
7. Continue this process until all the dice have "died" from heart disease.
8. Now determine the percentage of survivors for each generation using the following formula and record in Table 28-2:

$$\text{percentage surviving} = \frac{\text{number surviving}}{50} \times 100\%$$

Population 2

1. Start again with a full bucket of 50 dice. Assume that 1's die of heart disease and 2's die of cancer. Proceed as described for population 1, recording the counts in Table 28-2, until all dice are "dead."
2. Again determine the percentage of survivors for each generation with the following formula:

$$\text{percentage surviving} = \frac{\text{number surviving}}{50} \times 100\%$$

How does adding another cause of death affect the survivorship rate?

_____

_____

_____

Do you think this activity models human survivorship well? Why or why not?

_____

_____

_____

TABLE 28-2 Survival Rates of Dice Populations

Generation	Population 1 Survivors (Heart Disease Only)		Population 2 Survivors (Cancer & Heart Disease)	
	Number	Percentage	Number	Percentage
0	50	100%	50	100%
1	_____	_____	_____	_____
2	_____	_____	_____	_____
3	_____	_____	_____	_____
4	_____	_____	_____	_____
5	_____	_____	_____	_____
6	_____	_____	_____	_____
7	_____	_____	_____	_____
8	_____	_____	_____	_____
9	_____	_____	_____	_____
10	_____	_____	_____	_____
11	_____	_____	_____	_____
12	_____	_____	_____	_____
13	_____	_____	_____	_____
14	_____	_____	_____	_____
15	_____	_____	_____	_____
16	_____	_____	_____	_____
17	_____	_____	_____	_____
18	_____	_____	_____	_____
19	_____	_____	_____	_____
20	_____	_____	_____	_____
21	_____	_____	_____	_____
22	_____	_____	_____	_____
23	_____	_____	_____	_____
24	_____	_____	_____	_____
25	_____	_____	_____	_____

B. Soap Bubble Survivorship *(About 30 min.)*

Work in groups of four. One person blows bubbles, a second person times the "life" of the bubble, a third observes survivorship, and the fourth records data in Table 28-3.

Bubbles are "alive" as long as they are intact, but "die" when they burst. You will be following bubbles and timing how long they "live."

Different student groups in your lab section will create three different populations of soap bubbles. Your group will be assigned one population to work with during this exercise:

Population 1. Once the bubble leaves the wand, group members wave, blow, or fan in an effort to keep the bubble in the air and prevent it from breaking (dying) as long as possible.

Population 2. Group members do nothing to interfere with the bubbles or keep them up in the air once the bubbles leave the wand.

Population 3. This group uses a wand mounted on a wooden frame (Figure 28-1). The person blowing bubbles stands at the wand facing the frame, dips the wand in the soap solution, and tries to blow the bubbles through the opening in the frame. *Bubbles that break on the paper "hazard" or any other part of the frame rather than passing through the frame are timed and included in the data. Bubbles that fall to the side without passing through or breaking on the frame are ignored (not counted in the data).* **Do not** *attempt to manipulate the frame in any way so as to increase the chances that the bubbles will pass through it.*

1. Practice blowing bubbles for a few minutes until you can generate them with the single end of the wand. Even though you may generate several bubbles, follow only one at a time to measure its life span.
2. The timer should start the watch once the bubble is free of the wand. When the bubble bursts, the timer notes the time and checks the age at death in Table 28-3.
3. Measure the life span of 50 bubbles.
4. Summarize your data as follows:
   (a) Count the number of checks (number of bubbles dying) at each age. Record the number in the column marked "Total."
   (b) To determine the number surviving at each age, subtract the number dying at each age from the number surviving at the previous age. So, if 5 bubbles break (die) at age 1 second, then $50 - 5 = 45$ survive at least 1 second. If 10 more bubbles die at age 2 seconds, then $45 - 10 = 35$ survive at least 2 seconds.
   (c) Calculate the percentage surviving at each age. Since at birth (the moment the bubble leaves the wand) 50 bubbles are alive, 100% are alive at age 0. Use the following formula:

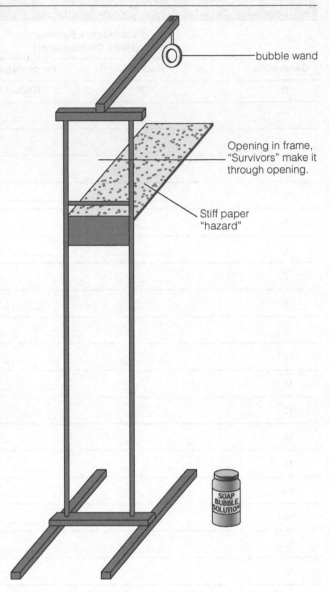

Figure 28-1 Frame for soap bubble population 3.

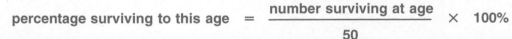

$$\textbf{percentage surviving to this age} \ = \ \frac{\textbf{number surviving at age}}{\textbf{50}} \ \times \ \textbf{100\%}$$

TABLE 28-3	Survivor Data for Soap Bubble Population			
	Bubbles Dying at This Age		**Bubbles Surviving to This Age**	
Age at Death (sec)	Number (✔ = 1)	Total	Number	Percentage
0	0	0	50	100%
1	_____	_____	_____	_____
2	_____	_____	_____	_____
3	_____	_____	_____	_____
4	_____	_____	_____	_____
5	_____	_____	_____	_____
6	_____	_____	_____	_____
7	_____	_____	_____	_____
8	_____	_____	_____	_____
9	_____	_____	_____	_____
10	_____	_____	_____	_____
11	_____	_____	_____	_____
12	_____	_____	_____	_____
13	_____	_____	_____	_____
14	_____	_____	_____	_____
15	_____	_____	_____	_____
16	_____	_____	_____	_____
17	_____	_____	_____	_____
18	_____	_____	_____	_____
19	_____	_____	_____	_____
20	_____	_____	_____	_____
21	_____	_____	_____	_____
22	_____	_____	_____	_____
23	_____	_____	_____	_____
24	_____	_____	_____	_____
25	_____	_____	_____	_____
26	_____	_____	_____	_____
27	_____	_____	_____	_____
28	_____	_____	_____	_____
29	_____	_____	_____	_____
30+	_____	_____	_____	_____

PROCEDURE

Work alone for this activity.

1. Use Figure 28-2 to make an *arithmetic plot* of survivorship of the dice and soap bubble populations. Note that the horizontal lines are the same distance apart on this graph.

Using open circles to indicate the percentage of *dice* surviving at each age, plot the percentage surviving data from Table 28-2. Using closed circles to indicate the percentage of *soap bubbles* surviving at each age, plot the percentage surviving data from Table 28-3. Use a ruler to connect the data points from each population.

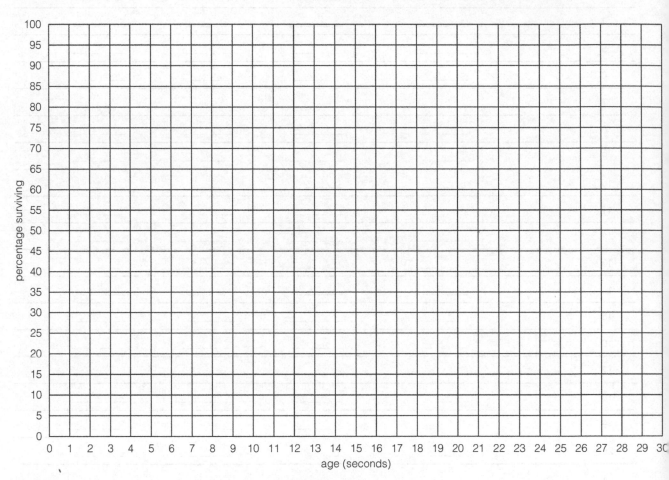

Figure 28-2   Arithmetic plot of survivorship.

2. Now you will use Figure 28-3 to make a *logarithmic plot* of survivorship. Note that on this *semilog* paper, the horizontal lines become closer together toward the top of the page.

If you've never used semilog paper, examine Figure 28-3 more closely. The scale on the vertical axis runs from 1 at the bottom to 9 near the middle, then from 1 to 9 near the top; there is a 1 at the topmost line. The lines are spaced according to the *logarithms* of the numbers.

The 1 at the top represents 100%, the 9 below it 90%, and so on. This means that the 1 in the middle of the page is 10%, while the 9 below it represents 9%; the 8, 8%; and so forth until the 1 at the bottom, which represents 1%.

Plot the survivorship of the dice and soap bubbles that your group generated, using the percentage surviving data from Tables 28-2 and 28-3. Open circles indicate the percentage of dice surviving and closed circles the percentage of soap bubbles surviving. Use a ruler to connect the points, forming a curve for each population.

After you've plotted your data, copy your plots onto the overhead transparency your instructor has for this purpose, using the pens provided. Different colored pens are available so that each population will have a separate color.

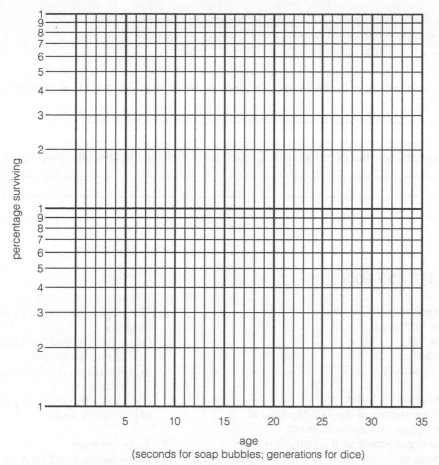

Figure 28-3 Logarithmic plot of survivorship.

Interpreting the Survivorship Curves (*About 15 min.*)

First examine the plots from the dice populations. In population 1 (heart disease only), a constant one-sixth, or 17%, of the population dies on average at each age. In population 2, two-sixths, or 33%, dies on average at each age. As you can see, on the *arithmetic plot* these data form a smooth curve, while on the *logarithmic plot* they form a straight line.

Both types of graphs provide useful information. A straight line on a logarithmic plot indicates the death rate is *constant*, but it's easier to see that more individuals die at a young age than at older ages in the arithmetic plot.

In natural populations, three basic trends of survivorship affecting population size have been identified. These are shown in Figure 28-4 and summarized in Table 28-4.

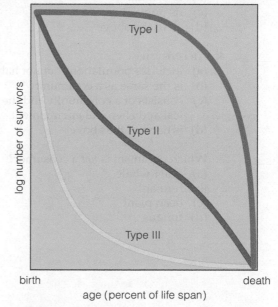

Figure 28-4 Three basic types of survivorship curves.

TABLE 28-4 Survivorship Curves

Type	Trend
I	Low mortality early in life, most deaths occurring in a narrow time span at maturity
II	Rate of mortality fairly constant at all ages
III	High mortality early in life

Now examine the logarithmic survivorship curves for the soap bubble population. How do they compare with the Type I, II, and III curves in Figure 28-4?

_____

_____

Do any of the soap bubble populations show a constant death rate for at least part of their life span? If so, for which population?

_____

_____

How did the treatments that populations 1, 2, and 3 were subjected to affect the shape of the curves?

_____

_____

_____

PRE-LAB QUESTIONS

_____ 1. The Greek word *oikos* means
   (a) environment
   (b) habitat
   (c) house
   (d) community

_____ 2. A population includes
   (a) all organisms of the same species in one place at a given moment
   (b) the organisms and their physical environment
   (c) the physical environment in which an organism exists
   (d) all organisms within the environment at a given moment

_____ 3. The place that an organism resides is
   (a) its community
   (b) its habitat
   (c) its population
   (d) all of the above

_____ 4. An ecosystem
   (a) includes populations but not habitats
   (b) is the same as a community
   (c) consists of a community and the physical and chemical environment
   (d) is none of the above

_____ 5. Which organism is *not* a consumer?
   (a) killer whale
   (b) human
   (c) bean plant
   (d) fungus

_____ 6. The trophic level that an organism belongs to
   (a) is determined by what it uses as an energy source
   (b) determines the population it belongs to
   (c) can be thought of as its feeding level
   (d) is both a and c

_____ 7. A decomposer
   (a) is exemplified by a fungus
   (b) is the same as a parasite
   (c) feeds on living organic material
   (d) feeds on primary consumers only

_____ 8. An herbivore
   (a) eats plant material
   (b) eats primary producers
   (c) is exemplified by a deer
   (d) is all of the above

_____ 9. Which term best describes a human being?
   (a) omnivore
   (b) herbivore
   (c) carnivore
   (d) parasite

_____ 10. A survivorship curve shows
   (a) the number of individuals surviving to a particular age
   (b) the cause of death of an individual
   (c) the place an organism exists in its environment
   (d) the organism's trophic level

EXERCISE 28

Ecology: Living Organisms in Their Environment

POST-LAB QUESTIONS

Introduction

1. Define the terms *herbivore, carnivore, omnivore.*

2. Define the terms *producer, primary consumer, secondary consumer.*

3. Characterize yourself, using the terms listed in questions 1 and 2. Justify your answer.

28.1 Food Webs

4. Examine the food web you constructed in Table 28-1. Suppose a severe outbreak of rabies occurs in the coyote population, resulting in the death of the entire population. Draw a rough graph that describes the impact of the loss of this predator on the rabbit, deer, and raspberry populations over time, and explain your reasoning.

5. You've probably read newspaper articles about the so-called ozone "hole" over Antarctica and other areas of the planet, as well as health-care cautions that destruction of the earth's protective ozone layer will allow harmful ultraviolet (UV) light to reach the surface. Already, a significant increase in skin cancer is being noticed, a disease known to be caused by excessive exposure to UV light (including that produced by tanning lights). Perhaps not so well known is the damage UV light causes to the photosynthetic machinery of green plants and protists.

    Explain what would happen over time to the populations of organisms of your food web (Table 28-1) if ozone depletion becomes massive.

28.2 Flow of Energy Through an Ecosystem

6. You may have heard the advice to "eat low on the food chain." Explain how choosing a soybean-protein "burger" rather than a beef burger for lunch is energy efficient with respect to energy flow and utilization in trophic levels.

28.3 Survivorship

7. Would you expect a population in which most members survive for a long time to produce few or many offspring? Which would be most advantageous to the population as a whole?

8. Suppose a human population exhibits a Type III survival curve. What would you expect to happen to the curve over time if a dramatic improvement in medical technology takes place?

9. Return to the survivorship model you created using dice: You found that the chance of dying from heart disease is approximately one-sixth for each die, indicating that survivorship is essentially the same for each age group. Relate what happened in your model with a realistic projection showing at what ages most humans die of heart disease.

Food for Thought

10. Ecologist Aldo Leopold, the father of American conservation, said, "The last word in ignorance is the man who says of an animal or plant: 'What good is it?' . . . If the biota, in the course of aeons, has built something we like but do not understand, then who but a fool would discard seeming useless parts? To keep every cog and wheel is the first precaution of intelligent tinkering."

    These are strong words indeed. How do they relate to what you've explored in this exercise?

Habitat Resource Partitioning in a Woodpecker Guild

Abstract

Elucidating how coexisting species partition resources is a fundamental aim of ecological research. Our research focused on determining how coexisting woodpecker species partition habitat space within a complex landscape mosaic. We found a statistically significant association between woodpecker species and habitat type, indicating that woodpecker species do differ in their use of habitat. Further studies focusing on more finely tuned data collection at the microhabitat level are discussed.

Introduction

A central goal within ecology is an assessment of the relationships between and among coexisting species. The competitive exclusion principle states that when resources are limited, coexisting species partition the resources such that the species never totally overlap in their use of the limiting resources (Diamond and Case 1986). Partitioning of resources may occur in several ways. Most simply, species may occupy different ranges of habitats with no co-occurrence at the same time. When species do have overlapping ranges, partitioning may occur on a local habitat (i.e., macrohabitat) level. For example, the thrashers of North America, *Mimidae*, all occupy some type of scrub habitat. However, each species is generally found to individually occupy a homogeneous vegetation type that has no overlap of use with the other species (Cody 1974). When species co-occur within a local habitat, partitioning may occur by species differing in their spatial and behavioral use of the habitat. This can occur in a vertical, horizontal, or temporal dimension. For example, MacArthur (1958) documented that separate foraging sites within a single habitat, the Maine spruce forests, were used by different species of *Dendroica* warblers. Yet another way in which co-occurring species may partition resources is through differences in the types of food taken. An example of this is provided by several species of European tits. All co-occur in a single local habitat type, oak woodland, but each species takes food items of slightly different sizes (Betts 1955).

Woodpeckers offer an excellent opportunity to explore the ways in which closely related species partition resources in order to coexist. Woodpeckers belong to the "trunk and branch foraging" guild, a predominantly insectivorous guild whose members primarily search for food on or in the trunks and branches of trees (De Graaf, Tilghman, and Anderson 1985). We chose to examine how woodpecker species coexist in the scrub-pine habitats of central Florida. The landscape is characterized by a complex mosaic of different vegetation associations. We quantified individual species' responses to the diversity of vegetation associations in order to understand how this woodpecker guild is able to coexist. Our objective was to determine whether woodpecker species differ in their use of *macrohabitat space* (herein defined as one of the major vegetation association types within the landscape).

Materials and Methods

Our study site was located at Archbold Biological Station (ABS), 12 kilometers south of Lake Placid in Highlands County, Florida. ABS includes the following vegetation associations: southern ridge sandhill (RS), sand pine scrub (SS), scrubby flatwoods (SF), flatwoods (FL), swale, bayhead, and seasonal ponds. Detailed descriptions of these habitats are given by Abrahamson et al. (1984). The vegetational structure and composition of these habitats is in large part determined by fire and drainage. As a result, a mosaic of habitats with different fire histories as well as different vegetation associations exists on the ABS property. To determine how the woodpecker species coexist within this mosaic of habitats, we chose three sites representing the predominant vegetation associations: RS, SS, and SF. The three sites chosen were of roughly the same size and shape (based on aerial photographs and GIS maps).

The species of woodpecker found on the ABS property include: the downy woodpecker, the red-cockaded woodpecker, the hairy woodpecker, the red-headed woodpecker, the red-bellied woodpecker, the pileated woodpecker, the yellow-bellied sapsucker, and the yellow-shafted flicker. As noted, these species belong to the trunk-and-branch foraging guild (De Graaf, Tilghman, and Anderson 1985). Their foraging behavior is predominately directed on and around the trunks and branches of trees. Of this guild, we observed and collected data on five species: the downy, flicker, hairy, red-bellied, and red-headed woodpecker species. We did not encounter the other members of the guild.

Our method of collecting data involved walking throughout the particular vegetation association site in such a manner as to ensure full coverage of the site. Each investigator covered a separate site so that all three sites were visited simultaneously. We spent a total of two hours canvassing each site. Data were recorded on each species encountered, and attempts were made to ensure that individuals were not counted more than once (by visual identification using binoculars). Data were compiled into a 3×5 contingency table and analyzed for habitat preferences of the five species using the chi-square test (Fowler, Cohen, and Jarvis 1998). Our null hypothesis stated that there was no association between habitat type and woodpecker species; our alternative hypothesis was that there is an association between habitat type and woodpecker species.

Results

We found a statistically highly significant association between woodpecker species and type of habitat ($X^2 = 44.59$, $P < 0.01$). We therefore rejected the null hypothesis and accepted the alternative hypothesis. To determine which woodpecker species are associated with which habitat type, we compared observed and expected frequencies of each species by habitat type. We found more red-headed woodpeckers than expected in scrubby flatwoods, more downy woodpeckers than expected in sandpine scrub, and more than expected numbers of the remaining three species (the yellow-shafted flicker and the hairy and red-bellied woodpeckers) in ridge sandhill (Table 1).

TABLE 1 Contingency Table of Frequencies for Woodpecker Species by Habitat Type

	Downy	Flicker	Hairy	Red-Bellied	Red-Headed	Total
Scrubby Flatwoods						
Observed	2	6	2	7	13	30
Expected	5.2	6.6	7.2	5.4	5.4	30
% within habitat	6.7	20.0	6.7	23.3	43.3	100
Sandpine Scrub						
Observed	22	10	16	8	11	67
Expected	11.7	14.8	16.2	12.1	12.1	67
% within habitat	32.8	14.9	23.9	11.9	16.4	100
Ridge Sandhill						
Observed	2	17	18	12	3	52
Expected	9.1	11.5	12.6	9.4	9.4	52
% within habitat	3.8	32.7	34.6	23.1	5.8	100

Discussion

This initial investigation into the ways in which closely related species partition resources has revealed that the woodpeckers co-occurring on the ABS property differ in the ways in which they use the habitat mosaic. Although all five species were found in all habitat types, they differed in their frequencies of occurrence within these habitats. None of the five species sampled was found to be using only one habitat category to the exclusion of the others. It may be that, by distributing their habitat use across more than one habitat, no two species come into severe enough competition for deleterious consequences to develop. Previous research has shown that woodpeckers are opportunistic foragers that use a variety of substrates (Short 1982). Habitat mosaics may allow enough partitioning of resources across the landscape such that the species avoid severe competition. One area for future research would be determining the size of the landscape mosaic of habitats that would be needed to support a large woodpecker species guild.

Alternatively, even though species co-occur within a macrohabitat category, they may actually be partitioning the resources at the level of the microhabitat. Studies could be designed that observe focal individuals. Data that could be collected include bird species, foraging behavior, substrate used (trunk, branch, or ground), tree species used, height on the tree, and whether the tree was live or dead. Bird foraging behavior could be divided into the following relevant categories: peering and poking, probing and drilling, scaling, and hawking. Additional investigations could examine the types of insects and other food sources used by species.

The results of studies such as the one reported here as well as those suggested will advance our understanding not only of how species coexist but also how species respond to environmental conditions such as food availability and habitat space. This type of data is necessary for making sound management decisions, especially decisions regarding landscape management and restoration activities.

Citations

Abrahamson et al. 1984

Betts, M. M. 1955. The food of titmice in oak woodland. *Journal of Animal Ecology* 24: 282–323.

Cody, M. L. 1974. *Bird communities.* Princeton, NJ: Princeton University Press.

De Graaf, R. M., N. G. Tilghman, and S. H. Anderson. 1985. Foraging guilds of North American birds. *Environmental Management* 9(6): 493–536.

Diamond, J., and T. J. Case (Eds.). 1986. *Community ecology.* New York: Harper & Row.

Fowler, J., L. Cohen, and P. Jarvis. 1998. *Practical statistics for field biology.* Chichester, UK: Wiley & Sons.

MacArthur, R. H. 1958. Population ecology of some warblers of northeastern coniferous forests. *Ecology* 39(4): 599–619.

Short, L. L. 1982. *Woodpeckers of the world.* Monographs Series No. 4. Greenville, DE: Delaware Museum of Natural History.

Human Impact on the Environment: Stream and Pond Ecology

OBJECTIVES

After completing this exercise, you will be able to

1. define *stream, pool, riffle, abiotic, biotic, watershed, point source pollution, nonpoint source pollution, pH, dissolved oxygen, nitrates, phosphates, chlorides, coliform bacteria, macroinvertebrate, larvae, Stream Quality Index;*

2. explain why most aquatic invertebrates are found in riffles;

3. describe how sediment pollution occurs and its major effects on the stream ecosystem;

4. describe the effects of organic pollutants and nutrient pollution on stream organisms;

5. explain the importance of dissolved oxygen in stream water;

6. describe how aquatic macroinvertebrates present in a stream can indicate water quality;

7. indicate the significance of finding coliform bacteria or bacteria identified via a PathoScreen test in a water sample;

8. discuss limitations and strengths of standard chemical water tests, as well as the Stream Quality Index;

9. discuss some of the ways in which human activity in a watershed can affect stream water quality.

INTRODUCTION

Global warming, extinction of species, damage to the stratospheric ozone layer, toxic chemical spills, ground water contamination, and deforestation. This partial list of environmental effects attributed to the activities of our exploding human population is diverse and alarming. Each of these issues is a complex problem with scientific, economic, philosophical, sociological, and political aspects, and no single exercise (or entire course!) can fully address any of these concerns. However, this exercise, which focuses on some of the ways humans affect streams and rivers, the arteries of our planet, will give you an idea of how scientists build the data base they use to assess environmental problems.

A **stream** is a body of running water flowing on the earth. Streams range from shallow, narrow brooks tumbling down mountainsides to broad, deep rivers like the Mississippi. Regardless of its size, every stream has distinct major habitat types that influence the types of organisms living there. Relatively slow-moving, deep waters are called **pools,** while **riffles** are shallow areas with fast-moving current bubbling over rocks (Figure 29-1).

a

b

(Photos by E. F. Benfield.)

Figure 29-1 Stream habitats. (a) Pool leading into a riffle, (b) closer look at a riffle.

Riffles are places where stream water tumbling over rocks and gravel bottom recharges the water with oxygen. Because of the higher dissolved oxygen levels and the rocks and gravel that provide habitat, most stream inhabitants live in riffles. The rapid current brings oxygen for cellular respiration as well as a steady supply of food from upstream. Pools, on the other hand, are still areas with lower oxygen content. Pools tend to have muddy beds, catch organic matter from upstream, and are areas of decomposition and nutrient recycling.

Gaseous, liquid, and solid substances enter the stream and either dissolve in the water or are carried along as suspended particles. The current thus transports these substances to where they may affect organisms downstream. The current speed, type of bottom structure, temperature, and dissolved substances are the **abiotic** components of the stream ecosystem. These abiotic components influence the life and habits of the organisms inhabiting the flowing water, the **biotic,** or living, components of the stream ecosystem.

The single most important factor influencing the abiotic and biotic components is the character of the **watershed,** all the surrounding land area that serves as drainage for a stream. A watershed may cover a few acres of land surrounding a very small stream, or include tens of thousands of square kilometers for a large river. You can think of a watershed as a bowl, with the stream flowing across the bottom. Water that falls on the higher elevations (the sides of the bowl) of a watershed eventually flows into the stream.

Your instructor will provide you with information about the boundaries and terrain of the watershed of the stream you will study.

The quality of a stream depends on its watershed and the human activities that occur there. In general, a stream is of the highest quality when the watershed and streamside vegetation remain in a natural state, while most human influences tend to be detrimental to stream quality. These human-induced effects can be lumped under the general term *pollution*. The most common types of water pollutants are

- *Human pathogens*, disease-causing agents, such as bacteria, viruses, and protists.
- *Toxic chemicals*, both organic (carbon-containing) and inorganic. Toxic chemicals can harm humans, and fish and other aquatic life. They include acids, heavy metals, petroleum products, pesticides, and detergents.
- *Organic wastes,* such as sewage, food-processing and paper-making wastes, and animal wastes. Organic wastes can be decomposed by bacteria, but the process uses up oxygen dissolved in the water. The decomposition of large quantities of organic wastes can deplete the water of oxygen, causing fish, insects, and other aquatic oxygen-requiring organisms to die.
- *Plant nutrients,* such as nitrates and phosphates, that can cause too much growth of algae and aquatic plants. When the masses of algae and plants die, the decay process can again deplete the critical dissolved oxygen in the water.
- *Sediments*, or suspended matter, the particles of soil or other solids that become suspended in the water. Sediment originates where soil has been exposed to rainfall, most often from construction sites, forestry operations, and agriculture. This is the biggest pollutant by weight, and it can reduce photosynthesis by clouding the water and can carry pesticides and other harmful substances with the soil particles. When sediment settles, it can destroy fish feeding and spawning areas and clog and fill stream channels, lakes, and harbors.
- *Heat*, or thermal pollution. Heated water holds less dissolved oxygen than colder water, so the addition of heated water by industries must be closely monitored to ensure that dissolved oxygen isn't driven to too low levels.

All these substances enter water as either "point" or "nonpoint" pollution. **Point source pollution** occurs at a specific location where the pollutants are released from a pipe, ditch, or sewer. Examples of point sources include sewage treatment plants, storm drains, and factory discharges. Point sources are relatively easy to identify, regulate, and monitor. Federal and state governments regulate the type and quantity of some types of pollutants for each point source discharge.

Nonpoint source pollution enters streams from a vast number of sources, with no specific point of entry contributing a large amount of pollutants. Most nonpoint source pollution results when rainfall and runoff carry pollutants from large, poorly defined areas into a stream. Examples of nonpoint source pollution include the washing of fertilizers and pesticides from lawns and agricultural fields, and soil erosion from construction sites. Because sources of nonpoint pollution are difficult to identify and define, it is much more difficult to regulate.

What human activities can you name that might be affecting your study stream? _____

_____

_____

_____

What activities might have affected your study stream in the past that are unlikely to occur today?

_____

_____

_____

A. Observations (About 20 min.)

MATERIALS

Per student group (4):

- clear glass jar or beaker
- map or aerial photo of watershed area (optional)

PROCEDURE

The first step in the quality analysis of your stream, as in so many scientific endeavors, is observation. Record your observations on the Stream Environment Characterization Data Sheet using the instructions and questions immediately following the data sheet for guidance.

B. Physical and Chemical Sampling (About 1 hr.)

MATERIALS

Per student:

- disposable plastic gloves (optional)

Per student group (4):

- nail clippers or scissors
- plastic beaker
- plastic sample bottles
- watch or clock

Per lab room:

- thermometer
- portable pH meter or pH indicator paper
- Hach or other dissolved oxygen test kit
- Hach or other nitrate test kit
- Hach or other chloride test kit
- Hach or other phosphate test kit
- dH_2O
- paper towels

PROCEDURE

In this section, you will take several streamside measurements, and collect water samples for analysis. Your class will divide into groups, with each group performing different tasks. As a class you will measure the temperature and pH of the water, plus determine several nutrient/pollutant levels. Additionally, you will collect macroinvertebrates for classification and sample the water for the presence of potentially pathogenic (disease-causing) bacteria.

1. **Temperature.** Stream organisms are typically more sensitive to higher temperatures. In addition to affecting the metabolic rate of the cold-blooded aquatic organisms, cooler water holds more dissolved oxygen than warmer water.

Measure the water temperature holding a thermometer about 10 cm below the water surface, if possible. Keep the thermometer in the water for approximately 1 minute so the temperature can equilibrate. Record the temperature measurements from 3 locations here, then total them and calculate the average temperature. Enter the average temperature in Table 29-1.

Temperature: _____; _____; _____

2. **pH.** pH is a measure of the concentration of hydrogen ions in the water. In essence, pH signifies the degree of acidity or alkalinity of water. The pH scale ranges from 0 (most acidic) to 14 (most basic), as indicated in Figure 29-2. A solution with a pH of 7 is neutral. A change of only 1 on the scale means a tenfold change in the hydrogen ion concentration of the water.

The bedrock in the watershed influences the pH of the stream water as does the pH of rainwater and discharges into the stream. pH values between 6.0 and 8.5 occur naturally in streams. A pH value below 6.0 indicates acid mine drainage, acid precipitation, or some other acid source, unless the stream drains a naturally acidic environment such as a swamp or bog.

pH that is too low can damage fish gills. Low water pH can also cause some minerals, such as aluminum, in the streambed to go into solution and become toxic. The higher the pH, the more carbonates, bicarbonates, and salts are dissolved in the water. Streams with moderate concentrations of these substances are nutritionally rich and support abundant life.

STREAM ENVIRONMENT CHARACTERIZATION DATA SHEET

(circle appropriate choice for each)

Riparian Zone

Predominant Surrounding Land Use:

Forest/Overgrown field   Agricultural/Pasture   Residential

Commercial   Industrial   Other

Streambank Erosion:

None   Moderate   Severe

Local Nonpoint Source Pollution:

No evidence   Potential sources   Obvious sources

Canopy Cover:

Open   Little shade (40%)   Moderate shade (40–80%)

Shady (>80%)

Stream Corridor

Sediment Odors:

Normal   Sewage   Petroleum

Chemical   Anaerobic   Other

Sediment Deposits:

Sludge   Sawdust   Sand   Relict shells   Other

Streambed:

Green filamentous algae   Blue-green algae

Yellow-orange coating

Water Odors:

Natural   Sewage   Petroleum

Chemical   Other

Water Surface Oils and Foams:

None   Flecks   Globs   Sheen   Slick

Turbidity:

Clear   Slightly turbid   Turbid   Opaque

Water Color: _____

Instructions and Background

STREAM ENVIRONMENT CHARACTERIZATION INSTRUCTIONS AND BACKGROUND

Predominant land use: Look up and downstream on both banks and observe how land adjacent to the stream is being used. Also note any minor uses (for example, small corn field, parking lot, unpaved road) that may have an impact on the stream.

Streambank erosion: Observe the condition of the land adjacent to the streambank. Is the land highly vegetated and stable? Are there any gullies eroding the streambank or is the bank being undercut?

Local watershed nonpoint source pollution: Observe any sources of pollution aside from sedimentation. Are there roads close to the stream bank? Are there farms near the stream? Is the site adjacent to a golf course or residential or commercial area?

Canopy cover: Stand in the center of the stream and look up at the sky. What percentage of the stream is shaded? Is the entire stream covered with tree canopy or is the vegetation predominantly shrubs or herbs? Shaded streams have cooler water than unshaded. Warmer water also contains less dissolved oxygen than colder, and so may limit the kinds of organisms that can survive.

Sediment odors: Carefully take a handful or sediment from the site. Can you detect any odors? Indicate any odor on the data sheet even if it is not listed. *Note:* Anaerobic (without oxygen) sediments contain hydrogen sulfide and smell like rotting eggs. Anaerobic conditions can occur naturally in a marsh but can also indicate the presence of sewage pollution in a stream.

Sediment deposits: Are any sediment deposits found at the site? Are the undersides of any loosely embedded rocks blackened? This indicates an anaerobic condition unfavorable to benthic (bottom-dwelling) invertebrates.

Streambed: Green filamentous algae may indicate the presence of organic pollutants or excessive nutrients if they colonize more than 10% of the streambed. Dark-colored blue-green algae (more properly called "cyanobacteria") are more tolerant of pollution and may indicate pollution by organic materials such as sewage, liquid released from sanitary landfills, and/or animal wastes. A yellow to orange-red coating on the streambed may indicate a high soil erosion rate, industrial pollution, or polluted water draining from a mine.

Water odors: Sniff the stream waters and record any detectable odor. If no odor is recognizable, answer "natural."

Water surface oils and foams: Note the presence of any oils on the surface of the water. In streams, oils are usually seen as a rainbowlike sheen when in sunlight. Are oils widespread? Or are they only in isolated areas, perhaps surrounding leaves? (Plants naturally release oils as they decay.)

    The presence of extensive white-colored foam more than 7 cm high indicates pollution from detergents.

Turbidity: Take a sample of water in a clear container and hold it up to a light background. Note the amount of turbidity (cloudiness) present. Turbidity is caused by the suspension of solid matter, including sediment and plankton. Determining turbidity will aid in evaluating water color.

Water color: Collect a water sample in a clear container and hold it up against a white background. What color does the water appear? Often a stream looks dark brown, tan, or green, but is actually clear, with the water color resulting from the stream bottom material. If any strange colors are observed, indicate whether they appear throughout the stream or in isolated areas, such as below a discharge pipe.

TABLE 29-1   Stream Measurements and Water Quality Standards

Physical/Chemical Measurement	Average Value	Water Quality Standards		Probable Cause of Degradation[a]
		General[b]	Trout Stream	
Temperature		< 32°C	< 20–23°C	Lack of shade
pH		6.0–8.5	6.5–8.5	Acid drainage
Dissolved oxygen (DO)		> 5 mg/L	> 6 mg/L	Organic and nutrient sources
Nitrate (NO_3^-)		< 2 mg/L	< 10 mg/L[3]	Organic and nutrient sources, fertilizers
Phosphate (PO_4^-)		< 0.1 mg/L		Organic and nutrient sources, fertilizers
Chloride (Cl^-)		< 30 mg/L		Road salt, sewage, agricultural chemicals
Total coliform bacteria			0 colonies[c]	Sewage, animal wastes
PathoScreen waterborne pathogens			0 colonies[c]	Sewage, animal wastes

[a] Probable cause(s) of values failing to meet standards.
[b] Standards for swimming and to protect general aquatic life. (*Source*: Wisconsin Department of Natural Resources.)
[c] Standards for drinking water. (*Source*: Wisconsin Department of Natural Resources.)

Use the pH meter or indicator paper to determine the pH of your stream water. Try to sample both riffle and pool areas. Determine and record the average pH in Table 29-1.

pH: _____ ; _____ ; _____

Perform each following chemical analysis on at least three locations, average your readings, and record the average concentration in Table 29-1. Follow the instructions provided with each test kit.

3. **Dissolved oxygen (DO).** Aquatic animals, like most living organisms, require oxygen for their metabolic processes. This oxygen is present in dissolved form in water. If much organic matter (for example, sewage) is present in water, microorganisms use the **dissolved oxygen** in the process of decomposing the organic matter. In that case, the level of dissolved oxygen might fall too low to be available for aquatic animals. How much DO an aquatic organism needs depends on its species, its physiology, water temperature, pollutants present, and other factors. However, many studies suggest that 4−5 parts per million (ppm) is the minimum needed to support a diverse fish population. The DO of good fishing waters generally averages about 9 ppm. When DO levels drop below 3 ppm, even pollution-tolerant fish species may die.

4. **Nitrate (NO_3^-).** Nitrogen (N) is a nutrient required for plant life. Nitrogen in the form of **nitrates** (NO_3^-) is a common component of the synthetic fertilizers used on lawns, golf courses, and farm fields to stimulate plant growth. Surface runoff and infiltration through the soil water from these sources is the most common route for nitrates to enter streams. Leaking sewage tanks, municipal wastewater treatment plants, manure from livestock and other animals, and discharges from car exhausts are all sources of stream nitrates. Excessive nitrates in streams can cause overgrowth of algae and other plant life. This plant growth may decrease the light available to other stream life, but eventually dies and decays. The decay process can deplete water of the vital dissolved oxygen.

5. **Phosphate (PO_4^-).** Phosphorous (P), another element crucial to plant life, is also a common component of commercial fertilizers in the form of **phosphates** (PO_4^-). When it rains, varying amounts of phosphates wash from farm fields and other fertilized areas into nearby waterways. Like nitrates, phosphate enrichment also causes nutrient problems in streams. Excessive levels indicate pollution by nutrient sources or organic substances (fertilizers, sewage, detergents, animal wastes).

6. **Chloride (Cl^-).** The chloride ion (Cl^-) has very different properties from chlorine gas (Cl_2), which is used as a disinfectant. **Chloride** is one of the major ions in sewage; it also enters waterways when washed from roadways that have been treated with salt ($NaCl$, sometimes $CaCl_2$) in wintertime to melt ice and snow. Agricultural chemicals are another source of chloride in water. Water with high chloride content is toxic to plant life.

C. Biological Sampling (About 1 hr.)

MATERIALS

Per student:

- hip boots or footwear that can get wet
- forceps or tweezers

Per student group (4):

- sterile water sample collection bottles or containers
- kickseine
- white enamel pan
- bottle of alcohol
- sample collecting jars

Per lab room:

- Presence/Absence ampoule
- 100-mL plastic beaker
- 2 Whirl-Pak™ bags
- PathoScreen powder pillow
- nail clipper

PROCEDURE

1. **Total coliform bacteria. Coliform bacteria** are tiny, rod-shaped bacteria found in soil and water. Certain coliform bacteria, notably *Escherichia coli,* live in the intestines of humans and all warm-blooded animals, and these fecal coliform bacteria pass out of the body with feces. They can cause human disease if they enter a drinking water supply such as a private well or a river, and they are also associated with other kinds of pathogenic bacteria. While municipal water treatment plants routinely treat water to kill many potential pathogens, people who rely on private wells don't usually have that protection. Other coliform bacteria are common soil inhabitants but tend to associate in water with the more dangerous fecal coliforms. You will sample the water of your stream to determine whether any coliform bacteria are present. The presence of coliform bacteria suggests contamination with sewage or animal wastes.

   (a) Break a Presence/Absence ampoule and pour the contents into one Whirl-Pak™ bag. Be careful not to touch the inside of the bag or otherwise contaminate the contents. See Figure 29-3.

   (b) Collect at least 100 mL of stream water for each sample. Pour the water carefully into the Whirl-Pak™ bag, again being careful not to contaminate the inside of the sterile container with anything other than stream water. Swirl to mix well. Record the appearance of the liquid.

   (c) Incubate the sample at 35–40°C for 24 hours. A color change from purple to orange indicates the presence of coliform bacteria.

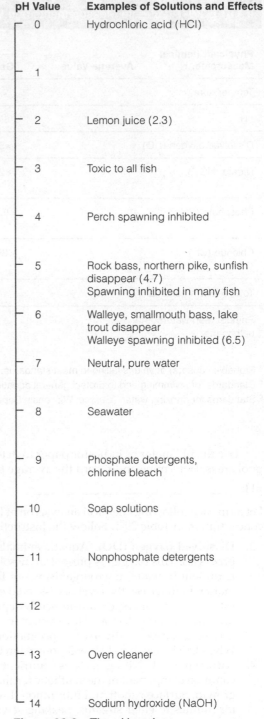

pH Value	Examples of Solutions and Effects
0	Hydrochloric acid (HCl)
1	
2	Lemon juice (2.3)
3	Toxic to all fish
4	Perch spawning inhibited
5	Rock bass, northern pike, sunfish disappear (4.7) Spawning inhibited in many fish
6	Walleye, smallmouth bass, lake trout disappear Walleye spawning inhibited (6.5)
7	Neutral, pure water
8	Seawater
9	Phosphate detergents, chlorine bleach
10	Soap solutions
11	Nonphosphate detergents
12	
13	Oven cleaner
14	Sodium hydroxide (NaOH)

Figure 29-2 The pH scale.

Figure 29-3 Use of sterile Whirl-Pak™ bag.

2. **PathoScreen for waterborne pathogens.** The PathoScreen test allows us to detect common human pathogens *other than* coliform bacteria. These bacteria may also be associated with fecal contamination, and include the organisms responsible for causing some kinds of dysentery, *Salmonella* poisoning, and other illnesses.

    (a) Clip the corner off a PathoScreen powder pillow and dump contents into a Whirl-Pak™ bag. Be careful not to touch the inside of the bag or otherwise contaminate the contents. (See Figure 29-3.)

    (b) Collect at least 100 mL of stream water for each sample. Pour the water carefully into the Whirl-Pak™ bag. Swirl to mix well. Record the appearance of the liquid.

    (c) Incubate the sample at 35–40°C for 24 hours. A color change from orange to black, accompanied by a strong sulfur odor, indicates the presence of potential pathogens.

3. **Aquatic Macroinvertebrates. Macroinvertebrates,** animals that are large enough to be seen with the naked eye and that lack a backbone, are an important part of the stream ecosystem. Nearly all of the macroinvertebrates present in the water are immature insects, called **larvae,** which mature and then spend their adult lives on land or in the air. Most aquatic larvae feed on decaying organic matter, but some graze on algae or pursue other animals. Other aquatic macroinvertebrates include mollusks (shelled organisms such as snails and clams), arthropods (relatives of shrimp and lobsters such as crayfish and scuds), and various types of worms.

Most aquatic macroinvertebrates, especially those that live on or in the bottom sediments of riffle areas, require a specific range of water quality conditions and thus are good indicators of water quality. Aquatic macroinvertebrates can be separated into three general groups on the basis of their tolerance to pollution. Some bottom dwellers like mayfly larvae need a high-quality, unpolluted environment and are placed in Group 1 (Figure 29-4). Dominance of Group 1 organisms in a stream signifies good water quality. Group 2 organisms can exist in a wide range of water quality conditions and include crayfish and dragonfly larvae. Group 3 organisms are tolerant of pollution and include midge and blackfly larvae and leeches.

By looking at the diversity of macroinvertebrates present, you will be able to get a general idea of the health of a stream. Typically, waters of higher quality support a greater diversity (more kinds) of pollution-intolerant aquatic insects. As water becomes more polluted, fewer kinds of organisms thrive. You will collect samples of aquatic insects and separate them in the laboratory into general types to determine the Stream Quality Index Value.

SAMPLING WITH A KICKSEINE. Try to sample from riffle areas only, since these areas usually have a higher quality environment for macroinvertebrates. Sample from pools only if most of the stream is pool area.

Work in groups.

1. Select a riffle area with a depth of 7–30 centimeters (cm), and stones that are 5–25 cm in diameter or larger if possible. Move into the sample area from the downstream side to minimize disturbance.
2. One or two students should hold the kickseine at the downstream edge of the riffle. Be sure that the bottom edge of the screen is held tightly against the stream bottom. (You might want to hold down the edge with a few rocks.) Don't allow any water to flow over the top of the seine or you might lose specimens.
3. The other group members should walk in front of the net and begin the collection procedure. Pick up all rocks fist-sized or larger and rub their surfaces thoroughly in front of the net to dislodge all organisms into the seine. Then disturb the streambed for a distance of 1 m upstream of the kickseine. Stir up the bed with hands and feet until the entire 1-m² area has been worked over. Kick the streambed with a sideways motion to allow bottom-dwelling organisms to be carried into the net.
4. Remove the seine from the water with a forward-scooping motion so that no specimens are washed from the surface. Carry the seine to a flat area of the stream bank for examination.
5. Rinse the organisms and debris collected to the bottom of the screen. Many aquatic macroinvertebrates are very small, so be certain to remove all of them from the screen. Use forceps to remove any that cling to the screen. Place the collection in a collecting jar; drain off as much water as possible and cover the material with alcohol to kill and preserve the organisms. Label each jar to identify the area sampled.

29.2   Laboratory Analysis—Evaluating Biological Tests and Samples

A. Total Coliform Bacteria

Note the reaction after 24 hours of incubation and record it in Table 29-1. A color change from reddish-purple to yellow or yellow-brown indicates the presence of coliform bacteria. If there is no change in 24 hours, incubate for an additional 24 hours and recheck for the color change. If after 48 ± 3 hours of incubation the sample still appears reddish-purple, record the test as negative for total coliform bacteria.

Note the reaction after 24 hours of incubation and record it in Table 29-1. A color change from orangish to black, accompanied by a strong smell of sulfur, indicates the presence of potentially pathogenic bacteria.

C. Macroinvertebrates *(About 90 min.)*

Identification of aquatic organisms can be difficult, to say the least. However, biological sampling results can be translated into a water quality assessment without actually identifying each insect specimen, but by assessing the number of different kinds of collected macroinvertebrates.

MATERIALS

Per student group:

- petri dishes
- white paper *or* white enamel pan

- forceps or tweezers
- dissection microscope *or* magnifying glass

PROCEDURE

1. Pour a portion of a kickseine sample into a petri dish. Place a piece of white paper onto the stage of a dissecting microscope to provide an opaque white background, and then observe the specimens in the petri dish. Alternatively, you might pour a sample into a white enamel pan and observe the specimens with a magnifying glass.

2. Use forceps or tweezers to sort all the organisms into "look-alike" groups, or taxa (Figure 29-4). Look primarily at body shape and number of legs and "tails" to aid in identification. Note the size range of the pictured taxa. Record each taxonomic group present in your sample under the appropriate group number in Table 29-2 (Group 1 = pollution-intolerant taxa; Group 2 = somewhat pollution-tolerant taxa; Group 3 = pollution-tolerant taxa). For purposes of this lab, disregard organisms that are not pictured in Figure 29-4.

TABLE 29-2 Aquatic Macroinvertebrate Survey and Worksheet

Taxa Present in Stream Samples		
Group 1 **(Pollution Intolerant)**	**Group 2** **(Somewhat Pollution Tolerant)**	**Group 3** **(Pollution Tolerant)**

3. Determine the Stream Quality Index (SQI) Value, a measure of stream ecosystem health, as follows:
   (a) The organisms in each group are assigned a group value as follows:
      Group 1 = group value 3
      Group 2 = group value 2
      Group 3 = group value 1
   (b) Count the number of taxa (*types* of organisms) in each group and multiply that number by the group value. Record the resulting group index value in worksheet space.

Example:	Group 1 Taxa	Group 2 Taxa	Group 3 Taxa
	Caddisfly	Damselfly	Blackfly
	Mayfly	Crane Fly	Midge
	Stonefly	Clam	
	3 taxa × group value 3 = 9	3 taxa × group value 2 = 6	2 taxa × group value 1 = 2

   (c) Add the respective group index values of the three groups to find the cumulative SQI value. (In the preceding example, 9 + 6 + 2 = 17.)

Number of taxa	×	group value	=	group index value
Group 1 _____	×	3	=	_____
Group 2 _____	×	2	=	_____
Group 3 _____	×	1	=	_____

Group 1 index value	+	Group 2 index value	+	Group 3 index value	=	SQI
_____	+	_____	+	_____	=	_____

The SQI value can be interpreted as follows:

Stream Quality Index Value	Water Quality
23 and above	Excellent
17–22	Good
11–16	Fair
10 or less	Poor

In addition to indicating general water quality, your aquatic macroinvertebrate survey can provide you with information regarding the types of pollutants that may be affecting the stream. If you found few taxa of pollution-tolerant invertebrates, but with great numbers of individuals of these taxa, the stream water may be overly enriched with organic matter or have severe organic pollution. Conversely, if you found relatively few taxa and only a few individuals of each kind (or if you found *no* macroinvertebrates but the stream appears clean), you might suspect the presence of toxic chemicals. A diversity of taxa, with an abundance of individuals for many of the taxa, indicates a healthy stream with good to excellent water quality.

29.3 Analysis of Results

Table 29-1 lists acceptable levels of a few chemicals in water to protect aquatic life, as well as the probable causes of test values that fail to meet those standards. Compare your test results with those values. Also consider your findings from biological observations.

Are your observations, chemical measurements, and biological observations consistent? (Refer to the Stream Environment Characterization Data Sheet, Table 29-1, and Table 29-2.) That is, do they *all* indicate that your stream is healthy, with good water quality? Or conversely, do they *all* point to a stream in trouble, with poor water quality? Or do your results conflict, with some indicators of good quality and some indicators of poor water quality? _____

_____

If your results aren't clear-cut, speculate on why this might be so.

_____

_____

What might you do to increase the consistency and reliability of your results?

_____

_____

(Continues on p. 660.)

Group 1 Taxa: Pollution-sensitive organisms found in good-quality water.

Stonefly Larvae: Order Plecoptera. 1.25–3.5 cm, 6 legs with hooked tips, antennae, 2 hairlike tails. Smooth (no gills) on lower half of body. (See arrow.)

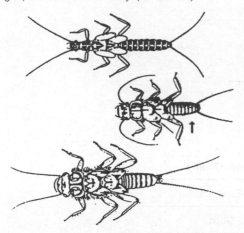

Caddisfly Larvae: Order Plecoptera. Up to 2.5 cm, 6 hooked legs on upper third of body, 2 hooks at back end. May be in a stick, rock or leaf case with its head sticking out. May have fluffy gill tufts on lower half.

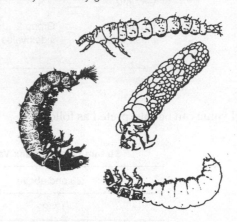

Water Penny Larva: Order Coleoptera. 0.5 cm, flat saucer-shaped body with a raised bump on one side and 6 tiny legs on the other side. Immature beetle.

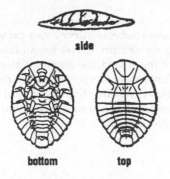

side

bottom   top

Dobsonfly (Hellgrammite) Larva: Family Corydalidae. 2–10 cm, dark-colored, 6 legs, large pinching jaws, eight pairs feelers on lower half of body with paired cottonlike gill tufts along underside, short antennae, 2 tails and 2 pairs of hooks at back end.

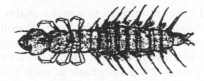

Gilled Snail: Class Gastropoda. Shell opening covered by thin plate called operculum. Shell usually opens on right.

Mayfly Larvae: Order Ephemeroptera. 0.5–2.5 cm, brown, moving, platelike or feathery gills on sides of lower body (see arrow), 6 large hooked legs, antennae, 2 or 3 long, hairlike tails. Tails may be webbed together.

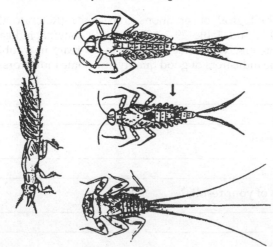

Riffle Beetle: Order Coleoptera. 0.5 cm, oval body covered with tiny hairs, 6 legs, antennae. Walks slowly underwater. Does not swim on surface.

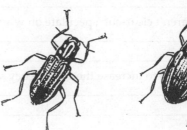

Figure 29-4 Stream insects and other macroinvertebrates. *Continues.*

Group 2 Taxa: Somewhat pollution-tolerant organisms can be in good- or fair-quality water.

Crayfish: Order Decapoda. Up to 15 cm, 2 large claws, 8 legs, resembles small lobster.

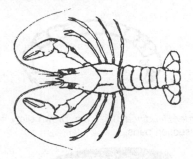

Sowbug: Order Isopoda. 0.5–2 cm gray oblong body wider than it is high, more than 6 legs, long antennae.

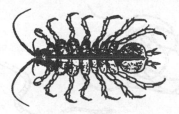

Scud: Order Amphipoda. 0.5 cm, white to gray, body higher than it is wide, swims sideways, more than 6 legs, resembles small shrimp.

Alderfly Larva: Family Sialidae. 2.5 cm long. Looks like small hellgrammite but has one long, thin, branched tail at back end (no hooks). No gill tufts underneath.

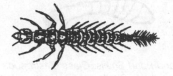

Fishfly Larva: Family Corydalidae. Up to 3.5 cm long. Looks like small hellgrammite but often a lighter reddish-tan color, or with yellowish streaks. No gill tufs underneath.

Clam. Class Bivalvia.

Damselfly Larvae: Suborder Zygoptera. 1.25–2.5 cm, large eyes, 6 thin hooked legs, 3 broad oar-shaped tails, positioned like a tripod. Smooth (no gills) on sides of lower half of body. (See arrow.)

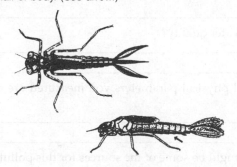

Watersnipe Fly Larva: Family Athericidae (Atherix). 0.5–2.5 cm, pale to green, tapered body, many caterpillarlike legs, conical head, feathery "horns" at back end.

Crane Fly Larva: Suborder Nematocera. 1.25–5 cm, milky, green, or light brown, plump caterpillarlike segmented body, 4 fingerlike lobes at back end.

Beetle Larva: Order Coleoptera. 0.5–2.5 cm, light colored, 6 legs on upper half of body, feelers, antennae.

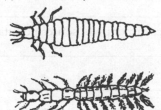

Dragon Fly Larvae: Suborder Anisoptera. 1.25–5 cm, large eyes, 6 hooked legs. Wide oval to round abdomen.

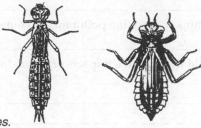

Figure 29-4 Stream insects and other macroinvertebrates. *Continues.*

Group 3 Taxa: Pollution-tollerant organisms can be in any quality of water.

Aquatic Worm: Class Oligochaeta. 0.5–5 cm, can be very tiny; thin wormlike body.

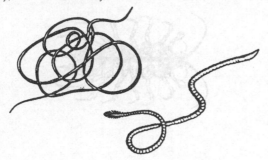

Midge Fly Larva: Suborder Nematocera. Up to 1 cm, dark head, wormlike segmented body, 2 tiny legs on each side.

Leech: Order Hirudinea. 0.5–5 cm, brown slimy body, ends with suction pads.

Blackfly Larva: Family Simulidae. Up to 0.75 cm, one end of body wider. Black head, suction pad on end.

Pouch Snail and Pond Snails: Class Gastropoda. No operculum. Breath air. Shell usually opens on left.

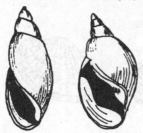

Other Snails: Class Gastropoda. No operculum. Breathe air. Snail shell coils in one plane.

Figure 29-4   Stream insects and other macroinvertebrates. *Continued.*

Which of your observed or measured factors indicate(s) higher quality water?

_____

_____

Which of your observed or measured factors indicate(s) lower water quality?

_____

_____

Why might you find a low SQI value, even if the chemical and physical parameters you measured are within limits for "good" water quality?

_____

_____

If you found indications of organic or nutrient pollution, what might be some of the sources for this pollution in your stream's watershed?

_____

_____

_____

Do you think most of the pollutants in your stream result from point source or nonpoint source pollutants?

_____

How could the negative impacts on your stream be reduced or eliminated?

_____

_____

_____

_____ 1. A riffle is
   (a) the land area surrounding a stream
   (b) an area of still, deep water
   (c) the living component of a stream
   (d) a fast-moving, shallow stream area

_____ 2. Which of these is *not* a type of water pollutant?
   (a) dissolved oxygen
   (b) toxic chemicals
   (c) organic wastes
   (d) sediment

_____ 3. Nonpoint source pollutants include
   (a) fertilizers from lawns
   (b) pesticides from farm fields
   (c) soil erosion from construction sites
   (d) all of the above

_____ 4. A yellow to orange-red coating on a stream bottom indicates
   (a) high dissolved oxygen
   (b) possible soil erosion or mine drainage
   (c) low nitrate concentration
   (d) high chloride content

_____ 5. pH is
   (a) a measure of phosphorous concentration
   (b) a measure of hydrogen ion concentration
   (c) measured by counting the number of different kinds of aquatic insects
   (d) measured on a scale from 1 to 100

_____ 6. Dissolved oxygen content in stream water is affected by
   (a) organic matter in the water
   (b) overgrowth of algae
   (c) water nutrient levels
   (d) all of the above

_____ 7. Which pollutant would most likely be the source of coliform bacteria in a stream?
   (a) fertilizers from a golf course
   (b) discharge from a factory
   (c) forestry activities
   (d) sewage or animal wastes

_____ 8. Most of the aquatic macroinvertebrates in a stream are
   (a) clams
   (b) insect larvae
   (c) worms
   (d) arthropods such as crayfish

_____ 9. Group 1 aquatic macroinvertebrates
   (a) require high-quality, unpolluted water
   (b) require polluted water
   (c) are tolerant of a wide range of water quality conditions
   (d) include leeches

_____ 10. Nitrates
   (a) are suspended sediments
   (b) measure hydrogen ion concentrations
   (c) are required for animal respiration
   (d) contain nitrogen, a necessary plant nutrient

EXERCISE 29

Human Impact on the Environment: Stream and Pond Ecology

POST-LAB QUESTIONS

Introduction

1. What is a watershed? What are some of the major human activities occurring in the watershed of the stream you sampled?

2. Distinguish between point and nonpoint sources of pollution. Provide several examples of each.

3. Why is it important for sewage treatment plants to remove nitrates and phosphates from sewage before discharging it into a river?

4. How can an algal bloom (population explosion of microscopic aquatic plants) affect fish in a pond?

5. Trout are fish that require relatively low-temperature water. Suppose a shaded trout stream has its over-hanging streamside trees removed. Explain the possible effects of this action on the trout population over the course of a summer.

29.1 Streamside Evaluation and Sampling

6. Here are the results of physical and chemical sampling of one stream. What do these results tell you about the suitability of this stream as fish habitat? What are likely causes for each of these values?

Parameter	Measurement
pH	5.2
Phosphate	0.15 mg/L
DO	3 mg/L

29.2 Laboratory Analysis—Evaluating Biological Tests and Samples

7. Here are the data from one stream's aquatic macroinvertebrate sample.

Taxa Present		
Group 1 (Pollution Intolerant)	Group 2 (Somewhat Pollution Tolerant)	Group 3 (Pollution Tolerant)
Caddisfly larvae	Crane fly larvae	Midge fly larvae
Mayfly larvae	Dragonfly larvae	Blackfly larvae
Riffle beetles	Crayfish	
Dobsonfly larvae		

Calculate the SQI value of this stream. What does this sample indicate about the water quality?

8. Referring to the stream in question 7, suppose you also know there are very few individuals of the Group 1 taxa present while midge fly larvae are extremely abundant. Does this information alter your assessment of the water quality of the stream? Why or why not?

Food for Thought

9. List several ways that sediment pollution could be/is being decreased in your community.

10. List several obstacles to improving stream quality in your community.

Estimating the Abundance of a Population of Aquatic Plants (*Lemna minor*, Duckweed)

Abstract

Soil, air, and water pollution can have serious impacts on the biotic component of ecosystems, including effects on human health and well-being. In the investigation reported here, I examine the population abundance of duckweed (*Lemna minor*) within two pond ecosystems on a local college campus. I estimated the population abundance of two separate duckweed populations that are assumed to experience different levels of chemical contamination because of parking lot surface runoff. I hypothesized that the pond closest to the potential pollution source would experience negative impacts on population dynamics. I predicted that the pond closest to the parking lot would support duckweed populations at a lower abundance than duckweed populations within ponds farther from the parking lot. I conducted field sampling using tea strainers to collect data for calculating estimates of population abundance based on frond counts. The results support the hypothesis and indicate that the pond closest to the parking lot supports a less abundant population than the pond farthest from the parking lot. Future studies investigating the effects of various compounds and compound concentrations on aquatic species and communities would be useful for improving our understanding of the ecotoxicological interactions caused by human behavior. Such studies would provide a sound empirical foundation for developing recommendations and strategies for sustainable interactions with the biosphere.

Introduction

Linkages between human actions and environmental inputs arising from those actions have effects on health, survival, and population growth or decline (McMichael 1993). Rapid growth of human populations combined with uncontrolled industrial growth, urban expansion, and agricultural land conversion has led to serious impacts on the environment and human quality of life (Molner and Molner 1999). The release of harmful chemicals into the environment continues, despite environmental activism and involvement from industry and the chemical and agricultural industries, as well as governmental regulations and oversight like that of the Environmental Protection Agency and the U.S. Department of Agriculture (Worldwatch Institute 2007).

An especially significant environmental impact from human actions is freshwater pollution. The Worldwatch Institute, a nonpartisan, independent, interdisciplinary research organization, reports that 81 percent of our nation's community water depends on groundwater sources, and that 53 percent of the population relies on groundwater as its source of drinking water (Worldwatch Institute 2007). Freshwater resources are critical for populations. Ninety-five percent of all freshwater on Earth exists as groundwater. It is imperative that we understand the sources and impact of pollution in order to develop recommendations and strategies for management, reduction, and prevention of pollution, as well as restoration and protection of freshwater resources.

The major sources of fresh water pollution can be classified as municipal, industrial, and agricultural. Freshwater contamination may occur by many routes from these sources, including surface runoff, liquid spills, wastewater discharges, littering, leaching to groundwater, leaking underground storage tanks, and municipal landfills. Among the most significant chemical contaminants are hydrocarbons, heavy metals, herbicides, pesticides, and chlorinated hydrocarbons.

The intensity of pollution and other disturbances can be assessed by understanding the biological responses of individual organisms, populations, and communities to environmental changes caused by freshwater pollution (Almar et al. 1998). These biological indicators can be used to encourage the ecologically sensitive management of freshwaters despite the conflicting demands of today's society.

My objective was to estimate the abundance of duckweed populations within two ponds on the campus of a local college. I hypothesized that duckweed population dynamics would be negatively impacted by surface runoff pollution such as from roads and parking lots. To test this hypothesis, I estimated population abundance of duckweed within two ponds, one adjacent to a large parking lot and the other isolated from the parking lot by both distance and elevation. I predicted that the pond adjacent to the parking lot would experience more surface pollutant runoff. This would lead to contamination of the pond, resulting in lower population abundance of duckweed.

Materials and Methods

Study Site

The study site is located on a local college campus in St. Louis, Missouri. Situated in the northeastern corner of the campus property are two ponds that were the focus of this investigation. These ponds are filled primarily by runoff coming from the areas around the pond that drain rainwater into the ponds. As the water flows into the ponds, it picks up minerals and soil particles. The type of rock and soil through which the water flows determines the type of minerals present in the water. Rocks exposed in this part of St. Louis County are usually limestone, made primarily of calcium carbonate. It is of significance to note that of the two ponds, the northernmost pond, surrounded by woodlands, is situated at a higher elevation and farther away from the large parking lot.

Study Species

The duckweeds (family *Lemnacea*) are a widely distributed group of 22 species found from the subpolar regions to the tropics. They are typically found within ponds, lakes, slow-moving streams and rivers, and ditches. The typical body structure consists of a few flattened, disc-like fronds or thalli that float on the surface. Reproduction is by budding, where new individuals grow from and break off of the "parent" plant, usually after six or so new fronds have developed.

Overall Procedure

For equipment, I obtained plastic tea strainers, white plastic trays, wax pencils, tape measures, and balances. I measured the surface area of the tea strainer and the surface area of ponds 1 and 2. I collected 12 samples from each pond by lowering the tea strainer carefully into the pond, allowing duckweed to redistribute itself once again. I then raised the tea strainer and allowed water to drain, and removed the duckweed attached to the *outside* of the tea strainer and returned them to the pond.

To calculate abundance, I marked the bottom of a shallow, rectangular plastic culture tray with a wax pencil to create a grid of squares 5 centimeters by 5 centimeters. I filled the tray with approximately 1 cm depth of water and emptied the fronds from the tea strainer into the tray. I used the standardized procedure of counting every visible frond. This included any tips of small new fronds that are just beginning to emerge from the pocket of the mother frond. I used a dissecting microscope (10×) for frond counts. I counted the number of fronds over 10 grid squares. To calculate the average population abundance (i.e., number of fronds) of duckweed in the two ponds, I used the following calculations:

1. the total estimated number of plants (N) collected in each tea-strainer sample = (the average number of plants counted in a grid square of the tray) × (total number of squares)

2. the estimated number of plants (P) on the pond = average N × (area of pond ÷ area of the tea strainer).

Statistical Analysis

Summary statistics were calculated for the data from both ponds. A *t*-test was conducted to compare the data.

Results

The results of the *t*-test revealed a significant difference between the two ponds in terms of duckweed abundance ($t = -5.16$, df $= 22$, $p < 0.0001$). Data are shown in Table 1, and Figure 1 presents the summary statistics for duckweed abundance. Pond 1, which is located closer to the parking lot, was predicted to exhibit lower abundance of duckweed because of assumed greater surface runoff of pollutants from the nearby parking lot. This was supported by the data gathered in May 2007, and therefore the hypothesis was accepted.

Discussion

The results of this investigation indicate that there is a statistically significant difference in the population abundances of duckweed within the two ponds located on the college campus.

The data support my hypothesis of a negatively impacted duckweed population in the lower pond resulting from elevated chemical contamination because of surface runoff from the parking lot.

The assumption that higher levels of chemical pollution are responsible for the decreased abundance of duckweed in the lower pond needs to be assessed. Further investigations into the mechanisms responsible for the observed differences in abundance are warranted. Artificial ponds may be used to conduct research investigations of this nature. Small artificial ponds can be created in the lab or field to investigate the impact of pollutants on pond ecosystems. Heimbach, Pflueger, and Ratte (1991) used 5-m³ artificial ponds to assess the influence of the potential hazards of chemicals to aquatic ecosystems. A study by Hughes et al. (1980) investigated the persistence and biological impact of organophosphorus insecticides within artificial ponds. In the laboratory, petri dishes or bowls may be used to create small artificial ponds under controlled conditions to determine the effect of independent variables. The strength of using such artificial ponds is that the investigator can ensure sufficient replication and control over relevant factors, increasing the strength of statistical evaluations of results (for example, see Rowe and Dunson 1994). The downside is that the artificial ponds may not accurately mimic larger natural ponds. Ideally, results from artificial ponds would be compared to natural ponds to check for similarity of results such as those done by Heimbach, Pflueger, and Ratte (1991).

TABLE 1 Summary data of duckweed abundance in the two ponds		
Sample Abundance	# of Fronds	# of Fronds
Estimate	Pond 1	Pond 2
1	82	115
2	72	111
3	89	99
4	79	78
5	98	102
6	90	122
7	96	112
8	92	104
9	86	118
10	82	128
11	93	120
12	88	109
Mean	87.2	109.8
SD	7.484833	13.16906
SE	2.160685	3.801581

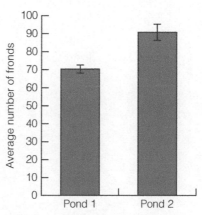

Figure 1 Average abundance of duckweed within the two ponds located on the college campus (values are means ± SE)

Of broader significance and potential application would be investigations into the uses of duckweed as a biological indicator for the intensity of pollution on aquatic ecosystems. Duckweeds are excellent model organisms because their growth rates can be measured by counting how many new fronds develop over a five-day period. Among the possible investigations are the effects of different chemical pollutants such as oil, gasoline, copper sulfate, various pesticides, herbicides, insecticides, and various fertilizers. Research such as this will continue to shed light on the linkages between human actions and environmental inputs as well as the consequences those actions and inputs have on health, survival, and population growth or decline. Armed with this knowledge, we may be better able to develop wise, sustainable, ecologically sensitive management and interaction with the world's freshwaters.

Citations

Almar, M., L. Otero, C. Santos, and J. González Gallego. 1998. Liver glutathione content and glutathione-dependent enzymes of two species of freshwater fish as bioindicators of chemical pollution. *Journal of Environmental Science & Health B* (Nov) 33(6): 769–83.

Heimbach, F., W. Pflueger, and H.-T. Ratte. 1991. Use of small artificial ponds for assessment of hazards to aquatic ecosystems. *Environmental Toxicology and Chemistry* 11(1): 27–34.

Hughes, D. N., M. G. Boyer, M. H. Papst, C. D. Fowle, G. A. V. Rees, and P. Baulu. 1980. Persistence of three organophosphorus insecticides in artificial ponds and some biological implications. *Archives of Environmental Contamination and Toxicology* 9(3): 269–79.

McMichael, A. J. 1993. *Planetary overload: Global environmental change and the health of the human species.* New York: Cambridge University Press.

Molner, S., and I. M. Molner. 1999. *Environmental change and human survival: Some Dimensions of human ecology.* Englewood Cliffs, NJ: Prentice Hall.

Rowe, C. L, and W. Dunson. 1994. The value of simulated pond communities in mesocosms for studies of amphibian ecology and ecotoxicology. *Journal of Herpetology* 28(3): 346–56.

Worldwatch Institute. 2007. Pollution. Online at http://www.worldwatch.org/taxonomy/term/109.

Animal Behavior

OBJECTIVES

After completing this exercise, you will be able to

1. define *instinctive behavior, learned behavior, phototaxis, geotaxis, thigmotaxis, posture*;

2. discuss the role of the genes in determining behavior;

3. create a list of major environmental stimuli;

4. describe what negative taxis and positive taxis mean;

5. list two aspects of frog behavior that do not require a forebrain and one that does not require a brain;

6. distinguish among the various environmental constituents that individuals can respond to.

INTRODUCTION

Animal behavior is the sum of the combined responses of individuals and groups to all of the stimuli they receive from external and internal environments. Receptors like the eyes, ears, olfactory epithelium, and taste buds gather information about various stimuli. Then, chemical messengers—hormones and the neurotransmitters of nervous tissues—direct glands and muscles as to how to respond, producing behavior. Responses result from reflexes, which occur below the level of consciousness, and reactions, which require a conscious decision.

Some behavior is in its entirety programmed in and directed by the genes. This **instinctive behavior,** initially at least, does not need any prior experience or learning. Other behavior does require some prior experience and learning. This **learned behavior** still has a genetic component, but the appropriate response or choice of responses must be learned. A clear example of learned behavior for us is the recognition, interpretation, and creation of spoken and written language. However, this behavior still requires adequate ears, a complex brain, and the correct vocal apparatus, the building of which is genetically programmed. Ultimately, it is the action of evolutionary agents over time that determines the genes animals possess and therefore the behaviors they will and can exhibit.

Not all learned behaviors require a conscious decision. Such behaviors include the imprinting of newborn goslings and many other animals on their mother and the classical conditioning of Pavlov's dogs to salivate when a bell is rung whether food is present or not. Other learned behaviors do require a conscious decision, such as braking your car after you observe a child's ball rolling into the street from between parked vehicles.

30.1   Instinctive Behavior *(40 min.)*

Examples of instinctive behavior include basic responses to major environmental stimuli such as gravity (**geotaxis**), light (**phototaxis**), and touch (**thigmotaxis**) and more complex actions such as the suckling, grasping, and smiling responses in human babies and food preference in newborn garter snakes. Responses can be positive (individuals are attracted to the stimulus) or negative (individuals avoid the stimulus). List up to five other major environmental stimuli that animals may respond to:

1. _____
2. _____
3. _____
4. _____
5. _____

MATERIALS

Per student pair:

■ mealworm beetle
■ plastic tray
■ wooden or plastic blocks

Per lab room:

■ snails in an algae-coated freshwater aquarium
■ cover for the left half of the aquarium

PROCEDURE

A. Response to Gravity

1. Work in pairs. Describe the feeding behavior of snails in an algae-coated freshwater aquarium.

_____

_____

_____

2. In the second column of Table 30-1, count and record the number of snails that orient with their heads pointed more upward than downward and those with their heads more downward than upward.

TABLE 30-1	Orientation of Snails in a Freshwater Aquarium	
Orientation	Number	Percentage
Heads upward		
Heads downward		
Total		

3. Total these two numbers, calculate the percentage of upwardly and downwardly pointing snails, and record them in the third column of Table 30-1.
4. Are most of the snails oriented with their heads directed downward (positive geotaxis) or upward (negative geotaxis)? _____

If the numbers are about equal, then the snails don't respond or are neutral in their response to gravity.

B. Response to Light

1. Count how many snails are in each half of the aquarium and record the numbers in the second column of Table 30-2.

TABLE 30-2	Snail Counts in the Light and Dark Halves of an Aquarium		
Side/Condition	Initial Count	Final Count	Number Gained (+) or Lost (−)
Right/Light			
Left/Dark			

2. At this time, your instructor will place a light over the right half of the aquarium and cover the left half.
3. At the end of the lab period, remove the lid and again count the snails in the right/light and left/dark sides of the aquarium. Record the numbers in third column of Table 30-2.
4. Record the number gained or lost from each side of the aquarium (fourth column).
5. Are snails phototactic?_____ If so, is their response negative or positive? _____

C. Response to Touch

1. Obtain a plastic tray and some wooden or plastic blocks and construct a simple maze.
2. Place a mealworm beetle in the center.
3. Describe its reaction to touching (walking into) an obstacle.

_____

_____

_____

_____

_____

_____

4. Is your beetle thigmotactic? _____ If so, is the response negative or positive? _____

_____

5. Start again and note if your beetle initially turns right or left in response to touching a wall by recording a check mark in the second or third column of Table 30-3. Continue to look until you have recorded a total of ten observations.
6. Does your beetle show any possible dominance for turning to the right or left? _____

TABLE 30-3 Response of Beetle to Touch

Observations	Right	Left
1		
2		
3		
4		
5		
6		
7		
8		
9		
10		
Totals		

30.2   Frog Behavior *(30 min.)*

It is often difficult to distinguish between instinctive and learned behavior in another animal and, for that matter, to recognize when a conscious decision is made. How other animals think and whether or not they think like humans are subjects of great controversy.

MATERIALS

Per student group (4):

- frog
- plastic tray
- small rectangular plastic tank filled with room-temperature H_2O

Caution

The secretions of amphibian skin may be irritating, and the skin itself may harbor bacteria. Use latex gloves whenever you handle a specimen and do not touch exposed parts of your body, especially your eyes.

PROCEDURE

1. Work in groups of four students. Put on latex gloves and place a frog in the middle of a plastic tray. If the frog jumps off the tray, catch and replace it. Once it sits still, describe the typical frog **posture**—the position of the body and its parts.

_____

_____

2. Place the frog on its back and let go. Describe its response.

_____

_____

Speculate as to the stimulus that initiates this righting response.

3. Put the frog in a tank of water and observe the frog's response to this stimulus. If the frog does not move, give a little push from the rear. Describe in some detail its response.

_____

_____

4. Hold the frog in your hand and stretch out a leg. Give a firm pinch to the frog's toes but not so hard that you risk breaking them. Describe the frog's response.

_____

_____

_____

5. Speculate as to how much of this behavior is instinctive or learned and how much requires conscious decisions on the part of the frog. Discuss your observations, come to a group consensus, and summarize them here.

_____

_____

_____

6. Likewise, speculate as to which of these behaviors can be done by a frog without a forebrain (no consciousness), without a brain, and without a central nervous system or CNS (no brain or spinal cord). Record your consensus predictions by placing a check mark in the Prediction columns of Table 30-4.

TABLE 30-4 Frog Behavior

Response	Without a Forebrain		Without a Brain		Without a CNS	
	Prediction	Result	Prediction	Result	Prediction	Result
Righting						
Swimming						
Withdrawal						

7. Your instructor may demonstrate, show a videotape, ask you to watch videoclips on a computer, or simply tell you about the correct results. Record them in Table 30-4.
8. Do the results affect the way you would answer the question, How much of human behavior is instinctive or learned and how much requires conscious decisions? If so, explain.

_____

_____

_____

30.3   Tropical Fish Behavior (30 min. or more)

In nature, animals don't live in isolation from the physical and chemical makeup of the environment or from other species—animals or plants. Behavior occurs between individuals of the same species and individuals of different species, as well as between individuals and nonliving environmental constituents.

MATERIALS
- aquarium with 5 species of tropical fish; sign that displays a picture, name, and code for each species represented

PROCEDURE
1. Go to the tropical fish aquarium. Note the sign with a picture of each species of fish present, its name, a code, and the number of individuals of each species in the tank. Transfer this information to Table 30-5.

TABLE 30-5	Fish Present in Aquarium		
Common Name	Species Name	Code	Number of Individuals

2. Sit and watch one of the five species of fish in the aquarium. Write the species name in Table 30-5.
3. Also note in Table 30-6 any behaviors you observe, their frequency, and, if appropriate, the living (same species, another species; use codes), or nonliving environmental constituent (plant, rock, bottom of tank, sides of tank, shady or open areas of tank, and so on) at which it is directed. Observations could include a brief description of feeding, display (taking a particular body orientation), mating, egg laying, aggression (such as fin nipping), location in tank, schooling (fish swim in synchrony), jumping, or relative swimming speed.

TABLE 30-6	Behavioral Observations of _____.			
Behavior	Frequency	Same Species	Another Species	Nonliving

4. Describe your chosen species' behavior under these conditions.
_____
_____
_____
_____
_____

30.4 Phototaxis Experiment *(60 min.)*

In this part of the exercise, you will test the hypothesis that *planarians exhibit negative phototaxis.*

MATERIALS

Per student group (4):

- petri dish
- fresh stream water (or aerated pond or spring water)
- planaria
- penlight

PROCEDURE

1. Work in a group of four. In Table 30-7, predict the planarian's response to a light pointed at the front of its head, the side of its head, and its tail when the planarian is placed in a petri dish of fresh stream water. As done previously, use "If . . . then . . . " formats for your statements.

TABLE 30-7	Presence or Absence, and Direction of Phototaxis in Planarians		
Predictions:			
Trial	**Light Directed at Front of Head**	**Light Directed at Side of Head**	**Light Directed at Tail**
1			
2			
3			
4			
5			
Most Common Response:			
Conclusion:			

2. Cooperate with other groups and darken the room. Determine the orientation of a living specimen in a glass culture dish filled with fresh stream water. Point a penlight at its head and turn it on. Describe briefly in the second column of Table 30-7 the specimen's response or lack of response (moves toward the light, moves away from the light, or no response). Repeat this trial for a total of five times, giving the specimen a 2-minute rest between each trial.
3. Repeat the procedure described in step 2, except shine the light at the side of the specimen's head and record the response in the third column of Table 30-7 (turns away from the light, turns toward the light, or no response).
4. Again, repeat the procedure described in step 2, except shine the light at the specimen's tail and record the response in the fourth column of Table 30-7 (moves toward the light, moves away from the light, or no response).
5. Note the most common response to each set of trials and write a conclusion as to whether the prediction is rejected or accepted in Table 30-7.
6. Share your results with the other groups and collect their information. Are your results consistent with their results?

_____

_____

7. If you were to look for planaria in a stream, where would you look for them? Base your answer on the results of this experiment.

_____

_____

_____

_____ 1. Behavior that is entirely controlled by genes and has no learning component is
(a) learned
(b) habituated
(c) instinctive
(d) classically conditioned

_____ 2. Learned behavior
(a) has a learning component
(b) has a genetic component
(c) has no genetic component
(d) has both a and b

_____ 3. All behavior
(a) has a genetic component
(b) involves learning
(c) requires a conscious decision
(d) both a and c

_____ 4. Braking your car after you observe a child's ball rolling into the street from between parked vehicles
(a) is instinctive
(b) requires a conscious decision
(c) is a reflex
(d) both a and b

_____ 5. An individual who avoids light is exhibiting
(a) negative phototaxis
(b) positive geotaxis
(c) negative thigmotaxis
(d) positive thigmotaxis

_____ 6. An individual who is attracted to light is exhibiting
(a) negative phototaxis
(b) positive phototaxis
(c) negative thigmotaxis
(d) positive thigmotaxis

_____ 7. An individual who responds to walking into an obstruction by pushing into it harder is exhibiting
(a) negative phototaxis
(b) positive geotaxis
(c) negative thigmotaxis
(d) positive thigmotaxis

_____ 8. An individual who avoids the pull of gravity is exhibiting
(a) negative phototaxis
(b) positive geotaxis
(c) negative geotaxis
(d) positive thigmotaxis

_____ 9. Schooling is
(a) a feeding behavior
(b) fin nipping
(c) fish swimming in synchrony
(d) a mating behavior

_____ 10. Fish can exhibit behaviors that are directed to
(a) members of the same species
(b) members of different species
(c) nonliving constituents of the environment
(d) all of these choices

EXERCISE 30

Animal Behavior

POST-LAB QUESTIONS

Introduction

1. Compare the contributions of genes and learning to instinctive behavior and learned behavior.

2. Does all learned behavior require a conscious decision before responding to a stimulus? Explain why or why not.

30.1 Instinctive Behavior

3. Match the terms in the left column to the definitions in the right column. Use each answer only once.

   _____ a. negative geotaxis   i. avoidance of light
   _____ b. positive geotaxis   ii. attracted to gravity
   _____ c. negative phototaxis   iii. attracted to light
   _____ d. positive phototaxis   iv. avoidance of touch
   _____ e. negative thigmotaxis   v. avoidance of gravity
   _____ f. positive thigmotaxis   vi. attracted to touch

30.2 Frog Behavior

4. Does a frog need a consciousness to behave like a frog? Explain.

30.3 Tropical Fish Behavior

5. Although a tropical fish aquarium is at best an approximation of a natural habitat, what constituents does the aquarium share with the natural habitat that are important to understanding the behavior of a particular species?

Food for Thought

6. What would you call an aquatic organism's avoidance of high concentrations of a chemical in the water?

7. Look up the fight–flight response in your textbook. Describe what it is and how it is still an advantageous part of human behavior.

8. You probably have had one if not several pet mammals—dogs, cats, and so on. Based on your interactions with them, hypothesize as to whether they can think like humans. If you believe that they can think, how and why might their thoughts be different from ours? (*Hint:* Consider the differences between the language of humans and other mammals.)

9. Search the World Wide Web for sites about animal thinking. List two sites and briefly summarize their contents.

   http://

   http://

10. Search the World Wide Web for sites that include animal cams. List and briefly describe the contents of two sites that you would recommend.

    http://

    http://

Female Mate Choice Based on Male Size in the Cichlid Fish, *Oreochromis mossambicus*

Abstract

Sexual selection is an important evolutionary process that can exert strong effects on species. We chose to investigate intersexual female mate choice in a haplochromine cichlid fish species, *Oreochromis mossambicus*. Our hypothesis was that females selected from among males based on size. We predicted that, given a choice between small and large males, females would choose the larger male. Our experimental design involved binary choice experiments in which females could select between two males of different sizes. We found no statistically significant difference between females choosing small versus females choosing large males. Although we found no statistically significant differences, possibly because of low sample size, there was a trend toward females choosing the larger male. We recommend further research with larger sample sizes. Research that investigates other potential factors that may be cues in sexual selection is suggested.

Introduction

Sexual selection is a microevolutionary process that can lead to the evolution of extravagant coloration or ornamentation among members of the competing or chosen sex (Starr, Evers, and Starr, 2008). Characteristics that provide an advantage in attracting or securing mates are typically favored. Sexual selection may occur when either males or females choose from among the other sex (intersexual selection) or when members of the same sex compete among themselves for access to mates (intrasexual selection). For example, collared lizards exhibit both types of sexual selection. Males compete with other males for access to females, and this has led to intrasexual selection for larger and larger male size, resulting in sexual size dimorphism within the species (Baird, Fox, and McCoy, 1997). Female collared lizards prefer males with brighter coloration, and this intersexual selection has led to sexual dichromatism within the species (McCoy et al., 1997).

The haplochromine cichlid fish species of Africa's great lakes provide an excellent system for studying the processes of sexual selection (Seehausen, 2000). We chose to investigate female mate choice in the cichlid species *Oreochromis mossambicus*. We were interested in the characteristics that females use to choose from among males. Females of this species are mouth brooders and invest more heavily in offspring production and caring than do males. We therefore expected that females would be choosy as to which males to mate with.

In laboratory experiments we tested the hypothesis that females would prefer large males over smaller males. The null hypothesis stated that females had no preference and both sizes of males were equally likely to be chosen.

Materials and Methods

Study Species

We obtained 20 adult female and 40 adult male *Oreochromis mossambicus*. Males of this species defend territories in which they dig spawning pits. These pits are used solely for mating and oviposition (Fryer and Iles, 1972). Females mouthbrood the eggs soon after they are fertilized at sites other than the mating sites. There are many potential factors that females may consider before choosing a mate: size of male, coloration of males, courtship behaviors, spawning pit characteristics, and so on. We chose to investigate male size.

Housing Conditions

In the laboratory, males were kept in pairs (one small and one large male in each pair) within aquariums containing two identical spawning pits at either end of the aquariums. Permanent opaque half-partitions were placed in the middle of the tank such that fish could swim to either side by moving around the partitions, but the partitions were staggered such that when fish were near the spawning pits they could not see each other. Males were allowed to acclimate to the aquariums for one week before conducting the experiments. Females were kept in separate communal aquariums from the males. All aquariums were maintained at about 28° C, illuminated with 15-watt daylight fluorescent lighting and exposed to a 12-hour by 12-hour light-to-dark regime. Fish were fed once a day with commercial cichlid pellets and commercial flakes.

Experimental Procedure

Binary choice experiments involved introducing a female to the aquariums containing the two males. A female was placed in the center of the aquariums between the two half-partitions. Ten females were used for these experiments. Once a day, spawning pits were examined for eggs. Experimental trials lasted until females spawned (1 to 28 days; median 2 days). Female preference was determined by the location of eggs. When a female spawned with both males or did not spawn, preference could not be attributed and the experimental trial was excluded from analysis.

Data Analysis

We used a two-tailed binomial test with a null hypothesis that stated the number of observations (female choices) in each category (of male size) is equal, and the alternative hypothesis is that the observed data are significantly different from the expected (Sokal and Rohlf, 1981).

Results

Eight out of ten females chose large versus small males to spawn with (Table 1 and Figure 1). A binomial test revealed that there is not a statistically significant preference for large versus small males (binomial test: $N = 10$, $P = 0.109$). These data are insufficient to provide evidence for a significant male size preference by females. Our data cannot be taken as different from what is expected for a population where there is no preference for male size. Therefore, we accepted the null hypothesis.

TABLE 1 Results of binary choice experiments in which female females ($N = 10$) chose to spawn with either large or small males

Females	Small Male	Large Male
1	No	Yes
2	No	Yes
3	Yes	No
4	No	Yes
5	Yes	No
6	No	Yes
7	No	Yes
8	No	Yes
9	No	Yes
10	No	Yes

Discussion

The results of our study do not reveal a significant difference between females choosing small versus large males. Our data were insufficient to indicate female mate choice based on male size. Although there was no significant difference, we did detect a statistically nonsignificant difference that bears further investigation. Because our sample size was limited to 10 binary experiment trials, our statistical test was more than likely insufficiently powerful enough to detect a significant difference in our data set. Future experiments with a larger sample size may be powerful enough to detect female mate choice.

Other studies indicate that female mate choice among cichlid species may be common, although the specific trait being chosen may differ between species. Seehausen and van Alphen (1998) report interspecific female mate choice based on male coloration in two closely related species of haplochromine cichlids from Lake Victoria. Territory quality (Dijkstra, van der Zee, and Groothuis, 2007) and male display behaviors (Fryer and Iles, 1972) have also been shown to affect female preference in cichlids. We suggest that future studies with sample sizes larger than ours could shed light on which, if any, male characters (such as size, coloration, courtship behavior) influence female mate choice in *Oreochromis mossambicus*.

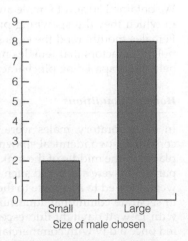

Figure 1 Results of binary choice experiments in which females ($N = 10$) chose to spawn with either small or large males.

Other valuable lines of inquiry involve investigating male mate choice and male intrasexual selection. Amorim and Almada (2004) were able to detect male–male encounters affected the subsequent social status and subsequent courtship behavior and mating success in *Oreochromis mossambicus*. In another species of cichlid, *Astatotilapia flaviijosephi*, Werner and Lotem (2003) detected male mate choice where males preferred to court larger rather than smaller females.

Citations

Amorim, M. C. P., and V. C. Almada. 2005. The outcome of male-male encounters affects subsequent sound production during courtship in the cichlid fish *Oreochromis mossambicus*. *Animal Behavior* 69: 595–601.

Baird, T. A., S. F. Fox, and J. K. McCoy. 1997. Population differences in the roles of size and coloration in intra- and inter-sexual selection in the collared lizard, *Crotaphytus collaris*: Influence of habitat and social organization. *Behavioral Ecology* 8(5): 506–17.

Dijkstra, P. D., E. M. van der Zee, and T. G. G. Groothuis. 2007. Territory quality affects female preference in a Lake Victoria cichlid fish. *Behavioral Ecology and Sociobiology* 62(5): 747–55.

Fryer, G., and T. D. Iles. 1972. *The cichlid fishes of the great lakes of Africa: Their biology and evolution*. Edinburgh: Oliver & Boyd.

McCoy, J. K., H. J. Harmon, T. A. Baird, and S. F. Fox. 1997. Geographic variation in sexual dichromatism in the collared lizard, *Crotaphytus collaris* (Sauria: Crotaphytidae). *Copeia* 1997(5): 565–71.

Seehausen, O. 2000. Explosive speciation rates and unusual species richness in haplochromine cichlid fishes: Effects of sexual selection. *Advances in Ecological Research* 31: 237–74.

Seehausen, O., and J. J. M. van Alphen. 1998. The effect of male coloration on female mate choice in closely related Lake Victoria cichlids (*Haplochromis nyererei* complex). *Behavioral Ecology and Sociobiology* 42(1): 1–8.

Sokal, R. R., and J. L. Rohlf. 1981. *Biometry*. New York: W. H. Freeman.

Starr, C., C. A. Evers, and L. Starr. 2008. *Biology: Concepts and application*, 7th ed. Belmont, CA: Cengage/Brooks Cole.

Werner, N. Y., and A. Lotem. 2003. Choosy males in a haplochromine cichlid: First experimental evidence for male mate choice in a lekking species. *Animal Behavior* 66: 293–8.

Measurement Conversions

Metric to American Standard	**American Standard to Metric**
Length	*Length*
1 mm = 0.039 inch	1 inch = 2.54 cm
1 cm = 0.394 inch	1 foot = 0.305 m
1 m = 3.28 feet	1 yard = 0.914 m
1 m = 1.09 yards	1 mile = 1.61 km
1 km = 0.622 miles	
Volume	*Volume*
1 mL = 0.0338 fluid ounce	1 fluid ounce = 29.6 mL
1 L = 4.23 cups	1 cup = 237 mL
1 L = 2.11 pints	1 pint = 0.474 L
1 L = 1.06 quarts	1 quart = 0.947 L
1 L = 0.264 gallon	1 gallon = 3.79 L
Mass	*Mass*
1 mg = 0.0000353 ounce	1 ounce = 28.3 g
1 g = 0.0353 ounce	1 pound = 0.454 kg
1 kg = 2.21 pounds	

The Scientific Method

To appreciate biology or, for that matter, any body of scientific knowledge, you need to understand how the **scientific method** is used to gather that knowledge. We use the scientific method to test the predictions of possible answers to questions about nature in ways that we can duplicate or verify. Answers supported by test results are added to the body of scientific knowledge and contribute to the concepts presented in your textbook and other science books. Although these concepts are as up-to-date as possible, they are always open to further questions and modifications.

One of the roots of the scientific method can be found in ancient Greek philosophy. The natural philosophy of Aristotle and his colleagues was mechanistic rather than vitalistic. A **mechanist** believes that only natural forces govern living things, along with the rest of the universe, while a **vitalist** believes that the universe is at least partially governed by supernatural powers. Mechanists look for interrelationships between the structures and functions of living things, and the processes that shape them. Their explanations of nature deal in **cause and effect**—the idea that one thing is the result of another thing (for example, fertilization of an egg initiates the developmental process that forms an adult.). In contrast, vitalists often use purposeful explanations of natural events (the fertilized egg strives to develop into an adult). Although statements that ascribe purpose to things often feel comfortable to the writer, try to avoid them when writing lab reports and scientific papers.

Aristotle and his colleagues developed three rules to examine the laws of nature: Carefully observe some aspect of nature; examine these observations as to their similarities and differences; and produce a principle or generalization about the aspect of nature being studied (for example, all mammals nourish their young with milk).

The major defect of natural philosophy was that it accepted the idea of *absolute truth.* This belief suppressed the testing of principles once they had been formulated. Thus, Aristotle's belief in spontaneous generation, the principle that some life can arise from nonliving things (say maggots from spoiled meat), survived over 2000 years of controversy before being discredited by Louis Pasteur in 1860. Rejection of the idea of absolute truth coupled with the testing of principles either by experimentation or by further pertinent observation is the essence of the modern scientific method.

Although there is not one universal scientific method, Figure A-1 illustrates the general process.

MATERIALS

Per lab room:

- blindfold
- plastic beakers with an inside diameter of about 8 cm stuffed with cotton wool
- four or five liquid crystal thermometers

Step 1. Observation. As with natural philosophy, the scientific method starts with careful observation. An investigator may make observations from nature or from the written words of other investigators, which are published in books or research articles in scientific journals and are available in the storehouse of human knowledge, the library. One subject we all have some knowledge of is the human body. The first four rows of Table 1 list some observations about the human body. The fifth row is blank so that you can fill in the steps of the scientific method for some observation about the human body, or anything else you and your instructor wish to investigate.

Step 2. Question. In the second step of the scientific method, *we ask a question* about these observations. The quality of this question will depend on how carefully the observations were made and analyzed. Table 1 includes questions raised by the listed observations.

Step 3. Hypothesis. Now we *construct a hypothesis*—that is, we derive by inductive reasoning a possible answer to the question. **Induction** is a logical process by which all known observations are combined and considered before producing a possible answer. Table 1 includes examples of hypotheses.

Step 4. Prediction. In this step we *formulate a prediction*—we assume the hypothesis is correct and predict the result of a test that reveals some aspect of it. This is deductive or "if-then" reasoning. **Deduction** is a logical process by which a prediction is produced from a possible answer to the question asked. Table 1 lists a prediction for each hypothesis.

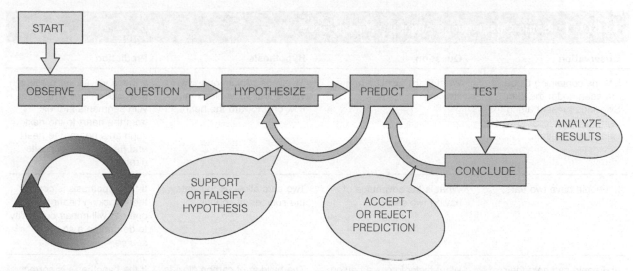

Figure A-1 The scientific method. Support or falsification of the hypothesis usually necessitates further observations, adjustment to the question, and modification of the hypothesis. Once started, the scientific method cycles over and over again, each turn further refining the hypothesis.

Step 5. Experiment or Pertinent Observations. In this step we *perform an experiment or make pertinent observations* to test the prediction. In an experiment of classical design, the individuals or items under study are divided into two groups: an **experimental group** that is treated with (or possesses) the independent variable and a **control group** that is not (or does not). Sometimes there is more than one experimental group. Sometimes subjects participate in both groups, experimental and control, and are tested both with and without the treatment.

In any test there are three kinds of variables. The **independent variable** is the treatment or condition under study. The **dependent variable** is the event or condition that is measured or observed when the results are gathered. The **controlled variables** are all other factors, which the investigator attempts to keep the same for all groups under study.

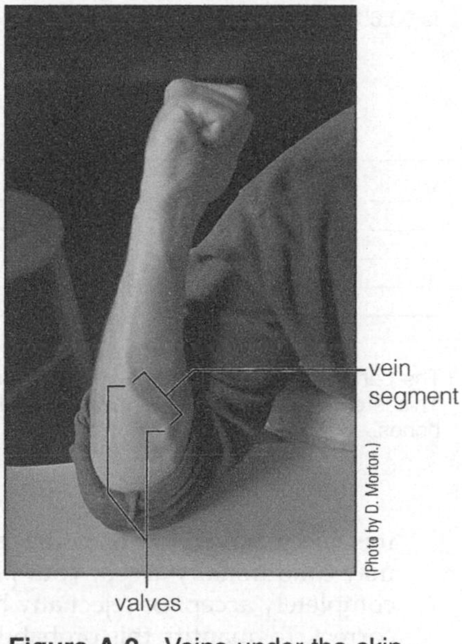

Note: The italicized statements show how the scientific method is applied to the predictions in Table 1 or give examples of the scientific method in practice.

To test the prediction that blocking hearing in one ear impairs our ability to point out a sound's source (row II in Table 1), a group of subjects is tested first with no ears blocked (control group) and then with one ear blocked (experimental group). The independent variable is the blocking of one ear; the dependent variable is the ability to point out a sound's source; and the controlled variables are the standard conditions used for each trial—same test sound, same background noise, same procedure for each subject (same blocked ear, same instructions, same sequence of trials, same time between trials), and recruitment of appropriate subjects.

Sometimes the best tests of the predictions of a hypothesis are not actual experiments but further pertinent observations. One of the most important principles in biology, Darwin's theory of natural selection, was developed by this nonexperimental approach. Although they are a little more difficult to form, the nonexperimental approach also has variables.

To test the prediction that a liquid crystal thermometer will record different temperatures on the forehead, back of neck, and forearm (row IV of Table 1), the independent variable is location; the dependent variable is temperature; and the controlled variables are using the same thermometer to measure skin temperature at the three locations and measuring all of the subjects at rest.

Step 6. Conclusion. *To make a conclusion*—the last step in one cycle of the scientific method—you use the results of the experiment or pertinent observations to evaluate your hypothesis. If your prediction does not occur,

Figure A-2 Veins under the skin.

— vein segment

valves

(Photo by D. Morton.)

TABLE 1 Some Observations About the Human Body

Observation	Question	Hypothesis	Prediction
I. Veins containing blood are seen under the skin. Swellings present along the vein are often located where veins join together.[a]	What is the function of the swellings?	Swellings contain one-way valves that allow blood to flow only toward the heart.	If these valves are present, then blood flows only from vein segments farther from the heart to the next segments nearer the heart and never in the opposite direction.
II. People have two ears.	What is the advantage of having two ears?	Two ears allow us to locate the sources of sounds.	If the hypothesis is correct, then blocking hearing in one ear will impair our ability to determine a sound's source.[b]
III. People can hold their breath for only a short period of time.	What factor forces a person to take a breath?	The buildup of carbon dioxide derived from the body's metabolic activity stimulates us to take a breath.	If the hypothesis is correct, then people will hold their breath a shorter time just after exercise compared to when they are at rest.
IV. Normal body temperature is 98.6°F.	Is all of the body at the same 98.6°F temperature?	The skin, or at least some portion of it, is not 98.6°F.	If the hypothesis is correct, then a liquid crystal thermometer will record different temperatures on the forehead, back of neck, and forearm.
V. _____	_____	_____	_____

[a] The portion of the vein between swellings is called a segment and is illustrated in Figure A-2.
[b] This is especially the case for high-pitched sounds because higher frequencies travel less easily directly through tissues and bones.

it is rejected and your hypothesis or some aspect of it is falsified. If your prediction does occur, you may conditionally accept your prediction and your hypothesis is supported. However, you can never completely accept or reject any hypothesis; all you can do is state a probability that one is correct or incorrect. To quantify this probability, scientists use a branch of mathematics called *statistical analysis*.

Even if the prediction is rejected, this does not necessarily mean that the treatment caused the result. A coincidence or the effect of some unforeseen and thus uncontrolled variable could be causing the result. For this reason, the results of experiments and observations must be *repeatable* by the original investigator and others.

Even if the results are repeatable, this does not necessarily mean that the treatment caused the result. *Cause and effect*, especially in biology, is rarely proven in experiments. We can, however, say that the treatment and result are correlated. A **correlation** is a relationship between the independent and the dependent variables.

Severe narrowing of a coronary artery branch reduces blood flow to the heart muscle downstream. This region of heart muscle gets insufficient oxygen and cannot contract and may die, resulting in a heart attack. The initial cause is the narrowing of the artery and the final effect is the heart attack. Perhaps the heart attack victim smoked cigarettes. Smoking cigarettes is one of several factors that make a person more likely to have a heart attack. This is based on a correlation between smoking and heart attacks in the general population but we cannot say for sure that the smoking caused the heart attack.

THEORIES and PRINCIPLES: When exhaustive experiments and observations consistently support an important hypothesis, it is accepted as a **theory**. A theory that stands the test of time may be elevated to the status of a **principle**. Theories and principles are always considered when new hypotheses are formulated. However, like hypotheses, theories and principles can be modified or even discarded in the light of new knowledge. Biology, like life itself, is not static but is constantly changing.

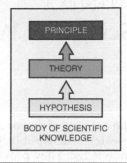

PRINCIPLE

THEORY

HYPOTHESIS

BODY OF SCIENTIFIC KNOWLEDGE

APPENDIX 3
Scientific Writing and Critiquing

The account of one or several related cycles of the scientific method is usually reported in depth in a research article published in a scientific journal. Writing such research articles allows scientists to share knowledge. They provide enough information so that other scientists can repeat the experiments or pertinent observations they describe. The journal *Science*, along with several others, presents its research articles in narrative form, and many of the details of the scientific method are understood and not stated. However, adherence to the modern scientific method is expected, and the scientific community understands that it is as important to expose mistakes as it is to praise new knowledge.

Most scientific journals present research reports in a standard format. This allows scientists to quickly and easily skim through the many new research articles published daily, reading in depth only those of particular interest. By understanding the purpose of each section of a research article, you not only will be able to locate the question, research context, main methods, results, and meaning of someone's research with speed and efficiency, but also learn how to design and communicate your own research investigations along the way! After all, the best way to learn science is to *do* science—reading research articles, exploring and trying out various experimental and observational methodologies that you are reading about, and eventually building on and extending previous research that you have read that particularly interested you. Scientists rarely investigate topics out of the blue; most often they become involved in research that they have either read about or observed that interested them. They then browse the available scientific knowledge base, most often the scientific journal literature, to see what is already known about the topic of interest. If they come across an area where nothing is known—an area of ignorance—then by using the scientific method they may choose to extend our knowledge.

The following sections are most commonly used in the scientific literature. The word counts are suggestions only and offered as guidelines to use when writing your own research reports.

Abstract (~200 words): The abstract contains a brief and concise summary of the main sections of the scientific report. Usually included are the research questions, the broad conceptual framework or relevance of the research topic, the hypothesis and prediction(s), the main methods, the main results and their interpretation, and a concluding statement.

Introduction (~200–300 words): The introduction is a short narrative that describes the research topic in some detail. It explains what is known regarding the research topic, based on previous research. This section concludes with what is not known regarding the research topic, the specific question that the research report focuses on, the hypothesis and prediction(s), and possibly a brief description of how the question will be answered.

Materials and Methods (~200 words): This section explains how the research was conducted. It should provide enough detail that someone could read the research report and repeat the experiment. The statistical analyses that will be used are also described in this section.

Results (~150 words): This section reports the results of the research but does not go into detail explaining or discussing what they mean (that goes in the next section). Here is where any tables, graphs, or figures would be located. The narrative of this section explains what the tables, graphs, or figures show.

Discussion (~200–300 words): In this section, the research question is restated, and the answer that was obtained from the research is given. Data reported in the results section are clearly related to the hypothesis and predictions. The results are interpreted, any unexpected or inconsistent results are explained, and a discussion of what the results mean is provided. The results are also integrated with the work of others, relating the research conducted to the larger body of work already completed and reported in the scientific literature. Finally, the research is related to the big picture or larger conceptual framework within which the research topic lies. New research questions or avenues of research are mentioned as well.

Citations or References: The sources of information that were used in the research report are cited in the text by author, and the details for how to locate the information is given.

It is recommended that you read several research articles to get an idea of how research is reported. This will increase your ability to read and interpret scientific literature as well as increase your ability to communicate through writing. The following is a general guide for critiquing scientific manuscripts.

Evaluating a Manuscript

When evaluating a manuscript, it is important to consider several things. Some of the following questions can only be answered by those who are intimately familiar with the organisms and techniques used. Others concern the structure, style, and soundness of a manuscript and can be answered by anyone with a peripheral knowledge of the subject matter. The ability to consider any of these problems will, of course, increase with experience.

ABSTRACT

Is it informative, concise, and complete?

INTRODUCTION

What is the main point of this study?

Is its relevance to scientific theory (a conceptual framework) apparent?

If it contains a hypothesis, is it stated in a way that is testable?

METHODS

Can the work be repeated on the basis of the methods given?

Are unwieldy technical terms defined?

Are the methods and organisms used appropriate for testing the hypothesis?

To what extent can conclusions drawn from laboratory experiments be extended to interpret events occurring in nature?

RESULTS

Are all important results given?

Were the results properly evaluated statistically?

Are the figures and tables clear?

Do the results contained in the figures and tables support the statements made in the text?

DISCUSSION

Are the data clearly related to the hypothesis?

Are the results properly interpreted?

Are the results integrated with those of other workers?

Are unexpected or inconsistent results explained?

Does the paper provide any new ideas or interpretations?

If the discussion contains any speculation, is it justifiable?

The following is a sample checklist that can be used for peer reviews of scientific reports.

Checklist for Journal Article Peer Review

Clearly critique how well the author sets his or her research topic in context with other scientific studies (the conceptual framework of the article and research).

Use the following outline for critiquing the introduction:

Background overview

Conceptual framework

Specific focus of article (hypothesis and objectives, or premise of a review article)

Significance of specific focus

Synthesis of literature on specific topic as provided by the article's author(s)

Broader impact and relevance of the research question

Evaluate the following:

a. Introduction

_____ Does the author of the article provide a sufficient context or background for the study, relating the specific focus to previous research or conceptual development?

_____ Does the author link his or her research topic to some overarching theory or conceptual framework?

_____ Are the hypothesis and objectives (the specific focus) clear?

_____ Is there a clear rationale and justification for the study (i.e., what is the significance of the study topic, in terms of advancing knowledge and understanding within its own field or across different fields of science)?

b. Methods, results, and discussion

_____ Critique the article (not just summarize), discussing such things as the development of ideas pertaining to the study topic, criteria for evidence, appropriateness and strength of research methodologies, and so on.

c. Conclusions

_____ Discuss the main conclusions (take-home message) of the article.

d. Broader impact

_____ Does the author clearly discuss how the article advances discovery and understanding within or across fields of scientific study (does the article move back to the big picture beyond the specific focus and back to the conceptual framework)?

_____ Does the author provide examples of application and use for the study's ideas, concepts, and findings?

Biological question(s) you came up with from reading the article and its conclusions—that is, questions that may lead to future research:

_____

_____

_____

_____

_____

_____

_____

_____

_____

_____

_____

APPENDIX 4
Statistics and Graphing with Excel

The following are references for commonly used statistical tests.

Nonparametric Tests

Nonparametric statistical tests do not require certain mathematical distributions of the data being analyzed. These tests can be used for measurement or count data as well as data that do not show a normal distribution.

Goodness-of-Fit Test (Chi-Square)

See discussion of chi-square in Exercise 9, section B.2.

Contingency Analysis (Cross-Tabulation of Frequency)

Contingency analysis is used with count data (frequencies) that are arranged in two or more rows, such as recording individuals by their species and their habitat location. This test is used to investigate the association between variables.

The Binomial Test

The binomial test is used when each individual in the sample is classified in one of two categories and the investigator wants to know if the proportion of individuals falling in each category differs from chance or from some previously specified probabilities of falling into those categories.

Parametric Tests

Parametric tests are based on certain assumptions of mathematical distribution of the data being analyzed. The most commonly used distribution is the normal distribution (other tests involve the binomial or Poisson distribution).

Comparison of Variances and Means for Two Samples: The t-Test

The *t*-test is one of the most widely used test procedures for comparing two samples to investigate whether significant differences exist between the two populations sampled.

Using Excel

Excel is an easy-to-use spreadsheet for creating data tables, calculating descriptive statistics, and graphing data. There are many useful online references that provide tutorials and walk you through using Excel. The following are recommended:

The Excel home page: http://office.microsoft.com/en-us/excel/default.aspx

An Excel tutorial: http://www.usd.edu/trio/tut/excel/

How to use Excel: http://serc.carleton.edu/introgeo/mathstatmodels/xlhowto.html

Using Inquiry-Based Module Reports

In the truest sense, science is apprenticeship-style learning by which the student learns from established scientists, often working with them on research projects and learning their "trade" while involved in authentic research investigations. Students locate and read numerous research articles, critiquing them so as to learn the elements of scientific research and communication.

The inquiry-based modules included in this lab manual are designed to teach students about scientific research by having them actually conduct research themselves. The inquiry-based modules not only contain information regarding scientific subject matter related to specific research topics, but also illustrate the scientific process: how science research is conducted and reported. They are designed to engage students in conducting their own research, stimulate higher-order cognitive thinking, and ask students to extend the research being reported.

The following guidelines are suggested for using these inquiry reports:

1. Have students critique the research report in the inquiry module (see sample rubric for peer reviews provided in Appendix 3).

2. Have students come up with their own way of extending the research reported in the module, coming up with their own experimental design (a detailed explanation of the process in the next section and a sample rubric provided later).

3. Have students conduct a background literature search to discover what is known about the research topic they are focusing on.

4. Have students describe their expected results as well as graph these expected results (if possible, also conduct a short pilot study).

5. Have students run their experiment, collect and analyze data, and graph their data.

6. Have students discuss their results and draw conclusions.

7. Finally, have students point to future research on the research topic, discussing ways of extending their research findings.

Creating a Research Proposal

One of the most challenging aspects of scientific research is synthesizing past work, current findings, and new hypotheses into research proposals for future investigation. Research proposals require the careful and thoughtful construction of the conceptual framework, the specific questions and hypotheses of the proposed research, a detailed experimental design and methods that will be used, the projected analysis of the data, and the significance of the proposed research.

The introduction provides background information and places the proposed study within a conceptual framework. Previous research is summarized, leading to the question the current research project proposes to answer. For purposes of clarity, explicit hypothesis/hypotheses are given, with a general statement regarding the main experimental design. (Think of an hourglass, leading from a wide section to a narrow neck. The wide section is the larger conceptual framework or context of the study; the narrow neck is the specific research focus or research question. Later, in the "significance of the proposed research" section, you will again lead to a large section, moving from the specific results of your study to relating these to the larger conceptual framework again.)

The methods section provides more details regarding the experimental design and methodologies you will use. This section is where specific predictions, and the underlying rationales justifying those predictions, are discussed. Background information can be given regarding the particular study site, focal organism, or process being investigated. How you will conduct the study—that is, the specific methods you will employ—are discussed in this section.

The results section for a proposal describes the type of analysis you will use: What type of statistics will be used to analyze the data?

The final section of a research proposal discusses anticipated results and the significance of the research findings. This section is where you expand back into the larger conceptual framework of the study. How will your specific results lead to advances in the field of study? Does your study address a topic that lacks quantitative data, or are their untested assumptions? This section should be pretty clear-cut if you did a thorough enough background investigation. The background review should have provided a clear path to an area within the field of study that requires more research.

Sample rubric for experimental design:

1. What is the research question you will investigate?

2. What is the significance of this question? How will the answer you discover contribute to and advance our understanding?

3. What is your hypothesis and prediction(s) (i.e., what do you *expect* to see *if* your hypothesis is accurate)? (Use the *If . . . , then . . .* format)

4. What is your experimental design? For example, control versus experimental group? Sample size in each group? What you will measure or take data on, your independent and dependent variables, and any statistical tests you will use.

CONTROL GROUP	EXPERIMENTAL GROUP

5. Describe the methods you will use so that someone could read your description and repeat your experiment exactly:

_____

_____

_____

_____

_____

_____

_____

_____

_____

_____

_____

_____

6. Create a data table of the data you *expect* to see if your hypothesis is accurate.

7. Create the *data-recording tables* you will use when you do your experiment; give a copy to your instructor.

8. Create a graph of your expected data and turn it in to instructor.

9. Make sure your equipment is assembled and ready and that your experiment will actually work! Do a *pilot study* run-through with a sample size of 1 in each group.

When you run your experiment, remember that you must be able to organize and graph your data, present your findings to the class, and interpret your results in terms of whether or not your hypothesis is supported or rejected. Be sure to explain your results.

Resources

Day, R. A. 1988. *How to Write and Publish a Scientific Paper* (3rd ed.). ISI Press, NY.

Friedland, A. J., and C. L. Folt. 2000. *Writing Successful Science Proposals.* Yale University Press, New Haven and London.

National Science Foundation. *Grant Proposal Guide.* NSF Web site.

Genetics Problems

You may find it helpful to draw your own Punnett squares on a separate sheet of paper for the following problems.

Monohybrid Problems with Complete Dominance

1. In mice, black fur (B) is dominant over brown fur (b). Breeding a brown mouse and a homozygous black mouse produces all black offspring.
    a. What is the genotype of the *gametes* produced by the brown-furred parent? _____
    b. What genotype is the brown-furred parent? _____
    c. What genotype is the black-furred parent? _____
    d. What genotype is the black-furred offspring? _____
    e. If two F_1 mice are bred with one another, what phenotype will the F_2 offspring be, and in what proportion?

    phenotype _____
    proportion _____

2. The presence of horns on Hereford cattle is controlled by a single gene. The hornless (H) condition is dominant over the horned (h) condition. A hornless cow was crossed repeatedly with the same horned bull. The following results were obtained in the F_1 offspring:

    8 hornless cattle

    7 horned cattle

    What are the parents' genotypes?

    cow _____

    bull _____

3. In fruit flies, red eyes (R) are dominant over purple eyes (r). Two red-eyed fruit flies were crossed, producing the following offspring:

    76 red-eyed flies

    24 purple-eyed flies
    a. What is the approximate ratio of red-eyed to purple-eyed flies? _____
    b. Based on your experience with previous problems, what two genotypes give rise to this ratio? _____
    c. What are the parents' genotypes? _____
    d. What is the genotypic ratio of the F_1 offspring? _____
    e. What is the phenotypic ratio of the F_1 offspring? _____

Monohybrid Problems with Incomplete Dominance

4. Petunia flower color is governed by two alleles, but neither allele is truly dominant over the other. Petunias with the genotype R^1R^1 are red-flowered, those that are heterozygous (R^1R^2) are pink, and those with the R^2R^2 genotype are white. This is an example of **incomplete dominance.** (Note that superscripts are used rather than upper- and lowercase letters to describe the alleles.)
    a. If a white-flowered plant is crossed with a red-flowered petunia, what is the genotypic ratio of the F_1 offspring? _____
    b. What is the phenotypic ratio of the F_1 offspring? _____
    c. If two of the F_1 offspring are crossed, what phenotypes will appear in the F_2 generation? _____
    d. What will be the genotypic ratio in the F_2 generation? _____

Monohybrid Problems Illustrating Codominance

5. Another type of monohybrid inheritance involves the expression of *both* phenotypes in the heterozygous situation. This is called **codominance.**

One well-known example of codominance occurs in the coat color of Shorthorn cattle. Those with reddish-gray roan coats are heterozygous (RR'), and result from a mating between a red (RR) Shorthorn and one that's white ($R'R'$). Roan cattle don't have roan-colored hairs, as would be expected with incomplete dominance, but rather appear roan as a result of having both red *and* white hairs. Thus, the roan coloration is not a consequence of pigments blending in each hair. Because the R and R' alleles are *both* fully expressed in the heterozygote, they are codominant.

a. If a roan Shorthorn cow is mated with a white bull, what will be the genotypic and phenotypic ratios in the F_1 generation?

genotypic ratio _____

phenotypic ratio _____

b. List the parental genotypes of crosses that could produce at least some

white offspring _____

roan offspring _____

Monohybrid, Sex-linked Problems

6. In humans, as well as in many other animals, sex is determined by special sex chromosomes. An individual containing two X chromosomes is a female, while an individual possessing an X and a Y chromosome is a male. (Rare exceptions of XY females and XX males have recently been discovered.)

I am a male/female (circle one).

a. What sex chromosomes do you have? _____

b. In terms of sex chromosomes, what type of gametes (ova) does a female produce? _____

c. What are the possible sex chromosomes in a male's sperm cells? _____

d. Which parent's gametes will determine the sex of the offspring? _____

7. The sex chromosomes bear alleles for traits, just like the other chromosomes in our bodies. Genes that occur on the sex chromosomes are said to be sex-linked. More specifically, the genes present on the X chromosome are said to be X-linked. Many more genes are present on the X chromosome than are found on the Y chromosome. Nonetheless, those genes found on the Y chromosome are said to be Y-linked.

The Y chromosome is smaller than its homologue, the X chromosome. Consequently, most of the loci present on the X chromosome are absent on the Y chromosome.

In humans, color vision is X-linked; the gene for color vision is located on the X chromosome but is absent from the Y chromosome.

Normal color vision (X^N) is dominant over color blindness (X^n). Suppose a color-blind man fathers the children of a woman with the genotype $X^N X^N$.

a. What genotype is the father? _____

b. What proportion of daughters will be color-blind? _____

c. What proportion of sons will be color-blind? _____

8. One daughter from the preceding problem marries a color-blind man.

a. What proportion of their sons will be color-blind? (Another way to think of this is to ask, What are the *chances* that their sons will be color-blind?) _____

b. Explain how a color-blind daughter might result from this couple.

_____

_____

Dihybrid Problems

Recall that pigmented eyes (P) are dominant to nonpigmented (p), and dimpled chins (D) are dominant to nondimpled chins (d).

9. A pigment-eyed, dimple-chinned man marries a blue-eyed woman without a dimpled chin. Their first-born child is blue-eyed and has a dimpled chin.

    a. What are the possible genotypes of the father? _____
    b. What genotype is the mother? _____
    c. What alleles may have been carried by the father's sperm? _____

10. Suppose a dimple-chinned, blue-eyed man whose father lacked a dimple marries a woman who is homozygous recessive for both traits.

    a. What is the expected genotypic ratio of children produced in this marriage? _____
    b. What is the expected phenotypic ratio? _____

11. In his original work on the genetics of garden peas, Mendel found that yellow seed color (YY, Yy) is dominant over green seeds (yy) and that round seed shape (RR, Rr) is dominant over shrunken seeds (rr). Mendel crossed pure-breeding (homozygous) yellow, round-seeded plants with green, shrunken-seeded plants.

    a. What will be the genotype and phenotype of the F_1 produced from such a cross?

       genotype _____
       phenotype _____
    b. If the F_1 plants are crossed, what will be the expected phenotypic ratio of the F_2 generation? _____

Multiple Alleles

12. The major blood groups in humans are determined by **multiple alleles;** that is there are *more than* two possible alleles, any one of which can occupy a locus.

    In this ABO blood group system, a single gene can exist in any of three allelic forms: I^A, I^B, or i. The alleles A and B code for production of antigen A and antigen B (two proteins) on the surface of red blood cells. Alleles A and B are codominant, while allele i is recessive.

    Four blood groups (phenotypes) are possible from combinations of these alleles (Table 1).

TABLE 1 The ABO Blood Groups			
Blood Type	**Anitgens Present**	**Antibody Present**	**Genotype**
O	Neither A nor B	A and B	ii
A	A	B	$I^A I^A$ or $I^A i$
B	B	A	$I^B I^B$ or $I^B i$
AB	AB	Neither A nor B	$I^A I^B$

    a. Is it possible for a child with blood type O to be produced by two AB parents? _____ (yes or no)
    Explain

    _____

    _____

    _____

    b. In a case of disputed paternity, the child is type O, the mother type A. Could an individual of the following blood types be the father? _____ Explain each possibility.

       O _____

       A _____

       B _____

       AB _____

Chi-Square Analysis

13. In fruit flies, red eyes (R) are dominant over white eyes (r). A student performs a cross between a heterozygous red-eyed fly and a white-eyed fly. The student counts the offspring and finds 65 red-eyed flies and 49 white-eyed flies.

   a. What is the expected phenotypic ratio of this cross? _____

   b. Using a χ^2 test, determine whether the deviation between the observed and the expected is the result of chance.

      $\chi^2 =$ _____

   c. Conclusion

_____

_____

14. In fruit flies, gray body (G) is dominant over ebony body (g).

   a. A red-eyed, gray-bodied fly known to be heterozygous for both traits is mated with a white-eyed fly that is heterozygous for body color.

      What is the expected phenotypic ratio for this mating? _____

   b. The observed offspring consist of 15 white-eyed, ebony-bodied flies; 31 white-eyed, gray-bodied flies; 12 red-eyed, ebony-bodied flies; and 38 red-eyed, gray-bodied flies.

      What is the χ^2 value for this cross? _____

   c. Is it likely that the observed results "fit" the expected values? _____

Terms of Orientation in and Around the Animal Body

Body Shapes

Symmetry. The body can be divided into almost identical halves:

Asymmetry. The body cannot be divided into almost identical halves (for example, many sponges).

Radial symmetry. The body is shaped like a cylinder (for example, sea anemone) or wheel (for example, sea star).

Bilateral symmetry. The body is shaped like ours in that it can be divided into halves by only one symmetrical plane (midsagittal).

Directions in the Body

Dorsal. At or toward the back surface of the body.

Ventral. At or toward the belly surface of the body.

Anterior. At or toward the head of the body—ventral surface of humans.

Posterior. At or toward the tail or rear end of the body—dorsal surface of humans.

Medial. At or near the midline of a body. The prefix *mid-* is often used in combination with other terms (for example, midventral).

Lateral. Away from the midline of a body.

Superior. Over or placed above some point of reference—toward the head of humans.

Inferior. Under or placed below some point of reference—away from the head of humans.

Proximal. Close to some point of reference or close to a point of attachment of an appendage to the trunk of the body.

Distal. Away from some point of reference or away from a point of attachment of an appendage to the trunk of the body.

Longitudinal. Parallel to the midline of a body.

Axis. An imaginary line around which a body or structure can rotate. The midline or *longitudinal axis* is the central axis of a symmetrical body or structure.

Axial. Placed at or along an axis.

Radial. Arranged symmetrically around an axis like the spokes of a wheel.

Planes of the Body

Sagittal. Passes vertically to the ground and divides the body into right and left sides. The *midsagittal* or *median plane* passes through the longitudinal axis and divides the body into right and left halves.

Frontal. Passes at right angles to the sagittal plane and divides the body into dorsal and ventral parts.

Transverse. Passes from side to side at right angles to both the sagittal and frontal planes and divides the body into anterior and posterior parts—superior and inferior parts of humans. This plane of section is often referred to as a cross section.

Illustration References

Abramoff, P., and R. G. Thomson. 1982. *Laboratory Outlines in Biology III*. New York: W. H. Freeman.

Boolootian, R. A., and K. A. Stiles Trust. 1981. *College Zoology*. Tenth Edition. New York: Macmillan.

Case, C. L., and T. R. Johnson. 1984. *Experiments in Microbiology*. Menlo Park, California: Benjamin/Cummings.

Fowler, I. 1984. *Human Anatomy*. Belmont, California: Wadsworth.

Gilbert, S. G. 1966. *Pictorial Anatomy of the Fetal Pig*. Second Edition. Seattle, Washington: University of Washington Press.

Glase, J. C., et al. 1975. *Investigative Biology*. Ithaca, New York.

Hickman, C. P. 1961. *Integrated Principles of Zoology*. Second Edition. St. Louis, Missouri: C. V. Mosby.

Hickman, C. P., et al. 1978. *Biology of Animals*. Second Edition. St. Louis, Missouri: C. V. Mosby.

Jensen, W. A., et al. 1979. *Biology*. Belmont, California: Wadsworth.

Kessel, R. G., and R. H. Kardon. 1979. *Tissues and Organs*. New York: W. H. Freeman.

Kessel, R. G., and C. Y. Shih. 1974. *Scanning Electron Microscopy in Biology*. New York: Springer-Verlag.

Lytle, C. F., and J. E. Wodsedalek, 1984. *General Zoology Laboratory Guide*. Complete Version. Ninth Edition. Dubuque, Iowa: Wm. C. Brown.

Patten, B. M. 1951. *American Scientist* 39: 225–243.

Scagel, R. F., et al. 1982. *Nanvascular Plants*. Belmont, California: Wadsworth.

Sheetz, M., et al. 1976. *The Journal of Cell Biology* 70:193.

Shih, C. Y., and R. G. Kessel. 1982. *Living Images*. Boston: Science Books International/Jones and Bartlett Publishers.

Stanier, R., et al. 1986. *The Microbial World*. Fifth Edition. Englewood Cliffs, New Jersey: Prentice-Hall.

Starr, C., and R. Taggart. 1984. *Biology*. Third Edition. Belmont, California: Wadsworth.

Starr, C., and R. Taggart. 1987. *Biology*. Fourth Edition. Belmont, California: Wadsworth.

Starr, C., and R. Taggart. 1989. *Biology*. Fifth Edition. Belmont, California: Wadsworth.

Starr, C., and R. Taggart. 2001. *Biology: The Unity and Diversity of Life*. Ninth Edition. Belmont, California: Wadsworth.

Starr, C. 1991. *Biology: Concepts & Applications*. Belmont, California: Wadsworth.

Starr, C. 2000. *Biology: Concepts and Applications*. Fourth Edition. Belmont, California: Wadsworth.

Steucek, G. L., et al. 1985. *American Biology Teacher* 471: 96–99.

Storer, T., et al. 1979. *General Zoology*. New York: McGraw-Hill.

Villee, C. A., et al. 1973. *General Zoology*. Fourth Edition. Philadelphia: W. B. Saunders.

Weller, H., and R. Wiley. 1985. *Basic Human Physiology*. Boston: PWS Publishers.

Wischnitzer, S. 1979. *Atlas and Dissection Guide for Comparative Anatomy*. Third Edition. New York: W. H. Freeman.

Wolfe, S. L. 1985. *Cell Ultrastructure*. Belmont, California: Wadsworth.